Lecture Notes in Computer Science 14427

The series Lecture Notes in Computer Science (LNCS), including its subseries Lecture Notes in Artificial Intelligence (LNAI) and Lecture Notes in Bioinformatics (LNBI), has established itself as a medium for the publication of new developments in computer science and information technology research, teaching, and education.

LNCS enjoys close cooperation with the computer science R & D community, the series counts many renowned academics among its volume editors and paper authors, and collaborates with prestigious societies. Its mission is to serve this international community by providing an invaluable service, mainly focused on the publication of conference and workshop proceedings and postproceedings. LNCS commenced publication in 1973.

Qingshan Liu · Hanzi Wang · Zhanyu Ma ·
Weishi Zheng · Hongbin Zha · Xilin Chen ·
Liang Wang · Rongrong Ji
Editors

Pattern Recognition and Computer Vision

6th Chinese Conference, PRCV 2023
Xiamen, China, October 13–15, 2023
Proceedings, Part III

Editors
Qingshan Liu
Nanjing University of Information Science and Technology
Nanjing, China

Zhanyu Ma
Beijing University of Posts and Telecommunications
Beijing, China

Hongbin Zha
Peking University
Beijing, China

Liang Wang
Chinese Academy of Sciences
Beijing, China

Hanzi Wang
Xiamen University
Xiamen, China

Weishi Zheng
Sun Yat-sen University
Guangzhou, China

Xilin Chen
Chinese Academy of Sciences
Beijing, China

Rongrong Ji
Xiamen University
Xiamen, China

ISSN 0302-9743 ISSN 1611-3349 (electronic)
Lecture Notes in Computer Science
ISBN 978-981-99-8434-3 ISBN 978-981-99-8435-0 (eBook)
https://doi.org/10.1007/978-981-99-8435-0

This Springer imprint is published by the registered company Springer Nature Singapore Pte Ltd.
The registered company address is: 152 Beach Road, #21-01/04 Gateway East, Singapore 189721, Singapore

Preface

Welcome to the proceedings of the Sixth Chinese Conference on Pattern Recognition and Computer Vision (PRCV 2023), held in Xiamen, China.

PRCV is formed from the combination of two distinguished conferences: CCPR (Chinese Conference on Pattern Recognition) and CCCV (Chinese Conference on Computer Vision). Both have consistently been the top-tier conference in the fields of pattern recognition and computer vision within China's academic field. Recognizing the intertwined nature of these disciplines and their overlapping communities, the union into PRCV aims to reinforce the prominence of the Chinese academic sector in these foundational areas of artificial intelligence and enhance academic exchanges. Accordingly, PRCV is jointly sponsored by China's leading academic institutions: the Chinese Association for Artificial Intelligence (CAAI), the China Computer Federation (CCF), the Chinese Association of Automation (CAA), and the China Society of Image and Graphics (CSIG).

PRCV's mission is to serve as a comprehensive platform for dialogues among researchers from both academia and industry. While its primary focus is to encourage academic exchange, it also places emphasis on fostering ties between academia and industry. With the objective of keeping abreast of leading academic innovations and showcasing the most recent research breakthroughs, pioneering thoughts, and advanced techniques in pattern recognition and computer vision, esteemed international and domestic experts have been invited to present keynote speeches, introducing the most recent developments in these fields.

PRCV 2023 was hosted by Xiamen University. From our call for papers, we received 1420 full submissions. Each paper underwent rigorous reviews by at least three experts, either from our dedicated Program Committee or from other qualified researchers in the field. After thorough evaluations, 522 papers were selected for the conference, comprising 32 oral presentations and 490 posters, giving an acceptance rate of 37.46%. The proceedings of PRCV 2023 are proudly published by Springer.

Our heartfelt gratitude goes out to our keynote speakers: Zongben Xu from Xi'an Jiaotong University, Yanning Zhang of Northwestern Polytechnical University, Shutao Li of Hunan University, Shi-Min Hu of Tsinghua University, and Tiejun Huang from Peking University.

We give sincere appreciation to all the authors of submitted papers, the members of the Program Committee, the reviewers, and the Organizing Committee. Their combined efforts have been instrumental in the success of this conference. A special acknowledgment goes to our sponsors and the organizers of various special forums; their support made the conference a success. We also express our thanks to Springer for taking on the publication and to the staff of Springer Asia for their meticulous coordination efforts.

We hope these proceedings will be both enlightening and enjoyable for all readers.

October 2023

Qingshan Liu
Hanzi Wang
Zhanyu Ma
Weishi Zheng
Hongbin Zha
Xilin Chen
Liang Wang
Rongrong Ji

Organization

General Chairs

Hongbin Zha	Peking University, China
Xilin Chen	Institute of Computing Technology, Chinese Academy of Sciences, China
Liang Wang	Institute of Automation, Chinese Academy of Sciences, China
Rongrong Ji	Xiamen University, China

Program Chairs

Qingshan Liu	Nanjing University of Information Science and Technology, China
Hanzi Wang	Xiamen University, China
Zhanyu Ma	Beijing University of Posts and Telecommunications, China
Weishi Zheng	Sun Yat-sen University, China

Organizing Committee Chairs

Mingming Cheng	Nankai University, China
Cheng Wang	Xiamen University, China
Yue Gao	Tsinghua University, China
Mingliang Xu	Zhengzhou University, China
Liujuan Cao	Xiamen University, China

Publicity Chairs

Yanyun Qu	Xiamen University, China
Wei Jia	Hefei University of Technology, China

Local Arrangement Chairs

Xiaoshuai Sun	Xiamen University, China
Yan Yan	Xiamen University, China
Longbiao Chen	Xiamen University, China

International Liaison Chairs

Jingyi Yu	ShanghaiTech University, China
Jiwen Lu	Tsinghua University, China

Tutorial Chairs

Xi Li	Zhejiang University, China
Wangmeng Zuo	Harbin Institute of Technology, China
Jie Chen	Peking University, China

Thematic Forum Chairs

Xiaopeng Hong	Harbin Institute of Technology, China
Zhaoxiang Zhang	Institute of Automation, Chinese Academy of Sciences, China
Xinghao Ding	Xiamen University, China

Doctoral Forum Chairs

Shengping Zhang	Harbin Institute of Technology, China
Zhou Zhao	Zhejiang University, China

Publication Chair

Chenglu Wen	Xiamen University, China

Sponsorship Chair

Yiyi Zhou	Xiamen University, China

Exhibition Chairs

Bineng Zhong	Guangxi Normal University, China
Rushi Lan	Guilin University of Electronic Technology, China
Zhiming Luo	Xiamen University, China

Program Committee

Baiying Lei	Shenzhen University, China
Changxin Gao	Huazhong University of Science and Technology, China
Chen Gong	Nanjing University of Science and Technology, China
Chuanxian Ren	Sun Yat-Sen University, China
Dong Liu	University of Science and Technology of China, China
Dong Wang	Dalian University of Technology, China
Haimiao Hu	Beihang University, China
Hang Su	Tsinghua University, China
Hui Yuan	School of Control Science and Engineering, Shandong University, China
Jie Qin	Nanjing University of Aeronautics and Astronautics, China
Jufeng Yang	Nankai University, China
Lifang Wu	Beijing University of Technology, China
Linlin Shen	Shenzhen University, China
Nannan Wang	Xidian University, China
Qianqian Xu	Key Laboratory of Intelligent Information Processing, Institute of Computing Technology, Chinese Academy of Sciences, China
Quan Zhou	Nanjing University of Posts and Telecommunications, China
Si Liu	Beihang University, China
Xi Li	Zhejiang University, China
Xiaojun Wu	Jiangnan University, China
Zhenyu He	Harbin Institute of Technology (Shenzhen), China
Zhonghong Ou	Beijing University of Posts and Telecommunications, China

Contents – Part III

Machine Learning

Vision Problems in Robotics, Autonomous Driving

Machine Learning

Loss Filtering Factor for Crowd Counting

Yufeng Chen(✉)

Uppsala University, Uppsala 75105, Sweden
yufeng.chen.4516@student.uu.se

Abstract. In crowd counting datasets, each person is annotated by a point, typically representing the center of the head. However, due to the dense crowd, variety of scenarios, significant obscuration and low resolution, label noise in the dataset is inevitable and such label noise has a negative impact on the performance of the model. To alleviate the negative effects of label noise, in this paper we propose the Loss Filtering Factor, which can filter out the losses assumed to be caused by label noise during the training process. By doing so, the model can prioritize non-noise data and focus on it during training and predicting. Extensive experimental evaluations have demonstrated that the proposed Loss Filtering Factor consistently improves the performance of all models across all datasets used in the experiments. On average, it leads to a 5.48% improvement in MAE and 6.43% in MSE. Moreover, the proposed approach is universal and it can be easily implemented into any neural network model architecture to improve performance.

Keywords: crowd counting · noisy labels · loss function

1 Introduction

Crowd counting is a fundamental computer vision task that uses computer vision techniques to estimate the number of people in unconstrained scene automatically. Recent years, crowd counting has been widely studied since its significance in various applications, such as crowd Analysis [2,7], video surveillance [14], etc. Crowd counting is challenging due to overlaps, occlusions and variations in perspective and illumination. In the past decade, many algorithms have been proposed. Early methods estimate crowd counts through individual detection. Currently, Methods that mainly cast crowd counting as a density map estimation problem and combines it with convolutional neural networks(CNNs) have made remarkable progresses [5,7,10]. By this methods, the values of crowd density map regressed by CNNs are summed to give the total size of the crowd.

The traditional approach is to convert these point annotations into density maps using a Gaussian kernel and treat the density map as the "ground truth". The model is then trained by regressing the value at each pixel in the density map. Currently, several methods (Bayesian, DMCC, ShanghaiTech) focus on improving the conversion process from point annotations to density maps and have achieved impressive results.

Q. Liu et al. (Eds.): PRCV 2023, LNCS 14427, pp. 3–15, 2024.
https://doi.org/10.1007/978-981-99-8435-0_1

Nowadays, widely used public datasets [2,3,15] only provide point annotations, i.e., only one pixel of each person is labeled (typically the center of the head). The traditional approach is to convert these point annotations into density maps using a Gaussian kernel and treat the density map as "ground truth". The model is then trained by regressing the value at each pixel in the density map. Currently, many methods [7,10,15] focus on improving the process that convert point annotations to density map and have achieved good performances.

However, it is important to note that the quality of datasets and the accuracy of point annotations play a critical role in developing effective crowd count estimators for crowd counting tasks. It should be acknowledged that it is human annotators who manually label the points for all datasets. Specifically, every person in each image of the dataset requires manual labeling. Due to factors like overlap, low resolution, and high crowd density near the vanishing point, mislabeled point annotations and spatial errors can occur. The small size of annotations compared to human heads leads to imperfect correspondence. These "unknown" mislabeled point annotations and spatial errors have a negative impact on crowd count estimator training.

To alleviate negative influences caused by annotation noises, we propose a Loss Filtering Factor (LFF) based on the hypothesis that, under certain conditions, the trained model predicts more correct signals than human-labeled annotations. During the model training process, the LFF helps filter out pixel-level losses that are likely caused by annotation noise. This allows the model to prioritize losses that are more meaningful and not affected by noise. Through extensive experimental evaluations on popular publicly available datasets including UCF-QNRF [3], ShanghaiTech [15], UCF_CC_50 [2] and RGBT-CC [6], we demonstrate that the proposed Loss Filtering Factor improves the performance of various backbone networks used for crowd counting tasks.

2 Related Work

Crowd Counting. Crowd counting is a key problem in computer vision and has been extensively studied. Methods for addressing this problem can be categorized into three groups: detection, direct count regression, and point supervision.

Early methods primarily detect individuals (persons, heads, or upper bodies) in images, but face challenges in detecting crowds with high density. Furthermore, Significant occlusion necessitates bounding box annotation, which is time-consuming and unclear. Later methods [3] overcome this issue by regressing to a density map based on point annotations, referred to as "ground truth". This method uses location information to learn a density map for each training sample. However, they assume crowd uniformity, whereas real-world crowd distribution in images is often uneven due to imaging conditions and camera angles. Recently, many works [7,12] have proposed that instead of generating density maps, point supervisions be used directly. Further works focus on optimal transport [8] and weak supervision [11] without depending on the assumption of Gaussian distribution.

Noisy Labels. The effectiveness of deep neural networks relies on having access to high-quality labeled training data. Label noise in the training data can significantly impair model's performance on clean test data [13]. Unfortunately, samples with faulty or incorrect labels are virtually always present in big training datasets. Recently, academics have increasingly focused on this issue. [1] proposed a Co-teaching model for combating with noisy labels and [4] offered a deeper understanding of deep learning using non-synthetic noisy labels.

These methods do not transfer well to crowd counting due to limited datasets and high computational expenses. On the contrary, deep learning models have shown superior performance to humans in tasks such as image classification [17], speech recognition [9], etc. Therefore, in crowd counting problem, especially high-density tasks, we may expect that, under certain conditions, existing deep learning models predict more true signals than human-labeled annotations.

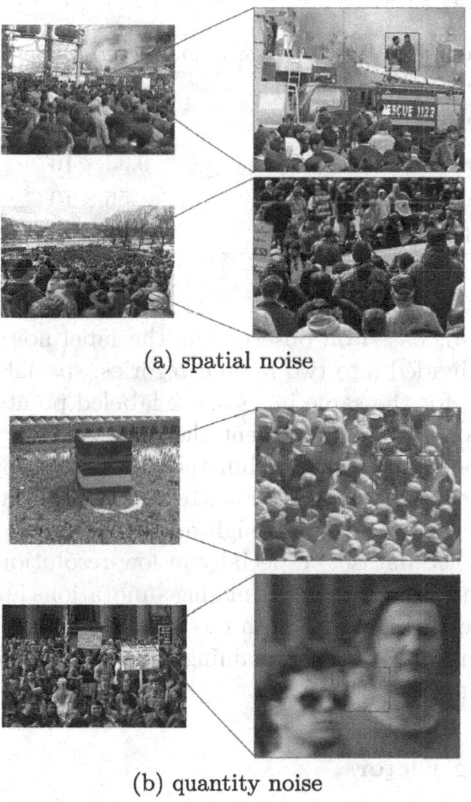

(a) spatial noise

(b) quantity noise

Fig. 1. This figure highlights the label noise issues present in existing dense crowd datasets. In (a), we observe cases where the annotations are located in other parts of the body rather than the center of the head, as well as annotations positioned outside of the body. On the other hand, (b) showcases examples of both duplicate and missing annotations, further emphasizing the challenges posed by label noise.

3 Proposed Method

3.1 Background and Motivation

In the realm of crowd counting tasks, the presence of label noise within datasets is an inescapable reality. This noise stems from the intricacies of dense crowds, a multitude of scenarios, and substantial occlusions. Widely-used benchmarks in the crowd counting domain exhibit notably high crowd densities, thereby introducing complexities in maintaining annotation consistency and accuracy for key points. The comprehensive data statistics for dense crowd counting across various multi-scene datasets are encapsulated in Table 1. Notably, a significant majority of these datasets feature an average of over a hundred individuals per image.

Table 1. Average Count and average density of popular crowd datasets

Dataset	Average Count	Average Density
UCF_CC_50	1279	2.02×10^{-4}
UCF-QNRF	815	1.12×10^{-4}
ShanghaiTech_A	501	9.33×10^{-4}
ShanghaiTech_B	123	1.56×10^{-4}
RGBT-CC	68	2.21×10^{-4}

On the other hand, based on observation, the label noise in popular crowd benchmarks can be divided into two main categories: spatial noise and quantity noise. In the dataset, for the same images, the labeled points may vary in their relative locations. Some points represent the pixel at the center of the head, while others are randomly placed within the person, such as on the chest or waist. These inaccuracies in annotation are referred as spatial noise, as depicted in Fig. 1(a). Moreover, the scarcity of high-resolution crowd images contributes to quantity noise in the dataset, especially in low-resolution images with high crowd density. This noise encompasses missing annotations and duplicated annotations, as illustrated in Fig. 1(b). Both of the aforementioned label noises will undoubtedly have an impact on the training and performance of deep learning models to some extent [13].

3.2 Loss Filtering Factor

To alleviate the negative effects of label noise, this paper introduces the Loss Filtering Factor. It is based on the assumption that, under certain conditions, the trained neural network model can generate more accurate predictions than human-labeled annotations. The proposed Loss Filtering Factor serves as a filtering mechanism to identify and discard losses that are likely caused by label noise, such as quantity noise and spatial noise, during the training process.

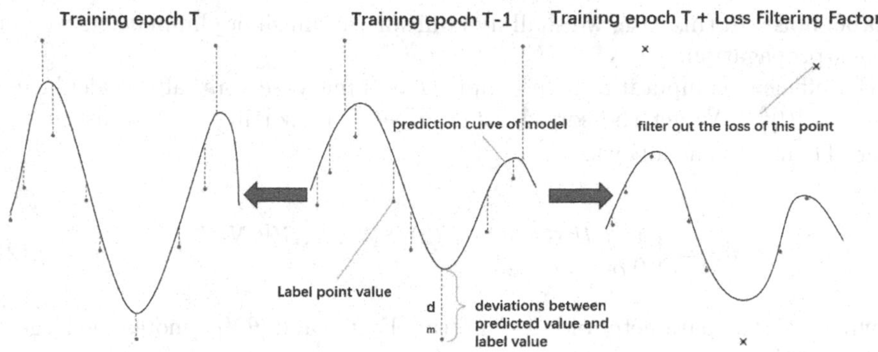

Fig. 2. Comparison of training with(right) and without(left) using loss filter factor

Figure 2 shows a simple (2-D)example of training with and without Loss Filtering Factor. During the training process, the predicted values of the model in epoch T will be closer to the label values than in epoch $T-1$, i.e., the total of losses of epoch T is less than epoch $T-1$. When training without using loss filter factor, the model will consider the losses caused by label noise as regular losses, therefore, even though the total losses are reduced, the non-noise losses are not always minimized; When training using Loss Filtering Factor, the Loss Filtering Factor can filter out the losses believed to be caused by label noise, as a result, both total losses and non-noise losses decrease.

The proposed method employs the mask $M = [m^j]_j^N$ to selectively supervise training of the neural network model. It can be defined as:

$$LFF = \sum_{j}^{N} m^j \cdot l^j \tag{1}$$

where $L = [l^j]_j^N$ is the deviations between the predicted value and label value. For example, for methods based on point supervision, l^j is the differences between each predicted point value and corresponding label point value; while for density-maps-based methods, the l^j is the differences between the predicted density maps and generated density maps. apparently, the value of sizes N is determined by the loss function. For example, in MSE Loss N equals to the size of density map, while in Bayesian Loss N equals to the number of annotated points.

Considering the inevitable presence of label noise within datasets, and the occasional instances where the trained model generates more accurate signals than the annotations themselves, the mask M serves as a mechanism to selectively filter losses. This, in turn, prevents these losses from participating in the back-propagation process during model training. We consider the deviation l^j between predictions and labels to be a type of label uncertainty. According to the foregoing argument, if the l^j is excessively large, it is most likely generated by label noise. In this case, we shall dynamically diminish or eliminate its weight in back-propagation.

For efficient computation, we adopt M as binary vectors, after calculating all losses $[l^j]_j^N$, We get the sorted list $S = [e^j]_j^N$ by sorting $[l^j]_j^N$ in ascending order. Then m^j is as follows:

$$m^j = \begin{cases} 1 \; if \; l^j \in \{S(1), S(2), S(3), ..., S([\theta N])\} \\ 0 \; otherwise \end{cases} \tag{2}$$

where θ is the parameter of Loss Filtering Factor and $[\theta N]$ denotes the largest integer no than θN.

Assuming $F = F(l^j)$ represents the model's loss function. Notably, when the loss filtering factor $\theta = 1.0$. it equates to the absence of loss filtering, implying the direct inclusion of all losses $[l^j]$ within the function F. Conversely, a loss filtering factor of $\theta = 0.85$ implies that 15% of label values are treated as potential noise, allowing only the 85% of label values exhibiting the least deviation from predictions to contribute to the supervision process. Ultimately, the comprehensive loss function incorporating the loss filtering factor can be defined as follows:

$$F_{LFF} = F(LFF) = F(\sum_j^N m^j \cdot l^j) \tag{3}$$

4 Experiments

4.1 Evaluation Metrics

Mean Absolute Error (MAE) and Mean Squared Error (MSE) are two extensively used metrics for evaluating crowd count estimate methods. They are defined as follows:

$$MAE = \frac{1}{K} \sum_{K=1}^{K} |N_k^{GT} - N_k| \tag{4}$$

$$MSE = \sqrt{\frac{1}{K} \sum_{K=1}^{K} |N_k^{GT} - N_k|^2} \tag{5}$$

where K is the number of test images, N_k^{GT} and N_k are the value of label count and the value of estimated count for the $k-th$ image, respectively.

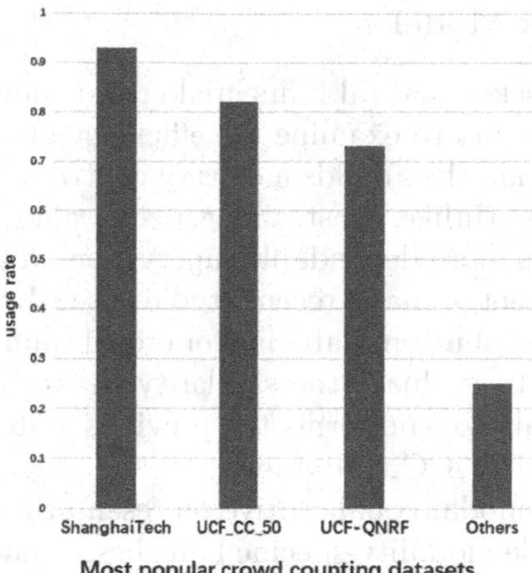

Fig. 3. Among the observed 200 papers, each utilized multiple datasets, averaging 2.73 per publication. Notably, the ShanghaiTech benchmark stands out with a usage rate exceeding 90% in pertinent research; the UCF_CC_50 benchmark follows closely with over 80% usage and the UCF-QNRF dataset commands attention at over 70% adoption within relevant research circles.

4.2 Datasets

Numerous public crowd counting datasets are currently accessible. To ensure representation, we randomly selected 200 recent publications from the past five years that utilized deep learning to address crowd counting. The results are shown in Fig. 3. Based on the popularity of crowd counting datasets, UCF-QNRF [3],ShanghaiTech [15], UCF_CC_50 [2] will be used in this paper. Furthermore, we will incorporate RGBT-CC [6], the pioneering large-scale RGBT Crowd Counting benchmark introduced in 2021.

ShanghaiTech [15] consists of part A and part B. Part A contains 300 training images and 182 testing images. while Part B includes 400 training images and 316 testing images. Part A has a significantly higher density than part B1.

UCF_CC_50 [2] includes 50 grayscale images with varying but high resolutions. Each image has an average resolution of 2013 × 2902. The average count for each image is 1,279.

UCF-QNRF [3] is one of the largest crowd counting datasets, with 1,535 images and 1.25 million point annotations. The training set contains 1,201 images, while the remaining 334 images are used for testing.

RGBT-CC [6] is the first publicly available RGBT dataset for crowd counting, containing 2,030 pairs of representative RGB-thermal images, 1,013 of which are captured in light and 1,017 of which are captured in darkness. Each image is the same size(640 × 480) and 1,030 pairs are used for training, 200 pairs are for validation and 800 pairs are for testing. A total of 138,389 persons are marked with point annotations, on average 68 people per image.

4.3 Neural Network Model

In this experiment, we select several influential crowd counting models and add different loss filtering factors to examine the efficiency of the proposed method.

BL [7] is a loss function that builds a density contribution probability model using point annotations. Unlike constraining pixel values in the density map, BL training loss employs more dependable supervision on the expected count at each marked point. Presently, many recent studies use BL as their loss function.

DMCC [10] uses Distribution Matching for crowd counting. It uses the Optimal Transport method to evaluate the similarity between the normalized predicted density map, it also outperforms the previous state-of-the-art results on the ShanghaiTech and UCF_CC_50 datasets.

IADM [6] is a crossmodal collaborative representation learning framework, which consists of multiple modality-specific branches, a modality-shared branch, and an Information Aggregation Distribution Module. It's universal for multimodal crowd counting problems.

CSCA [16] are modular building pieces that can be simply integrated into any modality-specific architecture. This method greatly improves performance across various backbone networks and outperforms the previous state-of-the-art results on RGBT-CC dataset.

Implementation Details. In this experiment, we use the standard image classification network VGG-19 as backbone and build the model respectively according to the official implementation of the methods in Sec.4.3. The backbone is pre-trained on ImageNet and the Adam optimizer with an learning rate 10^{-5} is used to update the parameters.

For training, images from various datasets are randomly cropped into different sizes. The crop size is 256×256 for ShanghaiTech Part A, UCF_CC_50 and RGBT-CC where images resolutions are smaller, and 512×512 for ShanghaiTech Part B and UCF-QNRF. And we perform five-fold cross validations to obtain the average test result of UCF_CC_50 dataset since it is a small-scale dataset with no data split designated for training and testing.

4.4 Experimental Evaluations

We compare the models with and without using proposed loss filtering factor on the benchmark datasets described in Sec.4.2. We set the model without Loss Filtering Factor($\theta = 1$) as the baseline and compare the performances of models at various loss filtering factor values. To make a fair comparison, for the same model, the same random seeds are used for each set of experiments as well as other parameters. The experimental results are shown in Table 2. For each model structure, we conduct five experiments on each dataset separately, including a model without the proposed method ($\theta = 1$), and four models using Loss Filtering Factor with different values of θ.

Quantitative Results. In all experiments, the proposed Loss Filtering Factor consistently improves the performances of all the four models in all datasets used in experiments by an average of 5.48% for MAE and 6.43% for MAE. For

Table 2. Benchmark evaluations on five benchmark crowd counting datasets using the MAE and MSE metrics.The best performance is shown in underlined.

Model	Dataset	Shanghai A		Shanghai B		UCF-QNRF		UCF-CC-50		RGBT-CC	
	θ	MAE	MSE	MAE	MSE	MAE	MSE	MAE	MSE	MAE	MSE
BL	1.00	61.16	100.85	8.53	13.29	93.92	166.63	230.27	343.57	18.85	32.05
	0.95	<u>58.98</u>	<u>95.33</u>	8.12	12.76	92.70	165.08	<u>226.26</u>	<u>326.64</u>	<u>17.97</u>	<u>30.20</u>
	0.90	62.86	99.77	<u>7.47</u>	12.99	<u>89.73</u>	<u>160.10</u>	239.23	348.85	18.41	30.79
	0.85	64.85	100.41	8.21	<u>12.52</u>	91.77	164.12	271.49	332.88	19.47	34.34
	0.80	66.23	100.82	8.61	13.60	93.71	166.23	229.74	333.83	19.48	33.15
	0.75	67.43	102.32	10.04	15.13	95.86	170.92	235.42	353.30	20.90	35.00
CSCA	1.00	–	–	–	–	–	–	–	–	14.81	27.25
	0.95	–	–	–	–	–	–	–	–	<u>14.35</u>	<u>24.87</u>
	0.90	–	–	–	–	–	–	–	–	14.57	25.68
	0.85	–	–	–	–	–	–	–	–	14.75	27.50
	0.80	–	–	–	–	–	–	–	–	15.51	28.14
	0.75	–	–	–	–	–	–	–	–	16.40	29.26
DMCC	1.00	62.08	98.59	7.91	14.97	90.88	157.71	229.89	313.79	18.42	31.88
	0.95	62.58	97.96	7.54	14.65	87.43	150.25	<u>208.66</u>	<u>289.20</u>	<u>17.36</u>	29.99
	0.90	<u>61.33</u>	<u>96.14</u>	<u>7.01</u>	<u>13.55</u>	<u>86.10</u>	<u>145.40</u>	211.15	290.19	17.55	<u>29.43</u>
	0.85	65.24	99.60	7.22	13.86	88.85	152.08	225.77	318.22	18.27	31.45
	0.80	67.88	104.02	7.24	13.75	92.12	156.71	226.90	330.53	19.18	32.81
	0.75	70.42	109.86	8.33	13.54	93.25	158.02	231.27	335.94	20.55	34.21
IADM	1.00	–	–	–	–	–	–	–	–	15.66	27.98
	0.95	–	–	–	–	–	–	–	–	15.52	26.24
	0.90	–	–	–	–	–	–	–	–	<u>15.20</u>	<u>25.94</u>
	0.85	–	–	–	–	–	–	–	–	15.36	26.38
	0.80	–	–	–	–	–	–	–	–	16.06	26.32
	0.75	–	–	–	–	–	–	–	–	17.18	28.05

most models and datasets, using Loss filtering factor with a θ of 0.9 or 0.95 may result in the best performance increase. When $\theta = 0.9$, it provides the greatest performance improvement, boosting the counting accuracy of DMCC on ShanghaiTechB dataset by 11.38% and 9.49% for MAE and MSE, respectively.

From the perspective of the model, using the appropriate Loss Filtering Factor makes around 6.81% improvements of DMCC and 5.27% of BL on all five datasets, and makes around 5.91% improvements of CSCA and 5.11% of IADM on RGBT-CC datasets. From the perspective of the dataset, using the appropriate Loss Filtering Factor makes average of 5.36% improvements of models trained on UCF-QNRF, 3.18% on ShanghaiTechA, 9.77% on ShanghaiTechB, 5.93% on UCF_CC_50 and 5.74% on RGBT-CC, respectively.

4.5 Key Issues and Discussion

Effect of θ. To analyze the impact of θ selection on model performance, we divide the results of each model with Loss Filtering Factor($\theta \neq 1$) by the corresponding result of the baseline model($\theta = 1$)and take the average. The result is shown in Fig. 4. Although the best value of θ vary for different models and datasets, the proposed Loss Filtering Factor increases model performance when the θ are between 0.85 and 1, and for most models, selecting the value of θ to 0.95 or 0.9 can maximize the model's performance. However, as θ is selected smaller, the performance of models decrease significantly.

The effect of θ can be explained by following ways:

- Approximately 10% of the dataset comprises point annotations that exert a detrimental impact on the model's counting performance when utilized in training. Stated differently, the dataset contains roughly 10% label noise, and the reduction of this noise during training can enhance the model's counting accuracy. This observation may also shed light on the varying optimal θ selections for the same model across different datasets, as the degrees of label noise tend to fluctuate among datasets.
- Utilizing the proposed Loss Filtering Factor results in the model abstaining from utilizing all data points for supervision in each training epoch, akin to the dropout layer in neural network models. Interestingly, even if the data removed from training due to the influence of factor θ isn't attributable to label noise, it can still contribute to enhanced generalization and a reduction in the model's tendency to overfit to the training data.

The explanations above demonstrate that an appropriate value of θ can significantly enhance model performance by mitigating label noise within the dataset during training. This, in turn, contributes to improved generalization and reduced overfitting of the model. However, when θ becomes excessively small ($\theta \leq 0.80$), it leads to diminished model accuracy due to inadequate utilization of ground truth, resulting in insufficient supervision for the model.

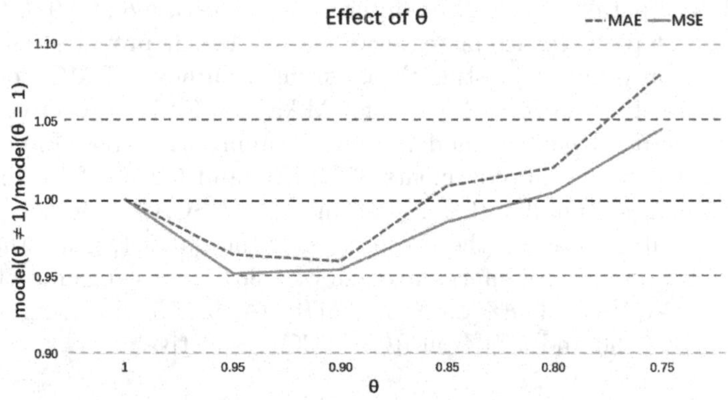

Fig. 4. Effect of Loss Filtering Factor θ.

Effect on Model Convergence Speed. Through our experiments, we have observed that upon the incorporation of the Loss Filtering Factor, the model's training time remains relatively stable across each epoch. Conversely, the rate of convergence varies considerably among models, with the model's overall training duration being approximately 5% longer when employing the Loss Filtering Factor compared to when it is not used.

Ablation Studies. We perform the ablation study on UCF-QNRF dataset by comparing the proposed LFF with removing the same number of points randomly during the training. Table 3 provides quantitative results of it.

Table 3. BL and DMCC is original model($\theta = 1$). We use $\theta = 0.9$ for model BL+LFF and DMCC+LFF, and randomly remove 10% of data supervision for model BL+random and DMCC+random each epoch in training

UCF-QNRF	MAE	MSE
BL	93.92	166.63
BL+random	93.25	165.98
BL+LFF	89.73	160.10
DMCC	90.88	157.71
DMCC+random	90.19	156.44
DMCC+LFF	86.10	145.40

When 10% of the data were randomly omitted from supervision during training, a slight improvement in the overall model performance was observed, albeit a marginal one. This suggests that the act of randomly reducing data supervision in each training epoch has only a modest impact on model performance. This observation underscores that the enhanced performance achieved by the proposed Loss Filtering Factor is not merely a result of random data omission during training. Rather, the LFF's filtering mechanism predominantly focuses on removing label noise that would otherwise detrimentally affect the model's counting performance.

5 Conclusions and Future Work

In this paper, we present the Loss Filtering Factor (LFF) method, which can be seamlessly integrated into various neural network model architectures. This method centers around the loss function and effectively filters out losses influenced by label noise during model training. Moreover, experimental findings uncover the presence of label noise in crowd counting datasets, adversely impacting the model's performance. The proposed approach successfully mitigates this negative influence of label noise, resulting in improved counting performance.

Currently, we have developed a straightforward yet effective implementation of the Loss Filtering Factor method and demonstrated its efficacy. In the future, we aim to explore more sophisticated strategies, such as the dynamic application of the Loss Filtering Factor in each training epoch, to further enhance the capabilities of our proposed method.

References

1. Han, B., et al.: Co-teaching: robust training of deep neural networks with extremely noisy labels. In: Advances in Neural Information Processing Systems 31 (2018)
2. Idrees, H., Saleemi, I., Seibert, C., Shah, M.: Multi-source multi-scale counting in extremely dense crowd images. In: Proceedings of the IEEE Conference on Computer Vision and Pattern Recognition (CVPR) (2013)
3. Idrees, H., et al.: Composition loss for counting, density map estimation and localization in dense crowds. In: Proceedings of the European Conference on Computer Vision (ECCV), pp. 532–546 (2018)
4. Jiang, L., Huang, D., Liu, M., Yang, W.: Beyond synthetic noise: deep learning on controlled noisy labels. In: International Conference on Machine Learning, pp. 4804–4815. PMLR (2020)
5. Lin, H., Ma, Z., Ji, R., Wang, Y., Hong, X.: Boosting crowd counting via multifaceted attention. In: Proceedings of the IEEE/CVF Conference on Computer Vision and Pattern Recognition, pp. 19628–19637 (2022)
6. Liu, L., Chen, J., Wu, H., Li, G., Li, C., Lin, L.: Cross-modal collaborative representation learning and a large-scale RGBT benchmark for crowd counting. In: Proceedings of the IEEE/CVF Conference on Computer Vision and Pattern Recognition, pp. 4823–4833 (2021)
7. Ma, Z., Wei, X., Hong, X., Gong, Y.: Bayesian loss for crowd count estimation with point supervision. In: Proceedings of the IEEE/CVF International Conference on Computer Vision, pp. 6142–6151 (2019)
8. Ma, Z., Wei, X., Hong, X., Lin, H., Qiu, Y., Gong, Y.: Learning to count via unbalanced optimal transport. In: Proceedings of the AAAI Conference on Artificial Intelligence, vol. 35, pp. 2319–2327 (2021)
9. Pham, N.Q., Nguyen, T.S., Niehues, J., Müller, M., Stüker, S., Waibel, A.: Very deep self-attention networks for end-to-end speech recognition. arXiv preprint arXiv:1904.13377 (2019)
10. Wang, B., Liu, H., Samaras, D., Nguyen, M.H.: Distribution matching for crowd counting. Adv. Neural. Inf. Process. Syst. **33**, 1595–1607 (2020)
11. Wang, M., Zhou, J., Cai, H., Gong, M.: CrowdMLP: weakly-supervised crowd counting via multi-granularity MLP. arXiv preprint arXiv:2203.08219 (2022)
12. Zand, M., Damirchi, H., Farley, A., Molahasani, M., Greenspan, M., Etemad, A.: Multiscale crowd counting and localization by multitask point supervision. In: ICASSP 2022–2022 IEEE International Conference on Acoustics, Speech and Signal Processing (ICASSP), pp. 1820–1824. IEEE (2022)
13. Zhang, C., Bengio, S., Hardt, M., Recht, B., Vinyals, O.: Understanding deep learning (still) requires rethinking generalization. Commun. ACM **64**(3), 107–115 (2021)
14. Zhang, S., Wu, G., Costeira, J.P., Moura, J.M.F.: Understanding traffic density from large-scale web camera data. In: Proceedings of the IEEE Conference on Computer Vision and Pattern Recognition (CVPR) (2017)
15. Zhang, Y., Zhou, D., Chen, S., Gao, S., Ma, Y.: Single-image crowd counting via multi-column convolutional neural network. In: Proceedings of the IEEE conference on computer vision and pattern recognition, pp. 589–597 (2016)

16. Zhang, Y., Choi, S., Hong, S.: Spatio-channel attention blocks for cross-modal crowd counting. In: Proceedings of the Asian Conference on Computer Vision, pp. 90–107 (2022)
17. Zoph, B., Vasudevan, V., Shlens, J., Le, Q.V.: Learning transferable architectures for scalable image recognition. In: Proceedings of the IEEE Conference on Computer Vision and Pattern Recognition, pp. 8697–8710 (2018)

Classifier Decoupled Training for Black-Box Unsupervised Domain Adaptation

Xiangchuang Chen[1], Yunhang Shen[2], Xuan Luo[3], Yan Zhang[4], Ke Li[2], and Shaohui Lin[1,5](✉)

[1] School of Computer Science and Technology, East China Normal University, Shanghai, China
shlin@cs.ecnu.edu.cn
[2] Tencent Youtu Lab, Shanghai, China
[3] Nanchang University, Nanchang, China
[4] Institute of Artificial Intelligence, Xiamen University, Xiamen, China
[5] KLATASDS-MOE, Shanghai, China

Abstract. Black-box unsupervised domain adaptation (B^{2}UDA) is a challenging task in unsupervised domain adaptation, where the source model is treated as a black box and only its output is accessible. Previous works have treated the source models as a pseudo-labeling tool and formulated B^{2}UDA as a noisy labeled learning (LNL) problem. However, they have ignored the gap between the "shift noise" caused by the domain shift and the hypothesis noise in LNL. To alleviate the negative impact of shift noise on B^{2}UDA, we propose a novel framework called Classifier Decoupling Training (CDT), which introduces two additional classifiers to assist model training with a new label-confidence sampling. *First*, we introduce a self-training classifier to learn robust feature representation from the low-confidence samples, which is discarded during testing, and the final classifier is only trained with a few high-confidence samples. This step decouples the training of high-confidence and low-confidence samples to mitigate the impact of noise labels on the final classifier while avoiding overfitting to the few confident samples. *Second*, an adversarial classifier optimizes the feature distribution of low-confidence samples to be biased toward high-confidence samples through adversarial training, which greatly reduces intra-class variation. *Third*, we further propose a novel ETP-entropy Sampling (E^{2}S) to collect class-balanced high-confidence samples, which leverages the early-time training phenomenon into LNL. Extensive experiments on several benchmarks show that the proposed CDT achieves 88.2%, 71.6%, and 81.3% accuracies on Office-31, Office-Home, and VisDA-17, respectively, which outperforms state-of-the-art methods.

Keywords: Domain adaptation · Adversarial learning · Noisy label

1 Introduction

Deep neural networks (DNNs) have achieved remarkable performance on various visual recognition tasks [11,14] with laborious large-scale labeled data. Obvi-

Q. Liu et al. (Eds.): PRCV 2023, LNCS 14427, pp. 16–30, 2024.
https://doi.org/10.1007/978-981-99-8435-0_2

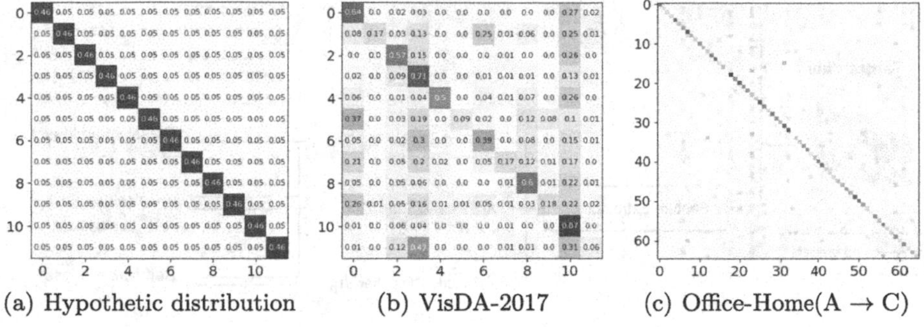

(a) Hypothetic distribution (b) VisDA-2017 (c) Office-Home(A $\rightarrow$ C)

Fig. 1. Confusion matrix for three different cases, the value in row r, column c denotes the proportion of samples belonging to category r that are labeled as c: (a) Uniformly distributed noise, widely adopted by Learning with Noisy Label (LNL) works [5, 10]; (b) Realistic noise matrix in VisDA-2017 [25] dataset based on the black-box source model f_s. The distribution of noise is imbalanced, and for some categories, the proportion of wrong labels is higher than that of correct labels, which seriously deteriorates the training; (c) Realistic noise matrix in Office-Home [27] (A $\rightarrow$ C), the distribution is similar to (b) with more classes.

ously, it is not only expensive but also not practical to label such massive data in various environments. Therefore, transfer learning [23] has received a great deal of research focus, especially on unsupervised domain adaptation (UDA) [4,8,13]. UDA aims to transfer the knowledge learned from a source domain to improve model performance on a target domain without labeled data, which has been widely used in natural image processing [11] and medical image analysis [7] to solve the domain shift problem. Existing UDA methods assume that the source domain data is available during transferring, however, it is not always practical in many situations due to data privacy, security issues, and the difficulty of data transmission.

As a solution, source-free unsupervised domain adaptation (SFUDA) [17] explores the model parameters to obtain the source domain knowledge without accessing source domain data during the adaptation. This approach is also known as white-box unsupervised domain adaptation (WBUDA). Although the source domain data is not used, WBUDA still suffers from data leakage because the knowledge inherent in model parameters may synthesize source domain data via generative methods [29,30]. To enhance data privacy, black-box unsupervised domain adaptation (B^2UDA) further limits the usage of the source domain model, where only the outputs of source domain models are exposed to learn the target domain model.

B^2UDA is more challenging than WBUDA due to the limited availability of source domain knowledge. Several recent attempts [19,24,31] take the black-box model as a pseudo labeling tool and introduce learning with noisy label (LNL) [5, 10] methods to train the target domain model with the noisy labels. For example, DINE introduces label smoothing and exponential moving average prediction to

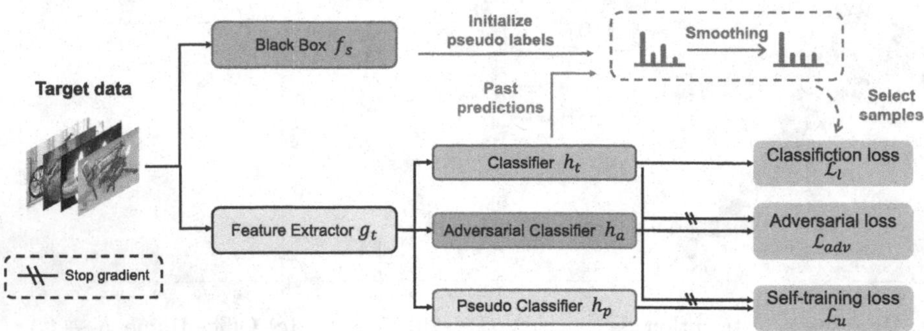

Fig. 2. The overall framework of the proposed method. The classifier h_t is trained with only the most confident and class-balanced samples, and pseudo classifier h_p and adversarial classifier h_a are introduced to decouple the training of high-confidence samples and low-confidence samples.

deal with noisy labels. However, these methods ignore the gap between the task setting of LNL and the B^{2}UDA problem. LNL commonly assumes a simple noise distribution, as shown in Fig. 1(a), where the noise is uniformly distributed across classes with pre-defined noise rates, which may fail to handle realistic noises. Under the B^{2}UDA setting, pseudo labels generated by the source model are subject to significant and imbalanced noises due to domain shift, as illustrated in Fig. 1(b). We refer to this noise caused by domain shift as **shift noise**, which is different from the noise distribution in general LNL tasks. Such shift noise seriously deteriorates model performance during the training process, and even dramatically drops the classification accuracies of some categories to zeros (see Fig. 4(b) in experiments).

In this paper, we focus on mitigating the negative impact caused by the shift noise to improve overall performance for B^{2}UDA. To this end, we propose a novel Classifier Decoupled Training (CDT) framework to identify and decouple the training of high-confidence and low-confidence data. Figure 2 presents our CDT framework. Firstly, we restrict the final target classifier for inference to be optimized only on high-confidence samples and introduce a pseudo classifier h_p to learn robust feature extractor from the remaining low-confidence data. Secondly, to avoid the bias caused by insufficient high-confidence data, an adversarial classifier h_a is designed to reduce the prediction error of the target model on low-confidence data via adversarial learning. Thirdly, we further introduce a new ETP-entropy sampling (E^{2}S) method to construct clean and class-balanced high-confidence training examples, which collects seldom high-confidence data based on the initialized labels at each training epoch. We find that clean samples at the start of training have lower entropy, which is also observed as the early-time training phenomenon (ETP) [20] in LNL tasks and allows us to use entropy as a metric for selecting samples. Thus we leverage ETP to improve the label quality by taking advantage of the model's tendency to effectively learn from clean samples rather than noisy ones.

The main contributions of our work are: (1) We propose a new training Classifier Decoupled Training CDT framework for black-box unsupervised domain adaptation (B^2UDA), which separates the classifier training of high-confidence data and low-confidence data to alleviate the influence of noise samples. (2) Based on the early-time training phenomenon (ETP) [20], we propose a ETP-entropy sampling (E^2S) method to obtain a clean class-balanced dataset. The size of the dataset is determined by the sample noise estimation, so there is no need to manually tune the hyper-parameters for various datasets. (3) Extensive experiments demonstrate that our proposed CDT consistently achieves significant performance gain on several widely-used benchmarks. The results of ablation experiments further demonstrate that CDT has an alleviating effect on shift noise.

2 Related Work

Unsupervised Domain Adaptation. (UDA) aims to transfer a model from a labeled source domain to an unlabeled target domain. Currently, UDA approaches can be divided into three categories: (1) Refine the feature representation of the target domain data to align with that of the source domain data [8,21]. The classical approach DANN [8] introduces an additional classifier to distinguish between the source domain data and the target domain data. Then the features of the target domain data are aligned with those of the source domain via adversarial training. (2) Generate target domain style data by leveraging labeled source domain data through generation models [4,6]. CrDoCo [4] introduces CycleGAN [33] to translate images between the source and target domains and proposes a cross-domain consistency method to enforce the adapted model to produce consistent predictions. However, generation-based methods usually lack control and change the semantic structure of the image. (3) Iteratively re-train the target domain samples with pseudo labels [13,17]. CVRN [13] exploits the complementarity of pseudo labels generated from different views of the same image to improve the quality of pseudo labels. Although UDA approaches have achieved significant success, they require to access source domain data, which raises privacy and portability issues and renders them inapplicable in the context of B^2UDA.

Black-Box Unsupervised Domain Adaptation. (B^2UDA) is first proposed in CRONUS [2], which is a subset problem of UDA and a new research area. B^2UDA is more restrictive than UDA in that models trained on source domain data are put into black boxes and only the outputs of these models are accessible. The pioneering work IterLNL [31] treats B^2UDA as an LNL problem, estimating the noise rate and gradually extracting clean labels from noisy labels. DINE [19] introduces knowledge distillation and mixup regularization to improve performance. RAIN [24] introduces frequency decomposition to capture the key objects and reduce background interference in the mixup strategy. These works focus on sample selection and data augmentation to enhance the generalization of the model and have achieved good results. However, they neglect the shift noise problem, which is our work's focus.

3 Method

3.1 Problem Definition

In B^{2}UDA setting, we are given access to a source model f_s trained on some inaccessible labeled source dataset $\mathcal{D}_\text{s} = \{x_i^s, y_i^s\}_{i=1}^{N_s}$ where x_i^s represents an image and y_i^s represents its associated label from the set of source classes $\mathcal{Y}_\text{s}$. Additionally, we have an unlabeled target dataset $\mathcal{D}_\text{t} = \{x_i^t\}_{i=1}^{N_t}$ with the same label space as the source domain, *i.e.*, $\mathcal{Y}_\text{t} = \mathcal{Y}_\text{s}$. However, the styles of the data in the target domain and the source domain are different. The objective is to utilize the source model f_s and target dataset $\mathcal{D}_\text{t}$ to train a model f_t that provides accurate predictions y_i^t for the target images x_i^t. It's worth noting that this setting is more restrictive than SFUDA [17], as B^{2}UDA does not have access to the parameters of the source model or the ability to extract feature vectors from the source model's middle outputs. To confront these constraints, a native solution is using knowledge distillation, which can be formulated as:

$$\mathcal{L}\left(f_\text{t}; \mathcal{D}_\text{t}, f_\text{s}\right) = \mathbb{E}_{x_t \in \mathcal{D}_\text{t}} \mathcal{L}_\text{kl}\left(f_\text{s}\left(x_t\right) \| f_\text{t}\left(x_t\right)\right), \tag{1}$$

where $\mathcal{L}_\text{kl}$ denotes the Kullback-Leibler (KL) divergence loss. However, the predictions from the source model f_s on target domain data $\mathcal{D}_\text{t}$ are subject to noise due to domain shift, resulting in an extremely imbalanced cross-class prediction error rate that we refer to as shift noise. Moreover, such shift noise increases with training, thus affecting the overall prediction accuracy.

3.2 Overall Framework

To mitigate shift noise during training, we propose a novel classifier decoupled training and an ETP-entropy sampling to learn the target dataset with noisy pseudo labels. The training pipeline is shown in Fig. 2. We formulate the overall objectives in two steps:

$$\mathcal{L} = \underbrace{\mathcal{L}_\text{warm}}_{\text{step 1}} + \underbrace{\left(\mathcal{L}_\text{c} + \mathcal{L}_\text{u} + \mathcal{L}_\text{adv}\right)}_{\text{step 2}}. \tag{2}$$

In step 1, a warmup scheme is adapted to initialize the target model on the target domain via $\mathcal{L}_\text{warm}$ loss. In step 2, we incorporate three losses, including classification loss $\mathcal{L}_\text{c}$, self-training loss $\mathcal{L}_\text{u}$ and adversarial loss $\mathcal{L}_\text{adv}$, to decouple the training of the target dataset with suspicious or confident labels estimated by ETP-entropy sampling. The optimization process is presented in Algorithm 1.

The warmup phase aims to ensure reliable pseudo labels at the start of training, which lasts for a few epochs. Specifically, we feed the entire target dataset into the source domain model to get initialized soft labels, which is the same approach used by DINE [19]. We use these labels to warm up g_t and pseudo classifier h_p so that the model initially adapts to the target domain and has a

Algorithm 1 Pseudocode of CDT for B²UDA.

[1]

1. Warmup Phase:
Require: Target data $\{x_t^i\}_{i=1}^{N_t}$, the number of epochs T_w, and iterations n_b in one epoch.
▷ Initialize the pseudo labels $P(x)$ via Eq. (3)
for $e = 1$ **to** T_w **do**
  for $i = 1$ **to** n_b **do**
    Train g_t and h_p by minimizing Eq. (3).
  end for
end for
2. Classifier Decoupled Training:
Require: Target data $\{x_t^i\}_{i=1}^{n_t}$, the number of epochs T_m, and iterations n_b in one epoch.
for $e = 1$ **to** T_m **do**
  ▷ Update the pseudo labels $P(x)$, and obtain high-confidence dataset $\mathcal{D}_l$ and low-confidence dataset $\mathcal{D}_u$ via Eq. (11).
  for $i = 1$ **to** n_b **do**
    ▷ Sample a batch B_i^1 from high-confidence dataset $\mathcal{D}_l$.
    ▷ Train g_t and h_t by minimizing Eq. (4) with B_i^1.
    ▷ Sample a batch B_i^2 from low-confidence dataset $\mathcal{D}_u$.
    ▷ Train g_t and h_p by minimizing Eq. (5) with B_i^2.
    ▷ Train g_t, h_t and h_a by minimizing Eq. (6) with B_1 and B_2.
  end for
end for

good feature space for the next step of training. The training objective of the warmup phase is:

$$\mathcal{L}_{\text{warm}}(g_t, h_p) = \mathbb{E}_{x_t \in \mathcal{D}_t} \mathcal{L}_{\text{kl}} \left((h_p \circ g_t)(x_t) \,\|P(x_t) \right),$$
$$P(x)_k = \begin{cases} f_s(x)_k, & k = \arg\max(f_s(x)) \\ (1 - \max(f_s(x)))/(K-1), & \text{otherwise} \end{cases}, \tag{3}$$

where $P(x_t)$ is the soft label initialized by the source model through adaptive label smoothing, and K is the number of classes.

3.3 Classifier Decoupled Training (CDT)

In addition to the target classifier p_t, the proposed CDT involves training extra pseudo classifier h_p and adversarial classifier p_a. The final target domain model f_t is the combination of the features extractor g_t and classifier h_t. *The adversarial classifier p_a and pseudo classifier h_p are discarded after training so that it does not increase inference time.*

Confident Learning. Different from the previous works [24,31] that employ the entire unlabeled target dataset for training the final classifier h_t, we segregate high-confidence subset $\mathcal{D}_l$ and low-confidence subset $\mathcal{D}_u$ from the target dataset

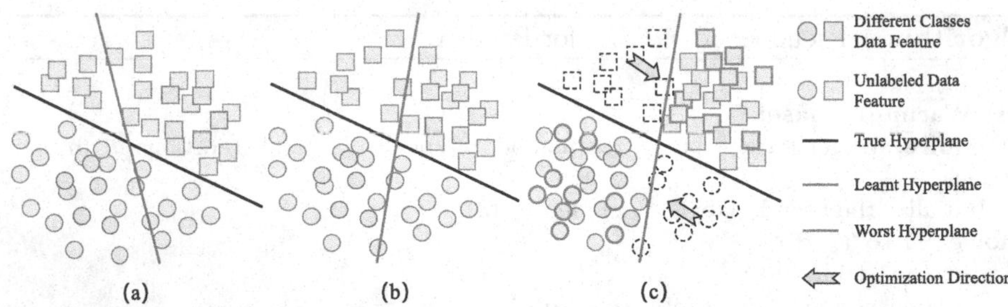

Fig. 3. The hyperplane is viewed as a representation of the classifier in the feature space. (a) Shift between the hyperplanes (*i.e.* h_t) learned on limited high-confidence data and the true hyperplanes. (b) The worst hyperplane (*i.e.* h_p) that correctly distinguishes high-confidence data while making as many mistakes as possible on low-confidence data. (c) Feature representations are optimized to improve the performance of the worst hyperplane.

by ETP-entropy sampling, which is detailed in the following section, and solely utilize high-confidence data to train h_t. The training objective of high-confidence dataset for target model f_t is as follows:

$$\mathcal{L}_{\mathrm{c}}(g_t, h_t) = \mathbb{E}_{x_l, y_l \in \mathcal{D}_{\mathrm{l}}} \mathcal{L}_{\mathrm{ce}} \left((h_t \circ g_t)(x_l), y_l \right). \tag{4}$$

As only high-confidence samples are utilized in this training and the classes are evenly distributed, thus h_t is less susceptible to shift noise. Here, y_l is a pseudo label of x_l in $\mathcal{D}_{\mathrm{l}}$ Eq. 11.

Pseudo-Label Learning. Since the size of $\mathcal{D}_l$ is relatively small compared to the entire dataset, we incorporated low-confidence data $\mathcal{D}_{\mathrm{u}}$ to prevent the target model from overfitting to few high-confidence data. As shown in Fig. 2, we introduce a pseudo classifier h_p on the top of feature extractor g_t and only optimize it with $\mathcal{D}_{\mathrm{u}}$. The training objective on the low-confidence dataset is

$$\begin{aligned} \mathcal{L}_{\mathrm{u}}(g_t, h_p, h_t) = \mathbb{E}_{x_u \in \mathcal{D}_{\mathrm{u}}} \mathbb{1}\left(\max\left(q_u\right) \geq \tau_u\right) \mathcal{L}_{\mathrm{ce}}\left(\left(h_p \circ g_t \circ \mathcal{A}\right)(x_u), \arg\max\left(q_u\right)\right), \\ q_u = \left(h_t \circ g_t \circ \alpha\right)(x_u), \end{aligned} \tag{5}$$

where $\mathcal{D}_u$ is the low-confidence dataset, τ_u is the threshold for filtering out clean samples, $\mathcal{A}$ is the strong data augmentation function, and α is the weak data augmentation function. We use the probability vector q_u, predicted by h_t for x_i, as a pseudo label to train h_p and g_t. Although q_u may contain wrong labels, the parameters of h_p are independent of h_t, which prevents h_t from directly accumulating the bias of noisy labels during training. This training step allows the feature extractor g_t to learn a better feature representation from $x^u \in \mathcal{D}_u$.

Adversarial Learning. Due to the scarcity of high-confidence samples $\mathcal{D}_l$, there may be significant variations in terms of the distance between supporting data points of each category and the true decision hyperplane, as illustrated in Fig. 3. Therefore, the learned decision hyperplane may have a bias towards certain categories, which could lead to wrong predictions on low-confidence data. To alleviate the bias, we propose the incorporation of an additional classification head h_a for the purpose of conducting adversarial learning as shown in Fig. 2. Firstly, we employ classifier f_t to make predictions on both high-confidence and low-confidence datasets. Then, we optimize classifier h_a to correctly predicts all high-confidence samples while incorrectly predicting low-confidence samples, thereby enabling inverse prediction between h_a and h_t. When classifier h_a makes poor predictions, the adversarial training imposes feature extractor g_t to map the low-confidence samples into a more effective feature space, as shown in Fig. 3. Thus, adversarial learning is formulated as:

$$\begin{aligned}\mathcal{L}_{\text{adv}}(g_t, h_t, h_a) =& \mathbb{E}_{x_l \in \mathcal{D}_\text{l}} \mathcal{L}_{\text{ce}}\left((h_a \circ g_t \circ \alpha)(x_l), \arg\max f_t(x_l)\right) - \\ & \mathbb{E}_{x_u \in \mathcal{D}_\text{u}} \mathcal{L}_{\text{ce}}\left((h_a \circ g_t \circ \alpha)(x_u), \arg\max f_t(x_u)\right).\end{aligned} \tag{6}$$

3.4 ETP-Entropy Sampling

In this section, we propose to sample high-confidence and low-confidence data from the target dataset. As shown in Fig. 1, the source model exhibits highly imbalanced cross-class prediction results when applied to target domain data due to the shift noise. Therefore, the conventional threshold-based sampling method may result in imbalanced category distribution and low data cleaning efficiency. To alleviate the negative impact of shift noise, we propose a balanced sampling method based on the early-time training phenomenon (ETP) [20].

Formally, for the k-th class, we select the N samples that are most likely to belong to the k-th class, regardless of which class the N samples are most likely to belong to, *i.e.*, $y \neq \arg\max(P(x))$. The formulation is constructed as follows:

$$\begin{gathered}P(x) \leftarrow \lambda P(x) + (1-\lambda) f_t(x), \\ \mathcal{D}_l^{k'} = \{(x, y=k) \mid P(x)_k \geq \tau^k\}, \quad |\mathcal{D}_l^{k'}| = N,\end{gathered} \tag{7}$$

where $P(x)_k$ is the probability that sample x belongs to the k-th class, pseudo labels $P(x)$ are updated in each iteration, λ is a momentum hyper-parameter and τ^k represents the N-th largest probability value that all samples belonging to k-th class.

Inspired by the ETP observed in LNL that classifiers have higher prediction accuracy during the early-training stage, we estimate the dataset size N based on current noise rate ϵ and training epoch n as follows:

$$N(n) = \left(1 - \min\left(\frac{n}{T}\epsilon, \epsilon\right)\right) |\mathcal{D}_t|, \tag{8}$$

where T is a hyper-parameter that is set to half of the maximum training epoch in the experiments. To estimate the rate of label noise ϵ, we calculate the predictive distribution $f_t(x)$ across all samples in the current epoch. Although isolated predictive probabilities may be overconfident and misleading, probability statistics are often sufficient for assessing overall classification accuracy as well as

Table 1. Accuracy (%) on Office-31 for B²UDA. H.Avg. represents the average accuracy of challenging transfer tasks, which includes the low half of all transfer tasks in terms of Source only accuracy. "A", "D", and "W" represent the three sub-datasets Amazon, DSLR, and Webcam on Office-31 dataset. The red numbers represent the best results achieved under the black-box setting, excluding those that source models are available.

Method	Type	A→D	A→W	D→A	D→W	W→A	W→D	Avg.	H.Avg.
Source only [11]	Pred.	79.9	76.6	56.4	92.8	60.9	98.5	77.5	64.6
LNL-OT [1]	Pred.	88.8	85.5	64.6	95.1	66.7	98.7	83.2	72.3
LNL-KL [31]	Pred.	89.4	86.8	65.1	94.8	67.1	98.7	83.6	73.0
HD-SHOT [16]	Pred.	86.5	83.1	66.1	95.1	68.9	98.1	83.0	72.7
SD-SHOT [16]	Pred.	89.2	83.7	67.9	95.3	71.1	97.1	84.1	74.2
DINE [19]	Pred.	91.6	86.8	72.2	96.2	73.3	98.6	86.4	77.4
CDT (Ours)	Pred.	94.0	88.9	75.1	96.4	75.3	99.6	88.2	79.8
BSP+DANN [3]	Data	93.0	93.3	73.6	98.	72.6	100.0	88.5	79.8
MDD [32]	Data	93.5	94.5	74.6	98.4	72.2	100.0	88.9	80.4
ATDOC [18]	Data	94.4	94.3	75.6	98.9	75.2	99.6	89.7	81.7

noise rates [9,12]. We calculate the proportion ρ of high-confidence samples in the target dataset:

$$\rho = 1 - \mathbb{E}_{x_t \in \mathcal{D}_t} \mathbb{1}\left[\max\left(f_t\left(x_t\right)\right) > \gamma\right]. \tag{9}$$

In order to mitigate overconfident predictions by the model, we rescale ρ to ρ':

$$\rho' = \begin{cases} -2\left(\rho - 0.5\right)^2 + 0.5, & \rho < 0.5 \\ 2\left(\rho - 0.5\right)^2 + 0.5, & \text{otherwise} \end{cases}. \tag{10}$$

This can be seen as a smoothing operation to avoid over-sampling and under-sampling. Finally, the noise rate is expressed as $\epsilon = 1 - \rho'$.

As the above sampling is only based on the estimated class probability, there may still exist noise in the high-confidence data. We calculate the cross entropy for all samples and select M samples with the minimum cross entropy for each class to construct a high-confidence dataset $\mathcal{D}_l$, while using the remaining samples to form a low-confidence dataset $\mathcal{D}_u$:

$$\begin{aligned} S_k &= \left\{\mathcal{L}_{\text{ce}}\left(x_l, y_l\right) \mid \left(x_l, y_l\right) \in \mathcal{D}_l^{k'}\right\}, \\ \mathcal{D}_l^k &= \{x_l \mid \mathcal{L}_{\text{ce}}\left(x_l, y_l\right) \in \text{top}M(S^k)\}, \\ \mathcal{D}_u^k &= \{x \mid x \in \mathcal{D}_t, x \notin \mathcal{D}_l^k\}, \end{aligned} \tag{11}$$

where S_k represents entropies of samples belonging to the k-th class and top$M(S_k)$ is a subset formed by the M smallest elements of S_k, and its proportion in S_k is a hyper-parameter denoted by $\mu = M/N$.

Table 2. Accuracy (%) on Office-Home for B^{2}UDA. (":" denotes "transfer to")

Method	Type	Ar:Cl	Ar:Pr	Ar:Re	Cl:Ar	Cl:Pr	Cl:Re	Pr:Ar	Pr:Cl	Pr:Re	Re:Ar	Re:Cl	Re:Pr	Avg.	H. Avg.
Source only [11]	Pred.	44.1	66.9	74.2	54.5	63.3	66.1	52.8	41.2	73.2	66.1	46.7	77.5	60.6	50.4
LNL-OT [1]	Pred.	49.1	71.7	77.3	60.2	68.7	73.1	57.0	46.5	76.8	67.1	52.3	79.5	64.9	55.6
LNL-KL [31]	Pred.	49.0	71.5	77.1	59.0	68.7	72.9	56.4	46.9	76.6	66.2	52.3	79.1	64.6	55.4
HD-SHOT [16]	Pred.	48.6	72.8	77.0	60.7	70.0	73.2	56.6	47.0	76.7	67.5	52.6	80.2	65.2	55.9
SD-SHOT [16]	Pred.	50.1	75.0	78.8	63.2	72.9	76.4	60.0	48.0	79.4	69.2	54.2	81.6	67.4	58.1
DINE [19]	Pred.	52.2	78.4	81.3	65.3	76.6	78.7	62.7	49.6	82.2	69.8	55.8	84.2	69.7	60.4
CDT (Ours)	Pred.	56.2	78.2	81.6	66.1	77.3	78.2	67.7	55.4	82.1	72.6	58.5	85.5	71.6	63.5
BSP+CDAN [3]	Data	52.0	68.6	76.1	58.0	70.3	70.2	58.6	50.2	77.6	72.2	59.3	81.9	66.3	58.1
MDD [32]	Data	54.9	73.7	77.8	60.0	71.4	71.8	61.2	53.6	78.1	72.5	60.2	82.3	68.1	60.2
CKB [22]	Data	54.7	74.4	77.1	63.7	72.2	71.8	64.1	51.7	78.4	73.1	58.0	82.4	68.5	60.7

Table 3. Accuracy (%) on VisDA-2017 for B^{2}UDA.

Method	Type	plane	bcycl	bus	car	horse	knife	mcycle	person	plant	sktbrd	train	truck	Avg.	H.Avg.
Source only [11]	Pred.	64.3	24.6	47.9	75.3	69.6	8.5	79.0	31.6	64.4	31.0	81.4	9.2	48.9	25.5
LNL-OT [1]	Pred.	82.6	84.1	76.2	44.8	90.8	39.1	76.7	72.0	82.6	81.2	82.7	50.6	72.0	67.2
LNL-KL [31]	Pred.	82.7	83.4	76.7	44.9	90.9	38.5	78.4	71.6	82.4	80.3	82.9	50.4	71.9	66.8
HD-SHOT [16]	Pred.	75.8	85.8	78.0	43.1	92.0	41.0	79.9	78.1	84.2	86.4	81.0	65.5	74.2	72.5
SD-SHOT [16]	Pred.	79.1	85.8	77.2	43.4	91.6	41.0	80.0	78.3	84.7	86.8	81.1	65.1	74.5	72.4
DINE [19]	Pred.	81.4	86.7	77.9	55.1	92.2	34.6	80.8	79.9	87.3	87.9	84.3	58.7	75.6	71.0
CDT (Ours)	Pred.	80.2	85.4	79.8	70.5	93.4	90.1	86.1	78.6	89.6	90.8	88.0	43.5	81.3	78.0
SAFN [22]	Dara	93.6	61.3	84.1	70.6	94.1	79.0	91.8	79.6	89.9	55.6	89.0	24.4	76.1	64.0
CDAN+E [21]	Data	94.3	60.8	79.9	72.7	89.5	86.8	92.4	81.4	88.9	72.9	87.6	32.8	78.3	69.1
DTA [15]	Data	93.7	82.2	85.6	83.8	93.0	81.0	90.7	82.1	95.1	78.1	86.4	32.1	81.5	73.5

4 Experiment

4.1 Setup

Datasets. Office-31 [26] is a popular benchmark for UDA, consisting of three different domains (<u>A</u>mazon, <u>W</u>ebcam, <u>D</u>SLR) with 31 categories. **Office-Home** [27] is a challenging medium-sized benchmark, consisting of four domains (<u>A</u>rt, <u>C</u>lip, <u>P</u>roduct, <u>R</u>eal World) in 65 categories. **VisDA-17** [25] is a large-scale benchmark for 12-class synthetic-to-real object recognition. The source domain contains 152 thousand synthetic images while the target domain has 55 thousand images.

Implementation Details. We implement our method via Pytorch and report the average accuracies among three runs. The source model f_s is trained using almost all samples in the source domain, leaving three samples per class as validation. We use ResNet50 as the backbone for Office-31 and Office-Home, and ResNet101 as the backbone for VisDA-2017. The parameters of the backbone are initialized using parameters pre-trained on ImageNet. An MLP-based classifier is added on the top of the backbone as the previous works [19,24,31]. The model is optimized by mini-batch SGD with the learning rate $1e-2$ for the new classifier and $1e-3$ for the pre-trained backbone. The warm-up epoch is empirically set to 5, and the training epoch is 50 except 10 for VisDA-2017. The learning scheduler is kept the same as DINE [19].

Baseline. We devise the following baseline methods to compare with our method. LNL-KL [31] and LNL-OT [1] are LNL methods and respectively adopt the diversity-promoting KL divergence and optimal transport to refine the noisy pseudo labels. HD-SHOT and SD-SHOT first warm up the target model by self-training and then exploit SHOT [17] by cross-entropy loss and weighted cross-entropy loss, respectively. We also compare with some traditional UDA methods, for example, MDD [32] and SAFN [28].

4.2 Performance Comparison

We show the results of cross-domain object recognition on Office-31, Office-Home, and VisDA-17 in Tables 1, 2, and 3, respectively. And type "Pred." denotes the black-box method, "Data" denotes the traditional UDA method. The proposed CDT achieves the best performance on all benchmarks. On average, our method outperforms the previous state-of-the-art DINE method by 1.8%, 1.9%, and 5.7% on Office-31, Office-Home, and VisDA-17, respectively. On hard transfer tasks (*i.e.*, VisDA-17), in which the predictions of the source model are extremely noisy, we achieve even more significant results. Specifically, we achieve 2.4%, 3.2%, and 7.0% improvement on Office-31, Office-Home, and Visda-2017, respectively. This indicates that our method handles transfer tasks with severe shift noise. In addition, we randomly selected 7 categories in transfer task Pr $\rightarrow$ Cl (Office-Home), whose source-only accuracy is shown in Fig. 4(a). The performance of DINE [19] on this task is illustrated in Fig. 4(b). Compared to the baseline method, although the overall accuracy of DINE [19] has improved, the accuracy of some categories decreases, where "soda" and "clipboard" dramatically drop to zero prediction accuracy. Our CDT achieves stable improvement across all categories, providing evidence that our proposed structure effectively alleviates shift noise.

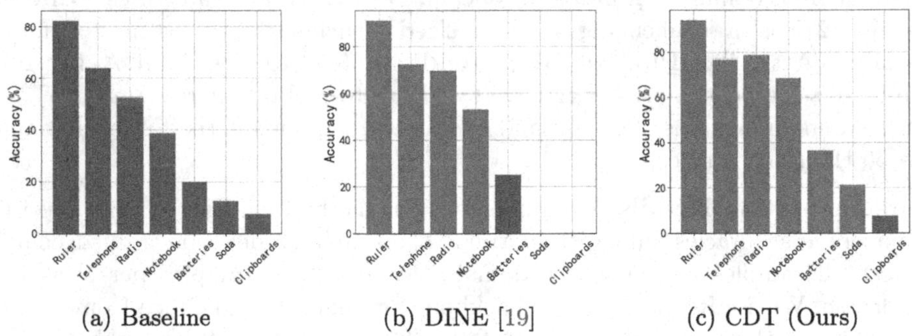

(a) Baseline (b) DINE [19] (c) CDT (Ours)

Fig. 4. Accuracy of different B^2UDA methods with 7 randomly selected categories on Office-Home (Pr $\rightarrow$ Cl). Affected by shift noise, DINE [19] significantly aggravates the prediction error of poorly-behaved classes, whereas our CDT effectively alleviates this phenomenon.

Table 4. Ablation. Results on Office-Home with different variants.

Method	Ar→Cl	Cl→Ar	Cl→Pr	Pr→Ar	Pr→Cl	Re→Cl	H.Avg.
Source only	44.1	54.5	63.3	52.8	41.2	46.7	50.4
CDT (Single Classifier)	50.1	57.2	66.2	58.3	47.4	49.7	54.8
CDT (w/o $\mathcal{L}_{\mathrm{u}}$)	50.4	60.6	72.2	62.4	51.5	53.3	58.4
CDT (w/o $\mathcal{L}_{\mathrm{adv}}$)	51.3	62.3	74.2	65.1	53.4	55.4	60.3
CDT (w/o etp)	55.5	65.7	75.8	66.7	54.7	57.8	62.7
CDT	56.2	66.1	77.3	67.7	55.4	58.5	63.5

4.3 Analysis

Ablation Study. We investigated the contribution of key components in CDT, as shown in Table 4. "Single Classifier" means that we use a native knowledge distillation method. "w/o $\mathcal{L}_{\mathrm{u}}$" means that after the warmup phase, the classifier h_p is discarded and only the high-confidence data is used to train the classifier h_t. The 5.1% drop in accuracy indicates the importance of low-confidence data in training. "w/o $\mathcal{L}_{\mathrm{adv}}$" means that we discard classifier h_a and the loss function $\mathcal{L}_{\mathrm{adv}}$. It can be seen that adversarial training contributes to 3.2% improvements. "w/o etp" means the selection of high-confidence data is based solely on confidence. It contributes to 1.2% improvements, which is better than the selection solely on confidence.

Hyper-parameter Sensitivity. We study the hyper-parameter μ and τ_u on Office-Home (Pr → Cl) across three runs. We select values between 0.3 and 0.9 for both μ and τ_u, as a too-small μ results in insufficient data, while a too-small τ_u leads to noise accumulation. As shown in Fig. 5(a), for μ the accuracies of CDT range from 46.1% to 55.4%, and the best result is achieved at 0.6. For τ_u, the accuracies of CDT rang from 49.2% to 55.4%, and the best result is achieved at 0.7.

Pseudo Labels for Poorly-Behaved Classes. To assess the effect of CDT in alleviating shift noise, we calculate the class imbalance ratio $I = \max_c N(c)/\max'_c N(c')$ for each iteration, where $N(c)$ represents the number of predictions assigned to class c. As shown in Fig. 5(b), DINE [19] results in a rapid increase of I to infinity, indicating that pseudo labels for poorly-behaved classes drop to zero, whereas our CDT approach avoids such an outcome. We also measure the quality of pseudo labels for the 10 and 20 worst-behaved classes and express it as average accuracy. As shown in Fig. 5(c), DINE exhibits an average accuracy of 0% and 13.1% on the poorly-behaved 10 and 20 classes. By effectively alleviating shift noise, CDT achieves a significant improvement of average accuracy by 16.3% and 12.1%.

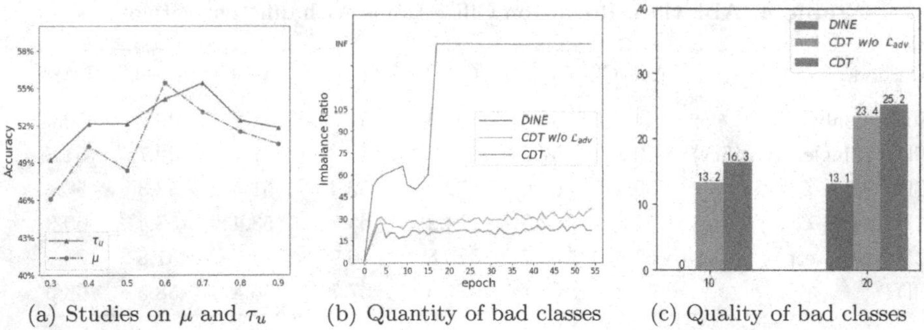

(a) Studies on μ and τ_u (b) Quantity of bad classes (c) Quality of bad classes

Fig. 5. The quantity and quality of pseudo labels and hyper-parameter sensitivity on Office-Home (Pr→Cl).

5 Conclusion

In this work, we propose CDT, a new training paradigm for B^{2}UDA. The target domain is divided into high-confidence and low-confidence subsets by our ES sampling method, and a classifier for unlabeled data training is introduced to alleviate the impact of shift noise on classifier training. In addition, to prevent the classifier from making poor predictions on low-confidence data due to the lack of labeled data, we introduce another classifier for adversarial training. Extensive experiments on various benchmarks show that CDT effectively suppresses the degradation of prediction accuracy for categories with high initial pseudo-label error rates, and achieves new state-of-the-art performance on several widely-used benchmarks.

Acknowledgments. This work is supported by the National Natural Science Foundation of China (NO. 62102151), Shanghai Sailing Program (21YF1411200), CCF-Tencent Open Research Fund, the Open Research Fund of Key Laboratory of Advanced Theory and Application in Statistics and Data Science, Ministry of Education (KLATASDS2305), the Fundamental Research Funds for the Central Universities.

References

1. Asano, Y.M., Rupprecht, C., Vedaldi, A.: Self-labelling via simultaneous clustering and representation learning. arXiv preprint: arXiv:1911.05371 (2019)
2. Chang, H., Shejwalkar, V., Shokri, R., Houmansadr, A.: Cronus: robust and heterogeneous collaborative learning with black-box knowledge transfer. arXiv preprint: arXiv:1912.11279 (2019)
3. Chen, X., Wang, S., Long, M., Wang, J.: Transferability vs. discriminability: batch spectral penalization for adversarial domain adaptation. In: ICML (2019)
4. Chen, Y.C., Lin, Y.Y., Yang, M.H., Huang, J.B.: CrDoCo: pixel-level domain transfer with cross-domain consistency. In: CVPR (2019)
5. Cheng, D., et al.: Instance-dependent label-noise learning with manifold-regularized transition matrix estimation. In: CVPR (2022)

6. Cui, K., Huang, J., Luo, Z., Zhang, G., Zhan, F., Lu, S.: GenCo: generative co-training on data-limited image generation. arXiv preprint: arXiv:2110.01254 (2021)
7. Dou, Q., Ouyang, C., Chen, C., Chen, H., Heng, P.A.: Unsupervised cross-modality domain adaptation of convnets for biomedical image segmentations with adversarial loss. In: IJCAI (2018)
8. Ganin, Y., et al.: Domain-adversarial training of neural networks. JMLR (2016)
9. Guo, C., Pleiss, G., Sun, Y., Weinberger, K.Q.: On calibration of modern neural networks. In: ICML (2017)
10. Han, B., et al.: Co-teaching: robust training of deep neural networks with extremely noisy labels. In: NeurIPS (2018)
11. He, K., Zhang, X., Ren, S., Sun, J.: Deep residual learning for image recognition. In: CVPR (2016)
12. Hendrycks, D., Gimpel, K.: A baseline for detecting misclassified and out-of-distribution examples in neural networks. arXiv preprint: arXiv:1610.02136 (2016)
13. Huang, J., Guan, D., Xiao, A., Lu, S.: Cross-view regularization for domain adaptive panoptic segmentation. In: CVPR (2021)
14. Krizhevsky, A., Sutskever, I., Hinton, G.E.: ImageNet classification with deep convolutional neural networks. In: NeurIPS (2012)
15. Lee, S., Kim, D., Kim, N., Jeong, S.G.: Drop to adapt: learning discriminative features for unsupervised domain adaptation. In: CVPR (2019)
16. Liang, J., Hu, D., Wang, Y., He, R., Feng, J.: Source data-absent unsupervised domain adaptation through hypothesis transfer and labeling transfer. PAMI (2020)
17. Liang, J., Hu, D., Feng, J.: Do we really need to access the source data? Source hypothesis transfer for unsupervised domain adaptation. In: ICML (2020)
18. Liang, J., Hu, D., Feng, J.: Domain adaptation with auxiliary target domain-oriented classifier. In: CVPR (2021)
19. Liang, J., Hu, D., Feng, J., He, R.: DINE: domain adaptation from single and multiple black-box predictors. In: CVPR (2022)
20. Liu, S., Niles-Weed, J., Razavian, N., Fernandez-Granda, C.: Early-learning regularization prevents memorization of noisy labels. In: NeurIPS (2020)
21. Long, M., Cao, Z., Wang, J., Jordan, M.I.: Conditional adversarial domain adaptation. In: NeurIPS (2018)
22. Luo, Y.W., Ren, C.X.: Conditional bures metric for domain adaptation. In: CVPR (2021)
23. Pan, S.J., Yang, Q.: A survey on transfer learning. TKDE (2010)
24. Peng, Q., Ding, Z., Lyu, L., Sun, L., Chen, C.: Toward better target representation for source-free and black-box domain adaptation. arXiv preprint: arXiv:2208.10531 (2022)
25. Peng, X., Usman, B., Kaushik, N., Hoffman, J., Wang, D., Saenko, K.: VisDA: the visual domain adaptation challenge. arXiv preprint: arXiv:1710.06924 (2017)
26. Saenko, K., Kulis, B., Fritz, M., Darrell, T.: Adapting visual category models to new domains. In: Daniilidis, K., Maragos, P., Paragios, N. (eds.) ECCV 2010. LNCS, vol. 6314, pp. 213–226. Springer, Heidelberg (2010). https://doi.org/10.1007/978-3-642-15561-1_16
27. Venkateswara, H., Eusebio, J., Chakraborty, S., Panchanathan, S.: Deep hashing network for unsupervised domain adaptation. In: CVPR (2017)
28. Xu, R., Li, G., Yang, J., Lin, L.: Larger norm more transferable: an adaptive feature norm approach for unsupervised domain adaptation. In: ICCV (2019)
29. Yin, H., Mallya, A., Vahdat, A., Alvarez, J.M., Kautz, J., Molchanov, P.: See through gradients: image batch recovery via gradinversion. In: CVPR (2021)

30. Yin, H., et al.: Dreaming to distill: Data-free knowledge transfer via deepinversion. In: CVPR (2020)
31. Zhang, H., Zhang, Y., Jia, K., Zhang, L.: Unsupervised domain adaptation of black-box source models. arXiv preprint: arXiv:2101.02839 (2021)
32. Zhang, Y., Liu, T., Long, M., Jordan, M.: Bridging theory and algorithm for domain adaptation. In: ICML (2019)
33. Zhu, J.Y., Park, T., Isola, P., Efros, A.A.: Unpaired image-to-image translation using cycle-consistent adversarial networks. In: ICCV (2017)

Unsupervised Concept Drift Detection via Imbalanced Cluster Discriminator Learning

Mingjie Zhao[1], Yiqun Zhang[1(✉)], Yuzhu Ji[1], and Yang Lu[2]

[1] School of Computer Science and Technology, Guangdong University of Technology, Guangzhou, China
2112205249@mail2.gdut.edu.cn, yqzhang@gdut.edu.cn
[2] Fujian Key Laboratory of Sensing and Computing for Smart City, School of Informatics, Xiamen University, Xiamen, China
luyang@xmu.edu.cn

Abstract. Streaming data processing has attracted much more attention and become a key research area in the fields of machine learning and data mining. Since the distribution of real data may evolve (called concept drift) with time due to many unforeseen factors and real data is usually with imbalanced cluster/class distributions during streaming data processing, drifts occurred in distributions with fewer data objects are easily masked by the larger distributions. This paper, therefore, proposes an unsupervised drift detection approach called Multi-Imbalanced Cluster Discriminator (MICD) to address the more challenging imbalance problem of unlabeled data. It first partitions data into compact clusters, and then learns a discriminator for each cluster to detect drift. It turns out that MICD can detect drift occurrence, locate where the drift occurs, and quantify the extent of the drift. MICD is efficient, interpretable, and has easy-to-set parameters. Extensive experiments on synthetic and real datasets illustrate the superiority of MICD.

Keywords: Streaming data · imbalanced distribution · concept drift · unsupervised drift detection · clustering

1 Introduction

With the popularization of data collection equipment, new data is generated and aggregated all the time to form streaming data. The processing and analysis of streaming data have thus become an important research topic [1–4] in the field of pattern recognition, machine learning, and data mining. The primary difference between streaming data and static data lies not only in the continuous arrival of data, but also in the change of data distribution that may occur at any time over time, which is also called concept drift [5–7].

When concept drift occurs, the previously learned knowledge may be outdated, leading to a decrease in the accuracy of data analysis tasks [8]. Therefore, it is crucial to detect drifts for streaming data [9]. Existing drift detection

Q. Liu et al. (Eds.): PRCV 2023, LNCS 14427, pp. 31–43, 2024.
https://doi.org/10.1007/978-981-99-8435-0_3

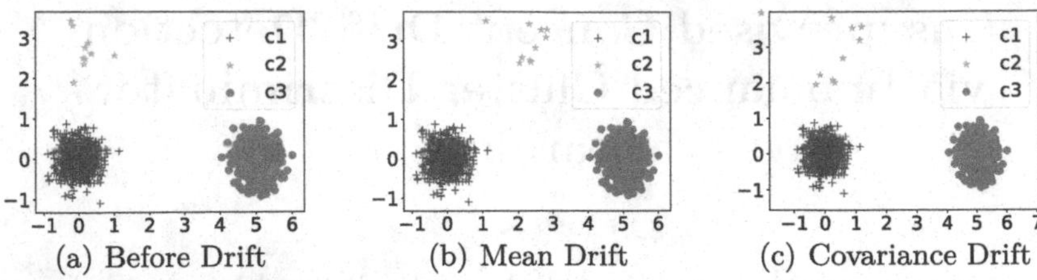

Fig. 1. A toy example of imbalanced clusters (a) before the occurrence of drift, (b) after mean drift, and (c) after covariance drift.

approaches are usually based on the analysis of a fixed number of data objects collected in continuous time (also called data chunks hereinafter) to judge the occurrence of drifts. The existing approaches can be roughly divided into (1) model performance change-based, and (2) distribution change-based, according to how they quantify the occurrence of drifts.

Model performance change-based methods are the predominant approach among existing concept drift detection methods. They monitor the statistics of the model output, and identify drift when a significant change occurs. Typical approaches include OCDD [5], D3 [6], DDM [10], EDDM [11], and ADWIN [12]. However, they focus on the overall detection performance rather than locating the exact drift and capturing the drift severity. Additionally, most of them require data labels for the drift detection.

Distribution change-based solutions detect drifts by quantifying the disparity between the historical chunk and the new chunk. Typical approaches include EI-KMeans (EIKM) [13], kdqTree [14], QuantTree (QTree) [15] and Mann-Whitney-Wilcoxon (MWW) [16]. They do not necessarily rely on labels and are thus more practical to real data. Although they have achieved promising drift detection performance, their effectiveness will be easily influenced by the common imbalance effect illustrated in Fig. 1, where the drifts in minor clusters are easy to be overlooked because this type of approach monitors the overall change.

By implementing EIKM, QTree, OCDD, and MWW on the two drift cases shown in Fig. 1, it can be found that although the two drift cases are intuitively obvious, all the four trial approaches fail to detect drift due to the imbalance issue. Furthermore, the distribution imbalance issue is more complex in reality, since there are usually multiple imbalanced clusters with changing distributions, which makes drift detection more challenging. To tackle this, it is imperative to develop a drift detector that can appropriately monitor the concept drift of multiple imbalanced distributions in an unsupervised dynamic environment.

This paper, therefore, proposes a new concept drift detection method called Multi-Imbalanced Cluster Discriminator (MICD), to detect if the distribution of multiple imbalanced clusters has changed in a newly arrived chunk. It first partitions a chunk into possibly imbalanced clusters and then trains a one-class classifier for each of them to achieve a detailed distribution description. It turns out that each classifier in MICD tracks the distribution of a certain cluster, and thus drifts can be detected accurately and timely, even if they occur in a minor cluster with a very small number of objects. Moreover, since the one-class classifiers are independent of each other, MICD can finely locate the drifts in different clusters according to their corresponding classifiers. To the best of our knowledge, this is the first attempt to specifically consider the imbalance issue in unsupervised concept drift detection. The main contributions are four-fold:

- A new paradigm for unsupervised drift detection has been proposed. It first partitions objects into probably imbalanced clusters, and then trains descriptors for each of them to avoid the overlook of drifts in minor clusters.
- To avoid the domination of major clusters, a partition and merging strategy is adopted, which initializes many extra seed points in a chunk to avoid the missing minor clusters, and then merges them to detect major clusters.
- Compared with the existing approaches that report drift occurrence based on the changes on global data statistics, our MICD can better understand concept drifts by finely locating the drift position and reflecting its severity.
- Comprehensive experiments have been conducted on 12 benchmark datasets. Experimental results illustrate that MICD achieves superior detection accuracy, and is robust to different degrees of imbalance and drift severity.

2 Related Works

This section overviews existing related works, including approaches proposed for concept drift detection and imbalanced data clustering.

2.1 Concept Drift Detection

Dasu et al. [14] utilizes kdqTree's spatial partitioning scheme to divide data into multiple uniformly dense subspaces. They construct histograms based on this partitioning and employ statistical tests to detect changes in data distributions. However, their partitioning approach relies heavily on the data itself and can significantly decrease efficiency when dealing with high-dimensional data. Similarly, the QuantTree method [15] employs a recursive binary partitioning scheme for individual covariates to divide the feature space, greatly improving partitioning efficiency for high-dimensional data [17,18]. However, both of them do not guarantee that the partitions are consistent with the true underlying distribution. As a result, these methods may introduce randomness and lack interpretability in the generated partitions. Lu et al. [13] proposed EIKM based on K-Means to partition and describe data distribution using histogram and then perform drift detection using the Chi-square test. However, since it forces each cluster to have enough number of objects to ensure the statistical meaning of the test, the partition result may not be entirely consistent with the underlying distribution.

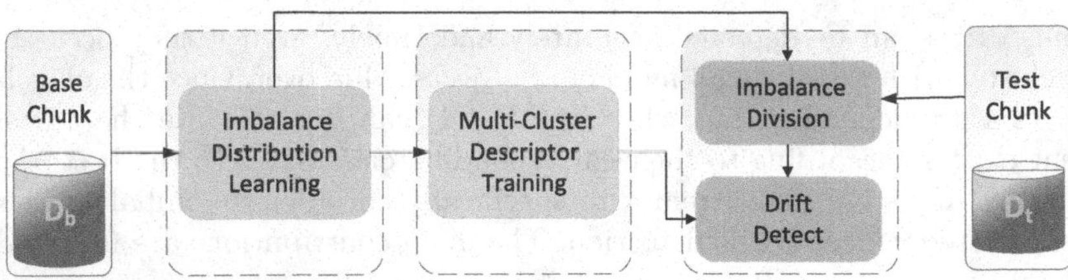

Fig. 2. Workflow of the proposed MICD.

2.2 Imbalance Data Clustering

In real complex data clustering analysis [19–21], it is common to encounter the imbalance issue or more generally the long-tail problem [22–24]. Most existing imbalanced data stream learning methods are based on the availability of labels [25]. Under unsupervised scenario [26], Multi-Exemplar Merging clustering (MEMC) utilizes an affinity propagation mechanism to merge the data objects into multiple sub-clusters. An approach combining bagging and boosting in an ensemble manner has also been proposed. They employ a new consensus function to generate combined clustering results and determine the cluster number k. On the other hand, Aksoylar et al. [27] utilize the spectral clustering method to reflect different levels of data imbalance, and then obtain the correct number of clusters accordingly. Most recently, SMCL [28] has been proposed adopting the "partition first, fusion late" strategy for learning imbalanced data. In the partition stage, it utilizes the competitive learning clustering algorithm to produce compact sub-clusters. When the sub-clusters are fine enough, it merges nearby sub-clusters according to their density connection, and finally forms the appropriate number of imbalanced clusters.

3 Propose Method

We first briefly overview the proposed Multi-Imbalanced Cluster Discriminator (MICD), then we present its details in the following subsections. Suppose data objects are collected in a chunk-wise manner. The first incoming chunk $D_b = \{\mathbf{x}_1, \mathbf{x}_2, ..., \mathbf{x}_w\}$ is the base chunk of length w, while a subsequent chunk $D_t = \{\mathbf{x}_1, \mathbf{x}_2, ..., \mathbf{x}_l\}$ will be tested in terms of if concept drift occurs. To detect the drifts, MICD first partitions D_b into k clusters, and trains k one-class classifiers accordingly to describe them. Then, MICD assesses if the distributions in D_t change based on the description. The overall MICD framework is shown in Fig. 2.

3.1 Imbalanced Distribution Learning

To describe the distribution of imbalanced data, we adopt an imbalanced data clustering process [28] to first partition D_b into z sub-clusters $\mathbf{M}' = \{M_1', M_2', ..., M_z'\}$ represented by seed points $S = \{\mathbf{s}_1, \mathbf{s}_2, ..., \mathbf{s}_z\}$, and then fuse the groups to form imbalanced clusters. During the partition, an object $\mathbf{x}_t$ is assigned to its closest seed point $\mathbf{s}_j$, which is selected by

$$j = \arg\min_{1 \le i \le z} d(\mathbf{x}_t, \mathbf{s}_i), \tag{1}$$

where $d(\mathbf{x}_t, \mathbf{s}_i)$ represents the Euclidean distance between $\mathbf{s}_i$ and $\mathbf{x}_t$. After partition all the objects according to Eq. (1), seed points are updated by

$$\mathbf{s}_j{}^{t+1} = \mathbf{s}_j + z\varepsilon \sum (\mathbf{x}_t - \mathbf{s}_j{}^t) q_{j,t} - z\beta_j \eta\varepsilon \sum (\mathbf{x}_t - \mathbf{s}_j{}^t)(1 - q_{j,t}) \tag{2}$$

where $q_{j,t} = 1$ when $\mathbf{x}_t \in \mathbf{s}_j$, otherwise, $q_{j,t} = 0$. β_j is a penalization factor

$$\beta_j = \exp\left(-\frac{d(\mathbf{s}_j, \mathbf{x}_t) - d(\mathbf{s}_v, \mathbf{x}_t)}{d(\mathbf{s}_v, \mathbf{x}_j)}\right), \tag{3}$$

where $\mathbf{s}_v$ is the winner seed point of $\mathbf{x}_t$, v is determined as j by Eq. (1), η and ε are two constants that control the strength of penalties in the competition process. The second term in Eq. (2) drags $\mathbf{s}_j$ towards the cluster center, while the third term penalizes rival seed points to be far from the current cluster, ensuring a better cluster exploration of the seeds.

If no seed points are displaced after convergence, it indicates that distributions are not adequately represented, and thus extra seed points should be added into cluster M_j' with higher winning count b_j and density gap σ_j, where

$$j = \arg\max_{r=\{1,\ldots z\}} \mathrm{b}_r \sigma_r \,. \tag{4}$$

Since multiple density peaks indicate the presence of multiple clusters within a cluster, the density gap can be assessed by examining whether there are multiple density peaks which can be computed by analyzing the ϵ-ball graph, which is constructed by joining together neighbors whose points are less than ϵ away from each other in the cluster [29]. After adding seed points and updating the number of sub-clusters z, the competition process is iteratively repeated until certain seed points are penalized to such an extent that they are expelled from all clusters. Once this happens, it means that each cluster is effectively represented by multiple sub-clusters, and the algorithm terminates.

Next, we fuse $\mathbf{M}'$ and determine the optimal number of clusters. To maximize the fusion of sub-clusters within the same cluster and avoid merging clusters from different groups, the algorithm employs a mixed one-dimensional binary Gaussian distribution to quantify the degree of separation between two sub-clusters M_i' and M_j', denoted as $h_{ij} = 1/\min f(\mathrm{A})$. The discrete function $f(A)$ is calculated by projecting points from M_i' and M_j' onto a one-dimensional plane.

After obtaining the separation degree h_{ij} between the sub-clusters, it is used to merge $\mathbf{M}'$ based on the ascending order of separation until only one cluster remains. Then, the optimal cluster number k, the final division $\mathbf{M} = \{M_1, M_2, ..., M_k\}$ of D_b and the corresponding sub-cluster fusion queue Q are obtained by using the internal evaluation metrics of clustering:

$$k = \arg\min_{1 \le j \le z-1} \left(\frac{sep_j}{\max_{j'}\{sep_{j'}\}} + \frac{com_j}{\max_{j'}\{com_{j'}\}}\right) \tag{5}$$

where sep_j and com_j represent the compactness and separation degree of the j-th sub-cluster, respectively. Once the partition of D_b is completed, we utilize the partition to train the cluster discriminators as discussed in the next section.

3.2 Multi-cluster Descriptor Training

After the partition of D_b, we obtain the partition results $\mathbf{M}$ and the cluster number k. Then we use it to train the multi-cluster discriminator $C = \{c_1(\mathbf{x}), c_2(\mathbf{x}), ..., c_k(\mathbf{x})\}$ composed of k one-class classifiers. One-class classifier possesses the capability to learn from the existing data and estimate the decision boundary based on the distribution of the data in the feature space. When dealing with data that exhibits an imbalanced distribution with multiple clusters, using a single one-class classifier to determine the decision boundary may lead to the drift changes of a specific cluster being obscured by other clusters.

To address this issue, we utilize the partition result $\mathbf{M}$ to train our multi-cluster discriminator. The one-class classifier used in our multi-cluster discriminator can be any type of classifier designed for one-class classification. For a cluster M_i in $\mathbf{M}$, we consider all the object points within M_i as positive examples and proceed to train the corresponding classifier $c_i(\mathbf{x})$ by $\min(f(\mathbf{x}))$, where $f(\mathbf{x})$ represents a cost function that quantifies the goodness of fit of the model to the positive examples. After completing the training of the multi-cluster discriminator, the decision boundaries of each cluster distribution in D_b can be effectively differentiated. This approach enables us to simultaneously and independently track the distribution changes of each cluster.

3.3 Concept Drift Detection Based on MCD

To detect drifts in the test chunk D_t, it should be first partitioned and then compared with the trained multi-cluster discriminator for detection. To address the potential imbalance issue in D_t, we still adopt the "partition first, fuse later" strategy, which has been applied to the base chunk D_b. It is worth noting that we use the seed points S obtained from D_b to "finely partition" D_t to achieve a higher drift sensitivity. Specifically, object-cluster affiliation $u_{i,j}$ is computed as

$$u_{i,j} = \begin{cases} 1, & \text{if } d(\mathbf{x}_i, \mathbf{s}_j) \leq d(\mathbf{x}_i, \mathbf{s}_t) \\ 0, & \text{otherwise} \end{cases}, \tag{6}$$

where $t = \{1, 2, ..., z\}$. $u_{i,j} = 1$ indicates that the i-th object is assigned to the j-th cluster. After the "fine partition" of D_t, the pre-partitioned clusters $\mathbf{O}' = \{O'_1, O'2, .., O'_z\}$ is obtained. We then use a fusion queue Q to merge them into the final clusters $\mathbf{O} = \{O_1, O_2, ..., O_k\}$, and the process can be written as

$$v_{i,j} = \begin{cases} 1, & \text{if } i \text{ in } Q[j] \\ 0, & \text{otherwise} \end{cases}, \tag{7}$$

where $j = \{1, 2, ..., k\}$, and $Q[j]$ represents the j-th element of fusion queue Q, which contains the indexes of the sub-clusters fused into M_j. $v_{i,j} = 1$ represents

Algorithm 1 Imbalance Concept Drift Detection

Input: Trained multi-cluster discriminator C, fusion queue Q, seed point list S, final cluster number k, drift threshold ρ, test chunk D_t
Output: Drift detection results $\mathbf{O}_{\text{drift}}$

$\mathbf{O}_{\text{drift}} = \emptyset$, $z = |S|$
Initialize pre-partition cluster set $\mathbf{O}' = \{O'_1, O'_2, ..., O_z\}$
Initialize Final cluster set $\mathbf{O} = \{O_1, O_2, ..., O_k\}$
for i $\leftarrow$ 1 to $|D_t|$ **do**
  Compute $u_{i,j}$ according to Eq.(6) ◁ *Pre-Partition* D_t
  if $u_{i,j} = 1$ **then**
    $O'_j = O'_j \cup \mathbf{x}_i$
  end if
end for
for i $\leftarrow$ 1 to z **do**
  Compute $v_{i,j}$ according to Eq.(7) ◁ *Fusion* D_t
  if $v_{i,j} = 1$ **then**
    $O_j = O_j \cup O'_i$
  end if
end for
for $j \leftarrow 1$ to k **do**
  Calculate α_j by Eq.(9) based on O_j and $c_j(\mathbf{x})$ ◁ *Drift detection of* D_t
  if $\alpha_j > \rho$ **then**
    $\mathbf{O}_{\text{drift}} = \mathbf{O}_{\text{drift}} \cup O_j$
  end if
end for

that the i-th sub-cluster in $\mathbf{O}'$ is fused into the j-th cluster of the final partition $\mathbf{O}$. Then each cluster in $\mathbf{O}$ is assigned to a corresponding one-class classification within the multi-cluster discriminator for subsequent drift detection.

For each object $\mathbf{x}_i \in O_j$, we employ one-class classification to classify it by

$$c_j(\mathbf{x}_i) = \begin{cases} 1, & \text{if } \mathbf{x}_i \text{ is from new concept} \\ 0, & \text{otherwise.} \end{cases}, \tag{8}$$

where $c_j(\mathbf{x}_i) = 1$ means object $\mathbf{x}_i$ is a normal point, and $c_j(\mathbf{x}_i) = 0$ means object $\mathbf{x}_i$ is an out-of-distribution data object. Accordingly, a threshold ρ is adopted to alarm the occurrence of drift in the corresponding cluster O_j. Specifically, if the proportion of out-of-distribution objects

$$\alpha_{\text{j}} = \frac{1}{|O_j|} \sum_{i=1}^{|O_j|} c_j(\mathbf{x}_i) \tag{9}$$

exceeds ρ, i.e., $\alpha_j > \rho$, then O_j is believed to have drift. It is worth noting that α_j is adopted to avoid over-sensitivity of the drift detection. After the detection, all the clusters with detect drifts are reported as $\mathbf{O}_{\text{drift}} = \{O_j | \alpha_j > \rho, j = 1, 2, ..., k\}$. We summarize the process of concept drift detection in Algorithm 1.

The whole MICD approach is described in Fig. 3.

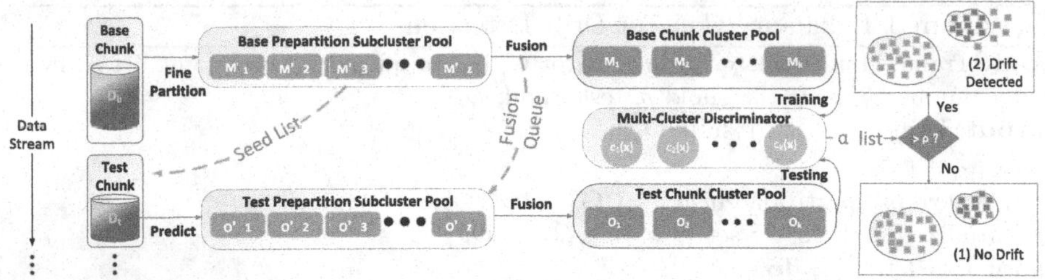

Fig. 3. Detailed working procedures of MICD.

4 Experiments

We first provide experimental setup in Sect. 4.1, and then demonstrate the results with observations in Sect. 4.2 – 4.4.

4.1 Experimental Setup

Experimental Design: We compare MICD with state-of-the-art methods on real and synthetic datasets, and also conduct ablation experiments to assess the importance of its modules. To test the robustness of MICD at different balance rates, we evaluate its performance on ten generated 2D-2G-Mean datasets with varying balance rates. To examine the drift-sensitivity of MICD, we also evaluate it on datasets with different degrees of drift severity. $Accuracy = (TP + TN)/(TP + TN + FP + FN)$, and False Negative Rate (FNR) are utilized as the evaluation metrics. Note that FNR refers to the rate of wrongly identifying a data chunk with drift to be without drift.

Datasets: We conduct experiments on seven synthetic and five real datasets. As we attempt to simultaneously consider imbalance and drift issues, a new data generation strategy is adopted to first generate data with imbalanced clusters, and then generate drifts following [13] and [30]. For each dataset, we use a base chunk for learning and generate 250 test chunks with the same distribution as the base chunk, and 250 test chunks with a different distribution for testing. For detailed dataset information and drift generation operations, please visit the link: https://drive.google.com/file/d/1o30zT6alN5ym9vMvl4dHiPLUNXeSyat4/view.

Counterparts: We compare our method with the following methods: K-Means plus our multi-cluster discriminator (KMCD), EI-KMeans (EIKM) [13], QuantTree (QTree) [15], OCDD [5], and Mann-Whitney-Wilcoxon (MWW) test [16]. Parameters of these methods were set according to the code and paper released by the authors. To ensure experimental reliability and minimize the introduction of additional factors, we employ the most basic form of SVM using the radial basis function as the kernel function as the one-class classifier for our multi-cluster discriminator. In all experiments, we set the hyper-parameter $\rho = 0.2$.

Table 1. FNR on synthetic datasets. The best result is highlighted in bold.

Dataset	MICD	KMCD	EIKM	QTree	OCDD	MWW
2D-2G-Mean	**6.8**	**6.8**	57.0	19.8	100.0	92.4
2D-2G-Cov	**3.6**	**3.6**	16.4	32.0	100.0	98.4
4D-4G-Mean	**0.4**	11.6	30.4	26.9	99.6	19.2
4D-4G-Cov	**12.4**	67.6	69.2	40.6	97.2	90.0
2D-8G-Mean	**2.8**	4.8	40.8	6.8	92.8	100.0
2D-8G-Cov	**0.4**	**0.4**	98.4	77.7	100.0	100.0
2D-Moon-N	**7.6**	100.0	73.6	72.1	100.0	100.0
Average FNR	**4.9**	27.8	55.1	39.4	98.5	85.7
Average Rank	**1.0**	2.0	3.9	3.0	5.4	4.7
Average Accuracy	**0.9**	0.8	0.6	0.7	0.5	0.6

Table 2. FNR on real datasets. The best result is highlighted in bold.

Dataset	MICD	KMCD	EIKM	QTree	OCDD	MWW
MFCCs	**26.8**	99.2	77.6	43.1	99.6	100.0
Avail	55.6	53.6	95.6	81.2	**53.2**	99.6
Posture	**73.6**	84.8	83.2	85.6	100.0	90.8
NOAA	**6.4**	21.6	96.4	62.8	29.2	17.6
Insect	**29.6**	46.8	84.4	63.6	100.0	96.4
Average FNR	**38.4**	61.2	87.4	67.3	76.4	80.9
Average Rank	**1.4**	2.8	4.0	3.6	4.4	4.8
Average Accuracy	**0.7**	0.7	0.6	0.6	0.6	0.6

4.2 Comparative Results

In this experiment, we employ multiple metrics to ensure a comprehensive evaluation. Tables 1 and 2 compare FNR and condensed summary of Accuracy results of different approaches on synthetic and real datasets, respectively. It can be seen that MICD outperforms the counterparts as it has the lowest average FNR and top ranking. According to the results shown in Table 1, MICD always performs the best among all the compared approaches on synthetic datasets. On real datasets in Table 2, It clearly indicates that MICD has the best FNR performance on four out of five datasets (expecting on Avail dataset). Moreover, KMCD performs worse than MICD because it ignores the imbalance effect when sketching the data distribution. But KMCD still performs better than the other counterparts because it can still avoid the domination of major cluster to a certain extent if the minor clusters can be discriminated by the adopted K-Means. Moreover, the condensed summary of Accuracy results reveals that our method also exhibits strong competitiveness in terms of FNR.

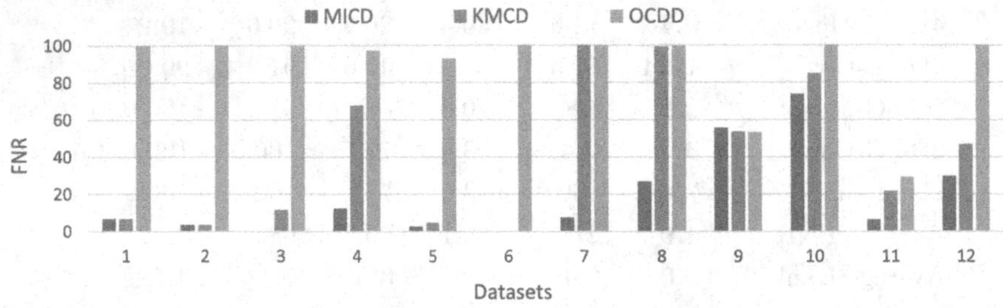

Fig. 4. Ablation Study of MICD.

4.3 Ablation Study

Detection performance of MICD and its two ablated versions are demonstrated in Fig. 4. MICD outperforms KMCD, validating the imbalanced partitioning mechanism it employs. This mechanism accurately describes data distribution, avoiding misclassification of minor clusters into major ones and forming a foundation for drift detection. Moreover, MICD's superiority over OCDD confirms the effectiveness of assigning a descriptor to each cluster. The use of independent descriptors for each cluster prevents dominance of detection results by large clusters, effectively identifying concept drift in imbalanced data. These results demonstrate the efficacy of MICD's core mechanisms: imbalanced partitioning and distribution description and detection using multiple one-class classifiers.

4.4 Study on Imbalance Rate and Drift Severity

As can be seen from Fig. 5(a), with the increasing of imbalance rate, MICD always has a lower FNR. It is because that MICD finely describes clusters with different imbalanced rates and can thus timely identify the occurrence of drifts. KMCD has the same performance as MICD because the synthetic data set has two separable clusters that can be easily partitioned by K-Means. By contrast, performance of the remainder four approaches deteriorates with the increasing of imbalance rate as they monitor the overall statistics of distributions to detect drifts, are thus less sensitive to the drifts occurring in very small clusters. In

Fig. 5(b), it can be seen that with the increasing of outer-cluster objects, all the approaches perform better because the drifts become more obvious and easier to detect. It is worth noting that MICD always outperforms the counterparts under different severity levels, which illustrates its superiority.

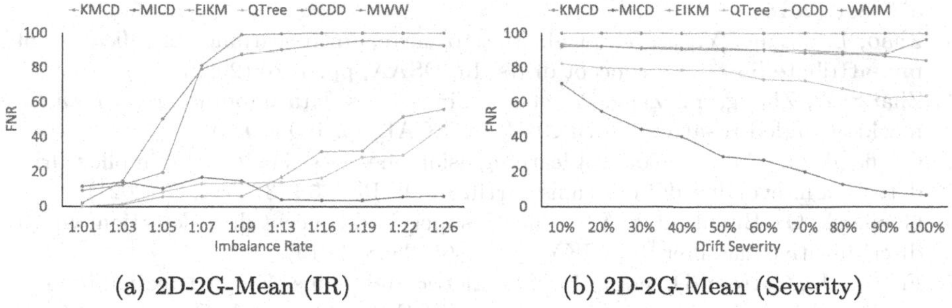

(a) 2D-2G-Mean (IR) (b) 2D-2G-Mean (Severity)

Fig. 5. FNR under different Imbalance Rates (IR) and varying drift-severity.

5 Conclusion

In this paper, we proposed a novel concept drift detection method called Multi-Imbalanced Cluster Discriminator (MICD). Compared with existing solutions, MICD further addresses the impact of data distribution imbalance in drift detection in an unsupervised environment, enabling MICD to operate in a real complex environment. It trains descriptors for the partitioned clusters of different sizes to determine whether subsequent objects conform to the description of the original distribution. Thanks to the reasonable partition of imbalanced clusters, and each cluster has an independent descriptor, MICD can accurately perceive the drifts that occur in minor clusters, and at the same time reveal the drift position and drift severity. Experimental results illustrate the promising performance of MICD, and observations on the experimental results shed light on the impact of imbalanced data on unsupervised concept drift detection.

Acknowledgements. This work was supported in part by the National Natural Science Foundation of China (NSFC) under grants: 62102097, 62002302, and 62302104, the Natural Science Foundation of Guangdong Province under grants: 2023A1515012855, 2023A1515012884, and 2022A1515011592, the Science and Technology Program of Guangzhou under grant 202201010548, and the China Fundamental Research Funds for the Central Universities under grant 20720230038.

References

1. Žliobaitė, I., Pechenizkiy, M., et al.: An overview of concept drift applications. In: Big Data Analysis: New Algorithms for a New Society, pp. 91–114 (2016)
2. Cheung, Y.m., Zhang, Y.: Fast and accurate hierarchical clustering based on growing multilayer topology training. IEEE Trans. Neural Netw. Learn. Syst. **30**(3), 876–890 (2018)
3. Zhao, L., Zhang, Y., Ji, Y., et al.: Heterogeneous drift learning: classification of mix-attribute data with concept drifts. In: DSAA, pp. 1–10 (2022)
4. Zhang, Z., Zhang, Y., Zeng, A., et al.: Time-series data imputation via realistic masking-guided tri-attention Bi-GRU. In: ECAI, pp. 1–9 (2023)
5. Gözüaçık, Ö., Can, F.: Concept learning using one-class classifiers for implicit drift detection in evolving data streams. Artif. Intell. Rev. **54**, 3725–3747 (2021)
6. Gözüaçık, Ö., Büyükçakır, A., et al.: Unsupervised concept drift detection with a discriminative classifier. In: CIKM, pp. 2365–2368 (2019)
7. Frittoli, L., Carrera, D., et al.: Nonparametric and online change detection in multivariate Datastreams using QuantTree. IEEE Trans. Knowl. Data Eng. **35**(8), 8328–8342 (2023)
8. Gemaque, R.N., Costa, A.F.J., et al.: An overview of unsupervised drift detection methods. Wiley Interdiscip. Rev. Data Min. Knowl. Discov. **10**(6), e1381 (2020)
9. Lu, J., Liu, A., et al.: Learning under concept drift: a review. IEEE Trans. Knowl. Data Eng. **31**(12), 2346–2363 (2018)
10. Gama, J., Medas, P., et al.: Learning with drift detection. In: Brazilian Symposium on Artificial Intelligence, pp. 286–295 (2004)
11. Baena-Garcıa, M., del Campo-Ávila, J., et al.: Early drift detection method. In: Fourth International Workshop Knowledge Discovery Data Streams, vol. 6, pp. 77–86 (2006)
12. Bifet, A., Gavalda, R.: Learning from time-changing data with adaptive windowing. In: SDM, pp. 443–448 (2007)
13. Liu, A., Lu, J., et al.: Concept drift detection via equal intensity k-means space partitioning. IEEE Trans. Cybern. **51**(6), 3198–3211 (2021)
14. Dasu, T., Krishnan, S., et al.: An information-theoretic approach to detecting changes in multi-dimensional data streams. In: Proceedings of 28th ISCAS (2006)
15. Boracchi, G., Carrera, D., et al.: QuantTree: histograms for change detection in multivariate data streams. In: ICML, pp. 639–648 (2018)
16. Friedman, J.H., Rafsky, L.C.: Multivariate generalizations of the Wald-Wolfowitz and Smirnov two-sample tests. Ann. Stat. **7**(4), 697–717 (1979)
17. Zhang, Y., Cheung, Y.m.: A fast hierarchical clustering approach based on partition and merging scheme. In: ICACI, pp. 846–851 (2018)
18. Zhang, Y., Cheung, Y.m., Liu, Y.: Quality preserved data summarization for fast hierarchical clustering. In: IJCNN, pp. 4139–4146 (2016)
19. Zhang, Y., Cheung, Y.M.: Graph-based dissimilarity measurement for cluster analysis of any-type-attributed data. IEEE Trans. Neural Netw. Learn. Syst. **34**, 6530–6544 (2022)
20. Zhang, Y., Cheung, Y.m., Zeng, A.: Het2Hom: representation of heterogeneous attributes into homogeneous concept spaces for categorical-and-numerical-attribute data clustering. In: IJCAI, pp. 3758–3765 (2022)
21. Zhang, Y., Cheung, Y.m.: Exploiting order information embedded in ordered categories for ordinal data clustering. In: ISMIS, pp. 247–257 (2018)

22. Shang, X., Lu, Y., et al.: Federated learning on heterogeneous and long-tailed data via classifier re-training with federated features. In: IJCAI, pp. 2218–2224 (2022)
23. Shang, X., Lu, Y., et al.: FEDIC: Federated learning on non-IID and long-tailed data via calibrated distillation. In: ICME, pp. 1–6 (2022)
24. Li, M., Cheung, Y.m., Lu, Y., et al.: Long-tailed visual recognition via Gaussian clouded logit adjustment. In: CVPR, pp. 6929–6938 (2022)
25. Lu, Y., Cheung, Y.m., et al.: Adaptive chunk-based dynamic weighted majority for imbalanced data streams with concept drift. IEEE Trans. Neural Netw. Learn. Syst. **31**(8), 2764–2778 (2020)
26. Tekli, J.: An overview of cluster-based image search result organization: background, techniques, and ongoing challenges. Knowl. Inf. Syst. **64**(3), 589–642 (2022)
27. Aksoylar, C., Qian, J., et al.: Clustering and community detection with imbalanced clusters. IEEE Trans. Signal Inf. Process Netw. **3**(1), 61–76 (2016)
28. Lu, Y., Cheung, Y.m., et al.: Self-adaptive multiprototype-based competitive learning approach: a k-means-type algorithm for imbalanced data clustering. IEEE Trans. Cybern. **51**(3), 1598–1612 (2019)
29. Rodriguez, A., Laio, A.: Clustering by fast search and find of density peaks. Science **344**(6191), 1492–1496 (2014)
30. Souza, V.M., dos Reis, D.M., et al.: Challenges in benchmarking stream learning algorithms with real-world data. Data Min. Knowl. Disc. **34**, 1805–1858 (2020)

Unsupervised Domain Adaptation for Optical Flow Estimation

Jianpeng Ding, Jinhong Deng, Yanru Zhang, Shaohua Wan, and Lixin Duan(✉)

Shenzhen Institute for Advanced Study, University of Electronic Science and Technology of China, Chengdu, China
{yanruzhang,shaohua.wan,lxduan}@uestc.edu.cn

Abstract. In recent years, we have witnessed significant breakthroughs of optical flow estimation with the thriving of deep learning. The performance of the unsupervised method is unsatisfactory due to it is lack of effective supervision. The supervised approaches typically assume that the training and test data are drawn from the same distribution, which is not always held in practice. Such a domain shift problem are common exists in optical flow estimation and makes a significant performance drop. In this work, we address these challenge scenarios and aim to improve the model generalization ability of the cross-domain optical flow estimation model. Thus we propose a novel framework to tackle the domain shift problem in optical flow estimation. To be specific, we first design a domain adaptive autoencoder to transform the source domain and the target domain image into a common intermediate domain. We align the distribution between the source and target domain in the latent space by a discriminator. And the optical flow estimation module adopts the images in the intermediate domain to predict the optical flow. Our model can be trained in an end-to-end manner and can be a plug and play module to the existing optical flow estimation model. We conduct extensive experiments on the domain adaptation scenarios including Virtual KITTI to KITTI and FlyingThing3D to MPI-Sintel, the experimental results show the effectiveness of our proposed method.

Keywords: Domain adaptation · Optical flow · Deep learning

1 Introduction

Optical flow estimation is the task of calculating the motion of each pixel between correlate frames in a video sequence. It's a core problem in computer vision and has a wide range of application scenarios such as video editing [15], behavior recognition [29] and object tracking [1]. With the thriving of deep convolution neural networks (DCNNs), we have witnessed significant breakthroughs and impressive performance on challenging benchmarks [18] in recent years. Nevertheless, the unsupervised method is still behind the supervised method

Q. Liu et al. (Eds.): PRCV 2023, LNCS 14427, pp. 44–56, 2024.
https://doi.org/10.1007/978-981-99-8435-0_4

due to it is the lake of effective supervision. Meanwhile, the supervised method usually assumes that the training data and the target data are drawn from the same distribution which can not be always held in practice. The model will encounter considerable domain shift caused by the illumination, background, and capture equipment, *etc.*

Unsupervised domain adaptation (UDA) is proposed to alleviate the domain shift problem by aligning the feature distributions between the source and target domain. It has been successful used in many computer vision tasks such as classification [34], semantic segmentation [22], object detection [5]. Most of them focus on adopting an adversarial learning, where a domain discriminator is employed to distinguish the target samples from the source samples, while the feature extractor tries to fool the discriminator by generating domain-invariant features.

Motivated by the UDA methods, we propose to develop an optical flow estimation model that can adapt to a new domain. To address domain shift, we design an extra simple yet effective domain adaptive autoencoder module to generate intermediate domain images which will be fed into the optical flow estimation model. To be specific, domain adaptive autoencoder has an encoder-decoder architecture where the encoder extract feature from image pairs to a latent space, and the decoder tries to reconstruct them. We hope the autoencoder can learn domain-invariant features in the latent space, so as to model adopts the reconstructed intermediate images and can accurately predict the optical flow for both source and target domain. Different from other UDA methods, they modify the architecture for specific tasks, our method does not need to remold the complex optical flow estimation model. Instead, our model can be a plug-and-play module to the existing optical flow estimation model and trained in an end-to-end manner.

2 Related Work

Supervised Learning: Top-performing optical flow estimation methods are mainly based on the recent advances of deep neural networks. As proposed by Doso-vitskiy et al. [6], FlowNet (FlowNetS and FlowNetC) is directly uses CNN (Convolutional Neural Networks) model for optical flow estimation. And then, FlowNet2 [13] stacks several basic FlowNet models into a large one, the performance has been further improved.

Recently, Recurrent All-Pairs Field Transforms (RAFT) model [31] propose to compute the cost volume between all pairs of pixels and iteratively updates a flow field through a recurrent unit that performs lookups on the correlation volumes. Many methods [16,27,36] extents the RAFT and further improve the performace. However, these method cannot generalize well on new domain when considerable domain shift exists between the training domain and test domain.

Unsupervised Learning: Unsupervised approaches do not need the annotations for training and they optimize the photometric consistency with some regularization [28]. This kind of methods focus on occlusion handling including DDFlow [20], SelFlow [21], and OAFlow [33]. Due to the lack of effective supervised information, there is still a big gap in performances between unsupervised and supervised methods. Therefore, in this paper, we attempt to leverage a labeled source domain to improve the performance on an unlabeled target domain.

Domain Adaptation: Transferring the learned knowledge from the labeled source domain to an unlabeled target domain becomes an appealing alternative in the high-level task, such as image classification and object detection. Domain adaptation (DA) addresses this problem by minimizing domain discrepancy between the source and target domains [37]. Many methods [4,7,12] employ a domain classifier to learning domain-invariant representation by adversarial training or a metric learning to minimize the domain disparity between the source and target domain.

To our best knowledge, the UDA for optical flow estimation has not been explored widely. Many methods focus on adapting synthetic-to-real images by adversarial learning or label transfer for pixel-level prediction tasks, such as semantic segmentation. Similarly, CyCADA [11] uses the CycleGAN [38] and transfers the source domain to the target domain with pixel-level alignment. However, these methods do not provide an improvement on the optical flow estimation task. Moreover, the optical flow estimation based on deep learning methods must compute the Cost Volumes [30], which makes it difficult to achieve feature alignment directly within the network framework for optical flow estimation. Therefore, we use the autoencoder module as an alternative to transform the images to a common intermediate domain images, which helps the optical flow estimation, so as to our approach can eliminate the domain discrepancy between source and target domains and have good generalization ability on the target domain.

3 Method

We have source domain $\mathcal{S}$ and the target domain $\mathcal{T}$. Formally, let us denote $D_S = \{(I_s^1, I_s^2)_i, y_s)|_{i=1}^{N_s}\}$ as the source domain training samples drawn from $\mathcal{S}$, each (I_s^1, I_s^2) is a source image pair, y_s is the corresponding optical flow label. Accordingly, we denote $D_T = \{(I_t^1, I_t^2)_i|_{i=1}^{N_t}\}$ as the target domain samples drawn from $\mathcal{T}$, and each (I_t^1, I_t^2) is a target image pair. It is noted that the optical flow label is unavailable for target domain. The objective of our method is to learn an optical flow estimation model, which has good generalization ability to the target domain $\mathcal{T}$.

3.1 Overview

To achieve this goal, we present an end-to-end trainable network which is composed of two modules: (1) the domain adaptive autoencoder (DAAE) module **A**

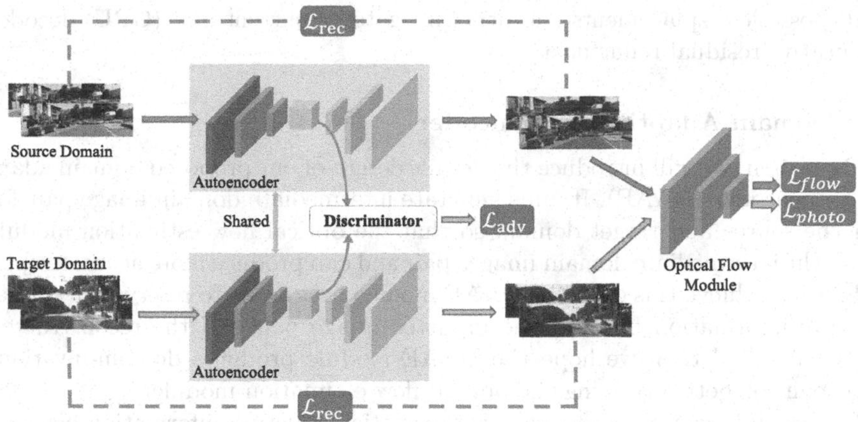

Fig. 1. Illustration of the proposed framework. It consists of two modules: a domain adaptive autoencoder (DAAE) and an optical flow module. The DAAE module consists of a encoder, a decoder and a discriminator, where the encoder adopts images and extract the feature to a latent space and the decoder reconstruct the image from the feature in latent space. The discriminator is to align the distribution between the source and target domain in the latent space based on adversarial learning. For source domain sample, we utilize optical flow module to calculate the flow loss $\mathcal{L}_{flow}$, while we use pthotometric loss $\mathcal{L}_{photo}$ to supervise the target domain samples.

and (2) the optical flow estimation network **F**. As illustrated in Fig. 1, the DAAE module adopts the source domain (target domain) image pairs and generate the intermediate image pair which is reconstructed from the domain-invariant latent space and fed into the optical flow estimation module to predict the corresponding optical flow. Through the collaboratively learning, the DAAE and the optical flow estimation module are capable of predicting accurate optical domain for the target domain. After training, we feed image pairs in the target domain to the DAAE module and optical flow estimation module to produce the final optical flow.

Specifically, the DAAE module consists of an encoder, a decoder and domain discriminator **D**. The encoder adopts the image pairs and extract the feature to a latent space, while the decoder reconstruct the original image pairs from the latent space. To align the distribution between the source domain and target domain, we apply a discriminator in latent space to learning domain-invariant representations. The encoder tries to extract domain-invariant features to fool the discriminator while the discriminator take efforts to figure out the inputted feature comes from the source domain or target domain. In this way, the DAAE module will produce the intermediate domain images that effectively help the learning of the optical flow estimation module. For the optical flow estimation module, we adopt RAFT [31] as the base optical flow estimator to improve its generalization ability due to its effectiveness and widely usage. RAFT introduce an all-pairs correlation volume, which explicitly models matching correlations for all possible displacements and utilizes a gated recurrent unit (GRU) decoder for iterative residual refinement.

3.2 Domain Adaptive Autoencoder

In this section, we will introduce the details design of our proposed domain adaptive autoencoder (DAAE). It aims generate intermediate domain image pair for both the source and target domain so that the optical flow estimation module adopts the intermediate domain images pair and can produce more accurate optical flow. To achieve this goal, the DAAE module first needs to reserve the spatial structure information between the inputted image pair and the reconstructed image pair. And then we hope the DAAE module produces domain-invariant image pair for better training the optical flow estimation module.

To this end, we propose to reserve the spatial structure information between the inputted image pair and the reconstructed image pair by defining a reconstruction loss, which can be defined as follows:

$$\mathcal{L}_{rec} = \alpha \cdot \mathcal{L}_{ss}(I, \hat{I}) + (1 - \alpha) \cdot \mathcal{L}_1(I, \hat{I}), \tag{1}$$

where $\alpha = 0.85$ is a trade-off parameter to balance the structural similarity loss $\mathcal{L}_{ss}$ and the l1-norm loss $\mathcal{L}_1 = ||I - \hat{I}||_1$, I and $\hat{I}$ represents the inputted image and the constructed image from decoder for both the source domain and target domain, respectively. The structural similarity is to ensure the pixel-level similarity using the Structural Similarity (SSIM) [35] term. The objective can be written as:

$$\mathcal{L}_{ss} = \frac{1 - SSIM(I, \hat{I})}{2}. \tag{2}$$

We expect the reconstructed image pair is domain-invariant and so that we can transfer the knowledge from the source domain to the target domain, and the optical flow estimation module can perform well on the target domain. To address the domain shift, we plug a discriminator into the autoencoder, as shown in Fig. 1. We can obtain the domain-invariant feature in latent space through adversarial learning in a min-max manner. On the one hand, the discriminator take efforts to figure out the inputted feature comes from the source or target domain. On the other hand, the encoder tries to fool the discriminator by extracting domain-invariant features. The adversarial loss can be written as:

$$\mathcal{L}_{adv} = -\sum_{h,w} \log(D(E(I_T))). \tag{3}$$

To train the discriminator network, we minimize the spatial crossentropy loss $\mathcal{L}_{\mathcal{D}}$ with respect to two classes using

$$\mathcal{L}_D = -\sum_{h,w} (1 - z) \cdot \log(D(E(I_s))) + z \cdot \log(D(E(I_t))), \tag{4}$$

where z is the domain label where $z = 0$ if the sample are from source domain and $z = 1$ for target domain, h and w is the height and width for the feature map, respectively.

3.3 Incorporating with RAFT

We utilize the DAAE module to generate the intermediate domain for both source domain images and target domain images for optical flow estimation. In this section, we illustrate how to incorporate the DAAE module with optical flow estimation method RAFT [31]. For the source domain, we have the ground truth label for RAFT to supervised the model.

Moreover, for the target domain, we cannot obtain the ground truth flow. Fortunately, many unsupervised approaches [17,20,21] have found that the photometric loss can be used as an alternative supervision signal for ground truth flow. However, the photometric loss can not work in the regions that have been occluded by other objects since their corresponding regions do not exist in another images. Thus, we apply forward-backward checking [25] estimate the occluded pixel. To this end, we can also provide some supervision for unlabeled target domain to some extent and the loss can be written as:

$$\mathcal{L}_{photo} = (1 - SSIM(\rho(I_t^2, f), I_t^1)) \odot M_t \tag{5}$$

where $\rho(.)$ is warp operate, the f represents the predicted of optical flow. M_t is the occlusion mask which is estimated by forward-backward checking, where 1 represents the non-occluded pixels in I_T^2 and 0 for those are occluded. I_T^1 and I_T^2 represents the image pair from the target domain.

3.4 Overall Objective

Combined with the loss of DAAE and RAFT on source samples and target samples, the overall object function can be written as:

$$\mathcal{L} = \mathcal{L}_{flow} + \mathcal{L}_{rec} + \lambda_1 \mathcal{L}_{adv} + \lambda_2 \mathcal{L}_{photo} \tag{6}$$

where λ_1 and λ_2 are the hyper-parameters used to control the relative importance of the respective loss terms. Based on the Eq. 6, we optimize the following min-max criterion:

$$\max_{D} \min_{E,F} \mathcal{L} \tag{7}$$

The ultimate goal is to optimize the optical flow loss for source images and the photometric loss for target images, while maximizing the probability of target features being considered as source features in latent space. Namely, to train our network using labeled source domain images pairs and unlabeled target domain image pairs, we minimize the optical flow loss $\mathcal{L}_{flow}$ and photometric loss $\mathcal{L}_{photo}$. The adversarial loss $\mathcal{L}_{adv}$ is optimized to align the feature distribution within the same domain.

3.5 Network Architecture

Our network architecture consists of two parts, including an optical flow network and an autoencoder network. The optical flow network has the same architecture as RAFT [31]. The autoencoder network is composed of encoder (we also call it generator) and decoder. The encoder contains three convolution layers and 9 residual blocks [10], each convolution layer is followed by an instance-normalization layer and a ReLU layer. To restore the input size and the overall image structure information, we use convolution layer and pixel-shuffle layer in the decoder. Our discriminator utilizes a fully-convolution layers. Similar to [32], it contains five convolution layers with a kernel size of 4×4 and stride of 2. Each convolution layer is followed by a ReLU layer. we jointly train the encoder-decoder network, discriminator, and optical flow estimation network in an end-to-end manner.

4 Experiments

In this section, we conduct our experiments on the cross domain scenarios from Virtual KITTI [3] to KITTI [3] and FlyingThings3D [24] to MPI-Sintel [2] to evaluate the proposed method. We also provide the quantitative results to show the effectiveness and study the ablated variants of our proposed method.

4.1 Experimental Setup

We conduct experiments on optical flow estimation public datasets: FlyingThings3D [24], Virtual KITTI [3], MPI-Sintel [2] and KITTI [8]. In both experiments, MPI-Sintel and KITTI are regarded as the target domain, while FlyingThings3D and Virtual KITTI are regarded as the source domain, respectively. For KITTI, we train on the multi-view extension on the KITTI 2015 [26] dataset, and following [17], we do not train on any data from KITTI 2012 [9] because it does not have moving objects.We implement our method with PyTorch on two RTX 3090 GPUs. During training, we pretrain on source domain for 100k iterations with out adaptation and use the largest possible mini-batch of 12 source image pairs. Then we train another 100k iterations for adaptive training both on the labeled source domain and the unlabeled target domain. Each mini-batch consists of 6 source image pairs and 6 target image pairs. We empirically set $\lambda_1 = 0.01$ and $\lambda_2 = 0.1$, respectively.

Following the conventions of the KITTI benchmark, average end-point error (EPE) and percentage of erroneous pixels (F1) are used as the evaluation metric of optical flow estimation. The EPE measure calculates the Euclidean distance between the endpoints of the GT flow and the estimated flow, the F1 measures the percentage of bad pixel averaged over all ground truth pixels of all images.

4.2 Experiment Results

We compare our method with existing methods on MPI-Sintel and KITTI-2015 optical flow benchmarks. The experimental results are shown in Table 1 and some qualitative comparison results are shown in Fig. 2 and Fig. 3.

Table 1. The performances of different methods on MPI-Sintel and KITTI 2015 benchmarks. '–' denotes the results is not reported in the original paper, "()" trained on the labeled evaluation set.

Method	Sintel(Clean)		Sintel(Final)		KITTI-15(train)		KITTI-15(test)
	train	test	train	test	EPE	F1-all	F1-all
FlowNetS [6]	4.05	7.42	5.45	8.43	–	–	–
PWC-Net [30]	3.33	–	4.59	–	13.20	–	–
RAFT [31]	1.43	–	2.71	–	5.04	17.4%	–
FlowNetS+ft [6]	(3.66)	6.96	(4.44)	7.76	–	–	–
PWC-Net+ft [30]	(1.70)	3.86	(2.21)	5.13	(2.16)	–	9.60%
RAFT+ft [31]	(0.77)	2.08	(1.20)	3.41	**(0.64)**	**(1.5%)**	5.27%
ARFlow [19]	(2.79)	4.78	(3.87)	5.89	2.85	–	11.80%
SimFlow [14]	(2.86)	5.92	(3.57)	6.92	5.19	–	13.38%
UFlow [17]	(2.50)	5.21	(3.39)	6.50	2.71	9.05%	11.13%
UpFlow [23]	(2.33)	4.68	(2.67)	5.32	2.45	8.90%	9.38%
Ours (Source-Only)	1.63	–	2.70	–	2.27	8.04%	–
Ours	**1.28**	**2.25**	**2.56**	**4.23**	**2.01**	**7.11%**	**7.64%**

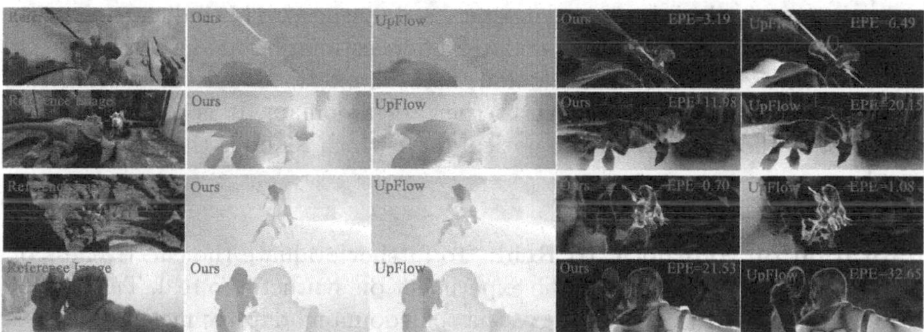

Fig. 2. Visual comparison of our method with the unsupervised state-of-the-art method UpFlow on MPI-Sintel benchmark. The resulting images is from the MPI-Sintel benchmark website (first two rows is Sintel Clean and last two rows is Sintel Final).

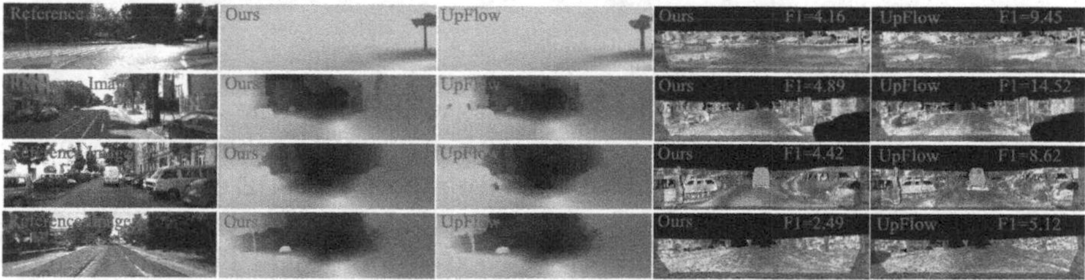

Fig. 3. Visual comparision of our method with the UpFlow on KITTI benchmark. The resulting images is from the KITTI benchmark website.

Table 2. Ablation studies of all the components. To evaluate each component of our method, we conduct the experiments on the scenario from Virtual KITTI to KITTI and from FlyingThings3D to Sintel.

L_{flow}	L_{rec}	L_{adv}	L_{photo}	Sintel (train)		KITTI-2015 (train)	
				EPE (Clean)	EPE (Final)	EPE	F1-all
✓	–	–	–	1.43	2.71	3.07	8.9%
✓	✓	–	–	1.63	2.70	2.27	8.04%
✓	–	✓	–	1.38	2.65	2.34	7.88%
✓	✓	✓	–	1.30	2.64	2.19	7.56%
✓	✓	✓	✓	1.28	2.56	2.01	7.11%

Adaptation on FlyingThings3D → MPI-Sintel: To evaluate the effectiveness of our method, we first conduct the experiment on FlyingThings3D → MPI-Sintel where they have a large visual mismatch. The experimental results are shown in Table 1, we can observe that the proposed model improves the source only by 0.35 and 0.14% EPE on the train set of train (test) set of Sintel (Clean) and Sintel (Final). Besides, our methods obtain 1.28(2.25), 2.56(4.23) EPE on the train (test) set of Sintel (Clean) and Sintel (Final), respectively, which outperform all the existing unsupervised methods. This indicates that we can improve the performance of the optical flow estimation model on the target domain by introducing a source domain with annotations and transferring the knowledge from the labeled source to the target domain. It is noted that the supervised method obtains better optical flow, which requires the ground truth optical.

Adaptation on Synthetic to Real: To further evaluate the effectiveness of our method, we also to conduct the experiment on synthetic to real. The results are shown in Table 1, we can observe that our domain adaptive model improves the baseline by 0.26 EPE and 0.93% F1-all scores on the training set of KITTI-15. Besides, our methods achieve 2.01 EPE, 7.11% F1-all score on the train set of KITTI-15, and 7.64% F1-all score on the test set of KITTI-15, which surpass all the existing unsupervised methods.

4.3 Ablation Study

To study the effectiveness of each component, we conduct the ablation experiments by removing the loss. The experiments are based on both the training set of KITTI-2015 and MPI-Sintel datasets. We report the EPE error and F1-all score to compare the performance of optical flow. It is worth noting that we only use frames 1–8 and 13–20 of the multi-view extension of KITTI-2015 datasets as the target domain during the training phase and evaluate on frames 10/11. The experimental results are shown in Table 2, taking the results on KITTI-2015 (train) as an example, the model that uses only the source samples to train the model can obtain 3.07 EPE and 8.9% F1-all score, respectively. We can observe that the model will achieve 2.19 EPE and 7.56% F1-all score which show the effectiveness of the proposed DAAE module. Besides, equipped with a photometric loss, the model performance is a boost to 2.01 and 7.11%. A similar observation can be obtained in the dataset Sintel.

5 Conclusion

In this work, we propose a novel approach for cross-domain optical flow estimation from the high demand in reality. We design an end-to-end framework based on adversarial learning to address domain shift problem. Specifically, we design a domain adaptive autoencode (DAAE) module to transform the source domain and target domain image into a common intermediate domain. We align the distribution between the source and the target domain in the latent space by a discriminator. And the optical flow estimation module adopts the images in the intermediate domain to predict the optical flow. For source domain samples, we calculate the flow loss to utilize the ground truth optical flow label. While we use the photometric loss to support to a extra supervision for the unlabeled target domain. Extensive experiments prove that our method is effective. Through collabaratively learning between DAAE and optical flow module, the model can adapt to the unlabeled target domain. The performances of our method outperform all unsupervised methods and some classic supervised methods.

Acknowledgements. This work is supported by the Major Project for New Generation of AI under Grant No. 2018AAA0100400, National Natural Science Foundation of China No. 82121003, and Shenzhen Research Program No. JSGG20210802153537009.

References

1. Behl, A., Hosseini Jafari, O., Karthik Mustikovela, S., Abu Alhaija, H., Rother, C., Geiger, A.: Bounding boxes, segmentations and object coordinates: how important is recognition for 3D scene flow estimation in autonomous driving scenarios? In: Proceedings of the IEEE International Conference on Computer Vision, pp. 2574–2583 (2017)
2. Butler, D.J., Wulff, J., Stanley, G.B., Black, M.J.: A naturalistic open source movie for optical flow evaluation. In: Fitzgibbon, A., Lazebnik, S., Perona, P., Sato, Y., Schmid, C. (eds.) ECCV 2012. LNCS, vol. 7577, pp. 611–625. Springer, Heidelberg (2012). https://doi.org/10.1007/978-3-642-33783-3_44
3. Cabon, Y., Murray, N., Humenberger, M.: Virtual KITTI 2 (2020)
4. Chen, J., Wu, X., Duan, L., Gao, S.: Domain adversarial reinforcement learning for partial domain adaptation. IEEE Trans. Neural Netw. Learn. Syst. (2020)
5. Deng, J., Li, W., Chen, Y., Duan, L.: Unbiased mean teacher for cross-domain object detection. In: Proceedings of the IEEE/CVF Conference on Computer Vision and Pattern Recognition, pp. 4091–4101 (2021)
6. Dosovitskiy, A., et al.: FlowNet: learning optical flow with convolutional networks. In: Proceedings of the IEEE International Conference on Computer Vision, pp. 2758–2766 (2015)

7. Duan, L., Tsang, I.W., Xu, D.: Domain transfer multiple kernel learning. IEEE Trans. Pattern Anal. Mach. Intell. **34**(3), 465–479 (2012)
8. Geiger, A., Lenz, P., Urtasun, R.: Are we ready for autonomous driving? The KITTI vision benchmark suite. IEEE Conf. Comput. Vis. Pattern Recogn. (CVPR) (2011)
9. Geiger, A., Lenz, P., Urtasun, R.: Are we ready for autonomous driving? The KITTI vision benchmark suite. In: 2012 IEEE Conference on Computer Vision and Pattern Recognition, pp. 3354–3361. IEEE (2012)
10. He, K., Zhang, X., Ren, S., Sun, J.: Deep residual learning for image recognition. In: Proceedings of the IEEE Conference on Computer Vision and Pattern Recognition, pp. 770–778 (2016)
11. Hoffman, J., et al.: CYCADA: cycle-consistent adversarial domain adaptation. In: International Conference on Machine Learning, pp. 1989–1998. PMLR (2018)
12. Hsu, H.K., et al.: Progressive domain adaptation for object detection. In: Proceedings of the IEEE/CVF Winter Conference on Applications of Computer Vision, pp. 749–757 (2020)
13. Ilg, E., Mayer, N., Saikia, T., Keuper, M., Dosovitskiy, A., Brox, T.: FlowNet 2.0: evolution of optical flow estimation with deep networks. In: Proceedings of the IEEE Conference on Computer Vision and Pattern Recognition, pp. 2462–2470 (2017)
14. Im, W., Kim, T.-K., Yoon, S.-E.: Unsupervised learning of optical flow with deep feature similarity. In: Vedaldi, A., Bischof, H., Brox, T., Frahm, J.-M. (eds.) ECCV 2020. LNCS, vol. 12369, pp. 172–188. Springer, Cham (2020). https://doi.org/10.1007/978-3-030-58586-0_11
15. Jiang, H., Sun, D., Jampani, V., Yang, M.H., Learned-Miller, E., Kautz, J.: Super SloMo: high quality estimation of multiple intermediate frames for video interpolation. In: Proceedings of the IEEE Conference on Computer Vision and Pattern Recognition, pp. 9000–9008 (2018)
16. Jiang, S., Campbell, D., Lu, Y., Li, H., Hartley, R.: Learning to estimate hidden motions with global motion aggregation. arXiv preprint arXiv:2104.02409 (2021)
17. Jonschkowski, R., Stone, A., Barron, J.T., Gordon, A., Konolige, K., Angelova, A.: What matters in unsupervised optical flow. arXiv preprint arXiv:2006.04902 1(2), 3 (2020)
18. Krizhevsky, A., Sutskever, I., Hinton, G.E.: ImageNet classification with deep convolutional neural networks. Adv. Neural. Inf. Process. Syst. **25**, 1097–1105 (2012)
19. Liu, L., et al.: Learning by analogy: Reliable supervision from transformations for unsupervised optical flow estimation. In: Proceedings of the IEEE/CVF Conference on Computer Vision and Pattern Recognition, pp. 6489–6498 (2020)
20. Liu, P., King, I., Lyu, M.R., Xu, J.: DDFlow: learning optical flow with unlabeled data distillation. In: Proceedings of the AAAI Conference on Artificial Intelligence, pp. 8770–8777 (2019)
21. Liu, P., Lyu, M., King, I., Xu, J.: SelFlow: self-supervised learning of optical flow. In: Proceedings of the IEEE/CVF Conference on Computer Vision and Pattern Recognition, pp. 4571–4580 (2019)
22. Liu, Y., Deng, J., Gao, X., Li, W., Duan, L.: BAPA-Net: boundary adaptation and prototype alignment for cross-domain semantic segmentation. In: Proceedings of the IEEE/CVF International Conference on Computer Vision, pp. 8801–8811 (2021)
23. Luo, K., Wang, C., Liu, S., Fan, H., Wang, J., Sun, J.: UPFlow: upsampling pyramid for unsupervised optical flow learning. arXiv preprint arXiv:2012.00212 (2020)

24. Mayer, N., et al.: A large dataset to train convolutional networks for disparity, optical flow, and scene flow estimation. In: Proceedings of the IEEE Conference on Computer Vision and Pattern Recognition, pp. 4040–4048 (2016)
25. Meister, S., Hur, J., Roth, S.: UnFlow: unsupervised learning of optical flow with a bidirectional census loss. In: Proceedings of the AAAI Conference on Artificial Intelligence (2018)
26. Menze, M., Geiger, A.: Object scene flow for autonomous vehicles. In: Proceedings of the IEEE Conference on Computer Vision and Pattern Recognition, pp. 3061–3070 (2015)
27. Poggi, M., Aleotti, F., Mattoccia, S.: Sensor-guided optical flow. In: Proceedings of the IEEE/CVF International Conference on Computer Vision, pp. 7908–7918 (2021)
28. Ren, Z., Yan, J., Ni, B., Liu, B., Yang, X., Zha, H.: Unsupervised deep learning for optical flow estimation. In: Proceedings of the AAAI Conference on Artificial Intelligence (2017)
29. Simonyan, K., Zisserman, A.: Two-stream convolutional networks for action recognition in videos. arXiv preprint arXiv:1406.2199 (2014)
30. Sun, D., Yang, X., Liu, M.Y., Kautz, J.: PWC-Net: CNNs for optical flow using pyramid, warping, and cost volume. In: Proceedings of the IEEE Conference on Computer Vision and Pattern Recognition, pp. 8934–8943 (2018)
31. Teed, Z., Deng, J.: RAFT: recurrent all-pairs field transforms for optical flow. In: Vedaldi, A., Bischof, H., Brox, T., Frahm, J.-M. (eds.) ECCV 2020. LNCS, vol. 12347, pp. 402–419. Springer, Cham (2020). https://doi.org/10.1007/978-3-030-58536-5_24
32. Tsai, Y.H., Hung, W.C., Schulter, S., Sohn, K., Yang, M.H., Chandraker, M.: Learning to adapt structured output space for semantic segmentation. In: Proceedings of the IEEE Conference on Computer Vision and Pattern Recognition, pp. 7472–7481 (2018)
33. Wang, Y., Yang, Y., Yang, Z., Zhao, L., Wang, P., Xu, W.: Occlusion aware unsupervised learning of optical flow. In: Proceedings of the IEEE Conference on Computer Vision and Pattern Recognition, pp. 4884–4893 (2018)
34. Wang, Z., Du, B., Shi, Q., Tu, W.: Domain adaptation with discriminative distribution and manifold embedding for hyperspectral image classification. IEEE Geosci. Remote Sens. Lett. **16**(7), 1155–1159 (2019)
35. Wang, Z., Bovik, A.C., Sheikh, H.R., Simoncelli, E.P.: Image quality assessment: from error visibility to structural similarity. IEEE Trans. Image Process. **13**(4), 600–612 (2004)
36. Zhang, F., Woodford, O.J., Prisacariu, V.A., Torr, P.H.: Separable flow: learning motion cost volumes for optical flow estimation. In: Proceedings of the IEEE/CVF International Conference on Computer Vision, pp. 10807–10817 (2021)

37. Zhao, S., Li, B., Xu, P., Keutzer, K.: Multi-source domain adaptation in the deep learning era: a systematic survey. arXiv preprint arXiv:2002.12169 (2020)
38. Zhu, J.Y., Park, T., Isola, P., Efros, A.A.: Unpaired image-to-image translation using cycle-consistent adversarial networks. In: Proceedings of the IEEE International Conference on Computer Vision, pp. 2223–2232 (2017)

Continuous Exploration via Multiple Perspectives in Sparse Reward Environment

Zhongpeng Chen[1,2] and Qiang Guan[2(✉)]

[1] School of Artificial Intelligence, University of Chinese Academy of Sciences, Beijing 100049, China

[2] Institute of Automation, Chinese Academy of Sciences, Beijing 100190, China

{chenzhongpeng2021,qiang.guan}@ia.ac.cn

Abstract. Exploration is a major challenge in deep reinforcement learning, especially in cases where reward is sparse. Simple random exploration strategies, such as ϵ-greedy, struggle to solve the hard exploration problem in the sparse reward environment. A more effective approach to solve the hard exploration problem in the sparse reward environment is to use an exploration strategy based on intrinsic motivation, where the key point is to design reasonable and effective intrinsic reward to drive the agent to explore. This paper proposes a method called CEMP, which drives the agent to explore more effectively and continuously in the sparse reward environment. CEMP contributes a new framework for designing intrinsic reward from multiple perspectives, and can be easily integrated into various existing reinforcement learning algorithms. In addition, experimental results in a series of complex and sparse reward environments in MiniGrid demonstrate that our proposed CEMP method achieves better final performance and faster learning efficiency than ICM, RIDE, and TRPO-AE-Hash, which only calculate intrinsic reward from a single perspective.

Keywords: Reinforcement Learning · Exploration Strategy · Sparse Reward · Intrinsic Motivation

1 Introduction

The goal of reinforcement learning is to train an effective policy, which drives the agent to obtain the maximum cumulative reward in the environment. The generation of the agent's policy primarily depends on the reward provided by the environment. Therefore, whether the agent can learn an effective policy is closely related to its interaction with the environment in order to capture rewards. Classical algorithms like DQN [1] and PPO [2] have demonstrated outstanding performance in the dense reward environment by using simple exploration strategies such as ϵ-greedy, entropy regularization, and Boltzman exploration. However, in the sparse reward environment, these simple exploration strategies face difficulties in guiding the agent to reach the target state and obtain effective reward.

Q. Liu et al. (Eds.): PRCV 2023, LNCS 14427, pp. 57–68, 2024.
https://doi.org/10.1007/978-981-99-8435-0_5

The agent will be unable to update the policy due to long-term inability to receive effective reward. Hence, this paper will concentrate on designing effective exploration strategies to help the agent better resolving hard exploration problem in the sparse reward environment.

In the sparse reward environment, the agent needs to continuously complete a series of correct decisions in order to obtain few effective rewards. However, a simple random exploration strategy is difficult for the agent to explore a trajectory that can complete the task. Exploration strategies based on intrinsic motivation have seen significant development in recent years and can effectively address the exploration challenges faced by the agent in the sparse reward environment. Intrinsic motivation comes from the concept of ethology and psychology [3]. During the learning process, higher organisms often spontaneously explore the unfamiliar and unknown environment without extrinsic stimuli to enhance their ability to survive in the environment. Exploration strategies based on intrinsic motivation formalize the concept of intrinsic motivation as an intrinsic reward by measuring the novelty of the state, thereby utilizing the intrinsic reward to drive the agent to spontaneously explore more unknown space in the sparse reward environment and increase the likelihood of the agent solving tasks. The novelty of a state is related to the number of times the agent accesses the state. Currently, there are two main perspectives for measuring the novelty of a state: global perspective and local perspective, as explained below.

Global Perspective: Use all historical samples collected by the agent from the environment to measure the novelty of a state.

Local Perspective: Evaluate the novelty of a state only using the samples collected in the current episode where the agent is interacting with the environment, without considering historical samples.

The drawback of measuring the novelty of a state from global perspective is that the novelty of the state and its corresponding intrinsic reward gradually decay during the training process. The decaying intrinsic reward cannot drive the agent to continuously explore in the environment. At the same time, measuring the novelty of a state from global perspective alone is not conducive to the agent discovering more novel trajectories. When a novel trajectory appears, it will be diluted by ordinary trajectories in history. Measuring the novelty of a state from local perspective can enable the agent to discover more novel trajectories and access more unknown states within the current episode. However, in the absence of global information, relying solely on local perspective to measure the novelty of a state may blindly and optimistically encourage the agent to explore unknown states.

Taking into account the characteristics of calculating intrinsic reward from different perspectives, we propose a method called CEMP (Continuous Exploration via Multiple Perspectives) that enables the agent to better explore in the sparse reward environment. CEMP calculates intrinsic reward from both global and local perspectives and integrates them to better drive the agent's exploration in the sparse reward environment. To summarize, the main contributions of our work are as follows:

1. We propose a method called CEMP that enables the agent to explore better in the sparse reward environment. This method combines intrinsic reward calculated from different perspectives to obtain a new intrinsic reward, which does not decay gradually during the training process and can continuously drive the agent's exploration in the sparse reward environment. Moreover, our proposed CEMP method can be easily integrated into various existing reinforcement learning algorithms.
2. The majority works that calculate intrinsic reward from local perspective only measure the novelty of a state within the current episode. In contrast, our proposed CEMP method can measure the novelty of a state from local perspective across multiple episodes and can flexibly control the range of the local perspective through parameters.
3. Experimental results in a series of complex and sparse reward environments in MiniGrid show that our proposed CEMP method achieves better final performance and faster learning efficiency compared to methods such as ICM, RIDE, and TRPO-AE-Hash that only calculate intrinsic reward from a single perspective.

2 Related Works

There are two main methods for calculating intrinsic reward from global perspective: state count and prediction error method.

The state count method extends UCB exploration strategy to the high dimensional state environment through pseudo-count or indirect count methods. DQN-PixelCNN [4] indirectly derives the pseudo-count of a state through the density probability model, which is modeled from the raw state space, but it is difficult to directly model a density probability model in the raw state space. To solve this issue, ϕ-EB [5] models a density probability model in a low dimensional feature space of the raw state. TRPO-AE-Hash [6] discretizes the raw state into low dimensional hash code using SimHash [7] and then computes the pseudo-count of a state based on its hash code. DORA [8] constructs an indirect count index E-value, which can become an effective generalization counter of (s_t, s_{t+1}). RND [9] uses the prediction error between the predictor network and the target network as an indirect measure of state count.

The prediction error method uses the prediction error between the next predicted state $\hat{s}_{t+1}$ output by a forward model and the ground-truth s_{t+1} as intrinsic reward. The key aspect of the prediction error method is how to construct a forward model. Dynamic-AE [10] reconstructs the raw state by using an autoencoder and constructs a forward model based on a low-dimensional feature space extracted by the middle layer of the autoencoder. In fact, using the features extracted by the middle layer of the autoencoder to train a forward model is highly susceptible to environmental noise. ICM [11] learns features of the raw state by utilizing an inverse model, which predicts the action a_t from the state (s_t, s_{t+1}) and extracts the intermediate layer output as low-dimensional features to construct a forward model. Inverse model only extracts feature from the raw state that is related to the agent's actions, which partially reduces the influence of noise from the raw state. Disagreement [12] trains multiple forward models based on random feature subspace and uses the variance of the predicted values of these multiple forward models as intrinsic reward.

The methods introduced above measure the novelty of a state from global perspective, while the methods below measure the novelty of a state from local perspective. DeepCS [13] incentivizes the agent to explore as many unknown states as possible within the same episode by setting intrinsic reward to 1 for unreachable states and 0 for visited states respectively. ECO [14] measures the novelty of a state by the reachability between states in the same episode. RIDE [15] directly uses the difference between two consecutive states as intrinsic reward.

3 Method

3.1 Continuous Exploration via Multiple Perspectives

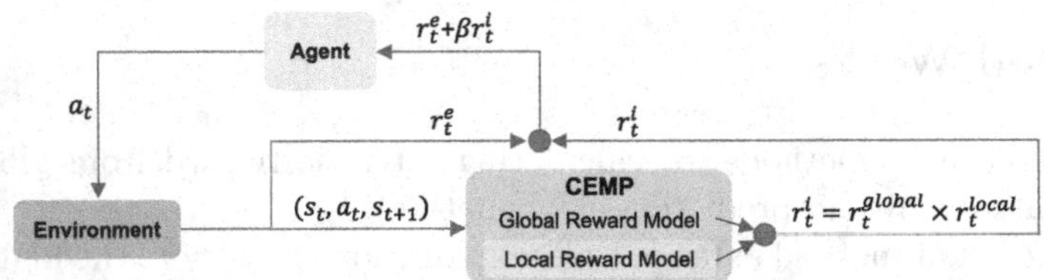

Fig. 1. Our proposed CEMP method: the design details of the Global Reward Model and the Local Reward Model can be found in Sect. 3.2 and 3.3, respectively.

As illustrated in Fig. 1, our proposed CEMP (Continuous Exploration via Multiple Perspectives) method calculates intrinsic reward from both global and local perspectives, and combines the reward from both perspectives by multiplication to obtain the new intrinsic reward r_t^i, as shown in Eq. (1). Next, the final reward that guides the update of the agent's policy is calculated by linearly weighting the new intrinsic reward r_t^i and the extrinsic reward r_t^e provided by the environment in terms of Eq. (2), where β is the weighting coefficient between both rewards.

$$r_t^i = r_t^{global} \times r_t^{local} \tag{1}$$

$$r_t = r_t^e + \beta r_t^i \tag{2}$$

We will use prediction error-based method to design the global intrinsic reward, and use state count method based on hash-discretization to design the local intrinsic reward. The specific reasons for these choices are as follows:

1. The prediction error-based method evaluates the novelty of a state from global perspective by using the prediction error of a forward model modeled with deep neural network, which is well-suited for processing large-scale data and has strong discriminative power when facing with large-scale states, thus, it is suitable for evaluating the novelty of a state from global perspective. However, it is not suitable for evaluating the novelty of a state from local perspective, because the forward model trained with a small dataset may have difficulty learning effective knowledge and distinguishing the novelty of a state due to insufficient data.
2. The state count method based on hash-discretization can quickly and accurately count different states in a small dataset by using the hash code of the raw state. However, in the case with large-scale states, it may not be able to distinguish different states due to the limited number of hash code bits.

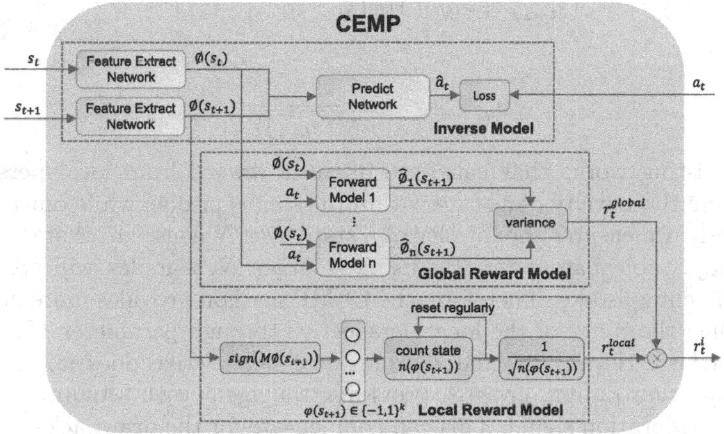

Fig. 2. The design details of the Global Reward Model and the Local Reward Model.

3.2 Global Reward Model

We first use the inverse model adopted in ICM [11] method to extract a low dimensional feature of the raw state, as shown in the green box in Fig. 2. Extracting a low-dimensional feature through the inverse model can capture the feature information related to the agent's actions as much as possible and reduce the impact of noise in the raw state. As shown in the red box in Fig. 2, we then use all the samples collected by the agent in the environment to train n forward models in the inverse feature space ϕ. The variance of the predicted values from these n forward models is then used as the intrinsic reward under global perspective, as shown in Eq. (3). The ensemble of these n forward models can further reduce the influence of state noise.

$$r_t^{global} = variance(\hat{\phi}_1(s_{t+1}), \hat{\phi}_2(s_{t+1}), ..., \hat{\phi}_n(s_{t+1})) \tag{3}$$

3.3 Local Reward Model

As shown in the blue box in Fig. 2, we use feature extract network of the inverse model to extract the inverse feature $\phi(s_{t+1})$ of the raw state s_{t+1}, and then use SimHash method to discretize $\phi(s_{t+1})$ into a k-dimensional binary hash code $\varphi(s_{t+1})$, as shown in Eq. (4), where M is a $k \times d$ matrix sampled from standard normal distribution, d is the dimension of $\phi(s_{t+1})$. Next, we count the state based on its hash code $\varphi(s_{t+1})$. In the counting process, we only count the access times of each state in N consecutive states, and reset the previous count information when the agent enters the next new N consecutive states. Equation (5) provides the relationship between the count of a state and corresponding intrinsic reward under local perspective. The local intrinsic reward r_t^{local} will incentivize the agent to frequently explore more novel states within the current N consecutive states, which is conducive to the agent discovering more novel trajectories.

$$\varphi(s_{t+1}) = sign(M\phi(s_{t+1})) \in \{-1, 1\}^k \tag{4}$$

$$r_t^{local} = \frac{1}{\sqrt{n(\varphi(s_{t+1}))}} \tag{5}$$

Most existing works that calculate intrinsic reward from local perspective only measure the novelty of a state within the same episode, while our proposed CEMP method measures the novelty of a state over N consecutive states, where N is an adjustable parameter. By using a larger N, samples will come from several different episodes. Therefore, the CEMP method provides more flexibility in controlling the range of the local perspective through parameter N.

The intrinsic reward calculated from local perspective does not gradually decay during the training process, providing the agent with a more continuous and stable exploration signal. This can compensate for the drawback of intrinsic reward calculated from global perspective that gradually decays over time, while also improving the agent's ability to discover more novel trajectories.

Our proposed CEMP method can be easily integrated into any reinforcement learning algorithm to enhance the agent's exploration ability in the sparse reward environment.

4 Experiment

4.1 Comparison Algorithms and Evaluation Metrics

We compared our proposed CEMP method with three intrinsic motivation-based exploration strategies: ICM [11], TRPO-AE-Hash [6] and RIDE [15]. ICM and TRPO-AE-Hash calculate intrinsic reward from global perspective, while RIDE calculates intrinsic reward from local perspective. Different reinforcement learning algorithms are used as baseline algorithms in the original papers of ICM, TRPO-AE-Hash, and RIDE. In order to ensure fairness in the comparison experiments, we use PPO as the baseline algorithm to reproduce the three exploration strategies and compare their performance with our proposed CEMP method. The CEMP method also uses PPO as the baseline algorithm.

We use the mean and standard deviation of the average reward over five different seeds as metrics to compare the performance of different exploration strategies. The mean of the average reward reflects the best performance that each exploration strategy can achieve, while the standard deviation of the average reward reflects the stability of the exploration strategy.

4.2 Network Architectures and Hyperparameters

PPO. In the PPO algorithm we reproduced, the Critic and Actor share a feature extract network, which consists of four linear layers with 256, 128, 64, and 64 nodes, respectively. Both the Critic and Actor consist of one linear layer with 64 nodes. The Critic and Actor take the output of the feature extract network as input and output value estimate and action probability distribution for the current state s_t. Other hyperparameters of the PPO algorithm we reproduced are shown in Table 1.

Table 1. Other hyperparameters of the PPO algorithm we reproduced.

Hyperparameter Name	Value
learning rate	0.0003
value loss weight	0.5
entropy loss weight	0.001
discount factor	0.99
λ, general advantage estimation	0.95
activation function	ReLU
optimizer	Adam

CEMP. As shown in Fig. 2, in our proposed CEMP method, the feature extract network in the inverse model consists of three linear layers with 64, 64, and 128 nodes, respectively. The network used to predict actions in the inverse model consists of one linear layer with 512 nodes. All forward models consist of two linear layers with 512 nodes. Other hyperparameters of CEMP method are summarized in Table 2.

ICM, RIDE, and TRPO-AE-Hash. ICM [11] trains a forward model based on the inverse feature space ϕ extracted by the inverse model. The difference between ICM and the Global Reward Model in CEMP lies in the number of forward model and the way intrinsic reward is calculated. ICM trains only one forward model and uses the prediction error of the forward model as the intrinsic

Table 2. Other hyperparameters of our proposed CEMP method.

Hyperparameter Name	Value
n, number of forward model	5
$step_{max}$, max step of one episode, determined by the environment	/
β, weighting coefficient of intrinsic and extrinsic reward	$10/step_{max}$
k, the dimension of hash code	16
N, adjust the range of local perspective	3200
m, batch size	320
learning rate	0.0003
T_{max}, max step of training	3×10^6
activation function	ReLU
optimizer	Adam

reward, as shown in Eq. (6), while the Global Reward Model in CEMP trains n forward models and uses the variance of the predicted values from n forward models as the intrinsic reward.

$$r_t^i = ||\phi(s_{t+1}) - f(\phi(s_t), a_t)||_2^2 \tag{6}$$

RIDE [15] directly uses the difference between the inverse features of two consecutive states as the intrinsic reward, as shown in Eq. (7). Here, ϕ represents the feature extract network in the inverse model, and $n(\varphi(s_{t+1}))$ is the pseudo-count of s_{t+1} in the current episode. The calculation method of $\varphi(s_{t+1})$ is consistent with the Local Reward Model used in CEMP.

$$r_t^i = \frac{||\phi(s_{t+1}) - \phi(s_t)||_2}{\sqrt{n(\varphi(s_{t+1}))}} \tag{7}$$

TRPO-AE-Hash [6] calculates the intrinsic reward in the same way as the Local Reward Model in the CEMP method. The only difference is that TRPO-AE-Hash omits the step of periodically resetting the count of a state, so it calculates the intrinsic reward from global perspective.

To ensure a fair comparison, all network architectures and hyperparameters used in the ICM, RIDE, and TRPO-AE-Hash methods that we reproduced are kept consistent with those used in our proposed CEMP method.

4.3 Experimental Results

Comparison with Other Methods. We select 9 sparse reward environments from MiniGrid for comparative experiments, which contain a total of 5 different tasks. Among the 5 different tasks, the tasks of DoorKey and LavaCrossing each have 3 environments of gradually increasing difficulty (difficulty ranking: DoorKey-5x5 < DoorKey-8x8 < DoorKey-16x16; LavaCrossingS9N1 < LavaCrossingS9N3 < LavaCrossingS11N5). The learning curves of

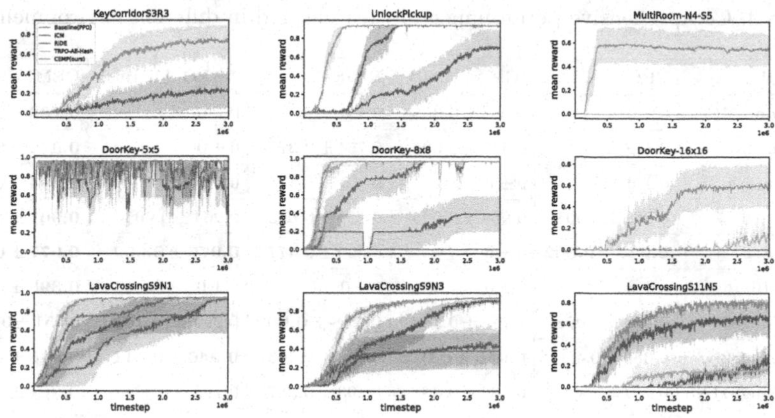

Fig. 3. Performance of each method in different sparse reward environments.

different methods are shown in Fig. 3, from which we can draw the following conclusions:

1. In some environments with lower difficulty levels, such as LavaCrossingS9N1 and DoorKey-5x5, ICM, RIDE, and TRPO-AE-Hash can achieve the same final performance as CEMP. However, our proposed CEMP method has a higher learning efficiency and more stable performance.
2. In the DoorKey and LavaCrossing tasks, as the difficulty increases, the performance of ICM, RIDE, and TRPO-AE-Hash gradually deteriorates, but our proposed CEMP method can still maintain good performance.
3. In more challenging environments, such as MultiRoom-N4-S5 and DoorKey-16x16, ICM, RIDE, and TRPO-AE-Hash are unable to complete the task, but our proposed CEMP method can still perform well and obtain a high average reward.

We evaluate the final policies of the agent obtained by various methods, and the evaluation results are summarized in Table 3. The evaluation metrics in the table are the mean and standard deviation of rewards obtained by each method in five different seeds. It can be seen from Table 3 that our proposed CEMP method has significantly better performance and is more stable than ICM, RIDE, and TRPO-AE-Hash that calculate intrinsic reward from only a single perspective.

Analysis of Local and Global Intrinsic Reward. From Fig. 4, we can see that as the training progresses, the global intrinsic reward gradually decays, while the local intrinsic reward remains in a relatively stable range and does not decay during the training process. Thus, the local intrinsic reward can drive the agent to continuously explore in the sparse reward environment, discover more novel states and trajectories, and compensate for the disadvantage of the global intrinsic reward, which gradually decays and cannot consistently drive the agent to explore.

Table 3. Comprehensive performance of each method in different environments.

mean ± std	PPO	ICM	RIDE	TRPO-AE-Hash	CEMP (ours)
KeyCorridorS3R3	0 ± 0	0.183 ± 0.365	0 ± 0	0 ± 0	**0.733 ± 0.367**
UnlockPickup	0 ± 0	0.947 ± 0.006	0.754 ± 0.377	0 ± 0	**0.949 ± 0.004**
MultiRoom-N4-S5	0 ± 0	0 ± 0	0 ± 0	0 ± 0	**0.594 ± 0.298**
DoorKey-5x5	0.962 ± 0.005	0.964 ± 0.005	0.964 ± 0.005	0.964 ± 0.005	**0.967 ± 0.003**
DoorKey-8x8	**0.977 ± 0.002**	0 ± 0	0.390 ± 0.477	**0.977 ± 0.002**	**0.977 ± 0.002**
DoorKey-16x16	0 ± 0	0 ± 0	0 ± 0	0 ± 0	**0.395 ± 0.484**
LavaCrossingS9N1	0.954 ± 0.008	0.766 ± 0.383	**0.958 ± 0.002**	0.950 ± 0.014	0.951 ± 0.002
LavaCrossingS9N3	0.380 ± 0.466	**0.946 ± 0.013**	0.378 ± 0.463	**0.946 ± 0.011**	**0.946 ± 0.015**
LavaCrossingS11N5	0 ± 0	0.762 ± 0.381	0.193 ± 0.387	0.192 ± 0.384	**0.961 ± 0.005**

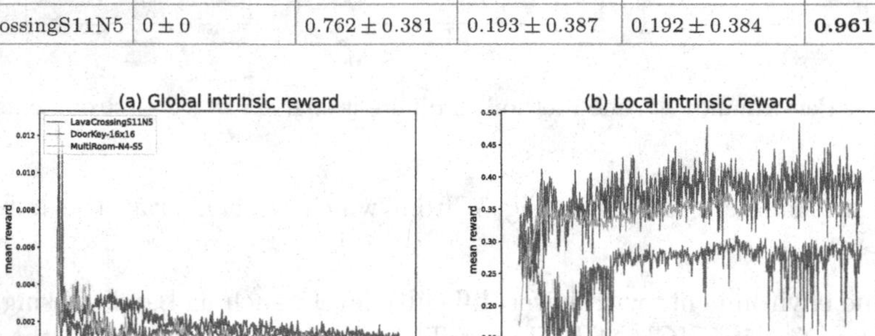

Fig. 4. The intrinsic reward calculated from both global and local perspectives in 3 environments: LavaCrossingS11N5, DoorKey-16x16, and MultiRoom-N4-S5.

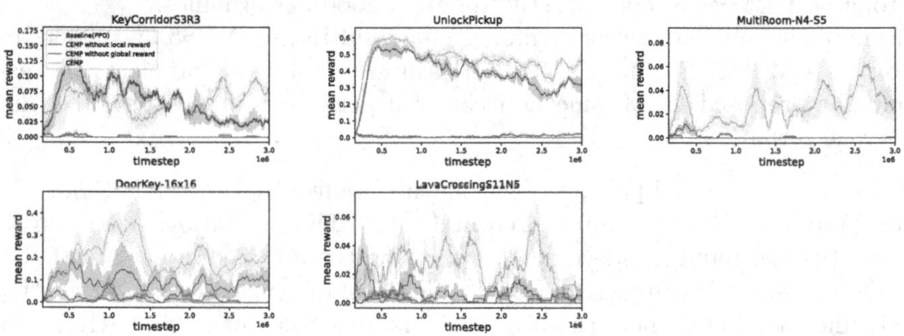

Fig. 5. Without extrinsic reward, only the intrinsic reward is used.

Learning with only Intrinsic Reward. To test whether the intrinsic reward designed from the local and global perspectives in our proposed CEMP method are effective, we only use the intrinsic reward from the local, global, and the fusion of both to guide the agent's learning. Based on the results shown in Fig. 5, we can conclude that our proposed CEMP method can achieve a certain level of performance in the sparse reward environment using only self-generated intrinsic reward. Moreover, the fusion of the intrinsic reward from the local and global perspectives can achieve better performance than a single intrinsic reward from either the global or local perspective.

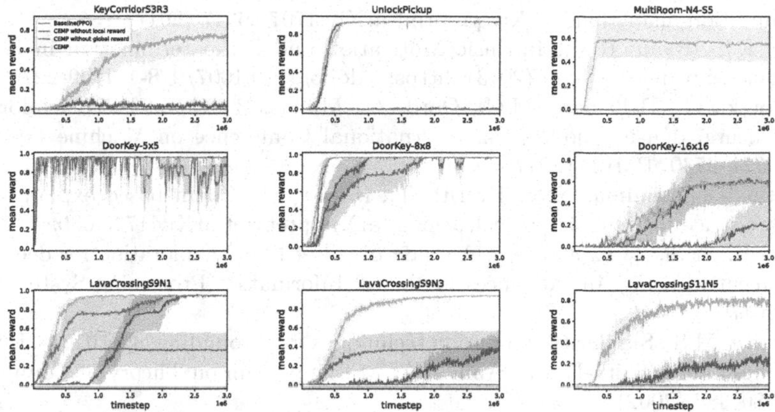

Fig. 6. The ablation experiment on each component of the intrinsic reward.

Ablation Experiment of the Intrinsic Reward. We conducted ablation experiment on each component of the intrinsic reward in our proposed CEMP method. From Fig. 6, we can draw the following conclusions:

1. In all environments, the intrinsic reward that integrates global and local intrinsic reward performs better than using only intrinsic reward calculated from a single perspective.
2. Using only global or local intrinsic reward can also achieve good performance in some environments.

5 Conclusion

This paper proposed a method called CEMP that enables the agent to explore better in the sparse reward environment. The CEMP method contributes a new framework for designing the intrinsic reward from multiple perspectives and can be easily integrated into various existing reinforcement learning algorithms. The experimental results in a series of complex sparse reward environments in MiniGrid demonstrate that our proposed CEMP method can achieve better final performance and faster learning efficiency than algorithms such as ICM [11], RIDE [15], and TRPO-AE-Hash [6] that calculate the intrinsic reward from only a single perspective.

Acknowledgments. This work was supported by the National Defense Science and Technology Foundation Reinforcement Program and the Strategic Priority Research Program of the Chinese Academy of Sciences, Grant No.XDA27041001.

References

1. Mnih, V., et al.: Human-level control through deep reinforcement learning. Nature **518**(7540), 529–533 (2015)
2. Schulman, J., Wolski, F., Dhariwal, P., Radford, A., Klimov, O.: Proximal policy optimization algorithms. arXiv preprint arXiv:1707.06347 (2017)
3. Deci, E.L., Ryan, R.M.: Intrinsic Motivation and Self-determination in Human Behavior. Springer, Cham (2013). https://doi.org/10.1007/978-1-4899-2271-7
4. Ostrovski, G., Bellemare, M.G., Oord, A., Munos, R.: Count-based exploration with neural density models. In: International Conference on Machine Learning, pp. 2721–2730. PMLR (2017)
5. Martin, J., Sasikumar, S.N., Everitt, T., Hutter, M.: Count-based exploration in feature space for reinforcement learning. arXiv preprint arXiv:1706.08090 (2017)
6. Tang, H., et al.: # exploration: A study of count-based exploration for deep reinforcement learning. In: Advances in Neural Information Processing Systems, vol. 30 (2017)
7. Charikar, M.S.: Similarity estimation techniques from rounding algorithms. In: Proceedings of the Thiry-Fourth Annual ACM Symposium on Theory of Computing, pp. 380–388 (2002)
8. Choshen, L., Fox, L., Loewenstein, Y.: Dora the explorer: directed outreaching reinforcement action-selection. arXiv preprint arXiv:1804.04012 (2018)
9. Burda, Y., Edwards, H., Storkey, A., Klimov, O.: Exploration by random network distillation. arXiv preprint arXiv:1810.12894 (2018)
10. Stadie, B.C., Levine, S., Abbeel, P.: Incentivizing exploration in reinforcement learning with deep predictive models. arXiv preprint arXiv:1507.00814 (2015)
11. Pathak, D., Agrawal, P., Efros, A.A., Darrell, T.: Curiosity-driven exploration by self-supervised prediction. In: International Conference on Machine Learning, pp. 2778–2787. PMLR (2017)
12. Pathak, D., Gandhi, D., Gupta, A.: Self-supervised exploration via disagreement. In: International Conference on Machine Learning, pp. 5062–5071. PMLR (2019)
13. Stanton, C., Clune, J.: Deep curiosity search: intra-life exploration improves performance on challenging deep reinforcement learning problems. corr abs/1806.00553 (2018) (1806)
14. Savinov, N., et al.: Episodic curiosity through reachability. arXiv preprint arXiv:1810.02274 (2018)
15. Raileanu, R., Rocktäschel, T.: Ride: rewarding impact-driven exploration for procedurally-generated environments. arXiv preprint arXiv:2002.12292 (2020)

Network Transplanting for the Functionally Modular Architecture

Quanshi Zhang[1(✉)], Xu Cheng[1], Xin Wang[1], Yu Yang[2], and Yingnian Wu[2]

[1] Shanghai Jiao Tong University, Shanghai, China
{zqs1022,xcheng8}@sjtu.edu.cn
[2] University of California, Los Angeles, USA

Abstract. This paper focuses on the problem of transplanting category-and-task-specific neural networks to a generic, modular network without strong supervision. Unlike traditional deep neural networks (DNNs) with black-box representations, we design a functionally modular network architecture, which divides the entire DNN into several functionally meaningful modules. Like building LEGO blocks, we can teach the proposed DNN a new object category by directly transplanting the module corresponding to the object category from another DNN, with a few or even without sample annotations. Our method incrementally adds new categories to the DNN, which do not affect representations of existing categories. Such a strategy of incremental network transplanting can avoid the typical catastrophic-forgetting problem in continual learning. We further develop a back distillation method to overcome challenges of model optimization in network transplanting. In experiments, our method with much fewer training samples outperformed baselines.

Keywords: Deep Learning · Network transplant

1 Introduction

Deep neural network (DNN) consisting of functionally meaningful modules has shown promise in various tasks. Such a functionally meaningful network architecture usually exhibits advantages in both learning and application.

Different from traditional learning DNNs from raw data, this paper proposes a new learning strategy, *i.e.*, network transplanting, to learn a DNN with functionally meaningful modules. Specifically, given an off-the-shelf DNN oriented to a specific task or a specific category, we transplant specific modules of this off-the-shelf DNN to a new DNN (namely *transplant net*), in order to enable the new DNN to accomplish a new task (*e.g.* classification or segmentation) or to deal with a new object category.

Q. Zhang and X. Cheng—Contributed equally to this paper.

Q. Liu et al. (Eds.): PRCV 2023, LNCS 14427, pp. 69–83, 2024.
https://doi.org/10.1007/978-981-99-8435-0_6

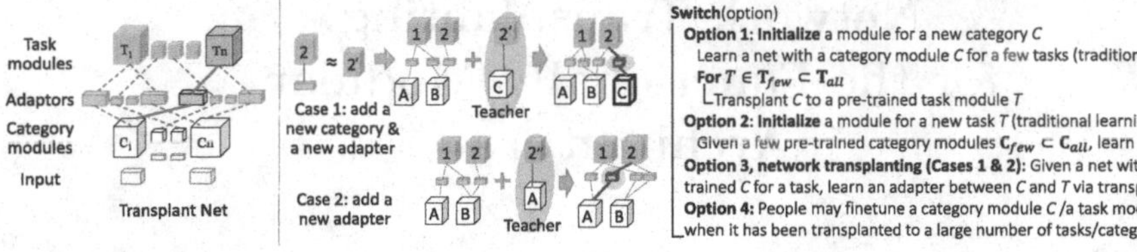

Fig. 1. A transplant net with a functionally modular architecture. We propose a theoretical solution to incrementally merging category modules from teacher nets into a transplant (student) net with a few or without sample annotations. (left) During the inference, people can manually activate the certain linkages (red) and deactivate others to conduct the task T_n on category C_1. (Middle) We show typical cases of network transplanting. Blue ellipses indicate teacher nets used for transplant. (right) Flowchart for network transplanting. (Color figure online)

To this end, the transplant net consists of three types of modules, including *category modules*, *task modules*, and *adapters*, as shown in Fig. 1(left). Specifically, each *category module* extracts general features for a specific category (*e.g.*, the dog), and each *task module* is learned for a certain task (*e.g.*, classification or segmentation). That is, the output feature of a specific category module is used for various tasks, while a task module is universal for all categories. The *adapter* projects the output feature of the category module into the input space of the task category, in order to enable the task category to deal with a new category. For example, for the segmentation of a dog, an adapter is learned to project output features of the dog-category module to the input space of the segmentation-task module, as shown in Fig. 2 (right).

Learning Adapters for Network Transplanting. We first follow the traditional scenario of multi-task and multi-category learning to train an initial transplant net for very few tasks and categories. We further train this initial transplant net for new categories or new tasks by gradually absorbing new modules, as shown in Fig. 1.

Specifically, given an initial transplant net, its task module g_S is trained to handle a specific task for many categories. When dealing with a new category, we collect another DNN (namely a *teacher net*) pre-trained for the same task on a new category, as shown in Fig. 2 (left). This teacher net has a category module f, which extracts generic features for the new category. Thus, in order to enable the task module g_S to deal with the new category, we transplant the new category module f to the transplant net by learning an adapter to project the output feature of f into the input space of g_S.

In this way, we may (or may not) have a few training annotations of the new category for the task to guide the transplanting. Besides, learning a small adapter to connect the new category module$f$ to the task module g_S **without** fine-tuning f and g_S preserves the generality of f and g_S, which helps avoid the catastrophic forgetting.

However, learning adapters but fixing category modules and task modules raises specific challenges to the optimization of DNNs. To overcome challenges, we propose a new algorithm, namely *back distillation*. In experiments, our back distillation method with much fewer training samples outperformed baseline methods.

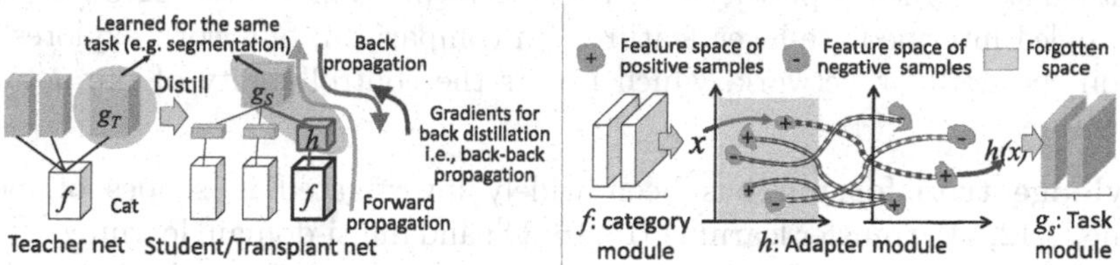

Fig. 2. Overview. (left) Given a teacher net and a transplant net (student net), the teacher net has a category module f for a single or multiple tasks, and the student net contains a task module g_S for other categories. To enable the task module g_S to deal with the new category, we transplant f to the student net by learning an adapter h via distillation. Blue ellipses shows that our method distills knowledge from the task modular g_T of a teacher net to the $h \circ g_S$ modules of the student net. Three red curves show directions of forward propagation, back propagation, and gradient propagation of back distillation. (right) During the transplant, the adapter h learns potential projections between f's output feature space and g_S's input feature space.

Distinct Advantages of Network Transplanting. As a new methodology, network transplanting exhibit distinct values in deep learning, because it builds a universal net for various categories and tasks, instead of learning various DNNs for different applications. First, the transplant net has compact representations, because each task/category module can be shared by various categories/tasks. Second, the network transplanting reduces the start-up cost of sample collection before the learning process, because transplanting can be conducted in an online manner by incrementally absorbing network modules of new categories. Third, network transplanting can be considered as a distributed-learning algorithm, which reduces the computational cost by distributing the heavy computation burden of modeling each category and each task. Fourth, network transplanting delivers trained DNNs to the centralized computation center, which is usually much cheaper than sending raw training data. Fifth, network transplanting can efficiently organize and use the knowledge in DNNs, because each module encodes the knowledge of a certain category or a certain task, which makes the transplant net controllable.

Contributions are summarized as follows. (1) We propose network transplanting, which learns with a few or even without additional training annotations. (2) We develop the back distillation to overcome specific challenges of network transplanting in model optimization. (3) Preliminary experiments verified the effectiveness of our method.

2 Related Work

Controllability. Learning a CNN with interpretable feature representations is a typical method to improve network interpretability. InfoGAN [7] and β-VAE [10] learned meaningful input codes of generative networks. In the capsule network [20], each capsule encoded different meaningful features. The ProtoPNet [5] was introduced to learn prototypical parts. Interpretable CNNs [22,31] learned disentangled intermediate-layer features. In comparison, our work explores the functionally modular network, which boosts the controllability of the network architecture.

Knowledge transferring has been widely investigated in scopes of meta-learning [2,12,14], transfer learning [15,16,30], and multi-domain learning [17,18, 28]. Especially, continual learning [1,4,19,29] has transferred knowledge from previous tasks to guide the new task.

However, network transplanting is **different** from previous studies from three perspectives, as summarized in Table 1. First, different from expanding the network architecture during the learning process [19,29], network transplanting has a better modular controllability, because it allows people to semantically control the knowledge to transfer, and requires much less network revision than that in multi-domain learning [17,18]. Second, network transplanting can physically avoid catastrophic forgetting, because it exclusively learns the adapter without affecting feature representations of existing category/task modules. Third, network transplanting uses the back distillation method to overcome the challenge of transferring upper modules, which is different from transferring low-layer features.

Table 1. Comparison between network transplanting and other studies. Note that this table only summarizes mainstreams in different research directions considering the huge research diversity.

	Annotation cost	Sample preparation	Interpretability	Catastrophic forgetting	Modular manipulate	Optimization
Directly learning a multi-task net	Massive	Simultaneously prepare samples for all tasks and categories	Low	—	Not support	Back prop
Transfer- / meta- / continual-learning	Some support weakly-supervised learning	Some learn a category/task after another	Usually low	Most algorithmically alleviate	Not support	Back prop
Transplanting	A few or w/o annotations	Learn a category after another	High	Physically avoid	Support	Back-back prop

3 Network Transplanting

Goal. Given an initial transplant net learned for multiple categories, the goal of network transplanting is to enable this transplant net to deal with a new category. To this end, we first collect another network (namely a *teacher net*) trained for a single or multiple tasks on the new category. This teacher net contains a category module f with m bottom layers to extract generic features for the new category, and a specific task module g_T made up of other layers. Thus, in order to enable the task module g_S of the transplant net to deal with the new category, we transplant the category module f to g_S by learning an adapter h with parameters θ_h to connect f and g_S, as shown in Fig. 2 (left).

Specifically, given an image I, let $x = f(I)$ denote the output of the category module f, $y_T = g_T(x)$ and $y_S = g_S(h(x))$ indicate outputs of the teacher net and the transplant net, respectively. To realize the goal of network transplanting, we use cascaded modules of h and g_S to mimic the task module g_T in the teacher net, *i.e.* $\max_{\theta_h} \text{similarity}(y_T, y_S)$. For convenience, the transplant net is termed a *student net* thereinafter.

Basic idea of the Gradient-Based Distillation. However, the above learning of the adapter h with parameters θ_h cannot be implemented as traditional distillation (*i.e.* directly pushing y_S towards y_T) for three reasons. First, g_S has vast forgotten spaces (introduced in Sect. 3.1). Second, y_T may has very few dimensions (*e.g.* a scalar output), which is usually not informative enough to force $g_S(h(x))$ to mimic g_T precisely, without lots of training samples. Third, except for the output y_T, features of other layers in g_T are not aligned with those of layers in g_S, which cannot be used for distillation.

As a solution, we propose to force the gradient (Jacobian) of the student net (transplant net) to approximate that of the teacher, which is a necessary condition of $y_T \approx y_S$.

$$\begin{aligned} y_T \approx y_S \quad &\Rightarrow \quad \forall \Delta x, \frac{\partial y_T}{\partial x} \approx \frac{y_T|_{x+\Delta x} - y_T|_x}{\Delta x} \approx \frac{y_S|_{x+\Delta x} - y_S|_x}{\Delta x} \approx \frac{\partial y_S}{\partial x} \\ &\Rightarrow \quad \forall \Delta x,\ \forall J,\ \ \frac{\partial J(y_S)}{\partial x} \approx \frac{\partial J(y_T)}{\partial x}. \end{aligned} \tag{1}$$

Equation (1) is supposed to be satisfied when $J(\cdot)$ is selected as any arbitrary linear or non-linear function of y that outputs a scalar. Thus, we use the following distillation loss to learn the adapter h with parameters θ_h.

$$\min_{\theta_h} Loss, \quad Loss = \mathcal{L}(y_S, y^*) + \lambda \cdot \| \frac{\partial J(y_S)}{\partial x} - \frac{\partial J(y_T)}{\partial x} \|_2^2 \tag{2}$$

where $\mathcal{L}(y_S, y^*)$ denotes the task loss of the student net (*e.g.* the CrossEntropy loss), and y^* indicates the ground-truth label. We omit $\mathcal{L}(y_S, y^*)$, if we learn the adapter without additional training labels. Note that Eq. (2) is similar to the Jacobian distillation in [24].

Avoiding Catastrophic Forgetting. Equation (2) exclusively revises θ_h without changing f or g_s. In this way, the transplant of a new category will not affect feature representations of all existing category modules f, task modules g_s, and all other adapters except for h, thereby avoiding catastrophic forgetting.

3.1 Space-Projection Problem of Standard Distillation and Jacobian Distillation

Problems of Vast Forgotten Spaces. In spite of using Jacobian distillation in Eq. (2), how to make an adapter h project the output feature space of f properly

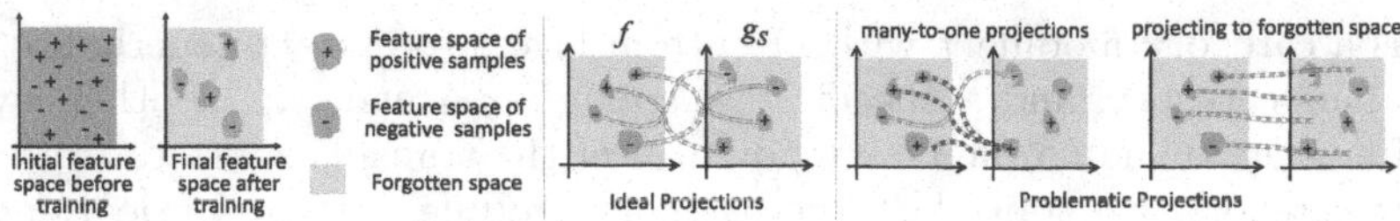

Fig. 3. An intermediate-layer feature space. (left) Features of an initialized DNN randomly cover the entire feature space. The learning process pushes the DNN towards typical feature spaces and produces vast forgotten space. (right) We show toy examples of space projection of adapters. Problematic space projections decrease both feature diversity and representation capability.

to the input feature space of g_S still remains a challenge, due to the vast forgotten space in Fig. 3 (left). Specifically, the information-bottleneck theory [21,27] reveals that a DNN *forgets* most intermediate-layer features and gradually limits its attention to very few discriminative features during the learning process, thereby yielding a vast forgotten space. That is, ***only very few features in the entire input space can pass through all ReLU layers in the task modular*** g_S. The output of f also has vast forgotten spaces, *i.e.*, the output of f is only distributed in a low-dimensional manifold inside the feature space.

More crucially, vast forgotten feature spaces block lots of paths for both the forward propagation and the backward propagation [26]. Specifically, in the forward propagation, most features of the adapter fall inside the forgotten space of g_S. Thus, as shown in Fig. 4, most feature information cannot pass through the task module g_S. Consequently, in backward propagation, the adapter will not receive informative gradients from enough diverse propagation paths.

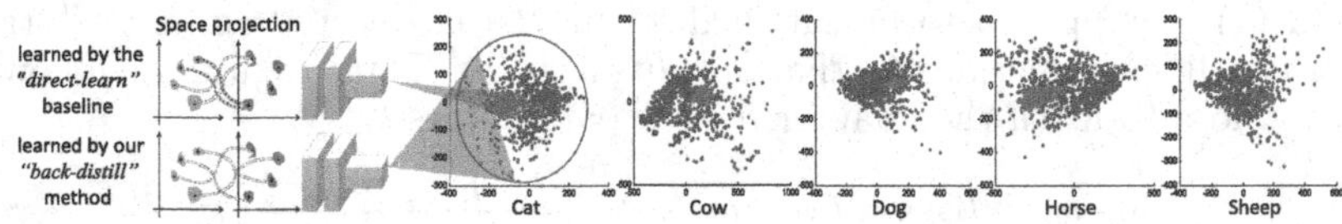

Fig. 4. Comparison of the projected feature spaces learned by our method (blue points) and those learned by the baseline method (red points). Each point represented two principal components of the *fc8* feature of an image. Features learned by the *direct-learn* baseline were more weakly triggered with lower diversity than those learned by our method due to problematic projections. (Color figure online)

Problems with the Jacobain Distillation. The above problem of forgotten feature spaces hampers most recent techniques, including the Jacobian distillation to optimize network parameters. Specifically, the vast forgotten input space of g_S blocks most propagation paths in the computation of feature gradients in Eq. (2). In other words, given a pre-trained task module g_S, ReLU layers in g_S tend to filter out most neural activations, if the input feature $h(x)$ has not been co-trained with g_S. Hence, without sufficient activations, gradients in

Eq. (2) carry very little information about network representations, which hurts the learning capacity of knowledge distillation.

3.2 Solution: Learning with Back-Distillation

To overcome the above optimization problem, in this section, we propose a back distillation method to reduce the forgotten space, *i.e.*, increasing the number of propagation paths during the computation of feature gradients.

Specifically, let $D_S = \frac{\partial J(y_S)}{\partial x}$ and $D_T = \frac{\partial J(y_T)}{\partial x}$ denote feature gradients, which both can be written in the form of back propagation $D(\mathbf{X}, \theta_h) = f'_{\text{conv}} \circ f'_{\text{relu}} \circ f'^{\text{max}}_{\text{pool}} \circ \cdots \circ f'_{\text{conv}}(G_y)$, where f'_i denotes the derivative of the layer-wise function f_i; $G_y \stackrel{\text{def}}{=} \frac{\partial J}{\partial y}$; $f'_i \circ f'_{i+1}(\cdot) = f'_i(f'_{i+1}(\cdot))$. To increase propagation paths, we revise ReLU layers and pooling layers in g_T, g_S, and h for distillation in Eq. (2) by proposing the following two pseudo-gradients $\hat{D}_S, \hat{D}_T$ to replace D_S, D_T in Eq. (1), respectively.

$$\hat{D}(\theta_h) \stackrel{\text{def}}{=\!=} f'_{\text{conv}} \circ f'_{\text{dummy}} \circ f'^{\text{avg}}_{\text{pool}} \circ \cdots \circ f'_{\text{conv}} \left(\hat{G}_y\right). \tag{3}$$

Eq. (3) contains three revisions to construct pseudo-gradients $\hat{D}_S$ and $\hat{D}_T$. First, we ignore dropout operations and replace derivatives of max-pooling $f'^{\text{max}}_{\text{pool}}$ with derivatives of average-pooling $f'^{\text{avg}}_{\text{pool}}$. Second, we revise the derivative of the ReLU to either $f^{\text{1st}}_{\text{dummy}}(\frac{\partial J}{\partial x^{(k)}}) = \frac{\partial J}{\partial x^{(k)}}$ or $f^{\text{2nd}}_{\text{dummy}}(\frac{\partial J}{\partial x^{(k)}}) = \frac{\partial J}{\partial x^{(k)}} \odot \mathbf{1}(x_{\text{rand}} > 0)$, where $x_{\text{rand}} \in \mathbb{R}^{s_1 \times s_2 \times s_3}$ denotes a random feature map, and $\odot$ indicates the element-wise product. Third, during the computation of both g_S and g_T for each image, we set the same values of x_{rand} and the random gradient $\hat{G}_y$ to make $\hat{D}_S$ and $\hat{D}_T$ comparable with each other.

Thus, the distillation loss based on $\hat{D}_S, \hat{D}_T$, namely *back distillation*, is formulated as $\min_{\theta_h} Loss = \mathcal{L}(y_S, y^*) + \lambda \|\hat{D}_S - \hat{D}_T\|^2$. Back distillation is eased due to that all above derivative functions $f'^{\text{1st}}_{\text{dummy}}$, $f'^{\text{2nd}}_{\text{dummy}}$, $f'^{\text{avg}}_{\text{pool}}$ do not block many activations and reduce the forgotten space. Moreover, back distillation can be optimized by propagating gradients of gradient maps to upper layers, which is considered as *back-back-propagation.*

Understanding the Back Distillation. Distillation loss based on pseudo-gradients $\hat{D}_S$ and $\hat{D}_T$ is a relaxation of the distillation loss in Eq. (2), where $\hat{D}_S$ and $\hat{D}_T$ are exclusively used for distillation and will **not** affect gradient propagation for the task loss $\mathcal{L}(y_S, y^*)$. More crucially, although $\hat{D}_S$ and $\hat{D}_T$ are not strictly equal to D_S and D_T, Tables 2 and 3 show that the solution based on $\hat{D}_S$ and $\hat{D}_T$ can well approximate the expected solution to Eq. (2).

Additionally, we can generate different values of $\hat{D}_S$ and $\hat{D}_T$ by enumerating $\hat{G}_y$ **even without any training samples**. It is because we can enumerate functions of J by generating different random gradients $\hat{G}_y$ according to Eq. (1), and each specific $\hat{G}_y$ can be used to compute a pair of $\hat{D}_S$ and $\hat{D}_T$ for back distillation. Specifically, for object segmentation task, the output is a tensor $y_S \in \mathbb{R}^{H \times W \times C}$, and we randomly sample $\hat{G}_y \in \mathbb{R}^{H \times W \times C}$ for each image. For single-category classification task, the output y_S is a scalar, and we generate a random matrix $\hat{G}_y \in \mathbb{R}^{S \times S \times 1}$ ($-1 \le \hat{G}_y^{ij1} \le +1$, $S = 7$ in experiments) for each image, which produces enlarged $\hat{D}_S, \hat{D}_T$ for back distillation.

4 Experiments

To simplify the story, in this section, we limited our attention to checking the effectiveness of network transplanting. We did not discuss other operations, *e.g.* fine-tuning or initializing category/task modules via traditional learning methods (see Fig. 1).

4.1 Implementation Details

We designed five experiments to evaluate the proposed method. Specifically, in *Experiment 1*, we learned toy transplant nets by inserting adapters between intermediate layers of pre-trained DNNs. *Experiments 2 and 3* learned transplant nets with two task modules (*i.e.* modules for object classification and segmentation) and multiple category modules in real applications. In *Experiment 4*, we conducted network transplanting to dissimilar categories. In *Experiment 5*, we transplanted category modules to generic task modules that had been learned for massive other categories.

Besides, we learned adapters with limited numbers of samples (*i.e.,* 10, 20, 50, and 100 samples) to check the back distillation strategy could decrease the demand for training samples. We further tested network transplanting **without any** training samples in *Experiment 1*, *i.e.* optimizing the distillation loss without considering the task loss.

Baselines. We compared our back distillation method (namely *back-distill*) with three baselines. All baselines exclusively learned the adapter without fine-tuning the task module for fair comparisons. The first baseline only optimized the task loss $\mathcal{L}(y_S, y^*)$ without distillation, namely *direct-learn.* The second baseline was the traditional distillation [11], namely *distill.* Its distillation loss $CrossEntropy(y_S, y_T)$ was applied to outputs of task modules g_S and g_T, because except for outputs, other layers in g_S and g_T did not produce features on similar manifolds. We tested the *distill* method in object segmentation, because unlike single-category classification with a single scalar output, segmentation outputs had sufficient correlations between soft output labels to enable the distillation. The third baseline was the Jacobian distillation [24], namely *Jacobian-distill,* where we replaced back distillation loss with MSE Jacobian-distillation loss.

Network Architectures. We transplanted category modules to a classification module in *Experiments 1, 2, 4, and 5*, and transplanted category modules to a segmentation module in *Experiment 3.* People usually extended the architecture of the widely used VGG-16 [23] to implement classification [31] and segmentation [13], as standard baselines of these two tasks. Thus, as shown in Fig. 5(a), we constructed a teacher net for both classification and segmentation as a network with a single category module and two task modules. The network branch for classification was exactly a VGG-16, and the network branch for segmentation was the FCN-8 s model [13].

Division of the Category Module and the Task Module. Because the first five layers of the FCN-8 s and those of the VGG-16 shared the same architecture,

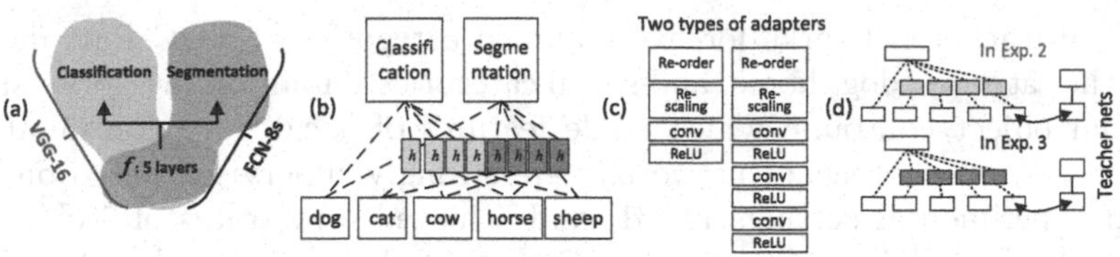

Fig. 5. (a) Teacher net; (b) transplant net; (c) adapters; (d) two sequences of transplanting.

we considered the first five layers (including two convolutional layers, two ReLU layers, and one pooling layer) as the shared category module, which encoded generic features of a specific category [3]. We regarded upper layers of the VGG-16 and the FCN-8s as two task modules, which encoded features to the task. Both branches were benchmark networks. Specifically, we followed the standard experimental settings in [13] to learn the FCN-8s for each category based on Pascal VOC 2011 dataset [9] for object segmentation. For object classification, we followed experimental settings in [31] to train DNNs on VOC Part dataset [6], CUB200-2011 dataset [25], and ILSVRC Animal-Part dataset [31] for binary classification of a single category from random images, respectively. We used SGD with learning rate 0.01 to train DNNs for 300 epochs.

Adapters. An adapter in Fig. 5(c) contained n convolutional layers, where we set $n = 1$ or 3 in experiments to analyze the influence of the layer number of adapters. Note that each convolutional layer in the adapter contained M filters. Each filter was a $3 \times 3 \times M$ tensor with *padding* $= 1$ in Experiment 1 or a $1 \times 1 \times M$ tensor without padding in Experiments 2 and 3, to avoid changing the size of feature maps, where M was the channel number of f. Additionally, we inserted a "reorder" layer and a "re-scaling" layer in the beginning of the adapter. The "reorder" layer is used to mimic feature states in real applications for fair evaluation. The "reorder" layer randomly reordered channels of the features x from the category module, which enlarged the dissimilarity between output features of different category modules. The "re-scaling" layer normalized the scale of features x from the category module, *i.e.* $x^{\text{out}} = \beta \cdot x$, for robust network transplanting. $\beta = \mathbb{E}_{I \in \mathbf{I}_S}[\|h(f(I))\|_F]/\mathbb{E}_{I \in \mathbf{I}_T}[\|f(I)\|_F]$ was a fixed scalar, where $\mathbf{I}_T$ and $\mathbf{I}_S$ denoted the image set of the new category and the image set of categories already modeled by the transplant net, respectively. $\|\cdot\|_F$ denoted the Frobenius norm.

4.2 Experimental Results and Analysis

- ***Experiment 1: Adding adapters to pre-trained DNNs.*** To examine effectiveness of network transplanting, we first conducted a toy test by inserting and learning an adapter between a category module and a task module. Here, we only considered networks with VGG-16 architectures for single-category classification. Note that in *Experiments 1 and 2*, we learned and merged teacher networks for five mammal categories of VOC Part dataset, *i.e.* the cat, cow, dog, horse, and sheep categories. Mammal categories shared similar object contours, which made features of a category transferable to other categories. Teacher networks were strongly supervised based on standard experimental settings in [31] and achieved error rates of 1.6%, 0.6%, 4.1%, 1.6%, and 1.0% for the classification of the cat, cow, dog, horse, and sheep categories, respectively. Then, we learned two types of adapters in Fig. 5(c) to analyze the influence of the layer number.

Specifically, we simply set $\hat{G}_y = 1$ without any gradient randomization, because the classification output y was a scalar without neighboring outputs to provide correlations. We used the revised dummy ReLU operations in Eq. (3) to ensure the value diversity of $\hat{D}_S, \hat{D}_T$ for learning. Moreover, we used $f'_{\text{dummy}} = f^{\text{2nd}}_{\text{dummy}}$ to compute expedient derivatives of ReLU operations in the task module, and used $f'_{\text{dummy}} = f^{\text{1st}}_{\text{dummy}}$ in the adapter. We set $\lambda = 10.0 / \left\| \frac{\partial \|\hat{D}_s - \hat{D}_t\|^2}{\partial \theta_h} \right\|$ for object classification in *Experiments 1 and 2.*

Figure 4 compared the space of *fc8* features, when we used *back distillation* and the *direct-learn* baseline to learn an adapter with three convolutional layers, respectively. We found that the adapter learned by our method passed much informative features to the *fc8* layer and yielded more diverse features. Which demonstrates that our method was more immune to problematic projections in Fig. 3 than the *direct-learn* baseline.

Table 2 reports the transplanting performance, when the adapter contained a single convolutional layer and when the adapter contained three convolutional layers. We discovered that a three-layer adapter was usually more difficult to optimize than the single-layer adapter. Table 2(left) shows that compared to the 9.79%–38.76% error rates of the *direct-learn* baseline, where our *back-distill* method yielded a significant lower classification error (1.55%–1.75%). *Even without any training samples, our method still outperformed the direct-learn method with 100 training samples.*

Note that given an adapter with three convolutional layers without any training samples, our *back-distill* method was not powerful enough to learn this adapter. Because deeper adapters with more parameters had more flexibility in representation, which required stronger supervision to avoid over-fitting. For example, in the last row of Table 2, our method successfully optimized an adapter with a single convolutional layer (the error rate was 1.75%), but was hampered when the adapter had three convolutional layers (the error rate was 49.9%, and our method produced a biased short-cut solution).

- ***Experiment 2: Operation sequences of transplanting to the classification module.*** As Fig. 5(b, d) shows, we could divide the entire network-transplanting procedure into an operation sequence of transplanting category modules to the classification module and another operation sequence of transplanting category modules to the segmentation module. Here, we evaluated the performance of transplanting category modules to the classification module, and conducted another sequence in *Experiment 3.*

Table 2. Error rates of classification when we inserted adapters with different convolutional layers. The last row corresponded to network transplanting without optimizing $\mathcal{L}(y_S, y^*)$. Our *back-distill* method yielded a significant lower classification error than the *direct-learn* baseline.

Insert one conv-layer	# of samples		cat	cow	dog	horse	sheep	**Avg.**
	100	direct-learn	12.89	3.09	12.89	10.82	9.28	9.79
		back-distill	**1.55**	**0.52**	**3.61**	**1.55**	**1.03**	**1.65**
	50	direct-learn	13.92	15.98	12.37	16.49	15.46	14.84
		back-distill	**1.55**	**0.52**	**3.61**	**1.55**	**1.03**	**1.65**
	20	direct-learn	16.49	26.80	28.35	32.47	25.77	25.98
		back-distill	**1.55**	**0.52**	**3.09**	**1.55**	**1.03**	**1.55**
	10	direct-learn	39.18	39.18	35.05	41.75	38.66	38.76
		back-distill	**1.55**	**0.52**	**3.61**	**1.55**	**1.03**	**1.65**
	0	direct-learn	–	–	–	–	–	–
		back-distill	**1.55**	**0.52**	**4.12**	**1.55**	**1.03**	**1.75**

Insert three conv-layers	# of samples		cat	cow	dog	horse	sheep	**Avg.**
	100	direct-learn	9.28	6.70	12.37	11.34	3.61	8.66
		back-distill	**1.03**	**2.58**	**4.12**	**1.55**	**2.58**	**2.37**
	50	direct-learn	14.43	13.92	15.46	8.76	7.22	11.96
		back-distill	**3.09**	**3.09**	**4.12**	**2.06**	**4.64**	**3.40**
	20	direct-learn	22.16	25.77	32.99	22.68	22.16	25.15
		back-distill	**7.22**	**6.70**	**7.22**	**2.58**	**5.15**	**5.77**
	10	direct-learn	36.08	32.99	31.96	34.54	34.02	33.92
		back-distill	**8.25**	**15.46**	**10.31**	**13.92**	**10.31**	**11.65**
	0	direct-learn	–	–	–	–	–	–
		back-distill	**50.00**	**50.00**	**50.00**	**49.48**	**50.00**	**49.90**

Specifically, we considered the classification module of the dog as a generic task modular, because the DNN for the dog was better learned due to the fact that the dog category contained more training samples. We transplanted category modules of other four mammal categories to this task module. We set the adapter to contain three convolutional layer, and used $f'_{\text{dummy}} = f^{\text{1st}}_{\text{dummy}}$ operation to compute derivatives of ReLU operations in both the task module and the adapter. The only exception was the lowest ReLU operation in the task module, for which we applied $f'_{\text{dummy}} = f^{\text{2nd}}_{\text{dummy}}|_{x_{\text{rand}}}$. We generated $x_{\text{rand}} = [x', x', \ldots, x']$ for each input image by concatenating x' along the third dimension, where $x' \in \mathbb{R}^{s_1 \times s_2 \times 1}$ contained 20%/80% positive/negative elements. The generation of $\hat{G}_y$ is introduced in Sect. 3.2. Here, $\mathbf{I}_S = \mathbf{I}_{\text{dog}}$ because the task module in the dog network was used as the generic task module g_S.

Table 3 presents the performance when we transplanted the category module to a task module oriented to categories with similar architectures using a few (10–100) samples. We discovered our *back-distill* method outperformed baseline methods.

- ***Experiment 3: Operation sequences of transplanting category modules to the segmentation module.*** Specifically, as introduced in Sect. 4.1, we used all training samples to learn five teacher FCNs for single-category segmentation with strong supervision. The pixel-level segmentation accuracies [13] for the cat, cow, dog, horse, and sheep categories were 95.0%, 94.7%, 95.8%, 94.6%, and 95.6%, respectively. We also considered the segmentation module of the dog as a generic task modular. We transplanted category modules of other four mammal categories to this task module. We set the adapter to contain one convolutional layer, and used the $f'_{\text{dummy}} = f^{\text{1st}}_{\text{dummy}}$ operation to compute derivatives of ReLU operations. We set $\lambda = 1.0/\mathbb{E}_I[\hat{D}_T]$ for all categories.

Table 3 compares the pixel-level segmentation accuracy between our *back-distill* method with baselines, where we tested our method with 10–100 training samples. We found our *back-distill* method exhibited 9.6%–11.5% higher accuracy than all baselines.

Table 3. Performance of transplanting to classification modules in Exp. 2 and transplanting to segmentation modules in Exp. 3. Our *back-distill* method outperformed baseline methods.

Error rate of single-category classification in Experiment 2 (transplanting to a classification module)													
# of samples		cat	cow	horse	sheep	**Avg.**	# of samples		cat	cow	horse	sheep	**Avg.**
100	direct-learn	20.10	12.37	18.56	11.86	15.72	20	direct-learn	31.96	37.11	39.69	35.57	36.08
	Jacobian-distill	19.07	**8.25**	9.79	11.86	12.24		Jacobian-distill	24.23	31.96	38.14	22.16	29.12
	back-distill	**8.76**	10.31	**8.25**	**4.12**	**7.86**		back-distill	**17.53**	**24.74**	**17.01**	**16.49**	**18.94**
50	direct-learn	22.68	19.59	19.07	14.95	19.07	10	direct-learn	41.75	37.63	44.33	33.51	39.31
	Jacobian-distill	14.95	19.07	21.13	13.40	17.14		Jacobian-distill	42.78	**28.35**	41.24	34.02	36.60
	back-distill	**11.86**	**13.40**	**15.98**	**12.37**	**13.40**		back-distill	**34.02**	33.51	**31.44**	**27.84**	**31.70**
Pixel accuracy of object segmentation in Experiment 3 (transplanting to a segmentation module)													
# of samples		cat	cow	horse	sheep	**Avg.**	# of samples		cat	cow	horse	sheep	**Avg.**
100	direct-learn	76.54	74.60	81.00	78.37	77.63	20	direct-learn	71.13	74.82	76.83	77.81	75.15
	distill	74.65	80.18	78.05	80.50	78.35		distill	71.17	74.82	76.05	78.10	75.04
	back-distill	**85.17**	**90.04**	**90.13**	**86.53**	**87.97**		back-distill	**84.03**	**88.37**	**89.22**	**85.01**	**86.66**
50	direct-learn	71.30	74.76	76.83	78.47	75.34	10	direct-learn	70.46	74.74	76.49	78.25	74.99
	distill	68.32	76.50	78.58	80.62	76.01		distill	70.47	74.74	76.83	78.32	75.09
	back-distill	**83.14**	**90.02**	**90.46**	**85.58**	**87.30**		back-distill	**82.32**	**89.49**	**85.97**	**83.50**	**85.32**

- ***Experiment 4: Transplant to dissimilar categories.*** We quantitatively evaluate the performance of transplanting to dissimilar categories to verify the effectiveness of network transplanting. In fact, identifying categories with similar architectures was still an open problem. People often manually defined sets of similar categories, *e.g.* learning a task module for mammals and learning another task module for different vehicles. In this way, we considered that the first four categories in VOC Part dataset, *i.e.* aeroplane, bicycle, bird, and boat, had dissimilar architectures with mammals. Thus, to transplant a mammal category (let us take the *cat* for example) to a classification/segmentation module, we learned a "leave-one-out" classification/segmentation module to deal with all four dissimilar categories and all mammal categories except the *cat*. Other experimental settings were the same as settings in *Experiments 2 and 3*.

Table 4 shows the performance of transplanting to a task module trained for both similar and dissimilar categories. We discover our method outperformed the baseline. Note that Table 4 did not report the result of the dog category, because we needed to compare with results in Table 3. Compared to the performance of transplanting to a task module oriented to similar categories in Table 3, transplanting to a more generic task modules for dissimilar categories in Table 4 hurt average segmentation performance but boosted the classification performance. It was because forcing a task module to handle dissimilar categories sometimes made the task module encode more robust representations, while it might also let the task module ignore details of mammal categories.

- ***Experiment 5: Transplanting to generic task modules learned for massive categories.*** Specifically, we followed experimental settings in [31] to learn DNNs on CUB200-2011 dataset and ILSVRC Animal-Part dataset [31]

Table 4. Transplanting to dissimilar categories. (top) Error rate of single-category classification when the classification module was learned for both mammals and dissimilar categories. (bottom) Pixel accuracy of object segmentation, the segmentation module was learned for both mammals and dissimilar categories. Our *back-distill* method outperformed the baseline method.

# of samples			cat	cow	horse	sheep	**Avg.**	# of samples		cat	cow	horse	sheep	**Avg.**
Classification	100	direct-learn	14.43	20.62	17.01	11.86	15.98	20	direct-learn	25.26	24.23	39.18	23.71	28.10
		back-distill	**5.67**	**3.61**	**6.70**	**2.58**	**4.64**		back-distill	**17.01**	**19.59**	**23.71**	**14.95**	**18.82**
	50	direct-learn	21.13	23.71	15.46	10.31	17.65	10	direct-learn	42.27	36.60	40.72	39.18	39.69
		back-distill	**7.22**	**9.28**	**8.76**	**5.67**	**7.73**		back-distill	**42.27**	**32.99**	**28.35**	**30.41**	**33.51**
Segmentation	10	direct-learn	64.97	69.65	80.26	69.87	71.19	20	direct-learn	68.69	81.02	71.88	72.65	73.56
		back-distill	**74.59**	**83.51**	**82.08**	**80.21**	**80.10**		back-distill	**73.34**	**84.78**	**81.40**	**81.04**	**80.14**

for binary classification, respectively. Target category modules were selected from DNNs for the last 10 categories of each dataset. These category modules were further transplanted to a generic task module, which was learned for all other categories in the dataset. Other experimental settings were the same as in *Experiment 2.* Table 5 shows that our *back-distill* method achieved the superior performance for transplanting to generic task modules.

Table 5. Transplanting to generic task modules for massive categories. We reported average error rates of single-category classification. Our *back-distill* method achieved the superior performance for transplanting to generic task modules.

	ILSVRC Animal-Part			ILSVRC Animal-Part			CUB200-2011	
# of samples	direct-learn	back-distill	# of samples	direct-learn	back-distill	# of samples	direct-learn	back-distill
100	12.73	**9.33**	20	27.99	**15.72**	20	14.17	**9.56**
50	20.67	**11.29**	10	32.99	**22.84**	10	19.91	**13.19**

- ***Discussion for network transplanting.*** First, as Table 2 shows, the back distillation method can significantly decrease the demand for training samples. Second, the comparison between Tables 3 and 4 indicates that a successful transplant requires target categories to share similar object shapes. Third, the transplant net can be applied to the multi-category classification by selecting the category module that yields the highest classification confidence as the classification result. Besides, we can learn a large transplant net for different categories and tasks by dividing it into lots of elementary operations of network transplanting in Fig. 1. Nevertheless, people can apply network transplanting to DNNs with different architectures for different tasks.

5 Conclusion

This paper proposes network transplanting to learn DNNs with a few or even without training annotations. We develop the back distillation to overcome challenges of network transplanting, and experiments verified the effectiveness of our method.

Acknowledgment. Quanshi Zhang and Xu Cheng contribute equally to this paper. Quanshi Zhang is the corresponding author. He is with the Department of Computer Science and Engineering, the John Hopcroft Center, at the Shanghai Jiao Tong University, China. This work is partially supported by the National Nature Science Foundation of China (62276165), National Key R&D Program of China (2021ZD0111602), Shanghai Natural Science Foundation (21JC1403800,21ZR1434600), National Nature Science Foundation of China (U19B2043).

References

1. Aljundi, R., Lin, M., Goujaud, B., Bengio, Y.: Gradient based sample selection for online continual learning. In NIPS
2. Andrychowicz, M., Denil, M., Colmenarejo, S.G., Hoffman, M.W., Pfau, D., Schaul, T., Shillingford, B., de Freitas, N.: Learning to learn by gradient descent by gradient descent. In NIPS (2016)
3. Bau, D., Zhou, B., Khosla, A., Oliva, A., Torralba, A.: Network dissection: Quantifying interpretability of deep visual representations. In CVPR (2017)
4. Buzzega, P., Boschini, M., Porrello, A., Abati, D., Calderara, S.: Dark experience for general continual learning: a strong, simple baseline. In NIPS
5. Chen, C., Li, O., Tao, D., Barnett, A., Rudin, C., Su, J.K.: This looks like that: deep learning for interpretable image recognition. In NIPS (2019)
6. Chen, X., Mottaghi, R., Liu, X., Fidler, S., Urtasun, R., Yuille, A.: Detect what you can: Detecting and representing objects using holistic models and body parts. In CVPR (2014)
7. Chen, X., Duan, Y., Houthooft, R., Schulman, J., Sutskever, I., Abbeel, P.: Infogan: Interpretable representation learning by information maximizing generative adversarial nets. In NIPS (2016)
8. Chou, Y.M., Chan, Y.M., Lee, J.H., Chiu, C.Y., Chen, C.S.: Unifying and merging well-trained deep neural networks for inference stage. In arXiv:1805.04980 (2018)
9. Hariharan, B., Arbelaez, P., Bourdev, L., Maji, S., Malik, J.: Semantic contours from inverse detectors. In ICCV (2011)
10. Higgins, I., Matthey, L., Pal, A., Burgess, C., Glorot, X., Botvinick, M., Mohamed, S., Lerchner, A.: β-vae: learning basic visual concepts with a constrained variational framework. In ICLR (2017)
11. Hinton, G., Vinyals, O., Dean, J.: Distilling the knowledge in a neural network. In NIPS Workshop (2014)
12. Hospedales, T., Antoniou, A., Micaelli, P., Storkey, A.: Meta-learning in neural networks: A survey. IEEE transactions on pattern analysis and machine intelligence **44**(9), 5149–5169 (2021)
13. Long, J., Shelhamer, E., Darrel, T.: Fully convolutional networks for semantic segmentation. In CVPR (2015)
14. Metz, L., Maheswaranathan, N., Cheung, B., Sohl-Dickstein, J.: Learning unsupervised learning rules. In ICLR (2019)
15. Neyshabur, B., Sedghi, H., Zhang, C.: What is being transferred in transfer learning? In NIPS

16. Raffel, C., Shazeer, N., Roberts, A., Lee, K., Narang, S., Matena, M., Zhou, Y., Li, W., Liu, P.J.: Exploring the limits of transfer learning with a unified text-to-text transformer. In JMLR
17. Rebuffi, S.A., Bilen, H., Vedaldi, A.: Learning multiple visual domains with residual adapters. In NIPS (2017)
18. Rebuffi, S.A., Bilen, H., Vedaldi, A.: Efficient parametrization of multi-domain deep neural networks. In CVPR (2018)
19. Rusu, A.A., Rabinowitz, N.C., Desjardins, G., Soyer, H., Kirkpatrick, J., Kavukcuoglu, K., Pascanu, R., Hadsell, R.: Progressive neural networks. In arXiv:1606.04671 (2016)
20. Sabour, S., Frosst, N., Hinton, G.E.: Dynamic routing between capsules. In NIPS (2017)
21. Schwartz-Ziv, R., Tishby, N.: Opening the black box of deep neural networks via information. In arXiv:1703.00810 (2017)
22. Shen, W., Wei, Z., Huang, S., Zhang, B., Fan, J., Zhao, P., Zhang, Q.: Interpretable compositional convolutional neural networks. In IJCAI (2021)
23. Simonyan, K., Zisserman, A.: Very deep convolutional networks for large-scale image recognition. In ICLR (2015)
24. Srinivas, S., Fleuret, F.: Knowledge transfer with jacobian matching. In ICML (2018)
25. Wah, C., Branson, S., Welinder, P., Perona, P., Belongie, S.: The caltech-ucsd birds-200-2011 dataset. Tech. rep, In California Institute of Technology (2011)
26. Wang, Y., Su, H., Zhang, B., Hu, X.: Interpret neural networks by identifying critical data routing paths. In CVPR (2018)
27. Wolchover, N.: New theory cracks open the black box of deep learning. In Quanta Magazine (2017)
28. Xiao, J., Gu, S., Zhang, L.: Multi-domain learning for accurate and few-shot color constancy (2020)
29. Yoon, J., Yang, E., Lee, J., Hwang, S.J.: Lifelong learning with dynamically expandable networks. In ICLR (2018)
30. Yosinski, J., Clune, J., Bengio, Y., Lipson, H.: How transferable are features in deep neural networks? In NIPS (2014)
31. Zhang, Q., Wu, Y.N., Zhu, S.C.: Interpretable convolutional neural networks. In CVPR (2018)

TiAM-GAN: Titanium Alloy Microstructure Image Generation Network

Zhixuan Zhang, Fusheng Jin(✉), Haichao Gong, and Qunbo Fan

Beijing Institute of Technology, Beijing, China
jfs21cn@bit.edu.cn

Abstract. The generation of titanium alloy microstructure images through mechanical properties is of great value to the research and production of titanium alloy materials. The appearance of GAN provides the possibility for image generation. However, there is currently no work related to handling multiple continuous labels for microstructure images. This paper presents a multi-label titanium alloy microstructure image generation network(TiAM-GAN). The TiAM-GAN proposed in this paper contains two sub-networks, a generation network for simple textures, which is based on the existing generation adversarial network, we reconstruct the loss function for multi-label continuous variables and deduce the error bound. Another microstructure image generation network for complex textures uses a mixture density network to learn the labels-to-noise mapping, and a deep convolution generation adversarial network is used to learn the noise-to-image mapping, then the noise output by the mixture density network will input to the deep convolution generation adversarial network to generate the image. Finally, we compared our method with existing methods qualitatively and quantitatively, which shows our method can achieve better results.

Keywords: Titanium Alloy Microstructure · Multi-Label · Image Generation · Generative Adversarial Network

1 Introduction

Titanium alloy is a material with excellent properties such as low density, high specific strength, high temperature resistance, corrosion resistance. [7], so it's widely used in aerospaces [4], medical [9], chemical industry [2] and other fields. The mechanical properties of titanium alloy are closely related to their microstructure characteristics. If the microstructure can be quickly given according to the mechanical performance parameters, it'll not only avoid the waste of materials caused by a large number of experiments and effectively guide the titanium alloy processing process, but also solve the limitations caused by existing equipment and technologies. Therefore, generating titanium alloy microstructure images from mechanical properties is of great significance for the theoretical foundation research and engineering application research.

Q. Liu et al. (Eds.): PRCV 2023, LNCS 14427, pp. 84–96, 2024.
https://doi.org/10.1007/978-981-99-8435-0_7

Most of the existing image generation tasks are for macro image datasets. However, the microstructure images of titanium alloy have some texture features. The difficulty is that the features among the images are similar, and the detail is closely related to its performance parameters, so it's more difficult for neural network to understand and learn. In addition, the current image generation tasks are almost all for discrete labels or single label [8]. However, the mechanical properties of titanium alloy are multiple continuous labels. The difficulty is that the neural network needs to learn the correlation among all the multiple labels, and there is no way to obtain the finite labels, so the labels are extremely unbalanced, and the learning effect will also be greatly reduced.

As far as we know, this paper is the first time to deal with the generation of titanium alloy microstructure images with multiple continuous labels. We propose a titanium alloy microstructure image generation network, TiAM-GAN, which is divided into two sub-networks, one is a continuous condition generative adversarial network based on feature fusion (Feature-Fusion CcGAN). The second sub-network is a generative adversarial network based on feature extraction and mapping (Feature-Extraction-Mapping GAN). Our contributions are as follows:

- We design a multi-label continuous condition generative adversarial network based on feature fusion (Feature-Fusion CcGAN). Based on the CcGAN(Continuous Conditional Generative Adversarial Network) [3], the network can accept multi-label input through label fusion, and we reconstruct the loss function of the generator and discriminator so that it can deal with multiple labels, and derive the error bound of the discriminator loss at the same time, it proves that the error is negligible. This network can generate microstructure images with simple texture and obvious grain boundaries.
- We design a generative adversarial network based on feature extraction and mapping (Feature-Extraction-Mapping GAN). The network learned label-to-noise mapping, and then sampled from its distribution to control the noise for input to generate a microstructure image that conforms to the selected labels. It solved the problem that Feature-Fusion CcGAN could not generate complex texture images.
- The ML-SFID(Multi-Label Sliding Fréchet Inception Distance) evaluation metric is proposed to quantitatively evaluate our proposed method and existing methods for multiple continuous labels.
- We conducted ablation experiments and comparative experiments on our titanium alloy Ti555 dataset, proved that our TiAM-GAN can achieve the best results through qualitative and quantitative evaluation.

2 Related Work

2.1 Image Generation

With the progress of deep learning, the field of image generation has developed rapidly. In 2014, Goodfellow et al. proposed a new network structure, GAN [5], based on the idea of competing against each other. For the original GAN, many variants have also appeared. In 2014, Mirza et al. proposed CGAN [8] for the uncontrollability of GAN, which solved the problem that the GAN cannot conditionally constrain the generated images. In 2015, Radford et al. proposed a CNN-based network, DCGAN [10] to make the quality of images better. In 2020, for the problem that traditional CGAN cannot handle continuous labels, Ding et al. proposed CcGAN [3], the first generative adversarial network for single continuous label. In addition, there are many other related works [1,11,13,14,16]. In 2020, the Denoising diffusion probabilistic models(DDPM) [6] has proposed which opens up a new path for image generation.

2.2 Mixture Density Network

Mixture density network is a model that uses neural networks to simulate data distribution. In theory, we can get arbitrary distributions by randomly combining model parameters. Therefore, Bishop [12] designed network: input a vector, through a designed neural network to obtain multiple parameters of the mixture Gaussian model, output a mixture probability distribution through these parameters. When we need to obtain the predicted value, we can directly randomly sample from this mixed probability distribution. In 2021, Yang et al. [15] applied MDN to the generation of material images. The optical absorption characteristics of the material obtained through physical simulation are used as the input of the network, then the parameters of the Gaussian mixture model are obtained through a fully connected neural network, finally the hidden variables are obtained through sampling, and it's used as the noise input of the generator.

3 Method

Then we'll introduce the framework of TiAM-GAN and its two subnetworks: Feature-Fusion CcGAN and Feature-Extraction-Mapping GAN.

3.1 Framework

Aiming at the generation of multi-label microstructure images, we propose a new network, the multi-label titanium alloy microstructure generation adversarial network TiAM-GAN, and the framework of the TiAM-GAN is shown in Fig. 1. The network consists of two subnetworks. We divide the microstructure images of titanium alloys into two part according to texture complexity. Microstructure images with simpler textures are trained and generated by Feature-Fusion CcGAN, while Microstructure images with more complex textures are trained and generated by Feature-Extraction-Mapping GAN.

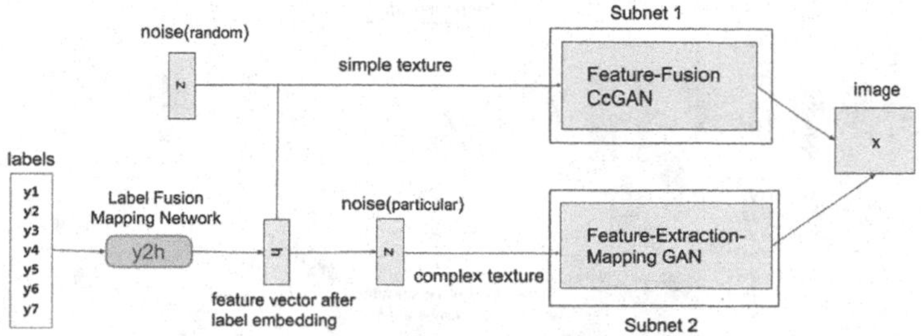

Fig. 1. Multi-label titanium alloy microstructure image generation network TiAM-GAN. The top is the first subnetwork (Feature-Fusion CcGAN) and the bottom is the second subnetwork (Feature-Extraction-Mapping GAN).

3.2 Feature-Fusion CcGAN

For the generator of Feature-Fusion CcGAN in Fig. 2, continuous conditional labels y1, y2,...,y7 are mapped to a 128-dimensional feature space through a label fusion mapping network to obtain the corresponding feature vector h, and then input it to the deconvolution layer together with a random noise z that is linearly mapped, and then though the deconvolution, batch normalization and activation function in turn, finally output the image x.

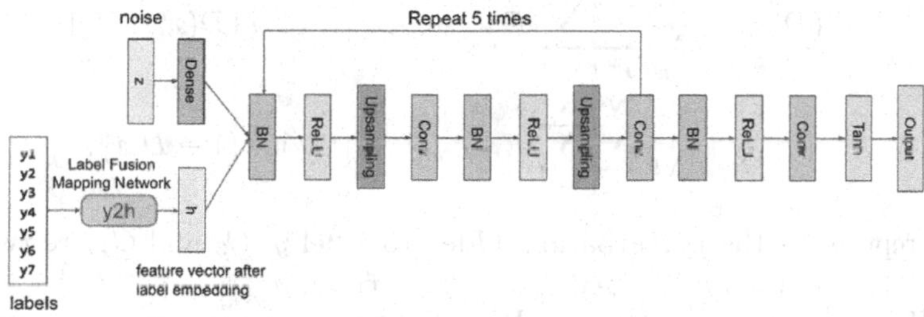

Fig. 2. Feature-Fusion CcGAN generator architecture.

For the discriminator of Feature-Fusion CcGAN in Fig. 3, input image x, though the convolution, normalization processing and activation function, and then do internal product and sum with the feature vector h mapped to feature space through a label fusion mapping network, finally output the discriminator result.

Reconstruction of Loss Function. Since the CcGAN [3] only targets single continuous label, does not extend to multiple labels. Therefore, on the basis of CcGAN, we reconstruct the loss function of the generator and discriminator in that paper to solve the problem of multi-label image generation.

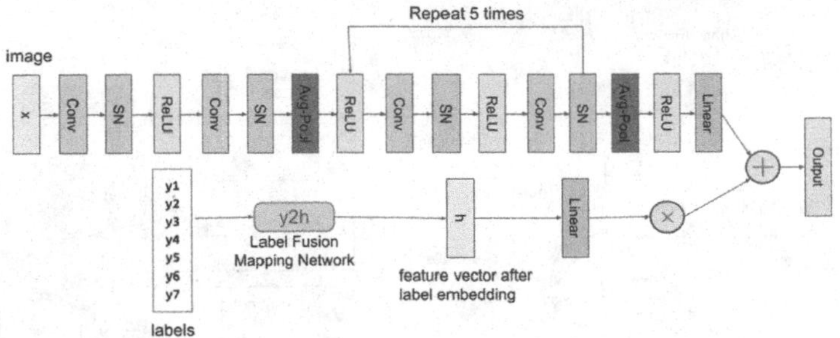

Fig. 3. Feature-Fusion CcGAN discriminator architecture.

The reconstruction loss function of generator is as follows:

$$L^{\varepsilon}(G) = -\frac{1}{N^g}\sum_{i=1}^{N^g} E_{\varepsilon^g \sim N(0,\sigma^2)} \log\left(D\left(G(z_i, \mathrm{h}_i^g), h_i^g\right)\right) \tag{1}$$

where h_i^g is the conditional labels $y1, ...y7$ is a 128-dimensional feature vector after label fusion and feature mapping. We performed the same perturbation ε on the seven conditional labels $y1, ...y7$.

The reconstruction loss function of discriminator: Multi Label Soft Vicinal Discriminator Loss (ML-SVDL) is:

$$\begin{aligned} L^{ML-SVDL}(D) = &-\frac{C_3}{N^r}\sum_{j=1}^{N^r}\sum_{i=1}^{N^r} E_{\varepsilon^r \sim N(0,\sigma^2)}\left[W_1 \log\left(D(x_i^r, h_j^r)\right)\right] \\ &-\frac{C_4}{N^g}\sum_{j=1}^{N^g}\sum_{i=1}^{N^g} E_{\varepsilon^g \sim N(0,\sigma^2)}\left[W_2 \log\left(1 - D(x_i^g, h_j^g)\right)\right] \end{aligned} \tag{2}$$

where, ε represents the perturbation added to label y, C_3 and C_4 are two constants, $W_1 = \frac{\prod_{k=1}^{7} w^r(y_{k_i}^r, y_{k_i}^r + \varepsilon^r)}{\sum_{i=1}^{N^r}\prod_{k=1}^{7} w^r(y_{k_i}^r, y_{k_i}^r + \varepsilon^r)}$, $W_2 = \frac{\prod_{k=1}^{7} w^g(y_{k_i}^g, y_{k_i}^g + \varepsilon^g)}{\sum_{i=1}^{N^g}\prod_{k=1}^{7} w^g(y_{k_i}^g, y_{k_i}^g + \varepsilon^g)}$, h_j^r represents the result of adding perturbation $y_j^r + \varepsilon^r$ to each real label and mapping to feature space, while h_j^g represents the result of adding perturbation $y_j^g + \varepsilon^g$ to each false label and mapping to feature space.

Error Bound Derivation. In this section, we refers CcGAN [3] to derive the error bound of the discriminator.

Firstly, assume that the range of all labels $y1, ...y7$ is $0 \sim 1$. D represents the hypothesis space of the discriminator. Let $p_w^r(y_i'|y_i) = \frac{w^r(y_i'|y_i)p^r(y_i')}{W^r(y_i)}$, $p_w^g(y_i'|y_i) = \frac{w^g(y_i'|y_i)p^g(y_i')}{W^g(y_i)}$, $W^r(y_i) = \int w^r(y_i', y_i)p^r(y_i')dy_i'$, $W^g(y_i) = \int w^g(y_i', y_i)p^g(y_i')dy_i'$, where, $i \in \{1, 2, 3, 4, 5, 6, 7\}$ represents seven types of mechanical property labels. Definition D^* represents the optimal discriminator, which minimizes $L^{ML-SVDL}$. Let $\tilde{D} = \arg\min L(D)$, $\hat{D}^{ML-SVDL} = \arg\min \hat{L}^{ML-SVDL}(D)$.

Definition 1: Define the Hölder Class of a function:

$$\sum(L) = \left\{ p : \forall t_1, t_2 \in y, \exists L > 0, st. \frac{|p'(t_1) - p'(t_2)|}{|t_1 - t_2|} \leq L \right\}$$

We make the following four assumptions:

(A1). All discriminators in D are measurable and bounded. let $U = \max\{\sup[-\log D], \sup[-log(1-D)]\}$, $U < \infty$
(A2) For any x and y_i, $\exists g^r(x) > 0, M^r > 0$, satisfies: when $\int g^r(x)dx = M^r$, $|p_r(x|y_i') - p_r(x|y_i)| \leq g^r(x)|y_i' - y|$. It means that the perturbation added to label y satisfies the Lipschitz condition, and $g^r(x)$ is called the Lipschitz constant of the function. That is to say, the perturbation added to label y can be ignored, and $g^r(x)$ is the minimum value of the slope.
(A3) For any x and y_i, $\exists g^g(x) > 0, M^g > 0$, satisfies: when $\int g^g(x)dx = M^g$, $|p_g(x|y_i') - p_g(x|y_i)| \leq g^g(x)|y_i' - y|$.
(A4) $p_r(y_i) \in \sum(L^r)$, $p_g(y_i) \in \sum(L^g)$.

We obtain Theorem 1: Assuming (A1) - (A4) holds, then for $\forall \delta \in (0,1)$, there is at least a probability of $1 - \delta$:

$$\begin{aligned} & L(D^{ML-SVDL}) - L(D^*) \\ & \leq 2U\left(\sqrt{\frac{C_{1,\delta}^{KDE}\log N^r}{N^r\sigma}} + L^r\sigma^2\right) + 2U\left(\sqrt{\frac{C_{2,\delta}^{KDE}\log N^g}{N^g\sigma}} + L^g\sigma^2\right) \\ & + 2U\sqrt{\frac{1}{2}\log\left(\frac{16}{\delta}\right)}\left(\frac{1}{\sqrt{N^r}}\prod_{i=1}^{7} E_{y_i \sim p_r^{KDE}(y_i)}\left[\frac{1}{W^r(y)}\right] + \frac{1}{\sqrt{N^g}}\prod_{i=1}^{7} E_{y_i \sim p_g^{KDE}(y_i)}\left[\frac{1}{W^g(y)}\right]\right) \\ & + U\left(M^r\prod_{i=1}^{7} E_{y_i \sim p_r^{KDE}(y_i)}\left[E_{y_i' \sim p_w^r(y_i'|y_i)}|y_i' - y_i|\right] + M^g\prod_{i=1}^{7} E_{y_i \sim p_g^{KDE}(y_i)}\left[E_{y_i' \sim p_w^g(y_i'|y_i)}|y_i' - y_i|\right]\right) \\ & + L(\tilde{D}) - L(D^*) \end{aligned} \quad (3)$$

3.3 Feature-Extraction-Mapping GAN

The framework of Feature Extraction Mapping GAN is shown in Fig. 4.

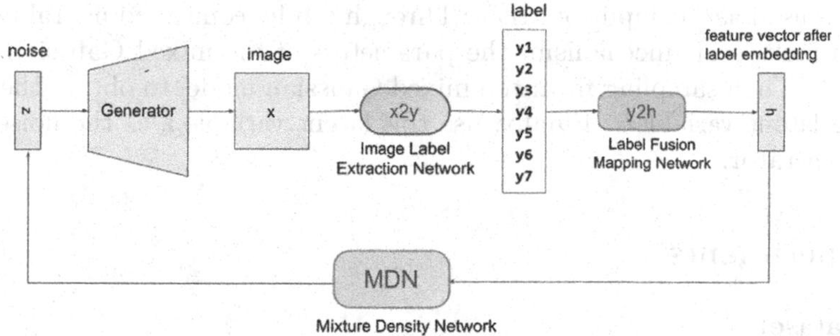

Fig. 4. Feature-Extraction-Mapping GAN architecture.

For Feature Extraction Mapping GAN, random noise z is fed to the trained generator to obtain a microstructure image. Then, obtain the corresponding label feature through the image label extraction network, and fuse these seven labels. After fusion, they are fed into the label fusion mapping network to obtain feature vectors h embedded in the feature space. Through a mixed density network, we can obtain the corresponding mapping distribution. Finally, the corresponding noise z can be sampled from the distribution to achieve the goal of generating images that satisfied the specified label conditions by controlling the noise.

Feature Extraction Mapping GAN mainly consists of three modules: multi-scale label fusion mapping network, image label extraction network, and MDN.

The label fusion mapping network uses the concat method in early fusion for feature fusion, which directly join these seven vectors in the first dimension, through a fully connected layer, finally mapping into a feature space. This mapping value h will input to the MDN network.

The image label extraction network is used to extract the label features of images. In order to make the entire network better learn the features of titanium alloy microstructure images, we need to add some prior knowledge to the network, that is, the extracted corresponding label features from the images. Since this process cannot be obtained through physical experiments, we transformed it into extracting features through training neural networks. The architecture for image label extraction network is shown in Fig. 5.

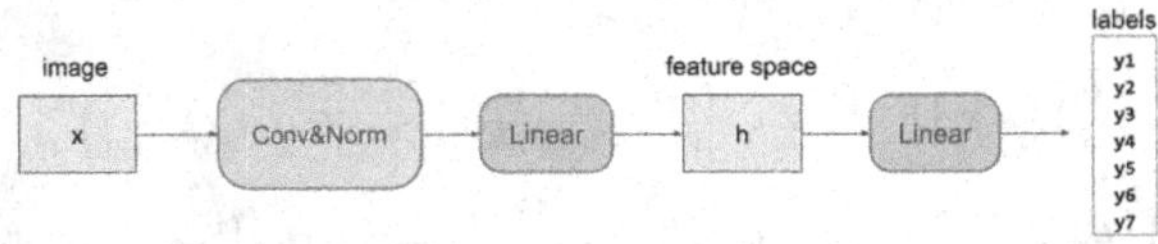

Fig. 5. image label extraction network architecture

The mixed density network in Feature Extraction Mapping GAN is partially based on Yang et al. [15]'s application in material image generation. The 32-dimensional feature vector h obtained through multi-scale label fusion mapping network is used as the input of MDN. Through a fully connected neural network based on multi-scale mechanism, the parameters of the mixed Gaussian model are output. Then sampling from the mixed Gaussian model to obtain the corresponding latent variable z. Finally, use this latent variable z as the noise input for the generator.

4 Experiments

4.1 Dataset

The experiments is based on real datasets of titanium alloy projects. Our titanium alloy category is Ti555, and the images were taken under optical and electron microscopes respectively. The microstructure images of titanium alloys can be divided into five classes: equiaxed structure, duplex structure, mixed structure, basket-weave structure, and Widmannstatten structure. Among them, equiaxed structure can be subdivided into five subclasses, namely equiaxed 1–5.

4.2 Metric

In order to better compare our model with existing models, we will evaluate both qualitatively and quantitatively. The quantitative metric is the new metric proposed in this paper: Multi label SFID (ML-SFID).

ML-SFID is an extension on SFID [3]. The smaller the ML-SFID value, the better the image generation effect. The calculation is shown in Eq. 4.

$$ML - SFID = \sum_{i=1}^{7} \frac{1}{N^c} \sum_{l=1}^{N^c} FID(i_l) \Big/ 7 \tag{4}$$

where, N^c represents the number of samples within the interval with center c, i represents the i-th label, and represents the i-th sample.

We use seven labels to generate images, so when calculating ML-SFID, it's necessary to divide the interval and radius of each label, the interval to each type of mechanical performance label is different. However, the number of centers for each type of mechanical performance label is unified to 100. The division of seven types of mechanical performance intervals is shown in Table 1. In addition, since the training labels of ACGAN are discrete labels, we need to redefine the intervals for ACGAN to calculate its ML-SFID value.

Table 1. Intervals of seven types of mechanical properties

Mechanical properties	Range	Interval length	Interval size
Tensile Strength	$400 \sim 1400$	1000	15
Yield Strength	$200 \sim 1100$	900	12
Elongation at Break	$0.17 \sim 0.3$	0.13	0.002
Rockwell Hardness	$20 \sim 70$	50	0.6
Impact Toughness	$30 \sim 100$	70	0.8
Dynamic Compressive Strength	$200 \sim 1600$	1400	20
Critical Fracture Strain	$1500 \sim 5000$	3500	40

4.3 Comparison Experiment

We compare TiAM-GAN with ACGAN. For ACGAN, we divide seven mechanical properties according to Table 2, and then each type of label is divided into multiple discrete intervals. At the same time, we modify the architecture and

Table 2. ACGAN Discrete Labels Based on the Range of seven Mechanical Properties

Mechanical properties	Range	Interval length	Interval size	Number of intervals
Tensile Strength	$400 \sim 1400$	1000	100	10
Yield Strength	$200 \sim 1100$	900	100	9
Elongation at Break	$0.17 \sim 0.3$	0.13	0.01	13
Rockwell Hardness	$20 \sim 70$	50	5	10
Impact Toughness	$30 \sim 100$	70	7	10
Dynamic Compressive Strength	$200 \sim 1600$	1400	200	7
Critical Fracture Strain	$1500 \sim 5000$	3500	500	7

loss function of ACGAN for multi-label. Finally, we calculate the ML-SFID. In order to make ACGAN perform well, we chose microstructure with simpler texture microstructure: equiaxed 1 with optical microscope and equiaxed 1 with electron microscope, that is more easily for neural network learning.

Qualitative. The qualitative results of the experiment are shown in Fig. 6. We can see that ACGAN performs very poorly, and even has no effect at all on equiaxed 1 with electron microscope. This is because we divide continuous variable labels into discrete values by interval, so the effect of learning will be worse. While our TiAM-GAN effect is very good.

Quantitative. The quantitative results is shown in Table 3. Our TiAM-GAN has the best ML-SFID results for images of each structure. Especially for equiaxed 1 with electron microscope, the ML-SFID of TiAM-GAN is over than 70% better than ACGAN. It's consistent of the quantitative evaluation results.

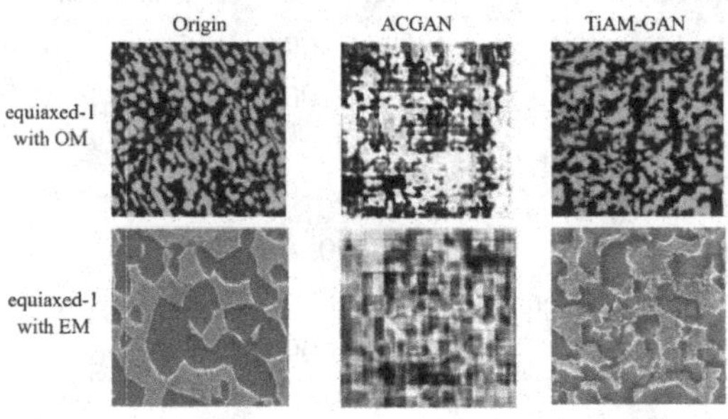

Fig. 6. Comparative experimental results between ACGAN and TiAM-GAN

Table 3. ML-SFID results of ACGAN and TiAM-GAN comparative experiments

model	equiaxed-1 with OM	equiaxed-1 with EM
ACGAN	0.7491	1.3516
TiAM-GAN	**0.4218**	**0.3071**

4.4 Ablation Experiment

The ablation experiment consists of two parts. The first part is a comparison between CcGAN and the first sub-network, Feature Fusion CcGAN; The second part is a comparison between CcGAN, Feature Fusion CcGAN, and TiAM-GAN.

For the first part of the ablation experiment, CcGAN was compared with Feature Fusion CcGAN. We chose three types of structures for the experiment: equiaxed-1 with optical microscopy(OM), duplex with optical microscopy, and equiaxed-1 with electron microscopy(EM).

Qualitative. The qualitative results of the experiment are shown in Fig. 7. We can see that CcGAN is not as good as our Feature Fusion CcGAN. For equiaxed-1 with electron microscopy, its generated structure is more blurry, and not as clear as Feature Fusion CcGAN.

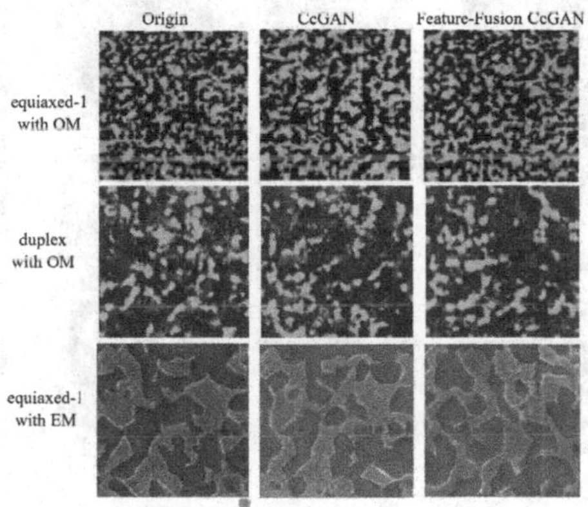

Fig. 7. Results of CcGAN and Feature Fusion CcGAN ablation experiments

Quantitative. The quantitative results is shown in Table 4. Our Feature Fusion CcGAN has the best evaluation results for each structure. The generation effect of duplex structure isn't as good as the other two, because duplex structure involve some complex textures, which is a huge test for the generator.

Table 4. ML-SFID in CcGAN and Feature Fusion CcGAN ablation experiments

model	equiaxed-1 with OM	duplex with OM	equiaxed-1 with EM
CcGAN	0.5057	0.9694	0.4041
Feature-Fusion CcGAN	**0.4218**	**0.6019**	**0.3071**

In the second part of the ablation experiment, we compare CcGAN, Feature Fusion CcGAN, and TiAM-GAN. In order to test the effectiveness of the TiAM-GAN for the microstructures with more complex textures, we selected three types of structure: duplex with electron microscopy(EM), basket-weave with optical microscopy(OM), and mixed structure with optical microscopy(OM).

Qualitative. The qualitative results of the experiment are shown in Fig. 8. We can see that CcGAN has the worst effect, followed by Feature Fusion CcGAN, and TiAM-GAN has the best effect. Especially for basket-weave, other two networks learned nothing, but our TiAM-GAN is similar to original image. It shows our second sub-network Feature Extraction Mapping GAN can process microstructure images of complex textures better.

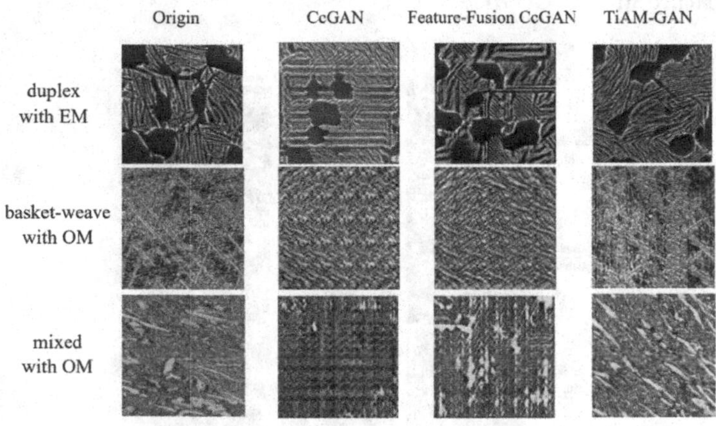

Fig. 8. Results of CcGAN, Feature Fusion CcGAN, TiAM-GAN ablation experiments

Quantitative. The quantitative results is shown in Table 5. Our TiAM-GAN has the best ML-SFID results for each structure image. Especially for Mixed structure, the ML-SFID of TiAM-GAN is over 80% better than other networks.

Table 5. ML-SFID results for second ablation experiments

model	duplex with EM	equiaxed-1 with EM basket-weave with OM	Mixed with OM
CcGAN	0.0929	0.33464	0.4487
Feature-Fusion CcGAN	0.0764	0.2826	0.1544
TiAM-GAN	**0.0302**	**0.1017**	**0.0255**

5 Conclusion

We proposes a new model, TiAM-GAN, which achieves the task of generating multi-label microstructure images of titanium alloys. The network is divided into two sub-networks according to the complexity of the microstructure. Then, an improved ML-SFID was proposed as a quantitative evaluation metric. Finally, we conduct comparative experiments and ablation experiments qualitatively and quantitatively, which found that our model achieved the best results in generating microstructure images of different types of titanium alloys.

References

1. Brock, A., Donahue, J., Simonyan, K.: Large scale GAN training for high fidelity natural image synthesis. arXiv preprint arXiv:1809.11096 (2018)
2. Chirico, C., Romero, A.V., Gordo, E., Tsipas, S.: Improvement of wear resistance of low-cost powder metallurgy β-titanium alloys for biomedical applications. Surf. Coat. Technol. **434**, 128207 (2022)
3. Ding, X., Wang, Y., Xu, Z., Welch, W.J., Wang, Z.J.: CcGAN: continuous conditional generative adversarial networks for image generation. In: International Conference on Learning Representations (2020)
4. Fu, Y.Y., Song, Y.Q., Hui, S.X.: Research and application of typical aerospace titanium alloys. Chin. J. Met. **30**, 850–856 (2006)
5. Goodfellow, I., et al.: Generative adversarial networks. Commun. ACM **63**(11), 139–144 (2020)
6. Ho, J., Jain, A., Abbeel, P.: Denoising diffusion probabilistic models. Adv. Neural. Inf. Process. Syst. **33**, 6840–6851 (2020)
7. Huang, Y., Wang, D.: Development and applications of titanium and titanium alloys for medical purposes. Rare Met. Mater. Eng. **34**, 658 (2005)
8. Mirza, M., Osindero, S.: Conditional generative adversarial nets. Comput. Sci. 2672–2680 (2014)
9. Munawar, T., et al.: Fabrication of dual Z-scheme TiO2-WO3-CeO2 heterostructured nanocomposite with enhanced photocatalysis, antibacterial, and electrochemical performance. J. Alloy. Compd. **898**, 162779 (2022)
10. Radford, A., Metz, L., Chintala, S.: Unsupervised representation learning with deep convolutional generative adversarial networks. arXiv preprint arXiv:1511.06434 (2015)
11. Salimans, T., Goodfellow, I., Zaremba, W., Cheung, V., Radford, A., Chen, X.: Improved techniques for training GANs. In: NIPS 2016: Proceedings of the 30th International Conference on Neural Information Processing Systems, pp. 2234–2242 (2016)

12. Silverman, B.W.: Density Estimation for Statistics and Data Analysis. Chapman & Hall (1986)
13. Wang, H., et al.: RGB-depth fusion GAN for indoor depth completion. In: 2022 IEEE/CVF Conference on Computer Vision and Pattern Recognition (CVPR), pp. 6199–6208 (2022)
14. Yang, S., Jiang, L., Liu, Z., Loy, C.C.: Unsupervised image-to-image translation with generative prior. In: 2022 IEEE/CVF Conference on Computer Vision and Pattern Recognition (CVPR), pp. 18311-18320 (2022)
15. Yang, Z., Jha, D., Paul, A., Liao, W.K., Choudhary, A., Agrawal, A.: A general framework combining generative adversarial networks and mixture density networks for inverse modeling in microstructural materials design. arXiv:2101.10553 (2021)
16. Zhu, J.Y., Park, T., Isola, P., Efros, A.A.: Unpaired image-to-image translation using cycle-consistent adversarial networks. In: 2017 IEEE International Conference on Computer Vision (ICCV) (2017)

A Robust Detection and Correction Framework for GNN-Based Vertical Federated Learning

Zhicheng Yang[1,2], Xiaoliang Fan[1,2](✉), Zheng Wang[1,2], Zihui Wang[1,2], and Cheng Wang[1,2]

[1] Fujian Key Laboratory of Sensing and Computing for Smart Cities, School of Informatics, Xiamen University, Xiamen 361005, People's Republic of China
{zcyang,zwang,wangziwei}@stu.xmu.edu.cn, {fanxiaoliang,cwang}@xmu.edu.cn
[2] Key Laboratory of Multimedia Trusted Perception and Efficient Computing, Ministry of Education of China, Xiamen University, Xiamen 361005, People's Republic of China

Abstract. Graph Neural Network based Vertical Federated Learning (GVFL) facilitates data collaboration while preserving data privacy by learning GNN-based node representations from participants holding different dimensions of node features. Existing works have shown that GVFL is vulnerable to adversarial attacks from malicious participants. However, how to defend against various adversarial attacks has not been investigated under the non-i.i.d. nature of graph data and privacy constraints. In this paper, we propose RDC-GVFL, a novel two-phase robust GVFL framework. In the detection phase, we adapt a Shapley-based method to evaluate the contribution of all participants to identify malicious ones. In the correction phase, we leverage historical embeddings to rectify malicious embeddings, thereby obtaining accurate predictions. We conducted extensive experiments on three well-known graph datasets under four adversarial attack settings. Our experimental results demonstrate that RDC-GVFL can effectively detect malicious participants and ensure a robust GVFL model against diverse attacks. Our code and supplemental material is available at https://github.com/zcyang-cs/RDC-GVFL.

Keywords: GNN-based Vertical Federated Learning · Adversarial attack · Robustness

1 Introduction

Graph Neural Network (GNN) has gained increasing popularity for its ability to model high-dimensional feature information and high-order adjacent information

Supplementary Information The online version contains supplementary material available at https://doi.org/10.1007/978-981-99-8435-0_8.

Q. Liu et al. (Eds.): PRCV 2023, LNCS 14427, pp. 97–108, 2024.
https://doi.org/10.1007/978-981-99-8435-0_8

on graphs [17]. In practice, privacy constraints prevent GNNs from learning better node representations when node features are held by different participants [6]. To enable collaborative improvement of node representation while protecting data privacy, GNN-based Vertical Federated Learning (GVFL) [2,3,12] employs GNNs locally for representation learning and Vertical Federated Learning (VFL) globally for aggregated representation.

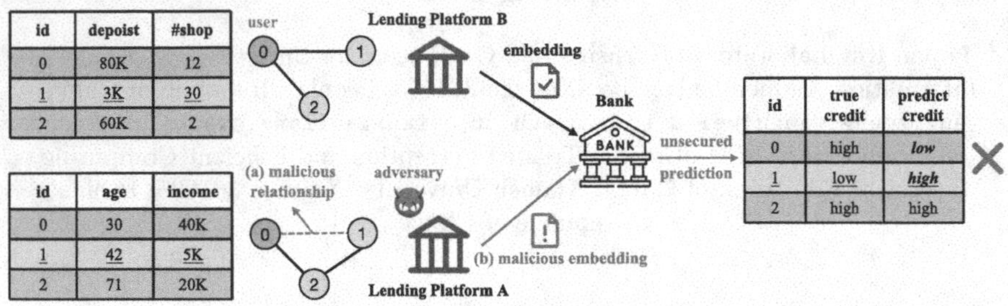

Fig. 1. An motivating example: an adversary can conduct various adversarial attacks: (a) manipulate the local connection relationship for fraudulent loan purposes by adding a connection from a low-credit user (user 1) to a high-credit user (user 0); and (b) upload the malicious embedding to deliberately misclassify the bank. To mitigate lending risks, the bank must implement effective defense measures.

In real-world applications, GVFL model is vulnerable to adversarial attacks [3]. For example, in Fig. 1, The adversary could compromise lending platform A and modify the uploaded embedding information to manipulate the bank's eventual prediction of the user's credit. Therefore, it is crucial for the bank to incorporate robust defensive measures, as failure to do so could result in loans being granted to users with lower credit.

However, neither graph defenses nor federated learning defenses against adversarial attacks can be directly applied in GVFL. On one hand, privacy restrictions prevent graph defenders from accessing training data and models in GVFL, which are essential for effectively implementing defense mechanisms [7,8]. On the other hand, federated learning defenses encounter challenges when dealing with inter-dependencies among node embeddings within a graph structure. Existing approaches, such as Copur's robust autoencoder [10,11], assume no dependency between embeddings of each sample, but the presence of inter-dependencies among node embeddings poses a hurdle for defenders to accurately learn the appropriate feature subspace.

To address the issues above, we propose a novel *Robust Detection and Correction Framework for GNN-based Vertical Federated Learning* (RDC-GVFL) to enhance the robustness of GVFL. RDC-GVFL has a detection phase and prediction phase. **In the detection phase**, the server first requires each participant to generate local embeddings to upload. Then, the server computes each participant's contribution to the validation set based on the Shapley Value, and identifies the participant with the lowest contribution as the malicious participant. **In the correction phase**, the server retrieves the relevant embeddings from historical embedding memory to correct malicious embeddings. Extensive experiments on three datasets against four adversarial attack settings confirm the effectiveness of RDC-GVFL.

The main contributions of our work can be summarized as follows:

- We propose RDC-GVFL, a novel robust framework for GVFL, which can enhance the robustness of GVFL under various attack scenarios. To the best of our knowledge, this is the first work dedicated to defending against adversarial attacks in GVFL.
- We present a Shapley-based detection method, enabling the effective detection of malicious participants in GVFL. Additionally, we propose a correction mechanism that utilizes historical embeddings to generate harmless embeddings, thereby obtaining accurate predictions.
- We conduct extensive experiments on three well-known graph datasets under four types of adversarial attacks. Experimental results demonstrate the effectiveness of the RDC-GVFL for a robust GVFL system.

2 Related Works

2.1 Attack and Defense in Graph Neural Networks

Extensive studies have demonstrated that GCNs are vulnerable to adversarial attacks [17]. These attacks can be classified into two categories based on the attack stage and goal. Evasion attacks (test-time) aim to deceive the GCN during inference, and they have been studied in works such as [20]. On the other hand, poisoning attacks (training-time) occur during the training process and aim to manipulate the GCN's learned representations, as explored in works like [19]. To enhance the robustness of GCNs, many defenses are proposed and they can be classified into three categories including improving the graph [8], improving the training [5], and improving the architecture [4] based on their strategy. However, all of these existing defenses require the defender to inspect either the training data or the resulting model, which is impractical in GVFL.

2.2 Attack and Defense in Vertical Federated Learning

Existing attacks on VFL have been shown to undermine model robustness [13]. For instance, the passive participant can launch adversarial attacks with different objectives. In targeted attacks, specific labels are assigned to triggered samples [14]. On the other hand, in non-targeted attacks, noise is added to randomly selected samples, or missing features are introduced to impair the model's utility [11]. For defense, RVFR [10] and Copur [11] is designed to defend against adversarial attacks. These defense approaches utilize feature purification techniques. However, they can not be directly applied to GVFL since the inter-dependencies among node embeddings in a graph structure makes it challenging to learn an appropriate feature subspace for defense purpose.

2.3 Attack and Defense in GNN-Based Vertical Federated Learning

GVFL has been found to be vulnerable to adversarial attacks, as highlighted in recent studies. One such attack method, called Graph-Fraudster, was proposed to perform evasion attacks in GVFL [3]. This attack assumes that a malicious participant can infer additional embedding information of a normal participant from the server. By inferring a perturbed adjacency matrix from normal embedding, the adversary can generate malicious embeddings. In practice, the adversary can employ various attack vectors to threaten GVFL and the defense against adversarial attacks in GVFL remains an open research issue.

3 Methodology

In this section, We first describe the GVFL system and threat models of GVFL. Next, we provide an overview of the proposed RDC-GVFL framework. Finally, we present the detection and correction methods of our framework. For convenience, the definitions of symbols used in this paper are listed in the Section A of Supplemental Material.

3.1 GNN-Based Vertical Federated Learning

As described in [2], GVFL involves M participants and a server that collaboratively train a model based on a graph dataset $G = \{G_1, ..., G_M, Y\}$. Here, $Y = \{y_j\}_{j=1}^{N}$ denotes labels of $|N|$ nodes held by the server, and $G_i = (V, E_i, \mathbf{X}_i)$ represents the subgraph held by participant i, where $\mathbf{X}_i = \{\mathbf{x}_j^i\}_{j=1}^{N}$ denotes the features associated with nodes V and edges E_i. The training process of GVFL is similar to VFL. Firstly, each participant i extract features and obtains the local embedding $\mathbf{h}_j^i = f_i(A_i, \mathbf{X}_i; \theta_i)$ of node j using its local Graph Neural Network f_i. Then, the server aggregates the node embeddings from all participants and generates the label prediction $\hat{y}_j$ for node j using its global classifier f_0.

3.2 Threat Model

Adversary. There are M participants which possess distinct subsets of features from the same set of N training instances, along with one adversary attempting to compromise the GVFL system. For simplicity, we assume the adversary is one of the participants in GVFL, referred to as the malicious participant.

Adversary's Goal. The malicious participant aims to induce the GVFL model to predict the wrong class. This may serve the malicious participant's interests, such as earning profits by offering a service that modifies loan application results at a bank or disrupting the normal trading system.

Adversary's Capability. We assume the training process is well-protected (attack-free). However, during the inference phase, the malicious participant can manipulate local data to send specific poisoning-embedded features to the server or send arbitrary embedded features to the server. As described in [3,11], we characterize the malicious participant's attack vector into the following four types and assume each malicious participant performs the same attack per round:

(a) **Graph-Fraudster attack** [3]: The malicious participant first steals normal embeddings from other participants. Then, the malicious participant adds noise to the stolen embeddings and computes an adversarial adjacency matrix $\hat{A}$ by minimizing the Mean Squared Error loss between the embedded features $f(\hat{A}, \mathbf{x})$ and the noise-added embedding $\mathbf{h}^m$. Finally, the malicious participant sends the poisoning embedded features based on the adversarial adjacency matrix.

$$\hat{A} = \arg\min_{\hat{A}} MSE(f(\hat{A}, \mathbf{x}), \mathbf{h}^m) \tag{1}$$

$$h_{gf} = f(\hat{A}, \mathbf{x}) \tag{2}$$

(b) **Gaussian feature attack**: The malicious participant sends malicious feature embedding which is added by Gaussian noise.

$$\mathbf{h}_{gauss} = \mathbf{h} + \mathcal{N}(\mu, \mathbf{\Sigma}) \tag{3}$$

(c) **Missing feature attack** [11]: The malicious participant don't send any embedded feature to the server. This can occur when the participant's device is damaged or disconnected.

$$\mathbf{h}_{miss} = \mathbf{0} \tag{4}$$

(d) **Flipping feature attack** [11]: The malicious participant sends adversarial feature embedding $\mathbf{h}_{flip}$ to mislead the server, whose magnitude λ can be arbitrarily large.

$$\mathbf{h}_{flip} = -\mathbf{h} * \lambda \tag{5}$$

3.3 Framework Overview

As shown in Fig. 2, we propose a two-phase framework called RDC-GVFL to enhance the robustness of GVFL. The Detection phase takes place between the model training and inference phases. During the Detection phase, the server collects the embeddings of participants from the validation dataset. Subsequently, the server calculates the contribution of each participant and identifies the one with the lowest contribution as the malicious participant. The Correction phase replaces the original model inference phase, and the server maintains an embedding memory that stores the embeddings of all training nodes. In this phase, the server utilizes normal embedding set as a query to retrieve the most similar embedding from historical embedding memory to correct the malicious embedding, thereby obtaining accurate predictions and sending them to all participants. The pseudo-code for RDC-GVFL is given in Algorithm 1.

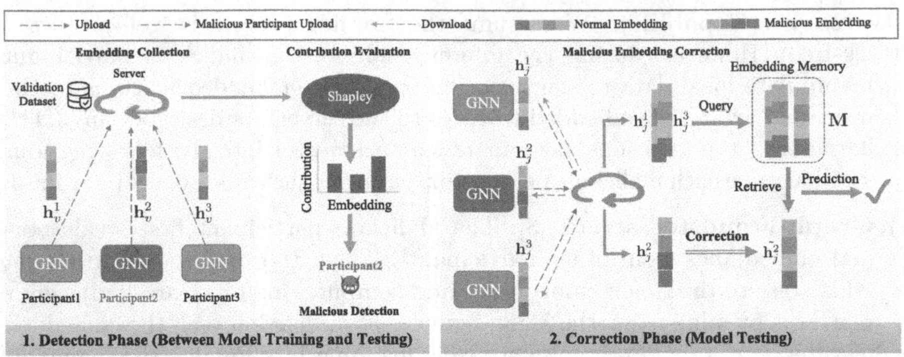

Fig. 2. Overview of the RDC-GVFL framework including two phases: (a) detection phase: the server using a validation dataset to evaluate contribution and identify the malicious participant; and (b) correction phase: the server leveraging the historical embeddings to retrieve the relevant embedding to correct the malicious embedding for accurate predictions sent to all participants.

3.4 Malicious Participant Detection

In addition to the training and testing processes of GVFL, we introduce a Detection phase to detect the malicious participant. In this phase, we propose a Shapley-based method to detect the malicious participant and it consists of three stages: embedding collection, contribution evaluation and malicious participant detection.

(a) **Embedding collection.** The server maintains a validation dataset $\mathcal{D}_{val} = \{V_{val}, Y_{val}\}$, which is used to mandate each participant generates a local embedding and sends it to the server. A normal participant i will send the embedding $\mathbf{h}_v^i$ of node $v \in V_{val}$ to the server while the malicious participant m, unaware of the server's detection process, will upload a poisoning embedding $\mathbf{h}_v^m$. As a result, the server collects a set of local embeddings aligned with its validation set.

(b) **Contribution evaluation.** We leverage Shapley Value, a fair contribution valuation metric derived from cooperative game theory, to evaluate contribution in GVFL. Specifically, we define the value function of a set of embeddings $F_v(S)$, which indicates whether the correct classification of node v can be achieved by making prediction using only the set S. The server then considers the total marginal value of each embedding $\mathbf{h}_v^i$ across all possible sets S as the contribution of participant i. This can be formalized as follows:

$$F_v(S) = \mathbb{I}(f_0(Agg(S), \theta_0) = y_v) \tag{6}$$

$$\phi_v^i(\{\mathbf{h}_v^i\}_{i=1}^M) = \sum_{S \subseteq (\mathbf{h}_v^1, \ldots \mathbf{h}_v^M) \setminus \mathbf{h}_v^i} \frac{|S|!(M - |S| - 1)!}{M!} (F_v(S \cup \mathbf{h}_v^i) - F_v(S)) \tag{7}$$

(c) **Malicious participant detection.** After the embedding collecting stage and contribution computation stage, the server accumulates the contributions of each participant on the validation set and identifies the participant with the lowest total contribution as a potential malicious participant $\hat{m}$. Based on this identification, the server may impose penalties on the malicious participant. This can be formalized as follows:

$$\hat{m} = \arg\min_{i} \sum_{v=1}^{|\mathcal{D}_{val}|} \phi_v^i(\{\mathbf{h}_v^i\}_{i=1}^M) \tag{8}$$

3.5 Malicious Embedding Correction

In the Correction phase, we leverage historical node embedding for correcting malicious embedding, which relies on the inter-dependency between nodes in the inference phase and nodes in the training phase. Specifically, during this phase, the server is already aware of the identity of the malicious participant, and it maintains a node embedding memory $\mathbf{M} \in \mathbb{R}^{N \times M}$ that stores the embeddings of all training nodes from all participants. When the GVFL model performs inference, i.e. each participant i sends its embedding $\mathbf{h}_j^i$ of node j to the server requesting prediction, the server will utilizes normal embedding set $\{\mathbf{h}_j^k\}$ where $k \in [M] \backslash m$ as a query to retrieve the most similar embedding $\mathbf{h}_\ell^m$ from the node embedding memory $\mathbf{M}$. Subsequently, the malicious embedding $\mathbf{h}_j^m$ is corrected or adjusted according to the $\mathbf{h}_\ell^m$. This process can be formulated as follows:

$$\ell = \arg\max_{\ell \in [N]} \{ \sum_{k \in [M] \backslash m} \frac{\mathbf{h}_j^k \cdot \mathbf{h}_\ell^k}{||\mathbf{h}_j^k|| \cdot ||\mathbf{h}_\ell^k||} \} \tag{9}$$

$$\mathbf{h}_j^m \leftarrow \mathbf{h}_\ell^m \tag{10}$$

4 Experiment

In this section, we carefully conduct comprehensive experiments to answer the following three research questions.

RQ1 Does the Shapley-based detection method within the RDC-GVFL framework have the capability to detect malicious participants?

RQ2 How does the performance of the RDC-GVFL framework compare to that of other defense methods?

RQ3 How does discarding the malicious embedding directly without correcting it affect the robustness?

4.1 Experiment Settings

Datasets. We use three benchmark datasets, i.e., Cora, Cora_ML [15] and Pubmed [16]. Dataset statistics are summarized in Table 1. The details of partition and evaluation strategies are described in the Section B of Supplemental Material.

Algorithm 1: RDC-GVFL

Input : Number of participants M, Validation Dataset $\mathcal{D}_{val}$, input embedding matrix $\mathbf{H}_j$ of node j

Output: Prediction label $\hat{y}_j$ of node j

// Model Training

Train the GVFL model to obtain the server model f_0 and local GNN models f_i;

Retrain the GVFL model to maintain the embedding memory $\mathbf{M}$;

$m =$ **ServerDetection**($\mathcal{D}_{val}$);

// Model Inference

$\hat{y}_j =$ **ServerCorrection**(m, $\mathbf{M}$, $\mathbf{H}_j$);

ServerDetection($\mathcal{D}_{val}$):

Initialize contribution list $C = [0] * M$;

for $v = 1, ..., |\mathcal{D}_{val}|$ **do**

  Randomly Sample $S \subset [|\mathcal{D}_{val}|]$;

  for *each participant i in parallel* **do**

    if *i belongs to malicious participant* **then**

      Malicious i computes and sends $\{\overline{\mathbf{h}_k^i}\}_{k \in S}$ to server accroding to (1)-(5);

    else

      participant i computes and sends $\{\mathbf{h}_k^i\}_{k \in S}$ to server;

  Server computes contribution c_i of each participant i according to (6)-(7);

  for $i = 1, ..., M$ **do**

    $C[i] \leftarrow C[i] + c_i$;

$m \leftarrow \arg\min_i C$;

return m;

ServerCorrection(m, $\mathbf{M}$, $\mathbf{H}_j$):

$\mathbf{H}_j = [\mathbf{h}_j^1 ... \mathbf{h}_j^m ... \mathbf{h}_j^M]$;

Server using normal embedding looks up in the node embedding memory $\mathbf{M}$ to retrieve the most relevant embedding $\mathbf{h}_\ell^m$ according to (9)-(10);

$\mathbf{h}_j^m \leftarrow \mathbf{h}_\ell^m$;

Server gets the prediction $\hat{y}_j = f_0(Agg(\mathbf{h}_j^1, ..., \mathbf{h}_j^m, ..., \mathbf{h}_j^M))$;

return $\hat{y}_j$;

Baseline Methods. We compare with 1) **Unsecured**: no defense. 2) **Krum** [1]: As no prior research has focused on robust GVFL, We borrow the idea of Krum which is the most well-known robust algorithm in FL to identify the participant with the highest pair-wise distance as the malicious participant, and discard its embedding for defense. 3) **RDC w/o Cor.**: We employ the RDC-GVFL framework to detect malicious participants, and subsequently, we directly discard the embedding associated with the identified malicious participant as a defense measure. 4) **Oracle**: where there is no attack. We use GCN [9] and SGC [18] as the local GNN model, the details of these models and parameter settings are shown in the Section B of Supplemental Material.

Table 1. The statistics of datasets.

Datasets	#Nodes	#Edges	#Features	#Classes	#Average Degree
Cora	2708	5429	1433	7	2.00
Cora_ML	2810	7981	2879	7	2.84
Pubmed	19717	44325	500	3	2.25

Table 2. Detection Rate(%) / Test Accuracy(%) against different attacks. 'V1/V2' represents scenarios where detection rate achieved 100% while 'V1/V2' represents scenarios where detection rate is below 100%.

Dataset	Attack	Local Model					
		GCN			SGC		
		M=2	M=3	M=4	M=2	M=3	M=4
Cora	Oracle	– –/80.13	– –/79.97	– –/78.97	– –/77.90	– –/77.73	– –/77.53
	GF	100/67.80	100/70.00	91.7/72.93	100/61.37	100/65.17	100/68.57
	Gaussian	100/69.93	89.0/68.97	83.3/69.00	100/37.20	100/52.97	100/54.97
	Missing	100/52.97	100/65.53	100/67.30	100/63.10	100/70.33	100/72.03
	Flipping	100/20.50	100/33.50	100/42.80	100/21.93	100/41.17	100/50.70
Cora_ML	Oracle	– –/84.90	– –/84.93	– –/84.87	– –/83.30	– –/83.30	– –/83.27
	GF	100/72.30	78.0/74.77	100/76.23	100/65.33	100/71.57	100/74.37
	Gaussian	100/64.47	78.0/63.50	100/62.93	100/35.73	100/35.57	100/35.87
	Missing	100/53.50	100/63.80	100/73.97	100/72.20	100/76.00	100/77.97
	Flipping	100/24.93	100/39.33	100/46.30	100/24.03	100/42.1	100/53.53
Pubmed	Oracle	– –/77.83	– –/78.33	– –/77.63	– –/75.80	– –/76.17	– –/75.63
	GF	100/44.13	100/53.17	100/51.80	100/35.30	100/36.97	100/42.47
	Gaussian	100/59.10	100/57.60	91.7/56.77	100/43.13	100/43.10	100/43.30
	Missing	100/60.57	100/62.47	100/67.13	100/64.37	100/68.33	100/67.40
	Flipping	100/35.03	100/42.63	100/51.97	100/34.77	100/51.80	100/55.67

4.2 Detection Performance(RQ1)

To answer **Q1**, we measured the detection rate of our detection method under the various attack settings, where the detection rate is calculated as the ratio of detected malicious participants to the total number of potential cases of malicious participants. We summarized the results in Table 2. Based on this results, we can draw several conclusions.

- **RDC-GVFL can effectively detects malicious participants across different attack scenarios**: Our detection method exhibits a high detection rate for identifying malicious participants in diverse scenarios. Specifically, our method achieves a 100% recognition rate against Missing attack and Flipping attack, while in other attack scenarios, the recognition rate remains above 78%.
- **RDC-GVFL is more effective against more significant attacks**: The effectiveness of identification increases as the attacks become more significant. For example, GF and Gaussian attacks exhibit higher effectiveness when the local model is SGC, and RDC-GVFL is capable of identifying the malicious participant in all these scenarios.

To better illustrate the effectiveness of the detection method, we visualize the contribution of each participant before and after the attack. These results can be found in Section C of Supplemental Material.

Table 3. The accuracy (%) of each defense under five adversarial attack scenarios on multiple datasets with a varying number of participants.

Local Model	Dataset	M	Attack	Defense			
				Unsecured	Krum	Ours w/o Cor.	Ours
GCN	Cora	2	Oracle	**80.10**	53.79	54.10	70.93
			GF	67.80	39.70	52.99	**70.76**
			Gaussian	69.93	47.60	52.99	**70.76**
			Missing	52.97	32.50	52.99	**70.76**
			Flipping	20.50	28.01	52.99	**70.76**
		3	Oracle	**80.00**	64.80	64.80	74.26
			GF	70.00	52.97	65.54	**74.23**
			Gaussian	68.97	65.54	64.20	**72.54**
			Missing	65.53	49.16	65.54	**74.23**
			Flipping	33.50	32.47	65.54	**74.23**
		4	Oracle	**79.00**	62.36	68.53	73.33
			GF	72.93	56.68	67.31	**73.86**
			Gaussian	69.00	67.32	66.29	**72.61**
			Missing	67.30	53.18	67.32	**74.35**
			Flipping	42.80	40.63	67.32	**74.35**
	Cora_ML	2	Oracle	**84.90**	51.55	56.59	78.09
			GF	72.30	42.75	53.46	**77.63**
			Gaussian	64.47	44.95	53.76	**71.10**
			Missing	53.50	32.17	53.46	**77.63**
			Flipping	24.93	28.67	53.46	**77.63**
		3	Oracle	**84.93**	54.41	71.60	82.29
			GF	74.77	47.27	64.45	**75.19**
			Gaussian	63.50	63.76	65.46	**75.00**
			Missing	63.80	32.15	63.76	**79.89**
			Flipping	39.33	23.64	63.76	**79.89**
		4	Oracle	**84.87**	69.64	76.89	82.69
			GF	76.23	60.43	74.07	**82.06**
			Gaussian	62.93	74.07	74.07	**82.06**
			Missing	73.97	53.56	74.07	**82.06**
			Flipping	46.30	28.64	74.07	**82.06**
	Pubmed	2	Oracle	**77.83**	61.00	64.20	71.56
			GF	44.13	46.68	60.35	**71.85**
			Gaussian	59.10	53.35	60.35	**71.85**
			Missing	60.57	47.26	60.35	**71.85**
			Flipping	35.03	37.18	60.35	**71.85**
		3	Oracle	**78.33**	56.43	70.00	74.76
			GF	53.17	54.30	62.29	**72.78**
			Gaussian	57.60	62.29	62.29	**72.78**
			Missing	62.47	54.70	62.29	**72.78**
			Flipping	42.63	34.02	62.29	**72.78**
		4	Oracle	**77.63**	63.76	70.70	76.16
			GF	51.80	56.50	67.66	**73.17**
			Gaussian	56.77	67.66	67.47	**72.15**
			Missing	67.13	57.27	67.66	**73.17**
			Flipping	51.97	41.74	67.66	**73.17**

4.3 Defense Performance(RQ2-RQ3)

To answer **Q2-Q3**, we conducted a series of experiments with different defense methods against various attack scenarios. The results based on GCN are summarized in Tabel 3 and the other results based on SGC can be found in Section D of Supplemental Material. Observations are concluded in this experiments.

- **RDC-GVFL exhibits a higher level of robustness compared to other defense methods**: RDC-GVFL consistently outperforms unsecured and Krum under various attack scenarios. For instance, on the GVFL based on two participants with Cora dataset, when a malicious participant performs Flipping attack, our RDC-GVFL framework demonstrates a remarkable accuracy improvement from 20.50% to 70.76%, while Krum only manages to enhance the accuracy to 28.01% from the same starting point of 20.50%.
- **Discarding malicious embeddings harms GVFL**: RDC-GVFL without correction does not generally improve accuracy compared to unsecured. This observation suggests that malicious embeddings contain information useful for model prediction, emphasizing the importance of their correction.
- **Negligible loss of accuracy with RDC-GVFL**: In the absence of an adversary, the use of the RDC-GVFL framework results in a maximum accuracy degradation of less than 10%. However, in the presence of an adversary, the RDC-GVFL framework can significantly enhance model accuracy and ensure the robustness of the GVFL system.

5 Conclusion

In this work, we introduced a novel robust framework of GVFL integrated with the Shapley-based detection and the correction mechanism to effectively defend against various attacks during inference in GVFL. In addition, the proposed detection method may have much wider applications due to the post-processing property. Through extensive experiments, we have demonstrated that our framework can achieve high detection rates for malicious participants and enhance the robustness of GVFL. For future work, we plan to extend our method to more scenarios and explore more effective defense methods for robust GVFL.

Acknowledgement. This work was supported by Natural Science Foundation of China (62272403, 61872306), and FuXiaQuan National Independent Innovation Demonstration Zone Collaborative Innovation Platform (No.3502ZCQXT2021003).

References

1. Blanchard, P., El Mhamdi, E.M., Guerraoui, R., Stainer, J.: Machine learning with adversaries: byzantine tolerant gradient descent. In: Advances in Neural Information Processing Systems, vol. 30 (2017)
2. Chen, C., et al.: Vertically federated graph neural network for privacy-preserving node classification. arXiv preprint arXiv:2005.11903 (2020)

3. Chen, J., Huang, G., Zheng, H., Yu, S., Jiang, W., Cui, C.: Graph-Fraudster: adversarial attacks on graph neural network-based vertical federated learning. IEEE Trans. Comput. Soc. Syst. **10**, 492–506 (2022)
4. Chen, L., Li, J., Peng, Q., Liu, Y., Zheng, Z., Yang, C.: Understanding structural vulnerability in graph convolutional networks. arXiv preprint arXiv:2108.06280 (2021)
5. Feng, W., et al.: Graph random neural networks for semi-supervised learning on graphs. Adv. Neural. Inf. Process. Syst. **33**, 22092–22103 (2020)
6. Fu, X., Zhang, B., Dong, Y., Chen, C., Li, J.: Federated graph machine learning: a survey of concepts, techniques, and applications. ACM SIGKDD Explor. Newsl. **24**(2), 32–47 (2022)
7. Jin, W., Ma, Y., Liu, X., Tang, X., Wang, S., Tang, J.: Graph structure learning for robust graph neural networks. In: Proceedings of the 26th ACM SIGKDD International Conference on Knowledge Discovery & Data Mining, pp. 66–74 (2020)
8. Jin, W., Zhao, T., Ding, J., Liu, Y., Tang, J., Shah, N.: Empowering graph representation learning with test-time graph transformation. arXiv preprint arXiv:2210.03561 (2022)
9. Kipf, T.N., Welling, M.: Semi-supervised classification with graph convolutional networks. arXiv preprint arXiv:1609.02907 (2016)
10. Liu, J., Xie, C., Kenthapadi, K., Koyejo, S., Li, B.: RVFR: robust vertical federated learning via feature subspace recovery. In: NeurIPS Workshop New Frontiers in Federated Learning: Privacy, Fairness, Robustness, Personalization and Data Ownership (2021)
11. Liu, J., Xie, C., Koyejo, S., Li, B.: CoPur: certifiably robust collaborative inference via feature purification. Adv. Neural. Inf. Process. Syst. **35**, 26645–26657 (2022)
12. Liu, R., Xing, P., Deng, Z., Li, A., Guan, C., Yu, H.: Federated graph neural networks: Overview, techniques and challenges. arXiv preprint arXiv:2202.07256 (2022)
13. Liu, Y., et al.: Vertical federated learning. arXiv:2211.12814 (2022)
14. Liu, Y., Yi, Z., Chen, T.: Backdoor attacks and defenses in feature-partitioned collaborative learning. arXiv preprint arXiv:2007.03608 (2020)
15. McCallum, A.K., Nigam, K., Rennie, J., Seymore, K.: Automating the construction of internet portals with machine learning. Inf. Retrieval **3**, 127–163 (2000)
16. Sen, P., Namata, G., Bilgic, M., Getoor, L., Galligher, B., Eliassi-Rad, T.: Collective classification in network data. AI Mag. **29**(3), 93–93 (2008)
17. Wu, B., et al.: A survey of trustworthy graph learning: reliability, explainability, and privacy protection. arXiv preprint arXiv:2205.10014 (2022)
18. Wu, F., Souza, A., Zhang, T., Fifty, C., Yu, T., Weinberger, K.: Simplifying graph convolutional networks. In: International conference on machine learning, pp. 6861–6871. PMLR (2019)
19. Zhang, S., Chen, H., Sun, X., Li, Y., Xu, G.: Unsupervised graph poisoning attack via contrastive loss back-propagation. In: Proceedings of the ACM Web Conference 2022, pp. 1322–1330 (2022)
20. Zügner, D., Akbarnejad, A., Günnemann, S.: Adversarial attacks on neural networks for graph data. In: Proceedings of the 24th ACM SIGKDD International Conference on Knowledge Discovery & Data Mining, pp. 2847–2856 (2018)

QEA-Net: Quantum-Effects-based Attention Networks

Juntao Zhang[1], Jun Zhou[1], Hailong Wang[1(✉)], Yang Lei[1], Peng Cheng[2], Zehan Li[3], Hao Wu[1], Kun Yu[1,3], and Wenbo An[1]

[1] Institute of System Engineering, AMS, Beijing, China
HailongWang_ISE@hotmail.com
[2] Coolanyp L.L.C, Wuxi, China
[3] University of Electronic Science and Technology of China, Chengdu, China

Abstract. In the past decade, the attention mechanism has played an increasingly important role in computer vision. Such an attention mechanism can be regarded as a dynamic weight adjustment process based on features of the input image. In this paper, we propose Quantum-Effects-based Attention Networks (QEA-Net), the simple yet effective attention networks, they can be integrated into many network architectures seamlessly. QEA-Net uses quantum effects between two identical particles to enhance the global channel information representation of the attention module. Our method could consistently outperform the SENet, with a lower number of parameters and computational cost. We evaluate QEA-Net through experiments on ImageNet-1K and compare it with state-of-the-art counterparts. We also demonstrated the effect of QEA-Net in combination with pre-trained networks on small downstream transfer learning tasks.

Keywords: Quantum mechanics · Attention mechanism · Image classification

1 Introduction

Quantum algorithms have been developed in the areas of chemistry, cryptography, machine learning, and optimization. Quantum computation uses the mathematic rules of quantum physics to redefine how computers create and manipulate data. These properties imply a radically new way of computing, using qubits instead of bits, and give the possibility of obtaining quantum algorithms that could be substantially faster than classical algorithms. There have been proposals for quantum machine learning algorithms that have the potential to offer considerable speedups over the corresponding classical algorithms, such as q-means [12], QCNN [13], QRNN [1]. However, more advanced quantum hardware is inevitable to translate such theoretical results into substantial advantages for real-world applications, which might still be years away. The near-future era of quantum

J. Zhang and J. Zhou—Contributed equally to this work.

Q. Liu et al. (Eds.): PRCV 2023, LNCS 14427, pp. 109–120, 2024.
https://doi.org/10.1007/978-981-99-8435-0_9

computing is limited due to small number of qubits that are also error prone. QUILT [20] is a framework for performing multi-class classification task designed to work effectively on current error-prone quantum computers. Another concept in literature is quantum logic, which refers to the non-classical logical structure and logical system originating from the mathematical structure of quantum mechanics. For example, Garg et al. [7] proposed a Knowledge-Base embedding inspired by quantum logic, which allows answering membership-based complex logical reasoning queries. Unlike the above works, Zhang et al. [26] proposed a novel method called Quantum-State-based Mapping (QSM) to improve the feature calibration ability of the attention module in transfer learning tasks. QSM uses the wave function describing the state of microscopic particles to embed the feature vector into the high dimensional space.

In a vision system, an attention mechanism can be treated as a dynamic selection process incorporating adaptive feature-weighing based on the importance of the input. Attention mechanisms benefit many visual tasks, e.g., image classification, object detection, and semantic segmentation. In this paper, we propose a new attention method called Quantum-Effects-based Attention Networks (QEA-Net). We evaluate the proposed method on large-scale image classification using ImageNet and compare it with state-of-the-art counterparts. Convolutional neural networks (CNNs) have been the mainstream architectures in computer vision for a long time until recently when new challengers such as Vision Transformers [5] (ViT) emerged. We train ResNet-50 with QEA-Net and adopt the advanced training procedure to achieve 79.5% top-1 accuracy on ImageNet-1K at 224×224 resolution in 300 epochs without extra data and distillation. This result is comparable to Vision Transformers under the same conditions.

Considering the high complexity of self-attention modules in the vision transformer, more simple architectures (e.g., MLP-Mixer [21], ResMLP [22]) that stack only multi-layer perceptrons (MLPs) have attracted much attention. Compared with CNNs and Transformers, these vision MLP architectures involve less inductive bias and have the potential to be applied to more diverse tasks. There is a long history of transfer learning and domain adaptation methods in computer vision [4,18]. Fine-tuning a pre-trained network model such as ImageNet on a new dataset is the most common strategy for knowledge transfer in the context of deep learning. We show the effectiveness of QEA-Net combined with the pre-trained MLP-Mixer in transfer learning tasks.

2 Related Works

2.1 Revisiting QSM

In order to highlight the plug-and-play ability of QSM, [26] use QSM only once at the appropriate layer in the module and do not change the data dimension. Specifically, assuming that the feature vector after pooling is $\mathbf{X} \in \mathbb{R}^{1\times d}$, and then the probability density function of one-dimensional particles under a coordinate representation in an infinite square well is used as a mapping:

$$\begin{cases} |\Psi(\mathbf{X})|^2 = QSM(\mathbf{X}) = \frac{2}{a}\sum_{n=1}^{N} c_n sin^2(\frac{n\pi}{a}\mathbf{X}) \\ a = max(|\mathbf{X}|) \end{cases} \tag{1}$$

This mapping does not change the dimension of $\mathbf{X}$. As analyzed in [26], training $|\Psi(\mathbf{X})|^2$ allows the neural network to exploit the probability distribution of the global information, which is beneficial for generating more effective attention. Clearly, this method is straightforward but preliminary. Instead of simply using the wave function as the mapping, QEA-Net in this paper uses the quantum effect between two identical particles to represent the spatial context information of feature map.

2.2 Attention Mechanism in CNNs

Spatial attention can be seen as an adaptive spatial region selection mechanism. Among them, GENet [10] represents those that use a sub-network implicitly to predict a soft mask to select important regions. SENet [11] proposed a channel attention mechanism in which SE blocks comprise a squeeze module and an excitation module. Global spatial information is collected in the squeeze module by global average pooling. The excitation module captures channel-wise relationships and outputs an attention vector using fully connected (FC) layers and non-linear layers (ReLU and sigmoid). Then, each channel of the input feature is scaled by multiplying the corresponding element in the attention vector. Some later work attempts to improve the outputs of the squeeze module (e.g., GSoP-Net [6]), improve both the squeeze module and the excitation module (e.g., SRM [15]), or integrate with other attention mechanisms (e.g., CBAM [24]). Qin et al. [19] rethought global information captured from the compression viewpoint and analyzed global average pooling in the frequency domain. They proved that global average pooling is a special case of the discrete cosine transform (DCT) and used this observation to propose a novel multi-spectral channel attention called FcaNet.

3 Quantum-Effects-based Attention Networks

In this section, we present a specific form of Quantum-Effects-based Attention Networks (QEA-Net). In order to highlight the role of quantum effects, we do not use a complex network structure. As shown in Fig. 1, the network consists of only one spatial attention module. The intermediate feature map is adaptively refined through our Network (QEA-Net) at every block of deep networks.

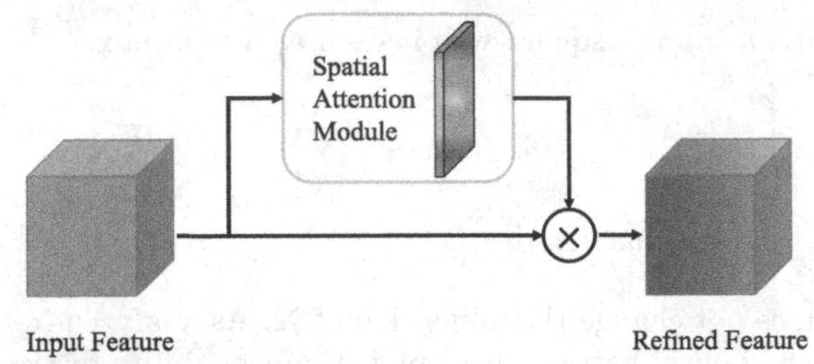

Fig. 1. The overview of the QEA-Net.

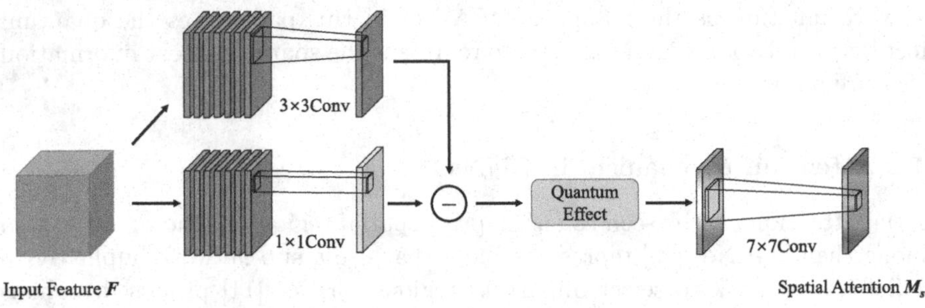

Fig. 2. Diagram of spatial attention module.

Given an intermediate feature map $\mathbf{F} \in \mathbb{R}^{C\times H\times W}$ as input, QEA-Net infers a $2D$ spatial attention map $\mathbf{M}_s \in \mathbb{R}^{1\times H\times W}$. The attention process can be summarized as:

$$\mathbf{F}' = \mathbf{M}_s(\mathbf{F}) \circ \mathbf{F} \tag{2}$$

where $\circ$ denotes element-wise multiplication. During multiplication, the attention values are broadcasted (copied) accordingly: spatial attention values are broadcasted along the channel dimension. $\mathbf{F}'$ is the refined output. Figure 2 depict the computation process of the attention map.

3.1 Spatial Attention Module Based on Quantum Effects

In the attention mechanism, average-pooling is usually used to aggregate information. But [24] argues that max-pooling gathers another important clue about distinctive object features to infer finer attention. However, CBAM [24] concatenates the feature maps after average-pooling and max-pooling operation, which ignores the intrinsic relationship between the two kinds of pooling. Therefore, we use the mutual constraints on the positions of two identical particles to aggregate information. In addition, the pooling operations ignore the distribution of spatial information in the channel dimension, so we use convolution with small-size kernels instead of pooling to preserve this information. Let the feature $\mathbf{F}$ after convolution be $\mathbf{F}^{1\times 1}_{Conv} \in \mathbb{R}^{1\times H\times W}$ and $\mathbf{F}^{3\times 3}_{Conv} \in \mathbb{R}^{1\times H\times W}$. Obviously, $\mathbf{F}^{3\times 3}_{Conv}$ focuses on the expression of the context between the pixels of the feature map, while $\mathbf{F}_{Conv}^{1\times1}$ is the weighted information of a single pixel in the channel dimension. We consider $\mathbf{F}_{Conv}^{1\times1}$ and $\mathbf{F}_{Conv}^{3\times3}$ as position vectors of two identical free particles, both of which are in momentum eigenstates (eigenvalues are $\hbar\boldsymbol{k}_\alpha$, $\hbar\boldsymbol{k}_\beta$). Without the exchange symmetry, the wave function of two free particles can be expressed as:

$$\psi\boldsymbol{k}_\alpha\boldsymbol{k}_\beta(\boldsymbol{r}_1,\boldsymbol{r}_2)=\frac{1}{(2\pi\hbar)^3}e^{i(\boldsymbol{k}_\alpha\cdot\boldsymbol{r}_1+\boldsymbol{k}_\beta\cdot\boldsymbol{r}_2)} \tag{3}$$

where $\boldsymbol{r}_1$ and $\boldsymbol{r}_2$ represent the spatial coordinates of the two particles respectively, and $\hbar$ is reduced Planck constant. In order to facilitate the study of the distribution probability of relative positions, let:

$$\begin{cases}\boldsymbol{r}=\boldsymbol{r}_1-\boldsymbol{r}_2\\ \boldsymbol{k}=(\boldsymbol{k}_\alpha-\boldsymbol{k}_\beta)/2\\ \boldsymbol{R}=\frac{1}{2}(\boldsymbol{r}_1+\boldsymbol{r}_2)\\ \boldsymbol{K}=\boldsymbol{k}_\alpha+\boldsymbol{k}_\beta\end{cases} \tag{4}$$

so Eq. 3 can be reduced to:

$$\psi\boldsymbol{k}_\alpha\boldsymbol{k}_\beta(\boldsymbol{r}_1,\boldsymbol{r}_2)=\frac{1}{(2\pi\hbar)^{\frac{3}{2}}}e^{i(\boldsymbol{K}\cdot\boldsymbol{R})}\phi_{\boldsymbol{k}}(\boldsymbol{r}) \tag{5}$$

where

$$\phi_{\boldsymbol{k}}(\boldsymbol{r})=\frac{1}{(2\pi\hbar)^{\frac{3}{2}}}e^{i\boldsymbol{k}\cdot\boldsymbol{r}} \tag{6}$$

we only discuss the probability distribution of relative positions. The probability of finding another particle around a particle in the spherical shell with radius $(r, r+dr)$ is:

$$4\pi r^2P(r)dr\equiv r^2dr\int|\phi_{\boldsymbol{k}}(\boldsymbol{r})|^2\,d\Omega=r^2dr\frac{4\pi}{(2\pi\hbar)^3} \tag{7}$$

so the probability density $P(r)$ is constant, independent of r. However, when we consider the exchange symmetry, the wave function is:

$$\psi^S\boldsymbol{k}_\alpha\boldsymbol{k}_\beta(\boldsymbol{r}_1,\boldsymbol{r}_2)=\frac{1}{\sqrt{2}}(1+P_{12})\psi\boldsymbol{k}_\alpha\boldsymbol{k}_\beta(\boldsymbol{r}_1,\boldsymbol{r}_2) \tag{8}$$

where P_{12} is the exchange operator. We directly give the probability density of relative positions under the condition that the exchange symmetry is satisfied:

$$P^S(r)=\frac{1}{(2\pi\hbar)^3}(1+\frac{sin2kr}{2kr}) \tag{9}$$

let x represent the relative distance between two particles and omit the constant factor $1/(2\pi\hbar)^3$, we get:

$$P(x)\propto 1+\frac{sinx}{x} \tag{10}$$

more elaborated derivation of the above equations can refer to most textbooks of quantum mechanics such as [9]. According to Eq. 10, the Quantum Effects (QE) can be expressed as:

$$QE(\mathbf{F}^{1\times1}_{Conv} - \mathbf{F}^{3\times3}_{Conv}) = \mathbf{ones}(H, W) + \frac{sin(\mathbf{F}^{1\times1}_{Conv} - \mathbf{F}^{3\times3}_{Conv})}{\mathbf{F}^{1\times1}_{Conv} - \mathbf{F}^{3\times3}_{Conv}} \quad (11)$$

where $\mathbf{ones}(H, W)$ represents a $H \times W$ matrix whose elements are all 1. As shown in Eq. 11, QE has no parameters to train, and the computation is negligible.

As shown in Fig. 2, we aggregate information of a feature map by using two convolution layers, generating two $2D$ maps: $\mathbf{F}^{1\times1}_{Conv} \in \mathbb{R}^{1\times H\times W}$ and $\mathbf{F}^{3\times3}_{Conv} \in \mathbb{R}^{1\times H\times W}$. The spatial attention is computed as:

$$\mathbf{M}_s(\mathbf{F}) = Sigmoid\Big(f^{7\times7}(QE(f^{1\times1}(\mathbf{F}) - f^{3\times3}(\mathbf{F})))\Big) \quad (12)$$

where $f^{1\times1}$, $f^{3\times3}$ and $f^{7\times7}$ represent convolutions with the filter size of 1×1, 3×3 and 7×7, respectively, and use padding to keep the size of the feature map unchanged.

4 Experiments

In this section, we first evaluate the QEA-Net on large-scale image classification using ImageNet, and compare it with state-of-the-art counterparts. Then, we show the effect of combining QEA-Net with a pre-trained MLP-Mixer [21] network in several downstream transfer learning tasks.

4.1 Implementation Details

To evaluate our QEA-Net on ImageNet-1K classification, we employ three widely used CNNs as backbone models, including ResNet-18, ResNet-34, and ResNet-50. To train ResNets with QEA-Net, we adopt the same data augmentation and hyperparameter settings. Specifically, the input images are randomly cropped to 224×224 with random horizontal flipping. The parameters of networks are optimized by stochastic gradient descent (SGD) with weight decay of 1e-4, a momentum of 0.9, and a mini-batch size of 256. All models are trained within 100 epochs by setting the initial learning rate to 0.1, which is decreased by a factor of 10 per 30 epochs.

As mentioned earlier, an attention module such as QEA-Net is inserted into the pre-trained network for feature recalibration to improve the model's performance in transfer learning tasks. In order to maintain the consistency of MLP-based architecture, Spatial Gating Unit (SGU) [16], Multi-Head Attention (MHA) [5] module and SE block [11] are selected as representatives of self-attention modules and channel attention modules, respectively. gMLP [16] is one of the SOTA MLP-based architectures, which the authors attribute to the effectiveness of SGU. The Multi-Head Attention (MHA) [5] module widely used in transformer-based architectures. These three modules rely on basic matrix multiplication routines, changes to data layout (reshapes and transpositions), and scalar nonlinearities. We fine-tune the model, which is reimplemented by MMClassification [17], using momentum SGD, learning rate of 0.06, batch size of 512, gradient clipping at global norm 1, and a cosine learning rate schedule with a linear warmup. No weight decay is used when fine-tuning. The above settings are consistent with MLP-Mixer, and the method of data augmentation is the same as above. All programs run on a server equipped with two RTX A6000 GPUs and two Xeon Gold 6230R CPUs.

Table 1. Comparison of different methods using ResNet-18 (R-18) and ResNet-34 (R-34) on ImageNet-1K in terms of network parameters (Param.), floating point operations per second (FLOPs), and Top-1/Top-5 accuracy (in %).

Method	CNNs	Years	Param.	GFLOPs	Top-1	Top-5
ResNet	R-18	CVPR 2016	11.15M	1.699	70.28	89.33
SENet [11]		CVPR 2018	11.23M	1.700	70.51	89.75
CBAM [24]		ECCV 2018	11.23M	1.700	70.60	89.88
QEA-Net(ours)			11.17M	1.700	70.55	89.92
ResNet	R-34	CVPR 2018	20.78M	3.427	73.22	91.40
SENet [11]		CVPR 2018	20.94M	3.428	73.69	91.44
CBAM [24]		ECCV 2018	20.94M	3.428	73.82	91.53
QEA-Net(ours)			20.82M	3.428	73.76	91.52

4.2 Comparisons Using Different Deep CNNs

ResNet-18 and ResNet-34. We compare different attention methods using ResNet-18 and ResNet-34 on ImageNet-1K. The evaluation metrics include both efficiency (i.e., network parameters, floating point operations per second (FLOPs)) and effectiveness (i.e., Top-1/Top-5 accuracy). We reimplement ResNet, SENet, as well as CBAM, and the results are listed in Table 1. Table 1 shows that QEA-Net improves the original ResNet-18 and ResNet-34 over 0.27% and 0.54% in Top-1 accuracy, respectively. Compared with SENet and CBAM, our QEA-Net achieves competitive performance with the least number of parameters.

ResNet-50. ResNet-50 is one of the most widely adopted backbones, and many attention modules report their performance on ResNet-50. We compare our QEA-Net with several state-of-the-art attention methods using ResNet-50 on ImageNet-1K, including SENet [11], GENet [10], SRM [15], CBAM [24], A^2-Nets [3], GCT [25] and FcaNet [19]. For comparison, we report the results of

Table 2. Comparison of different methods using ResNet-50 (R-50) on ImageNet-1K in terms of network parameters (Param.), floating point operations per second (FLOPs), and Top-1/Top-5 accuracy (in %). †: GENet was trained for 300 epochs. ‡: GCT uses warmup method. *: FcaNet uses more advanced training schedules, such as cosine learning rate decay and label smoothing. We reimplement it with the same settings as the other methods for a fair comparison.

Method	CNNs	Years	Param.	GFLOPs	Top-1	Top-5
ResNet	R-50	CVPR 2016	24.37M	3.86	75.20	92.52
SENet [11]		CVPR 2018	26.77M	3.87	76.71	93.39
GENet† [10]		NeurIPS 2018	25.97M	3.87	77.30	93.73
A^2-Nets [3]		NeurIPS 2018	25.80M	6.50	77.00	93.69
CBAM [24]		ECCV 2018	26.77M	3.87	77.34	93.69
SRM [15]		ICCV 2019	25.62M	3.88	77.13	93.51
GCT‡ [25]		CVPR 2020	24.44M	3.86	77.30	93.70
FcaNet-LF* [19]		ICCV 2021	26.77M	3.87	77.32	93.55
QEA-Net(ours)			24.52M	3.86	77.19	93.53

other compared methods in their original papers, except FcaNet. Compared with state-of-the-art counterparts, QEA-Net obtains better or competitive results, and achieves a balance between performance and efficiency (Table 2).

Table 3. Comparison of different architectures on ImageNet-1K in terms of network parameters (Param.), and Top-1 accuracy (in %). *: The result was reimplemented using the code provided by the open source toolbox MMClassification [17].

Method	Years	Resolution	Param.	Top-1
ViT-S/16-SAM [2]	ICLR 2022	224 × 224	22M	78.1
ViT-B/16 SAM [2]	ICLR 2022	224 × 224	87M	79.9
ResNet50 (rsb-A2)* [23]	NeurIPS 2021 Workshop	224 × 224	26M	79.2
QEA-ResNet50(ours)		224 × 224	26M	79.5

Recently, most of the state-of-the-art vision models have adopted the Transformer architecture. To boost the performance, these works resort to large-scale pre-training and complex training Settings, resulting in excessive demands of data, computing, and sophisticated tuning of many hyperparameters. Therefore, it is unfair to judge an architecture by its performance alone. We employ rsb-A2 [23], a strong training procedure, to train QEA-ResNet50 and ResNet50. Due to hardware limitations, we adjusted the minibatch to 512 according to the Linear Scaling Rule [8], which inevitably led to performance reduction. All models were trained for 300 epochs without extra data. As shown in Table 3, the strong training procedure reduces the performance gap between QEA-ResNet50

Table 4. Comparison of different models based on MLP-Mixer in terms of network parameters (Param.), floating point operations per second (FLOPs), and Top-1 accuracy.

	Mixer-B/16				
	base	+ SGU	+ SE	+ MHA	+ QEA
C-10	93.58%	94.55%	94.36%	93.95%	**94.61%**
C-100	78.54%	82.48%	82.29%	82.37%	**82.50%**
GFLOPs	12.61	12.68	12.61	12.61	12.61
Param.(M)	59.12	59.45	59.19	61.55	59.12
	Mixer-L/16				
	base	+ SGU	+ SE	+ MHA	+ QEA
C-10	91.03%	**94.54%**	94.39%	91.50%	93.88%
C-100	76.54%	78.60%	78.74%	**81.91%**	81.87%
GFLOPs	44.57	44.69	44.57	44.57	44.57
Param.(M)	207.18	207.75	207.31	211.47	207.18

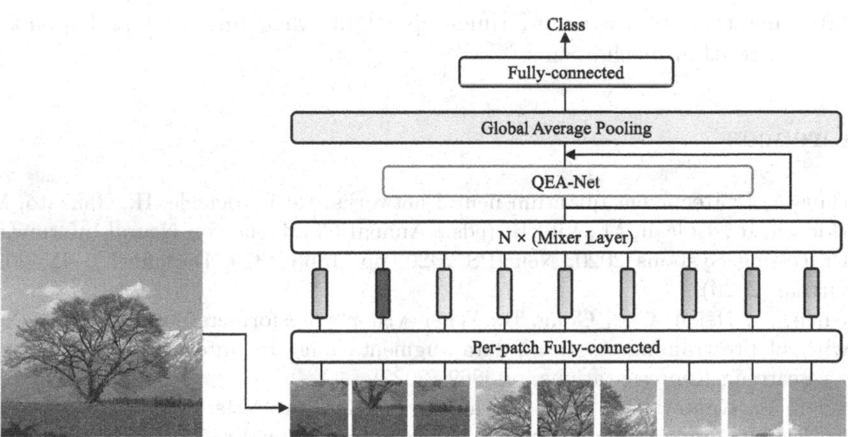

Fig. 3. The MLP-Mixer architecture with QEA-Net.

and ResNet50, but QEA-ResNet50 still outperforms ResNet50. In addition, the performance of QEA-ResNet50 is competitive with ViT-SAM, especially when the number of parameters is close.

4.3 Comparisons Using MLP-Mixer

We fine-tune the model and validate QEA-Net performance on small downstream tasks. Specifically, as shown in Fig. 3, we replace the classifier head to fit different tasks and add a QEA-Net before global average pooling.

Kornblith et al. [14] point out that ImageNet pre-training can greatly accelerate convergence. So we trained 10 epochs on CIFAR-10 (C-10) and 15 epochs on CIFAR-100 (C-100). The fewer epochs, the less pre-trained network changes. It is beneficial to verify the feature recalibration ability of the QEA-Net intuitively. As shown in Table 4, SGU and SE blocks have little difference in effect, but the QEA-Net is better, especially in deeper networks and datasets with more classes. MHA is also suitable for performing high-level feature recalibration tasks in MLP architectures but is more complex than other modules. To sum up, our method has lower model complexity than state-of-the-art while achieving very competitive performance.

5 Conclusion

This paper proposes an efficient channel attention method, call QEA-Net. QEA-Net uses the quantum effect between two identical particles to represent the spatial context information. The experimental results show that our QEA-Net is a lightweight plug-and-play module, which can improve the performance of various backbone networks and is suitable for various tasks. Future work will focus on exploring the application of time-dependent wave functions and quantum effects to attention mechanisms.

References

1. Bausch, J.: Recurrent quantum neural networks. In: Larochelle, H., Ranzato, M., Hadsell, R., Balcan, M., Lin, H. (eds.) Annual Conference on Neural Information Processing Systems 2020, NeurIPS 2020, pp. 1368–1379, December 6–12, 2020, virtual (2020)
2. Chen, X., Hsieh, C.J., Gong, B.: When vision transformers outperform ResNets without pre-training or strong data augmentations. In: International Conference on Learning Representations, pp. 869 (2022)
3. Chen, Y., Kalantidis, Y., Li, J., Yan, S., Feng, J.: A2-Nets: double attention networks. In: Annual Conference on Neural Information Processing Systems 2018, NeurIPS 2018, December 3–8, 2018, Montréal, Canada, pp. 350–359 (2018)
4. Csurka, G.: A comprehensive survey on domain adaptation for visual applications. In: Csurka, G. (ed.) Domain Adaptation in Computer Vision Applications, pp. 1–35. Springer, Advances in Computer Vision and Pattern Recognition (2017)

5. Dosovitskiy, A., et al.: An image is worth 16×16 words: transformers for image recognition at scale. In: International Conference on Learning Representations (2021)
6. Gao, Z., Xie, J., Wang, Q., Li, P.: Global second-order pooling convolutional networks. In: IEEE Conference on Computer Vision and Pattern Recognition, CVPR 2019, Long Beach, CA, USA, June 16–20, 2019, pp. 3024–3033. Computer Vision Foundation / IEEE (2019)
7. Garg, D., Ikbal, S., Srivastava, S.K., Vishwakarma, H., Karanam, H.P., Subramaniam, L.V.: Quantum embedding of knowledge for reasoning. In: Wallach, H.M., Larochelle, H., Beygelzimer, A., d'Alché-Buc, F., Fox, E.B., Garnett, R. (eds.) Annual Conference on Neural Information Processing Systems 2019, NeurIPS 2019, December 8–14, 2019, Vancouver, BC, Canada, pp. 5595–5605 (2019)
8. Goyal, P., et al.: Accurate, large minibatch SGD: training ImageNet in 1 hour. CoRR abs/1706.02677 (2017). http://arxiv.org/abs/1706.02677
9. Griffiths, D.J.: Introduction to quantum mechanics. Am. J. Phys. **63**(8) (2005)
10. Hu, J., Shen, L., Albanie, S., Sun, G., Vedaldi, A.: Gather-excite: exploiting feature context in convolutional neural networks. In: Annual Conference on Neural Information Processing Systems 2018, NeurIPS 2018, pp. 9423–9433
11. Hu, J., Shen, L., Sun, G.: Squeeze-and-excitation networks. In: 2018 IEEE Conference on Computer Vision and Pattern Recognition, CVPR 2018, Salt Lake City, UT, USA, June 18–22, 2018, pp. 7132–7141. Computer Vision Foundation / IEEE Computer Society (2018)
12. Kerenidis, I., Landman, J., Luongo, A., Prakash, A.: q-means: a quantum algorithm for unsupervised machine learning. In: Wallach, H.M., Larochelle, H., Beygelzimer, A., d'Alché-Buc, F., Fox, E.B., Garnett, R. (eds.) Annual Conference on Neural Information Processing Systems 2019, NeurIPS 2019, December 8–14, 2019, Vancouver, BC, Canada, pp. 4136–4146 (2019)
13. Kerenidis, I., Landman, J., Prakash, A.: Quantum algorithms for deep convolutional neural networks. In: 8th International Conference on Learning Representations, ICLR 2020, Addis Ababa, Ethiopia, April 26–30, 2020. OpenReview.net (2020)
14. Kornblith, S., Shlens, J., Le, Q.V.: Do better ImageNet models transfer better? In: IEEE Conference on Computer Vision and Pattern Recognition, CVPR 2019, Long Beach, CA, USA, June 16–20, 2019. pp. 2661–2671. Computer Vision Foundation / IEEE (2019)
15. Lee, H., Kim, H., Nam, H.: SRM: a style-based recalibration module for convolutional neural networks. In: 2019 IEEE/CVF International Conference on Computer Vision, ICCV 2019, Seoul, Korea (South), October 27 - November 2, 2019, pp. 1854–1862. IEEE (2019)
16. Liu, H., Dai, Z., So, D.R., Le, Q.V.: Pay attention to MLPs. In: Advances in Neural Information Processing Systems 34: Annual Conference on Neural Information Processing Systems 2021, NeurIPS 2021, December 6–14, 2021, virtual, pp. 9204–9215 (2021)
17. MMClassification Contributors: Openmmlab's image classification toolbox and benchmark. https://github.com/open-mmlab/mmclassification (2020)
18. Pan, S.J., Yang, Q.: A survey on transfer learning. IEEE Trans. Knowl. Data Eng. **22**(10), 1345–1359 (2010)
19. Qin, Z., Zhang, P., Wu, F., Li, X.: FcaNet: frequency channel attention networks. In: 2021 IEEE/CVF International Conference on Computer Vision (ICCV), pp. 763–772 (2021)

20. Silver, D., Patel, T., Tiwari, D.: QUILT: effective multi-class classification on quantum computers using an ensemble of diverse quantum classifiers. In: Thirty-Sixth AAAI Conference on Artificial Intelligence, AAAI 2022, pp. 8324–8332 (2022)
21. Tolstikhin, I.O., et al.: MLP-mixer: an all-MLP architecture for vision. In: Ranzato, M., Beygelzimer, A., Dauphin, Y., Liang, P., Vaughan, J.W. (eds.) Advances in Neural Information Processing Systems, vol. 34, pp. 24261–24272. Curran Associates, Inc. (2021)
22. Touvron, H., et al.: ResMLP: feedforward networks for image classification with data-efficient training. CoRR abs/2105.03404 (2021). https://arxiv.org/abs/2105.03404
23. Wightman, R., Touvron, H., Jegou, H.: ResNet strikes back: an improved training procedure in timm. In: NeurIPS 2021 Workshop on ImageNet: Past, Present, and Future (2021)
24. Woo, S., Park, J., Lee, J.-Y., Kweon, I.S.: CBAM: convolutional block attention module. In: Ferrari, V., Hebert, M., Sminchisescu, C., Weiss, Y. (eds.) ECCV 2018. LNCS, vol. 11211, pp. 3–19. Springer, Cham (2018). https://doi.org/10.1007/978-3-030-01234-2_1
25. Yang, Z., Zhu, L., Wu, Y., Yang, Y.: Gated channel transformation for visual recognition. In: Proceedings of the IEEE/CVF Conference on Computer Vision and Pattern Recognition (CVPR) (2020)
26. Zhang, J., et al.: An application of quantum mechanics to attention methods in computer vision. In: ICASSP 2023–2023 IEEE International Conference on Acoustics, Speech and Signal Processing (ICASSP) (2023)

Learning Scene Graph for Better Cross-Domain Image Captioning

Junhua Jia[1], Xiaowei Xin[1], Xiaoyan Gao[1], Xiangqian Ding[1], and Shunpeng Pang[2(✉)]

[1] Faculty of Information Science and Engineering, Ocean University of China, Shandong 266000, China
{jiajunhua,xinxiaowei,gxy9924}@stu.ouc.edu.cn, dingxq@ouc.edu.cn

[2] School of Computer Engineering, Weifang University, Shandong 261061, China
pangshunpeng@gmail.com

Abstract. The current image captioning (IC) methods achieve good results within a single domain primarily due to training on a large amount of annotated data. However, the performance of single-domain image captioning methods suffers when extended to new domains. To address this, we propose a cross-domain image captioning framework, called SGCDIC, which achieves cross-domain generalization of image captioning models by simultaneously optimizing two coupled tasks, *i.e.*, image captioning and text-to-image synthesis (TIS). Specifically, we propose a scene-graph-based approach SGAT for image captioning tasks. The image synthesis task employs a GAN variant (DFGAN) to synthesize plausible images based on the generated text descriptions by SGAT. We compare the generated images with the real images to enhance the image captioning performance in new domains. We conduct extensive experiments to evaluate the performance of SGCDIC by using the MSCOCO as the source domain data, and using Flickr30k and Oxford-102 as the new domain data. Sufficient comparative experiments and ablation studies demonstrate that SGCDIC achieves substantially better performance than the strong competitors for the cross-domain image captioning task.

Keywords: Image Captioning · Scene Graph · Text-to-Image Synthesis · Dual Learning

1 Introduction

With the development of deep learning, image captioning (IC) models have achieved increasingly impressive performance. However, it is worth noting that these IC models are solely trained and tested exclusively on single-domain datasets. The lack of training and testing these models on datasets from various domains limits their ability to generate novel and contextually relevant information. Especially when dealing with multi-object complex scenes in cross-domain image captioning tasks, the current mainstream image captioning methods perform poorly.

Q. Liu et al. (Eds.): PRCV 2023, LNCS 14427, pp. 121–137, 2024.
https://doi.org/10.1007/978-981-99-8435-0_10

To address the above problems, we propose a scene graph-based cross-domain image captioning method, dubbed SGCDIC, which fully leverages both the source and target domains. This is because scene graphs can effectively encode different regions, their attributes, and relationships in an image. To be specific, we propose SGAT (Scene Graph-based Approach for Image-to-Text Conversion), a method that excels at transforming images into text across various target scenes. Additionally, we utilize a GAN variant called DFGAN [22] to synthesize realistic images based on the text descriptions generated by SGAT. By comparing the rendered images with real images, we enhance the performance of IC model SGAT in novel domains.

Our contributions are as follows: (1) We leverage scene graphs to encode main regions, their attributes, and relationships in an image. (2) We propose a novel dual-learning mechanism to simultaneously proceed with two coupled tasks, *i.e.*, image captioning, and text-to-image synthesis. Cross-domain extension of image captioning is achieved based on the text-to-image synthesis task. (3) We conduct extensive experiments to evaluate the performance of SGCDIC. We employ MSCOCO as the source domain, and two other datasets, namely Flickr30k and Oxford-102, as the target domains. Through comprehensive evaluation, our proposed method, SGCDIC, demonstrates significant improvements compared to the existing methods in terms of performance and effectiveness.

2 Related Work

2.1 Scene Graph

Scene graphs capture the structured semantic information of images, encompassing the knowledge of objects, their attributes, and pairwise relationships. Consequently, scene graphs serve as a structured representation for various vision tasks, including VQA [6], image generation [10,14], and visual grounding [19]. Yang et al. [24] proposed a scene-graph-based solution for unpaired IC to solve the problem of a lack of paired data by learning a shared dictionary by including these language inductive biases. Recognizing the potential of leveraging scene graphs in vision tasks, several approaches have been proposed to improve scene graph generation from images. Conversely, some researchers have explored the extraction of scene graphs solely from textual data. In our study, we employ CogTree [27] for parsing scene graphs from images and SPICE [1] for parsing scene graphs from captions. These techniques enable us to effectively utilize scene graphs. By incorporating scene graphs into our framework, we enhance the understanding of the image content and facilitate the generation of more descriptive and contextually-rich captions.

2.2 Cross Domain Image Captioning

Cross-domain IC has received significant attention and has achieved remarkable results, as highlighted by several notable studies [5,7,29,30]. The work in

[5] proposes an adversarial training strategy to generate cross-domain captions by leveraging labeled and unlabeled data. Zhao et al. [30] introduce a cross-modal retrieval-based method for cross-domain IC. Their approach utilizes a cross-modal retrieval model to generate pseudo pairs of images and sentences in the target domain. This aids in the adaptation of the captioning model to the target domain. Yuan et al. [29] present a novel approach called Style-based Cross-domain Image Captioner for IC. Their method integrates discriminative style information into the encoder-decoder framework and treats an image as a unique sentence based on external style instructions. Dessi et al. [7] introduce a straightforward finetuning approach to enhance the discriminative quality of captions generated by cross-domain IC models. In our work, we propose a scene-graph-based dual-learning mechanism to render cross-domain IC models to achieve excellent performance.

3 Methods

To address the limitations of image captioning and achieve cross-domain generalization of the proposed IC model, we introduce a novel scene-graph-based learning mechanism SGCDIC based on the IC and TIS processes. SGCDIC is a multitask system consisting of two key components: (i) Image Captioning Task: we proposed a scene-graph-based image captioning model SGAT to train the image captioning task, which generates textual descriptions for input images. Specifically, we encode the input raw image into image patches (vector representations) with positional information using ViT [8], capture visual features using ResNet-101 [9], incorporate scene graph information obtained through CogTree [27], and feed them into GRU. The initial textual description of the image is generated by the GRU decoder. Subsequently, we input the generated text description into SeqGAN [28] with a back-translation module, utilizing external knowledge from a CLIP-based knowledge base and scene graph information obtained earlier to further refine the text description and align it with human cognition. (ii) Text-to-Image Syntheis Task: We employ DFGAN [22] to generate images based on the corresponding text descriptions. In DFGAN, a generative model G synthesizes realistic images based on the text descriptions, while a discriminative model D attempts to distinguish between real images from the training data and images generated by G. The optimization of G and D through a minimax game enables G to generate realistic and high-quality images. This adversarial process ultimately guides G to generate realistic and high-quality images. Finally, the generated images are fine-tuned using CLIP-based knowledge priors to align with human perceptual preferences.

When the image-to-text conversion task starts with an image description M, the following process is performed: (1) Using M to represent the image description task, an intermediate text description t_{mid} is generated for each image i in the target domain. This is achieved by the LSTM decoder $P(t_{mid}|i;\theta_M)$, where θ_M denotes the parameters of M. (2) The text-to-image synthesis task is then applied using the intermediate text description t_{mid} to generate the image

$i\prime$ using the text-to-image generator $G_{\theta_N^g}(z_{noise}, t_{mid})$. Here, θ_N^g represents the parameters involved in the N process. (3) The losses computed based on the two-generation processes are used to update the image description M and the text-to-image synthesis N. The entire process is repeated multiple times until the parameters of both tasks converge to a threshold.

Similarly, when the image-to-text conversion task starts with text-to-image synthesis N, the process is as follows: (1) The generator $G_{\theta_N^g}(z_{noise}, t)$ in DFGAN is used to generate an intermediate image i_{mid} based on the text description t. (2) Then, the description $t\prime$ is re-generated from the intermediate image i_{mid} using $P(y|i_{mid}; \theta_M)$. (3) The losses computed based on the above generation process are used to update the image description M and the text-to-image synthesis N. The entire process is repeated multiple times until the parameters of both tasks converge to a threshold.

In our study, we employ reinforcement learning techniques, specifically policy gradient, to maximize the long-term rewards of the image captioning (IC) task M and the text-to-image synthesis (TIS) task N. The following components of the reinforcement learning process are introduced: state, policy, action, and reward. State: The state represents the information describing the system or environment. In our tasks, the distributed representation of an image $i \in I$ can be viewed as the state s_i for the IC process M, while the distributed representation of a text description $t \in T$ can be viewed as the state s_t for the TIS process N. The selection of states is crucial for the performance and effectiveness of the tasks. Policy: The policy is the probability distribution of actions taken by the subtask model given a state. In the context of this chapter, we adopt the stochastic policy gradient approach, which approximates the policy through an independent function approximator. Specifically, we use the LSTM decoder $P(t \mid i; \theta_M)$ as the policy for M and the generator $G_{\theta_N^g}(z_{noise}, t_{mid})$ in DFGAN as the policy for N. The goal of the policy is to generate the best actions based on the current state. The input to the policy is the state, and the output is the action. Action: The action is the action or decision taken by the model given a state. For the image captioning task, the generated description t conditioned on the input image i is considered as the action for M. For the TIS task, the synthesized image i generated based on the input description t is considered as the action for N. Reward: The reward is the feedback signal obtained by the model based on the actions taken and the current state. The reward can be designed based on specific objectives of the tasks. Policy gradient algorithms are reinforcement learning techniques that optimize the policy through gradient descent to obtain long-term rewards. We will discuss the main factors contributing to the rewards for image captioning M and image synthesis N. In the image captioning task, the reward can be evaluated based on the similarity between the generated description and the ground truth description. In the TIS task, the reward can be evaluated based on the similarity between the generated image and the ground truth image.

3.1 The Principle of SGCDIC

When the image-text transformation task starts with M, the reward r_M for the image captioning M is calculated as follows:

Reconstruction Reward. We use the reconstruction reward to evaluate the intermediate description t_{mid} generated by the image captioning model $P(t_{mid} \mid i; \theta_M)$. By using the generator $G_{\theta_N^g}(z_{noise}, t_{mid})$ in DFGAN, we generate an image $i\prime$ from t_{mid} (referred to as reconstructed image). To compute the reconstruction reward, we measure the similarity between the original image i and the reconstructed image $i\prime$ as the reconstruction reward for M. In these processes, the scene graph serves as an intermediate bridge. Prior to generating the intermediate description t_{mid}, the image i needs to be transformed into a scene graph. And before generating the image $i\prime$ from the intermediate description t_{mid}, the intermediate description t_{mid} needs to be transformed into a scene graph, as shown in Fig. 1. In this study, we use the negative squared difference to measure the similarity between the images i and $i\prime$:

$$r_M^{rec} = -\left\|i - i'\right\|^2 \tag{1}$$

Evaluation Metric Reward. Given an image-text pair (i, t) in the target domain, which belongs to the dataset D^{w_p}, the evaluation metric is considered as the evaluation metric reward for M, and it is calculated using the following formula:

$$r_M^{eval} = \frac{BLEU\left(t_{mid}, t\right) + CIDEr\left(t_{mid}, t\right)}{2} \tag{2}$$

We combine the reconstruction reward r_M^{rec} in a linear manner and the evaluation metric reward r_N^{eval} as the final reward for M in each iteration.

$$r_M = \begin{cases} \lambda_M^1 r_M^{rec} + \lambda_M^2 r_M^{eval}, & \text{if } (x, y) \in D^{w_p} \\ \text{where } \lambda_M^1 = \lambda_M^2 = 0.5 & \\ \lambda_M^1 r_M^{rec}, & \text{if } (x, y) \in D^{w_u} \\ \text{where } \lambda_M^1 = 1.0. & \end{cases} \tag{3}$$

When the task is initiated by the subtask N, the reward r_N for the image description N is calculated as follows:

Reconstruction Reward. Given the intermediate image i_{mid} generated by the generator $G_{\theta_N^g}(z_{noise}, t)$ of DFGAN (conditioned on the caption t), we calculate the logarithm probability of reconstructing the caption from i_{mid}, denoted as $P\left(t \mid i_{mid}; \theta_M\right)$, as the reconstruction reward. The reconstruction reward for the image generation process N is defined as:

$$r_N^{rec} = \log P\left(t \mid i_{mid}; \theta_M\right) \tag{4}$$

Similarity Reward. Given the text-image pair (i, t) in the target domain, we calculate the similarity between the intermediate image i_{mid} and the original image i. The similarity is defined as the negative value of the squared difference between the two images:

$$r_N^{sim} = -\left\|i - i^{mid}\right\|^2 \tag{5}$$

Finally, the reward r_N for TIS task N is calculated as a combination of the reconstruction reward r_N^{rec} and the similarity reward r_M^{sim}:

$$r_N = \begin{cases} \lambda_N^1 r_N^{rec} + \lambda_N^2 r_N^{eval}, & \text{if } (x, y) \in D^{w_p} \\ \text{where } \lambda_N^1 = \lambda_N^2 = 0.5 & \\ \lambda_N^1 r_N^{rec}, & \text{if } (x, y) \in D^{w_u} \\ \text{where } \lambda_N^1 = 1.0 & \end{cases} \tag{6}$$

3.2 Parameters Updating

M and N can be learned to maximize the expected long-term rewards $\mathbb{E}[r_M]$ and $\mathbb{E}[r_N]$. First, at the beginning of the task, we update the model of M by computing the gradient of the expected reward $\mathbb{E}[r_M]$ with respect to the parameter set θ_M and θ_N, and update it according to the policy gradient theorem:

$$\nabla_{\theta_M} \mathbb{E}[r_M] = \mathbb{E}\left[r_M \nabla_{\theta_M} \log P\left(t_{mid} \mid i; \theta_M\right)\right] \tag{7}$$

$$\nabla_{\theta_N^g} \mathbb{E}[r_M] = \mathbb{E}\left[2\lambda_M^1 (i - i') \nabla_{\theta_N^g} G_{\theta_N^g}\left(z_{noise}, t_{mid}\right)\right] \tag{8}$$

Next, when the task starts with N, we update the model by computing the gradient of the expected reward $\mathbb{E}[r_N]$ with respect to the parameter sets θ_N^g and θ_M, and update them based on the policy gradient theorem, as follows:

$$\nabla_{\theta_N^g} \mathbb{E}[r_N] = \mathbb{E}\left[r_N \nabla_{\theta_N^g} G_{\theta_N^g}\left(z_{noise}, i_{mid}\right)\right] \tag{9}$$

$$\nabla_{\theta_M} \mathbb{E}[r_N] = \mathbb{E}\left[\lambda_N^1 \nabla_{\theta_M} \log P\left(t_{mid} \mid i; \theta_M\right)\right] \tag{10}$$

Once we have learned a high-quality generator $G_{\theta_N^g}$, the discriminator model $D_{\theta_N^d}$ is retrained as follows:

$$\begin{aligned} \max_{\theta_N^d} J\left(\theta_N^d\right) = \max_{\theta_N^d} \Big[& \mathbb{E}\left[\log D_{\theta_N^d}(i, t)\right] + \mathbb{E}\left[\log\left(1 - D_{\theta_N^d}\left(i, t_{mis}\right)\right)\right] \\ & + \mathbb{E}\left[\log\left(1 - D_{\theta_B^d}\left(G_{\theta_N^g}\left(z_{noise}, t\right), t\right)\right)\right]\Big] \end{aligned} \tag{11}$$

We divide the input to the discriminator D into three categories: synthesized images with random descriptions $D_{\theta_N^d}\left(G_{\theta_N^g}\left(z_{noise}, t\right), t\right)$, real images matching the description $D_{\theta_N^d}(i, t)$, and real images with mismatched descriptions $D_{\theta_B^d}\left(i, t_{mis}\right)$.

The gradient computation for the objective function $J(\theta_N^d)$ with respect to θ_N^d is as follows:

$$\begin{aligned}\nabla_{\theta_N^d} J\left(\theta_N^d\right) = \mathbb{E}\left[\nabla \log D_{\theta_N^d}(i,t)\right] + \mathbb{E}\left[\nabla \log\left(1 - D_{\theta_N^d}\left(i, t_{mis}\right)\right)\right] \\ + \mathbb{E}\left[\nabla \log\left(1 - D_{\theta_N^d}\left(G_{\theta_N^g}\left(z_{noise}, t\right), t\right)\right)\right]\end{aligned} \tag{12}$$

When computing the stochastic gradients of $\mathbb{E}[r_M]$ and $\mathbb{E}[r_N]$ with respect to the parameters θ_M, θ_N^g, and θ_N^d, we use the Adam optimizer to update these parameters.

The two components of SGCDIC are iteratively performed multiple times. In each iteration, we select an image and use the generator of image description M to generate its corresponding text description. Then, we use N to reconstruct the image from the text description. Similarly, we select a text description and use the text-to-image generator N to generate the corresponding image, and then reconstruct the text description using the image description M. Subsequently, the parameters of M and N are updated through iteration to further improve their performance. This process is repeated multiple times until a predetermined stopping condition is met.

3.3 Object Geometry Consistency Losses and Semantic Similarity

We can optimize the IC process by calculating the object geometry consistency losses between the generated image I' from the text description T and the real image I. This can be represented as $I \longrightarrow T_{mid} \longrightarrow I'$. For 2D static images, this paper calculates the object geometry consistency losses as follows:

$$\begin{aligned}\mathcal{L}_{seg} = \ell_{static} + \ell_{smooth} = \frac{1}{N}\left\|\left(\sum_{k=1}^{K} o_k * \left(\mathcal{T}_k \circ \boldsymbol{i}\right)\right) - (\boldsymbol{i} + \boldsymbol{m})\right\|_2 \\ + \frac{1}{N}\sum_{k=1}^{K}\left(\frac{1}{H}\sum_{h=1}^{H} d\left(\boldsymbol{o}_{i_n}, \boldsymbol{o}_{i_n^h}\right)\right)\end{aligned} \tag{13}$$

Given an input image set $\boldsymbol{I}$, the generated image $\boldsymbol{I}+\boldsymbol{M}$, object mask $\boldsymbol{O}$, and deviation $\boldsymbol{M}$, this paper introduces the following objectives to achieve geometric consistency between the input image $\boldsymbol{I}$ and the text-based generated image $\boldsymbol{I}+\boldsymbol{M}$. It is important to note that the initial object mask $\boldsymbol{O}$ is meaningless at the beginning and needs to be optimized appropriately. For the *k-th* object, we need to estimate its transformation matrix $\mathcal{T}_k \in \boldsymbol{R}^{4\times 4}$.

For the text description T and image I, we can optimize the TIS process by calculating the semantic similarity between the generated text description T' and the real text description T based on cosine similarity. This can be represented as $T \longrightarrow I_{mid} \longrightarrow T'$. In this paper, the semantic similarity of text is computed using cosine similarity as follows:

$$\mathcal{L}_{sim} = 1 - \mathcal{L}_{cos}, \mathcal{L}_{cos} = \frac{T \cdot T'}{\|T\|_2 \|T'\|_2} \tag{14}$$

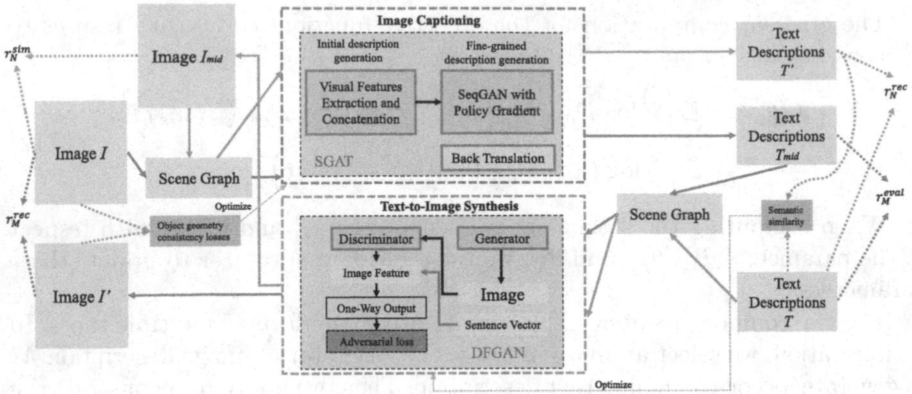

Fig. 1. The Principle of SGCDIC. SGAT can be seen in Fig. 2. DFGAN can be seen in research [22]. Pink represents images, scene graphs, and text descriptions. Orange solid arrows represent the dual learning process starting with the IC process, while blue solid arrows represent the dual learning process starting with TIS process. Please refer to Sect. 3.1 for the dashed arrows in these two colors. Specifically, the real image I is transformed into a scene graph, undergoes SGAT, and generates a text description T_{mid}. T_{mid} is then transformed into a scene graph, undergoes DFGAN, and generates the image $I^{'}$. The real text description T is transformed into a scene graph, undergoes DFGAN, and generates the image I_{mid}. I_{mid} is then transformed into a scene graph, undergoes SGAT, and generates the text description $T^{'}$. Purple represents the computation of object geometry consistency losses for different images and semantic similarity for different text descriptions, which optimize SGAT and DFGAN respectively. Please refer to Sect. 3.3 for detailed computational methods.

4 Experiments and Results Analysis

4.1 Datasets and Implementation Details

We employ MSCOCO [18] as the source domain, while Flickr30k [26] and Oxford-102 [21] were considered as the target domains. We provide a detailed description of these three datasets below: (1)MSCOCO: It consists of 2,783 training images, 40,504 validation images, and 40,775 image-text pairs for testing. Each image is associated with five captions manually created by Amazon Mechanical Turk workers. To ensure a fair comparison with other methods, we adopted the commonly used data split strategy proposed in [15]. We used 5,000 images for validation and testing, respectively, and the remaining images were used for training. (2)Flickr30k: It contains a total of 31,783 images with five textual descriptions per image. We followed the same data split as [15], where we used 29,000 images for training and 1,000 images for validation and testing. (3)Oxford-102: This dataset consists of 8,189 flower images from 102 different categories. Following the strategy proposed in [15], we used 1,000 images for validation and testing, and the remaining images were used for training. During the preprocessing

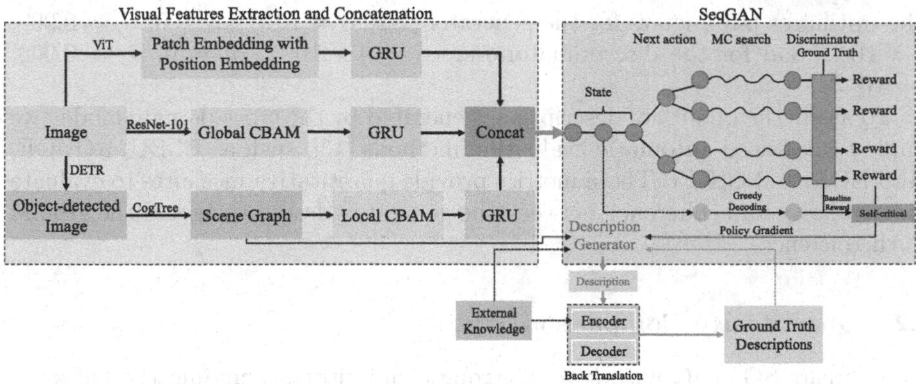

Fig. 2. Vanilla Architecture of the proposed model SGAT. For simplicity, CBAM refers to convolutional block attention [23]. Patch embedding with position embedding can be obtained through Vision Transformer [8]. Using ResNet-101 [9] to capture visual features. Then through Global CBAM, we get the informative visual features of the main objects of the image. DETR [3] is used to detect objects for scene graph generation. Then, we employ Cogtree [27] to generate scene graphs. The Local CBAM module is used to capture the relationship of different scene graphs. All the above information needs to be input into the simplified LSTM (GRU) for further concatenate preparation. Finally, the results are fed into the SeqGAN module. Among them, scene graphs provide explicit guidance in semantic structure design and refinement for description generation.

stage, we utilized publicly available code[1] for text description preprocessing. All descriptions were converted to lowercase, and non-alphanumeric characters were discarded. For these three datasets, we pruned the vocabulary by filtering out words with a frequency of less than 4, including special sentence start (BOS) and end (EOS) tokens. If a word was not present in the vocabulary, it was replaced with the "UNK" token.

All experiments are conducted on a server equipped with four NVIDIA Tesla V100 GPUs, Ubuntu 16.04.6 LTS operating system, 64GB of memory, Python 3.8.3, CUDA 11.2, and PyTorch 1.7.0. The parameters of the SGAT network were initialized using a Gaussian distribution $\mathcal{N}(0, 0.01)$, while other parameters were initialized using a uniform distribution [-0.01, 0.01]. The word embedding size was set to 300. Batch normalization was applied to all convolutional layers with a decay rate of 0.9 and ϵ of 1×10^{-5}. For the DFGAN network, Adam optimizer was used with $\beta_1 = 0$ and $\beta_2 = 0.9$. The learning rate was set to 0.0001 for the generator and 0.0004 for the discriminator according to the Two Timescale Update Rule [11]. During the large-scale training process of SGCDIC, the batch size was set to 100, and the learning rate was set to 2×10^{-4}. To prevent overfitting, dropout, and L2 regularization were employed with a dropout rate of 0.2 and weight decay of 0.001. The optimization algorithm used was Adam.

[1] https://github.com/karpathy/neuraltalk.

The optimizer parameters for the generator were $\alpha = 10^{-4}$, $\beta_1 = 0$, $\beta_2 = 0.999$, $\epsilon = 10^{-6}$, and for the discriminator were $\alpha = 4 \times 10^{-4}$, $\beta_1 = 0$, $\beta_2 = 0.999$, $\epsilon = 10^{-6}$.

To assess the quality of descriptions generated by the cross-domain model, we employ rule-based automatic evaluation methods [13], such as BLEU, ROUGE, METEOR, and CIDEr. These metrics provide quantitative measures to evaluate the similarity and coherence between the generated descriptions and the ground truth references.

4.2 Quantitative Comparison

We compare SGCDIC with several strong competitors, including the following methods: (1) GCN-LSTM model [25]: This model utilizes GCN to encode contextual information for visually related regions in structured semantic/spatial graphs. It then feeds the relation-aware region-level features learned from each semantic/spatial graph into an LSTM decoder with attention mechanism for sentence generation. (2) Bottom-Up and Top-Down attention (BUTD) model [2]: This model combines bottom-up and top-down attention mechanisms to compute attention scores at the levels of objects and salient image regions. The bottom-up attention mechanism determines the image regions, while the top-down attention mechanism determines the feature weights. (3) SAT model [5]: SAT utilizes adversarial learning to leverage unpaired image-text data in the target domain. SAT employs a critic network to evaluate whether the generated descriptions can be distinguished from real descriptions, and uses another critic network to evaluate whether the image and its generated description form a valid combination. (4) SGAE model [24]: By leveraging the complementary advantages of symbolic reasoning and end-to-end multimodal feature mapping, the SGAE model introduces inductive bias into the encoder-decoder architecture of language generation. It achieves this by narrowing the gap between visual perception and language composition through scene graphs, thus transferring the inductive bias from the purely linguistic domain to the visual-linguistic domain. (5) SubGC [31]: This model decomposes the scene graph into a series of subgraphs and decodes the target sentence by selecting important subgraphs and using an LSTM with an attention mechanism. (6) SG2Caps [20]: This model leverages the spatial positions of entities and person-object interaction labels to bridge the semantic gap between the image scene graph and the corresponding textual description scene graph. (7) SGCC [17]: This model utilizes the semantic attributes of objects and the 3D information of images (*i.e.*, object depth, relative 2D image plane distance, and relative angles between objects) to construct a scene graph. It employs graph convolutional networks to aggregate nodes and decomposes the constructed scene graph into attribute subgraphs and spatial subgraphs. (8) SGGC [4]: This model introduces a unique approach to the image captioning process by leveraging the semantic associations generated from the scene graph and visual features. It addresses the issue of neglecting semantic relationships among key regions in image description within the encoder-decoder architecture. (9) ICDF model [7]: This is a simple fine-tuning method that combines cross-domain retrieval with reinforcement learning to make the generated descriptions by the cross-domain IC model more discriminative. In this comparison, it is only included as the best-performing model among the fine-tuned ones, and the fine-tuned results (denoted as ICDF) are compared against the proposed method. (10) GCN-LSTM-dual, SAT-dual, SGAE-dual, SubGC-dual, SG2Caps-dual, SGCC-dual and SGGC-dual. These methods are embedded into the proposed dual learning framework individually. Specifically, they share a framework with SGCDIC, where the IC module in the framework is replaced by each of them. (11) L-Verse [16] and UMT [12] also focuses on intra-domain text-to-image synthesis. In this study, we pretrain UMT on MSCOCO dataset (source domain) and then test its performance on two target domains, namely Oxford-102 and Flickr30k. Please note that we do not compare SGCDIC with BUTD using the dual learning framework because SGCDIC is based on scene graphs and extends BUTD by leveraging the dual learning framework.

We employ the fine-tuning method proposed in [7], which involves fine-tuning the models listed in Table 1 using all paired image-text data in the target domain. We consider the best evaluation metric achieved after fine-tuning as a reliable reference for model comparison (denoted as ICDF result - fine-tuning on the best-performing metric among the compared models), utilizing all paired image-text data. In experiments, all training samples in the target domain are treated as unpaired image-text data. Specifically, for the models listed in Table 1, we first pretrain the models on the source domain (MSCOCO dataset) training set. Then, we fine-tune these models using the unpaired image and text data from the target domain (*e.g.*, Oxford-102 and Flickr30k). Finally, we evaluate the performance of all models on the target domain test set.

From Table 1, it can be observed that SGCDIC consistently outperforms its competitors with a significant advantage on the Oxford-102 and Flickr30k datasets. Specifically, on the Oxford dataset, SGCDIC performs better than other methods on all evaluation metrics (even better than the ICDF result). As expected, models embedded in the dual learning framework (models with "dual") show better performance than the individual IC models (models without "dual") in cross-domain tasks, as they leverage unpaired data from the target domain for training. All baseline methods using the dual learning framework outperform the original models, validating the effectiveness of the dual learning framework.

It is worth noting that SGCDIC shows more significant improvement on the Oxford-102 dataset compared to Flickr30k. This could be because there is a larger domain shift between MSCOCO and Oxford-102 (*i.e.*, a larger semantic gap) compared to MSCOCO and Flickr30k. The Oxford-102 dataset focuses on images of flowers, while MSCOCO and Flickr30k datasets contain multi-object images in daily life scenes, which have some similarities. Hence, the domain shift between MSCOCO and Oxford-102 is smaller (*i.e.*, a smaller semantic gap). Therefore, in cross-domain scenarios, generating semantically meaningful descriptions for complex scenes becomes more challenging, and the differences in generalization capabilities among different methods become more apparent. The

Table 1. In the absence of labeled image-text pairs, we performed the automatic evaluation of cross-domain image captioning tasks on two target domains, namely Oxford-102 and Flickr30k. We conducted pre-training of all methods on MSCOCO, which served as the source domain.

Method	Target	BLEU-1	BLEU-2	BLEU-3	BLEU-4	METEOR	ROUGE	CIDEr
GCN-LSTM	Oxford-102	0.574	0.335	0.261	0.219	0.229	0.447	0.253
BUTD	Oxford-102	0.541	0.253	0.159	0.104	0.161	0.364	0.105
SAT	Oxford-102	0.843	0.771	0.679	0.612	0.358	0.726	0.275
SGAE	Oxford-102	0.852	0.789	0.684	0.627	0.371	0.736	0.283
SubGC	Oxford-102	0.892	0.794	0.695	0.641	0.383	0.756	0.345
SG2Caps	Oxford-102	0.912	0.809	0.706	0.653	0.392	0.776	0.357
SGCC	Oxford-102	0.935	0.893	0.736	0.671	0.424	0.782	0.368
SGGC	Oxford-102	0.924	0.881	0.724	0.663	0.405	0.779	0.415
GCN-LSTM-dual	Oxford-102	0.763	0.547	0.394	0.358	0.426	0.568	0.362
SAT-dual	Oxford-102	0.863	0.781	0.689	0.629	0.375	0.754	0.349
SGAE-dual	Oxford-102	0.874	0.797	0.694	0.635	0.387	0.759	0.355
SubGC-dual	Oxford-102	0.913	0.817	0.715	0.673	0.396	0.784	0.394
SG2Caps-dual	Oxford-102	0.925	0.823	0.728	0.682	0.405	0.793	0.379
SGCC-dual	Oxford-102	$0.946^{\dagger}$	$0.914^{\dagger}$	$0.745^{\dagger}$	$0.683^{\dagger}$	$0.437^{\dagger}$	$0.794^{\dagger}$	0.385
SGGC-dual	Oxford-102	0.935	0.906	0.741	0.676	0.428	0.789	0.439
L-Verse	Oxford-102	–	–	–	0.373	0.285	0.567	$0.674^{\dagger}$
UMT	Oxford-102	–	–	–	0.356	0.261	0.546	0.652
ICDF	Oxford-102	0.949◇	0.921◇	0.751◇	0.688◇	0.442◇	0.806◇	0.677◇
SGCDIC (Ours)	Oxford-102	0.957♡	0.e928♡	0.763♡	0.694♡	0.459♡	0.815♡	0.683♡
GCN-LSTM	Flickr30k	0.631	0.356	0.274	0.225	0.259	0.463	0.271
BUTD	Flickr30k	0.615	0.403	0.249	0.181	0.157	0.439	0.232
SAT	Flickr30k	0.676	0.421	0.294	0.283	0.365	0.478	0.331
SGAE	Flickr30k	0.779	0.574	0.395	0.431	0.492	0.582	0.463
SubGC	Flickr30k	0.793	0.685	0.486	0.457	0.534	0.613	0.532
SG2Caps	Flickr30k	0.814	0.726	0.595	0.583	0.652	0.679	0.627
SGCC	Flickr30k	0.835	0.748	0.624	0.608	0.692	0.725	0.718
SGGC	Flickr30k	0.826	0.735	0.617	0.596	0.674	0.709	0.735
GCN-LSTM-dual	Flickr30k	0.794	0.546	0.457	0.468	0.427	0.552	0.483
SAT-dual	Flickr30k	0.689	0.445	0.316	0.394	0.473	0.528	0.475
SGAE-dual	Flickr30k	0.804	0.706	0.514	0.623	0.648	0.704	0.603
SubGC-dual	Flickr30k	0.815	0.765	0.562	0.675	0.655	0.743	0.674
SG2Caps-dual	Flickr30k	0.833	0.783	0.569	0.692	0.667	0.758	0.658
SGCC-dual	Flickr30k	$0.865^{\dagger}$	$0.794^{\dagger}$	$0.663^{\dagger}$	$0.712^{\dagger}$	$0.686^{\dagger}$	$0.778^{\dagger}$	0.679
SGGC-dual	Flickr30k	0.852	0.791	0.586	0.704	0.673	0.764	$0.813^{\dagger}$
L-Verse	Flickr30k	–	–	–	0.356	0.254	0.532	0.647
UMT	Flickr30k	–	–	–	0.334	0.237	0.522	0.628
ICDF	Flickr30k	0.874◇	0.803◇	0.674◇	0.732◇	0.691◇	0.782◇	0.827◇
SGCDIC (Ours)	Flickr30k	0.883♡	0.827♡	0.754♡	0.776♡	0.723♡	0.806♡	0.845♡

results of cross-domain image captioning generation demonstrate that SGCDIC can adapt well to the target domain without the need for fine-tuning paired image-text data in the target domain. Notably, the reason why SGCDIC achieves competitive results compared to other models such as SGAE, SubGC, SG2Caps, SGCC, SGGC, and their corresponding "dual" models (*i.e.* SGAE-dual, SubGC-dual, SG2Caps-dual, SGCC-dual, SGGC-dual) is that scene graphs can encode objects, attributes, and their relationships in the image, while the dual learning mechanism can capture fine-grained visual features. The reason why most metrics of SGCC and SGCC-dual are higher than other baseline models may be that 3D information is more discriminative for identifying multi-object dense scenes in images. 2D information is more prone to occlusion and blocking phenomena, where one object is obstructed by others and is not easily identifiable. As for L-Verse and UMT, they demonstrate the advantages of the Transformer architecture in capturing fine-grained cross-modal (image-text) features. However, when faced with target domains such as Oxford-102 and Flickr30k, their performance is relatively limited. This is because the Transformer architecture fails to learn the crucial aspect of domain transfer, *i.e.*, the commonality of different domain data (*e.g.* shared fine-grained features).

4.3 Qualitative Comparison

In this section, we present some representative descriptions and image samples generated by SGCDIC and other comparative methods.

Table 2. Qualitative examples of different methods on the Oxford-102 dataset. Each example consists of an input image on the left, the corresponding ground truth, and generated descriptions on the right.

Images	Ground truth descriptions	Generated descriptions
	**1**:this flower has a large number of red petals close together surrounding a group of tall yellow stamen. **2**:this flower is pink and yellow in color, with petals that are rounded. **3**:this flower has round, dark magenta petals and a big collection of yellow stamens. **4**:this dark pink flower has several layers of petals, it also has yellow stamen protruding from the center. **5**:this flower has thin yellow stamen and very bright pink petals with rounded tips.	BUTD:this flower has red petals with yellow stamens. SAT:this flower has petals that are pink and has red dots. SGAE: this flower has petals that are yellow and has green pedicel. SGCDIC:this flower has red petals surrounding yellow stamen, and has green pedicel.
	**1**:this flower is purple in color, with petals that are ruffled and wavy. **2**:this flower has a simple row of blue petals with the stamens at the center. **3**:this flower has petals that are purple with black stamen. **4**:this purple flower is trumpet shaped and has deep purple vertical lines coming out from the inside. **5**:this flower has large light purple petals with a wrinkled texture and rounded edges.	BUTD: this flower has purple wavy petals. SAT: this flower has purple petals. SGAE: a purple flower has wavy petals. SGCDIC: the flower has purple petals with wavy and ruffled shapes.
	**1**:the flower petals are oval shaped and are pink in color with the outer part colored yellow. **2**:the flower has pink petals with yellow outer edges. **3**:this flower is red and yellow in color, with petals that are multi colored. **4**:this flower is pink and yellow in color, and has petals that are yellow on the tips. **5**:this flower has many long red with yellow tip petals, and a wide red and yellow pistil.	BUTD:a flower has red and yellow petals. SAT:this flower has pink petals with yellow tips. SGAE:this flower has red petals with yellow tips. SGCDIC:this flower has wide red petals with yellow outer edges.
	**1**:the flower has long stamen that are pink with pink petals. **2**:this flower is red and black in color, with petals that are spotted. **3**:this flower has long red stamen surrounded by red petals with maroon spots. **4**:this flower has petals that are red with black dots. **5**:a large red flower with black dots and a very long stigmas.	BUTD:the flower has red petals that are spotted. SAT:this flower has petals that are red and has black dots. SGAE: a large red flower with black dots. SGCDIC:the flower has red petals with long stigmas, black dots in petals.
	**1**:pedicel is green in color,petals are white in cplor with yellow anthers. **2**:this flower is white and yellow in color, with petals that are wavy and curled. **3**:a flower with white petals accompanied by yellow anther filaments and pistils. **4**:the flower shown has several white petals, and a green center. **5**:this flower has pure white petals that curve out slightly at the top with a bright orange stamen.	BUTD:this flower has white petals and yellow stamen. SAT:a flower with white petals and yellow pistil. SGAE: this flower is white and yellow in color, petals are white. SGCDIC:this flower has white petals with yellow stamen and light green pistil.
	**1**:this flower is pink and yellow in color, with petals that are wavy and curled. **2**:a large purple pedaled flower with a sunflower looking center. **3**:the flower shown has several pink petals, and a yellow center. **4**:this pink flower has rounded petals and orange stamen and anthers. **5**:this flower has light purple petals that look like scoops and a thick layer of yellow stamen.	BUTD:a flower has pink petals and a yellow center. SAT:a flower with pink petals and yellow stamen. SGAE: this flower has several pink petals with yellow stamen SGCDIC:this flower has pink curved petals with yellow stamens.

On the Oxford-102 dataset, representative descriptions generated by SGCDIC and other comparative methods are shown in Table 2. It is easy to observe from Table 2 that all methods are capable of generating reasonable and relevant image descriptions. However, the domain adaptive model (SGCDIC) tends to produce more accurate and detailed descriptions. In most cases, SGCDIC achieves the best performance. For example, in the fourth sample, the description generated by SGCDIC, "the flower has red petals with long stigmas, black dots in petals," accurately describes the color, shape, and content of the flower (see Table 2).

Table 3. Qualitative examples of different methods on the Flickr30k dataset. Each example consists of the input image on the left, and the ground truth and generated descriptions on the right.

Images	Ground truth descriptions	Generated descriptions
	1:a boy leaping in the air towards a soccer ball. **2**:a boy in midair trying to kick a soccer ball. **3**:boy leaning backward to kick a soccer ball in midair. **4**:a child falling back to kick a soccer ball. **5**:a boy in a stripy shirt is playing with a soccer ball in a grassy park.	BUTD:a man in a field playing with a frisbee. SAT:a young boy playing with a soccer ball in a field. SGAE:a boy in a stripy shirt is playing with a soccer ball. SGCDIC:a boy in a stripy shirt is playing with a soccer ball in a grassy park.
	1:A man in a blue baseball cap and green waders fumbles with a fishing net in a blue boat docked beside a pier. **2**:A bright blue fishing boat and fisherman at dock preparing nets. **3**:A man in a small boat readies his net for the day ahead. **4**:A lone fisherman is on his boat checking his net. **5**:Man in blue boat holding a net.	BUTD: a man in boat is preparing nets. SAT: a man in boat is preparing nets. SGAE: a man in a small boat is checking his net. SGCDIC: A fisherman in a blue boat is holding a net.
	1:A brown beast of burden is suspended in midair by the cart it's pulling which is carrying various white packages and toppled over in the sand. **2**:An overloaded wagon full of white boxes tips backwards and pulls the mule attached to the wagon into the air. **3**:A cart that is so heavy it tipped over backwards and put the donkey that was pulling it in the air. **4**:A donkey suspended in air after a cart it is attached to over turns. **5**:A cart has been overloaded lifting the animal attached.	BUTD:An animal suspended in air after a cart. SAT: A beast of burden is suspended in midair. SGAE: A brown animal of burden is suspended in midair by the art. SGCDIC: A donkey of burden is suspended in midair by the cart which is carrying various white packages.
	1:little boy looking through a telescope on playground equipment with the sun glaring off the pole his hand is on. **2**:A child is looking through a pretend telescope on playground equipment in front of a blue sky. **3**:A little blond boy is looking through a yellow telescope. **4**:A child looking through a telescope on a playground. **5**:A kid with blond-hair using a telescope.	BUTD:a boy is looking through a telescope. SAT: a child is looking through a telescope. SGAE:A kid is looking through a yellow telescope. SGCDIC:A little boy with blond hair is looking through a yellow telescope.
	1:A horn and percussion band playing their instruments in public. **2**:A band of men and women are playing music in the street. **3**:A group of people are playing instruments. **4**:The band is playing music in this city. **5**:A band entertaining the crowd.	BUTD:many people are playing instruments. SAT: a lot of people are playing instruments. SGAE:a group of people are playing instruments. SGCDIC:A group of men and women are playing musical instruments in the street.
	1:Girls in white shirts and blue skirts shoot arrows from bows. **2**:5 children japanese with bows practicing in front of sensei. **3**:Five woman all dressed in uniform are in an archery class. **4**:A group of five women in blue skirts practicing archery. **5**:Five women are practicing bow and arrow skills indoors.	BUTD:several girls are playing arrows. SAT:many girls are practicing archery. SGAE:five girls are practicing archery. SGCDIC:five girls in white shirts and blue skirts are practicing archery indoors.

Furthermore, we provide several representative image descriptions in Flickr30k, as shown in Table 3. In the first example of Table 3, SGCDIC correctly and specifically describes what the boy is wearing as "a boy wearing a striped shirt," while other comparative methods simply state "a boy." The third and fourth examples demonstrate SGCDIC's ability to capture fine-grained features such as "white packages" and "blond hair." It is worth noting that in the example of the last row, all methods, including SGCDIC, fail to identify the man standing behind the five girls wearing similar-styled clothing. This means cross-domain IC models show lower accuracy than non-cross-domain models on test images from the target domain, particularly when encountering challenges like shadows, occlusions, and low lighting conditions that impact the IC process.

5 Conclusion

In this paper, we propose a dual-learning framework called SGCDIC, which is a scene graph-based multi-task learning algorithm for cross-domain image captioning. In this regard, we propose a scene-graph image captioning method called SGAT and leverage DFGAN for text-to-image synthesis. SGCDIC employs a dual-learning mechanism to optimize two coupled tasks (*i.e.* image captioning and text-to-image synthesis) simultaneously to transfer knowledge from the source domain to the target domain. SGCDIC follows a two-stage strategy. Firstly, we learn the shared knowledge of cross-modal generation from the source domain (MSCOCO), such as aligning visual representations with natural language descriptions, by accessing a large-scale dataset of text-image pairs. This enables fine-grained feature alignment between the visual and language modalities. Secondly, we transfer the learned shared knowledge to the target domain (Flickr30k, Oxford-102) and fine-tune the model on limited or unpaired data using the dual-learning mechanism. Extensive experiments validate our proposal and analysis. More remarkably, our method outperforms other methods, achieving the best cross-domain image captioning performance.

References

1. Anderson, P., Fernando, B., Johnson, M., Gould, S.: SPICE: semantic propositional image caption evaluation. In: Leibe, B., Matas, J., Sebe, N., Welling, M. (eds.) ECCV 2016. LNCS, vol. 9909, pp. 382–398. Springer, Cham (2016). https://doi.org/10.1007/978-3-319-46454-1_24
2. Anderson, P., et al.: Bottom-up and top-down attention for image captioning and visual question answering. In: Proceedings of the IEEE Conference on Computer Vision and Pattern Recognition, pp. 6077–6086 (2018)
3. Carion, N., Massa, F., Synnaeve, G., Usunier, N., Kirillov, A., Zagoruyko, S.: End-to-end object detection with transformers. In: Vedaldi, A., Bischof, H., Brox, T., Frahm, J.-M. (eds.) ECCV 2020. LNCS, vol. 12346, pp. 213–229. Springer, Cham (2020). https://doi.org/10.1007/978-3-030-58452-8_13
4. Chen, H., Wang, Y., Yang, X., Li, J.: Captioning transformer with scene graph guiding. In: 2021 IEEE International Conference on Image Processing (ICIP), pp. 2538–2542. IEEE (2021)
5. Chen, T.H., Liao, Y.H., Chuang, C.Y., Hsu, W.T., Fu, J., Sun, M.: Show, adapt and tell: Adversarial training of cross-domain image captioner. In: Proceedings of the IEEE International Conference on Computer Vision, pp. 521–530 (2017)
6. Damodaran, V., et al.: Understanding the role of scene graphs in visual question answering. arXiv preprint arXiv:2101.05479 (2021)

7. Dessì, R., Bevilacqua, M., Gualdoni, E., Rakotonirina, N.C., Franzon, F., Baroni, M.: Cross-domain image captioning with discriminative finetuning. In: Proceedings of the IEEE/CVF Conference on Computer Vision and Pattern Recognition, pp. 6935 6944 (2023)
8. Dosovitskiy, A., et al.: An image is worth 16×16 words: transformers for image recognition at scale. arXiv preprint arXiv:2010.11929 (2020)
9. He, K., Zhang, X., Ren, S., Sun, J.: Deep residual learning for image recognition. In: Proceedings of the IEEE Conference on Computer vision and Pattern Recognition, pp. 770–778 (2016)
10. Herzig, R., Bar, A., Xu, H., Chechik, G., Darrell, T., Globerson, A.: Learning canonical representations for scene graph to image generation. In: Vedaldi, A., Bischof, H., Brox, T., Frahm, J.-M. (eds.) ECCV 2020. LNCS, vol. 12371, pp. 210–227. Springer, Cham (2020). https://doi.org/10.1007/978-3-030-58574-7_13
11. Heusel, M., Ramsauer, H., Unterthiner, T., Nessler, B., Hochreiter, S.: GANs trained by a two time-scale update rule converge to a local nash equilibrium. In: Advances in Neural Information Processing Systems, vol. 30 (2017)
12. Huang, Y., Xue, H., Liu, B., Lu, Y.: Unifying multimodal transformer for bi-directional image and text generation. In: Proceedings of the 29th ACM International Conference on Multimedia, pp. 1138–1147 (2021)
13. Jia, J., et al.: Image captioning based on scene graphs: a survey. Expert Syst. Appl. **231**, 120698 (2023)
14. Johnson, J., Gupta, A., Fei-Fei, L.: Image generation from scene graphs. In: Proceedings of the IEEE Conference on Computer Vision and Pattern Recognition, pp. 1219–1228 (2018)
15. Karpathy, A., Fei-Fei, L.: Deep visual-semantic alignments for generating image descriptions. In: Proceedings of the IEEE Conference on Computer Vision and Pattern Recognition, pp. 3128–3137 (2015)
16. Kim, T., et al.: L-verse: bidirectional generation between image and text. In: Proceedings of the IEEE/CVF Conference on Computer Vision and Pattern Recognition, pp. 16526–16536 (2022)
17. Liao, Z., Huang, Q., Liang, Y., Fu, M., Cai, Y., Li, Q.: Scene graph with 3D information for change captioning. In: Proceedings of the 29th ACM International Conference on Multimedia, pp. 5074–5082 (2021)
18. Lin, T.Y., et al.: Microsoft coco: common objects in context (2014). arXiv preprint arXiv:1405.0312 (2019)
19. Lyu, F., Feng, W., Wang, S.: vtGraphNet: learning weakly-supervised scene graph for complex visual grounding. Neurocomputing **413**, 51–60 (2020)
20. Nguyen, K., Tripathi, S., Du, B., Guha, T., Nguyen, T.Q.: In defense of scene graphs for image captioning. In: Proceedings of the IEEE/CVF International Conference on Computer Vision, pp. 1407–1416 (2021)
21. Nilsback, M.E., Zisserman, A.: Automated flower classification over a large number of classes. In: 2008 Sixth Indian Conference on Computer Vision, Graphics & Image Processing, pp. 722–729. IEEE (2008)
22. Tao, M., Tang, H., Wu, F., Jing, X.Y., Bao, B.K., Xu, C.: DF-GAN: a simple and effective baseline for text-to-image synthesis. In: Proceedings of the IEEE/CVF Conference on Computer Vision and Pattern Recognition, pp. 16515–16525 (2022)
23. Woo, S., Park, J., Lee, J.Y., Kweon, I.S.: CBAM: convolutional block attention module. In: Proceedings of the European conference on computer vision (ECCV), pp. 3–19 (2018)

24. Yang, X., Tang, K., Zhang, H., Cai, J.: Auto-encoding scene graphs for image captioning. In: Proceedings of the IEEE/CVF Conference on Computer Vision and Pattern Recognition, pp. 10685–10694 (2019)
25. Yao, T., Pan, Y., Li, Y., Mei, T.: Exploring visual relationship for image captioning. In: Proceedings of the European Conference on Computer Vision (ECCV), pp. 684–699 (2018)
26. Young, P., Lai, A., Hodosh, M., Hockenmaier, J.: From image descriptions to visual denotations: new similarity metrics for semantic inference over event descriptions. Trans. Assoc. Comput. Linguist. **2**, 67–78 (2014)
27. Yu, J., Chai, Y., Wang, Y., Hu, Y., Wu, Q.: CogTree: cognition tree loss for unbiased scene graph generation. arXiv preprint arXiv:2009.07526 (2020)
28. Yu, L., Zhang, W., Wang, J., Yu, Y.: SeqGAN: sequence generative adversarial nets with policy gradient. In: Proceedings of the AAAI Conference on Artificial Intelligence, vol. 31 (2017)
29. Yuan, J., et al.: Discriminative style learning for cross-domain image captioning. IEEE Trans. Image Process. **31**, 1723–1736 (2022)
30. Zhao, W., Wu, X., Luo, J.: Cross-domain image captioning via cross-modal retrieval and model adaptation. IEEE Trans. Image Process. **30**, 1180–1192 (2020)
31. Zhong, Y., Wang, L., Chen, J., Yu, D., Li, Y.: Comprehensive image captioning via scene graph decomposition. In: Vedaldi, A., Bischof, H., Brox, T., Frahm, J.-M. (eds.) ECCV 2020. LNCS, vol. 12359, pp. 211–229. Springer, Cham (2020). https://doi.org/10.1007/978-3-030-58568-6_13

Enhancing Rule Learning on Knowledge Graphs Through Joint Ontology and Instance Guidance

Xianglong Bao[1], Zhe Wang[2], Kewen Wang[2(✉)], Xiaowang Zhang[1], and Hutong Wu[1]

[1] College of Intelligence and Computing, Tianjin University, Tianjin, China
{baoxl,xiaowangzhang,wht}@tju.edu.cn

[2] School of Information and Communication Technology, Griffith University, Brisbane, Australia
{zhe.wang,k.wang}@griffith.edu.au

Abstract. Rule learning is a machine learning method that extracts implicit rules and patterns from data, enabling symbol-based reasoning in artificial intelligence. Unlike data-driven approaches such as deep learning, using rules for inference allows for interpretability. Many studies have attempted to automatically learn first-order rules from knowledge graphs. However, existing methods lack attention to the hierarchical information between ontology and instance and require a separate rule evaluation stage for rule filtering. To address these issues, this paper proposes a **O**ntology and **I**nstance Guided **R**ule **L**earning (OIRL) approach to enhance rule learning on knowledge graphs. Our method treats rules as sequences composed of relations and utilizes semantic information from both ontology and instances to guide path generation, ensuring that paths contain as much pattern information as possible. We also develop an end-to-end rule generator that directly infers rule heads and incorporates an attention mechanism to output confidence scores, eliminating the need for rule instance-based rule evaluation. We evaluate OIRL using the knowledge graph completion task, and experimental results demonstrate its superiority over existing rule learning methods, confirming the effectiveness of the mined rules.

Keywords: Knowledge graph · Rule learning · First-order Logic

1 Introduction

Knowledge graph (KG) is a structured and semantic knowledge representation model and serves as a graph-based knowledge repository. In a KG, entities are represented as nodes in the graph, and the relationships between entities are expressed as edges between nodes. Each triple (h, r, t) represents the relationship r between entities h and t. For example, $(Alice, nationality, China)$ represents that Alice's nationality is China. Knowledge graphs are capable of capturing

Q. Liu et al. (Eds.): PRCV 2023, LNCS 14427, pp. 138–150, 2024.
https://doi.org/10.1007/978-981-99-8435-0_11

and expressing complex relationships and attributes between entities, enabling various applications in artificial intelligence such as question-answering systems [31] and recommendation systems [26]. However, real-world knowledge graphs are often incomplete, with numerous unknown entities and relationships. Knowledge graph completion is an important task that aims to infer missing knowledge based on existing entities and relationships.

Currently, the field of machine learning predominantly focuses on using data-driven approaches to complete missing facts in knowledge graphs. TransE [1] serves as a representative translation model, which models the triple (h, r, t) as $h + r \approx t$, where + represents vector addition and $\approx$ denotes similarity. RotatE [22] defines the relationship between entities as a rotation from the source entity to the target entity in the complex plane. These Knowledge Graph Embedding (KGE) methods aim to explore potential semantic information between different entities by mapping entities and relations to a low-dimensional vector space, thereby facilitating downstream tasks. However, these data-driven methods have limitations in terms of interpretability and inductive reasoning.

Unlike KGE methods, rule-based reasoning employs symbolic logic and offers the ability to interpret predictions through rules, thus providing interpretability. Additionally, rule-based reasoning does not target specific entities during inference, enabling predictions to be made for previously unseen entities. This property demonstrates its inductive nature. For instance, consider the rule $nationality(x, y) \leftarrow bornin(x, z) \wedge isCityof(z, y)$, which implies that if entity x is born in city z and z is a city in county y, we can infer that the nationality of x is y.

Rule learning is an important research direction in the field of machine learning, aiming to extract useful rules from data. Recently, various rule mining methods have been proposed in the literature. For example, AMIE [8] is a top-down search-based rule learning method that performs first-order rule mining based on an incomplete knowledge base. It starts with empty rule bodies and expands the rule bodies while retaining candidate rules with support greater than a threshold. RLvLR [17] utilizes embeddings in representation learning to address the limitations of search-based rule mining algorithms that are not applicable to large-scale knowledge graphs. It requires counting the number of ground truth instances across the entire knowledge graph to evaluate the quality of rules. NCRL [2]utilizes the compositional property of rule bodies to derive rule heads and is the best performing rule mining algorithm at present. However, it obtains training paths through simple random walks, which leads to high time complexity. Moreover, it cannot guarantee the quality of the sampled paths. Note that hierarchical information between ontology and instance contains different pattern information for each relation. Figure 1 shows the domain and range constraint information among relationships. However, existing rule mining methods lack attention to this hierarchical information.

In this paper, we explore a solution that involves a two-stage rule mining approach. The first module is a path sampling module that utilizes the ontology level type information from the TBox and the entity information from the

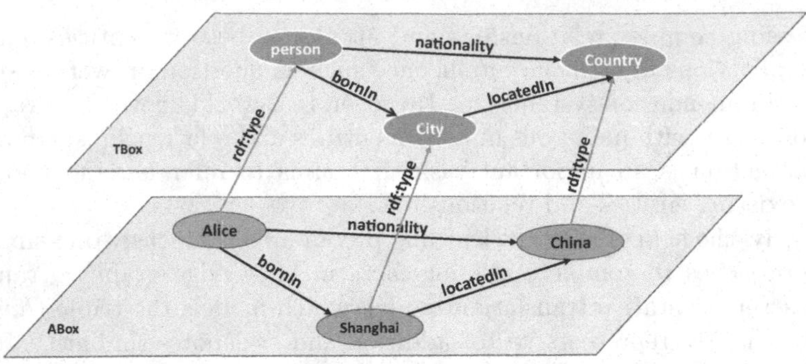

Fig. 1. The TBox contains ontology level information, while the ABox contains instance level information. Additionally, the TBox specifies the domain and range of different relations.

instance level ABox. Due to the focus of first-order rules on the logical relationships between relations rather than the path connections between entities, we not only consider entity information but also pay attention to the rich type information in the graph during the path sampling process. By utilizing the domain and range information of types in the TBox, we incorporate more pattern information into the sampled paths. The second part is the path synthesis module based on the Gate Recurrent Unit (GRU). This module synthesizes paths and generates rules along with their corresponding confidence scores in an end-to-end manner. It is important to note that our model avoids the rule evaluation process used in traditional rule mining algorithms. Instead, it directly outputs confidence scores, thus avoiding the computational cost and time consumption associated with the rule evaluation phase. We have implemented the prototype system OIRL and evaluated its performance on benchmark datasets. The experimental results demonstrate that our model outperforms existing rule mining methods. Our contributions can be summarized in the following three points:

- We propose a rule learning model that decouples the rule mining process into two stages and simultaneously incorporates ontology-level type information from the TBox and entity-level information from the ABox during the path sampling stage. This enables the sampled paths to capture more pattern information.
- Our model achieves end-to-end output of rule confidence, eliminating the need for a separate rule evaluation stage as found in previous methods.
- We conducted extensive experiments, and the results demonstrate that our model outperforms existing rule mining methods.

This paper is organized as follows. Section 2 introduces related work. Section 3 presents our method, OIRL, and describes the technical details. In Sect. 4, we will present the experimental results of OIRL on benchmark datasets. Finally, we conclude this paper in Sect. 5.

2 Related Work

2.1 Reasoning with Embeddings

In recent years, with the development of deep learning, many methods have represented entities and relationships as embeddings to accomplish reasoning tasks. As a representation learning approach, embedding methods aim to learn latent semantic information by mapping entities and relationships into a low-dimensional vector space. TransE [1] is one of the earliest methods that introduced embeddings into knowledge graphs. The core idea of TransE is to approximate the sum of the embedding of the head entity and the relationship to be close to the embedding of the tail entity. This allows it to capture relationships between entities. However, TransE represents relationships as unique vector offsets, which limits its ability to model many-to-one and many-to-many relationships. TransH [25] introduced a projection matrix to address the limitations of TranE by mapping entities and relationships onto hyperplanes. Subsequent models such as TransR [12], TransD [11], TransG [28], and TransAt [18] are all translation-based models. RESCAL [15] represents relationships as an association matrix between entities. DistMult [29] models relationships using tensor decomposition, where the matrix is constrained to be a diagonal matrix compared to RESCAL. ComplEx [24] extends DistMult by introducing complex embeddings, enhancing the modeling capability for asymmetric relationships. However, it cannot infer compositional patterns. RotatE [22] addresses the challenges of capturing relationship polysemy and symmetry in previous embedding methods by introducing complex vector space rotation operations. However, these embedding methods focus on existing triples and cannot predict unseen entities, lacking generalization. Additionally, embedding methods make predictions based on vectors, lacking symbolic meaning and interpretability. On the other hand, rule-based reasoning methods can provide human-interpretable explanations for predictions based on rules. They can also reason about unseen entities since rules are not specific to particular entities, exhibiting inductive reasoning capabilities.

2.2 Reasoning with Rules

Rule learning allows for the automatic mining of logical rules from knowledge graphs. These rules can be utilized for knowledge inference and completion. Inductive Logic Programming (ILP) [14] induces logical rules from basic facts. However, ILP's requirement for negative examples limits its applicability. AMIE [8] and AnyBURL [13] are both symbolic-based methods. AMIE treats rules as a series of atoms and extends them by adding three types of operators: Dangling Atom, Instantiated Atom and Closing Atom. It evaluates the quality of rules using Head Coverage (HC) and Standard Confidence (SC). AMIE+ [7] improves upon the rule expansion process of AMIE by only extending rules that can be closed within a set maximum length and optimizing the search process. However, due to the enormous search space, these ILP-based methods still face

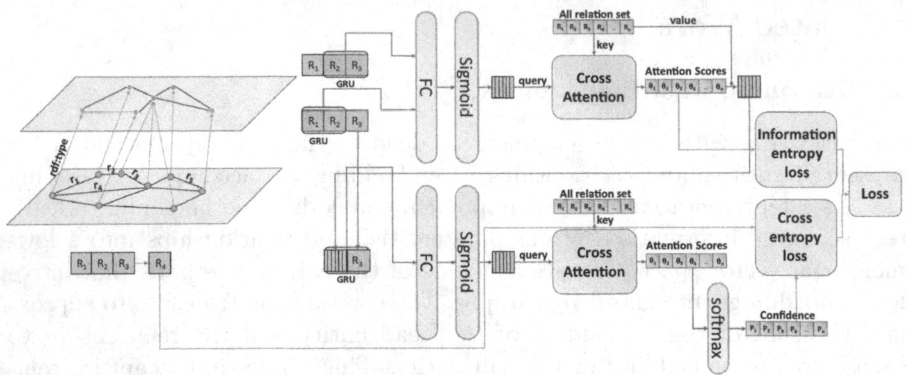

Fig. 2. Framework overview. OIRL consists of two modules: the Path Sampling Module and the Path Composition Module. The Path Sampling Module samples paths by capturing pattern information of different relations from both ontology and instance. The relation r_4 serves as the closing relation r_c. The closed paths are fed into the Path Composition Module for further processing. Note that the orange arrows represent the training process and do not appear during the inference process. (Color figure online)

challenges when applied to large-scale knowledge graphs. RLvLR [17] introduces a rule learning approach that combines embeddings. Unlike ILP methods that use refinement operators to search the rule space, RLvLR samples relevant subgraphs from the knowledge graph, containing only entities and facts related to the target predicate. It then utilizes the embedding model RESCAL [15] to prune the search space. TyRule [27] mines typed rules through type information. The above rule mining method requires a separate stage where pre-defined metrics are used to score the quality of the rules. There are also methods that employ neural-driven symbolic reasoning, such as KALE [9], RUGE [10], IterE [32], UniKER [4], and RPJE [16]. These methods require the prior mining of high-quality rules to drive the neural approaches, making the upstream rule mining methods crucial. Tensorlog [5] and Neura-LP [30] use differentiable processes for reasoning. NCRL [2] utilizes the compositional properties of atoms in rule bodies to infer rule heads, but it does not consider the quality of sampled paths in the path sampling phase, and the sampling process is also time-consuming.

3 Methodology

In this section, we will introduce a novel two-stage rule mining algorithm called OIRL. We first formally define ABox, TBox, and Horn rules.

ABox & TBox. The knowledge graph is formally defined as $\mathcal{G} = (E, R, O)$, E, R, and O represent the sets of entities, relationships, and triples. Our approach considers facts as the ABox and TBox parts, as shown in Fig. 1. In the ABox, we have instance-level facts, such as a triple $(Alice, bornin, Shanghai)$ which represents the connection between the entity *Alice* and the entity *Shanghai* through the relationship *bornin*. The ABox also indicates the type information of entities. For example, $(Alice, rdf : type, Person)$ describes the type of the entity *Alice* as a person. Unlike the ABox, the TBox describes ontology-level knowledge. For example, the two triples $(bornin, rdfs : domain, Person)$ and $(bornin, rdfs : range, City)$ indicate that the head entity type for the relation *bornin* should be *Person*, and the tail entity type should be *City*. The domain and range of a relation represent the types of the head entity and tail entity, respectively. Note that the domain and range of a relation are not necessarily unique.

Horn Rule. In this paper, we focus on mining Horn rules, which are a special form of first-order rules. Horn rules are composed of a body of conjunctive predicates, i.e., relations in the knowledge graph. The form of the rules mined by our method is as follows:

$$r_h(x, y) \leftarrow r_{b_1}(x, z_1) \wedge \cdots \wedge r_{b_n}(z_{n-1}, y) \tag{1}$$

In the mentioned rule, $r_h(x, y)$ represents the head of the rule, while $r_{b_1}(x, z_1) \wedge \cdots \wedge r_{b_n}(z_{n-1}, y)$ represents the body of the rule. Here, x, y, and z_i are variables that can be replaced with entities. Each individual $r(u, v)$ is referred to as an atom. When the body of a rule is satisfied, it allows for the inference of the facts in the head of the rule. Due to the fact that this form of rules manifests as a closed path in a graph, they are also referred to as closed-path rules.

3.1 Framework Details

The overall framework of our proposed approach is illustrated in Fig. 2. We combine the ontology-level information from the TBox with the instance-level information from the ABox to guide path searching. The generated paths simultaneously consider both type information and entity information. We consider the paths in the graph as a sequence and use GRU to iteratively combine the paths. Then, we input this combined representation into an attention module to compute the probability of each relation being the head of a rule. At each step, we calculate the entropy loss to optimize the probability distribution. Finally, we select the relation with the highest probability as the head of the rule.

Path Sampling Module. In our approach, we treat rules as sequence composed of relations, and therefore, we extract paths from the graph and use them as sequences for training. Most existing methods randomly select entities as anchors in the graph and start random walks from these anchors to obtain paths. Since Horn rules focus more on the combination of relations rather than entity level connections, in the process of path sampling, we sample based on relations rather than entities. In the knowledge graph, relations follow a heavy-tailed distribution, so we independently sample each relation and allocate a varying number of anchors based on the weight of each relation to reflect their importance in capturing rule patterns. We assign different sampling weights to each relation based on two aspects: (1) the weight of the instance level triplets in the ABox, and (2) the weight of the domain and range types at the ontology level in the TBox. The assignment of weights is as follows:

$$w_r = \lambda \frac{n_r}{\sum_{i=1}^{|R|} n_i} + (1-\lambda) \frac{m_r^{dom} + m_r^{ran}}{\sum_{i=1}^{|R|} m_i^{dom} + \sum_{i=1}^{|R|} m_i^{ran}} \tag{2}$$

where λ is a parameter that adjusts the degree of attention to the ABox and TBox. n_r represents the number of triples of relation r in the ABox, while m_r^{dom} and m_r^{ran} represent the number of domain and range types of relation r in the TBox. The weight metric w_r is used to assess the importance of relation r in terms of both instance and ontology levels, guiding the allocation of sampling weights. We can calculate the number of anchors for relation r by $a_r = \lceil w_r a \rceil$, where a is the total number of anchors, which is an adjustable parameter. After obtaining the anchor relation, a path search will be performed. Setting the sampling length to n, if the i-th relation in the sampling sequence is denoted as x_i, then the probability distribution of x_i can be defined as follows:

$$p(x_i = r_i | x_{i-1} = r_j) = \begin{cases} \frac{1}{|S(r_j)|}, & \text{if } r_j \text{ has connected relations} \\ 0, & \text{otherwise} \end{cases} \tag{3}$$

where $S(r_j)$ represents the number of relations connected to the relation r_j, relation r_i and r_j are connected if there exist triples $(e_x, r_j, e_y) \in O$ and $(e_y, r_i, e_z) \in O$. By sampling, we will obtain a relation sequence $s = [r_1, r_2, ..., r_n]$. We can close this path by adding a relation r^c that connects the head entity of r_1 with the tail entity of r_n. Please note that this step can be achieved with a time complexity of $O(n)$ and our method does not require searching the entire graph to obtain paths with rich pattern information.

Path Composition Module. Our method aims to generate the most probable rule heads for rule bodies and provide their confidence scores. We employ GRU to process the relation sequence s of length n, as shown in Fig. 2, the sliding window size for the GRU is set to 2, and h_i represents the combined representation of the two relations within the sliding window. After the sliding process, we obtain $n-1$ hidden states, which are then fed into a fully connected layer (fc). We select the h_i with the highest probability as the query for the attention unit:

$$Q = max([fc(h_1), fc(h_2), \ldots, fc(h_{i-1})]) \tag{4}$$

We concatenate each relation together to form a matrix $R \in \mathbb{R}^{|N| \times d}$ as the content. Here, $|N|$ represents the number of relations, and d represents the dimension of relation embeddings. We perform a dot product between the query and key, divided by the square root of the embedding dimension d, to scale the attention scores. Then, we multiply the result by the value to obtain the attention for each relation. By using attention as the weight for each relation, we can obtain a new representation through the composition of the two relations:

$$\mathbf{r_i} \wedge \mathbf{r_{i+1}} = softmax\left(\frac{QK^T}{\sqrt{d}}\right)V \tag{5}$$

where $\mathbf{r_i} \wedge \mathbf{r_{i+1}}$ represents the vector representation of the composition of two relations, r_i and r_{i+1}. We add the representation of $\mathbf{r_i} \wedge \mathbf{r_{i+1}}$ to the relation sequence s and repeat the aforementioned combination process.

Note that in previous methods, standard confidence (SC) and head coverage (HC) are often calculated to measure the quality of rules. We have avoided the need for a separate rule quality evaluation process. Instead, we take the attention scores obtained from the final combination as the probabilities for each relation to be the rule head. The relation with the highest probability is selected as the rule head, and the probability serves as the confidence score for the rule. This allows us to achieve an end-to-end evaluation of the rule's quality:

$$P\left(r_h = r_i | r_b = [r_1, r_2, \cdots, r_n]\right) = \max\left[softmax\left(\frac{QK^T}{\sqrt{d}}\right)\right] \tag{6}$$

Optimization. Our loss function consists of two parts: (1) the cross-entropy loss between the predicted result and the actual closed rule body path relation r^c, and (2) the entropy loss of the attention scores at each step:

$$L = \sum_{i=1}^{|N|} \mathbf{r}_i^c \log \boldsymbol{\theta}_i + \sum_{j=1}^{|n-2|} \boldsymbol{\theta}_j \log \boldsymbol{\theta}_j \tag{7}$$

where $|N|$ is the number of relations, $\mathbf{r}^c$ represents the embedding of the relation that closes the path. For example, given a path $s = [r_1, r_2, ..., r_n]$, if there exists r^c that connects the head entity and the tail entity of this path, we say that r^c closes this path. $\boldsymbol{\theta}_i$ represents the attention scores from the last round of output, which are the predicted values of the model. The first part of the loss function calculates the difference between the predicted values of the model and the true labels. In the second part, $\boldsymbol{\theta}_j$ represents the attention scores obtained at each step. By calculating the entropy of $\boldsymbol{\theta}_j$, we measure the degree of dispersion in the distribution of attention scores. n represents the length of the rule body sequence $[r_1, r_2, \cdots, r_n]$. Since the relations are combined pairwise, there will be n-1 steps in the intermediate combination process. The attention score from the last step will be directly used as the predicted value. Therefore, n-2 entropy losses will be calculated.

4 Experiment

In this section, we conducted experiments on benchmark datasets to demonstrate the effectiveness of our method and explored the impact of using rules of different lengths on the experimental results. We first provided the experimental settings and then presented a comparison of the experimental results between OIRL and other baselines.

4.1 Experiment Settings

Datasets. In our experiments, we evaluated our method using two widely used benchmark datasets: FB15K-237 [23] and WN18RR [6]. Both of these datasets are large-scale knowledge graphs. FB15K-237 is a dataset built upon Facebook's Freebase knowledge base. It is an improved version of the original FB15K dataset. The dataset consists of various relationship types, such as birthplace, nationality. WN18RR is a dataset constructed based on WordNet, which includes semantic relationships such as hypernyms and hyponyms. FB15K-237 includes explicit type information, which is extracted from the original dataset and not external knowledge. WN18RR does not have explicit type information added.

Evaluation Metrics. We use rules to complete the KG completion task, and for the learned rules, we can use forward chaining [21] to deduce missing facts. For each test triple, we will construct two queries $(h, r, ?)$ and $(t, r^{-1}, ?)$ to predict the missing entity. We use three evaluation metrics: Mean Reciprocal Rank (MRR), Hit@1, and Hit@10, which are commonly used in most existing studies.

Compared Algorithms. In our experiments, we compared the following algorithms: (1) Embedding-based methods: This category includes TransE [1], DistMult [29], ConvE [6], and ComplEx [24]. (2) Rule learning-based methods: This category includes Neural-LP [30], DRUM [20], RNNLogic [19], RLogic [3], and NCRL [2]. By incorporating embedding methods for comparison, we aim to demonstrate that despite rule mining methods not being specifically designed for KG completion tasks, our approach can achieve comparable results to embedding methods.

4.2 Results

Comparison Against Existing Methods. Table 1 shows the experimental results of our method on two large-scale datasets, FB15K-237 and WN18RR. Our method achieves state-of-the-art (SOTA) performance when compared to existing rule learning approaches. This demonstrates the effectiveness of the rules mined by OIRL. Existing methods fail to handle relations differently based on the varying amount of information contained in TBox and ABox, which leads to an inability to fully exploit the pattern information in the entire graph. Experimental results indicate the effectiveness of introducing ontology and instance guidance. Compared to embedding-based reasoning, OIRL exhibits inductive properties. Inductive properties refer to the ability to extract features and patterns from training data and apply the acquired knowledge to unknown data.

Table 1. Results of reasoning on FB15k-237 and WN18RR. H@k is in %. The bold numbers represent the best performances among all methods.

Methods	Models	FB15K-237			WN18RR		
		MRR	Hits@1	Hits@10	MRR	Hits@1	Hits@10
KGE	TransE	0.29	18.9	46.5	0.23	2.2	52.4
	DistMult	0.22	13.6	38.8	0.42	38.2	50.7
	ConvE	0.32	21.6	50.1	0.43	40.1	52.5
	ComplEx	0.24	15.8	42.8	0.44	41.0	51.2
Rule Learning	Neural-LP	0.24	17.3	36.2	0.38	36.8	40.8
	DRUM	0.23	17.4	36.4	0.38	36.9	41.0
	RNNLogic	0.29	20.8	44.5	0.46	41.4	53.1
	NCRL	0.30	20.9	47.3	0.67	56.3	85.0
	OIRL	**0.32**	**21.9**	**49.7**	**0.68**	**58.0**	**85.2**

Logic rules, being independent of entities, naturally possess inherent inductive capabilities and can generalize to unseen entities. Logic rules also allow for tracing the steps of reasoning and the facts upon which the reasoning process relies, providing interpretability.

Table 2. Performance w.r.t. Length of the rule. H@k is in %.

Length of the rule	FB15K-237			WN18RR		
	MRR	Hits@1	Hits@10	MRR	Hits@1	Hits@10
OIRL(Length of 2)	0.30	20.7	49.1	0.66	56.3	85.1
OIRL(Length of 3)	0.27	17.3	46.6	0.64	51.1	84.7
OIRL(Lengths of 2 and 3)	0.32	21.9	49.7	0.68	58.0	85.2

Performance W.r.t. Length of the Rule. According to our research findings, we discovered a correlation between rule length and experimental performance. Specifically, we investigated rules of length 2 and 3, comparing their impact on the experimental results. As shown in Table 2, we observed that rules of length 2 made the most significant contribution to performance. This indicates that shorter rules can effectively capture pattern information within the dataset. However, in comparison to exclusively using rules of length 2, we also found that employing rules of multiple lengths further enhances the system's performance.

Case Study of Logic Rules. Table 3 shows the Horn rules mined by OIRL in FB15k-237. It can be observed that logical rules unlike neural network methods,

Table 3. Case study of the rules generated by OIRL.

Confidence	Horn Rule
0.964	/people/person/nationality ← /base/biblioness/bibs_location/country ∧ /people/person/places_lived./people/place_lived/location
0.942	/people/person/languages ← /film/film/language ∧ /film/actor/film./film/performance/film
0.933	/film/film/language ← /location/country/official_language ∧ /film/film/country
0.901	/location/location/contains ← /location/hud_county_place/place ∧ /location/country/second_level_divisions

possess interpretability and provide transparent and understandable audit trails for prediction results. Furthermore, in comparison to KGE methods, logical rules are not specific to particular entities.

5 Conclusions

This paper investigates learning logical rules for knowledge graph reasoning. We propose a method called OIRL, which enhances rule learning on knowledge graphs through Joint Ontology and Instance Guidance. OIRL treats rules as a sequential model and uses information in both the TBox and ABox to guide path sampling. Different weights are assigned to different relations to incorporate more pattern information into the paths. By employing an end-to-end rule generator, we eliminate the need for a separate rule quality evaluation stage. Experimental results demonstrate that OIRL outperforms existing rule learning methods.

References

1. Bordes, A., Usunier, N., Garcia-Duran, A., Weston, J., Yakhnenko, O.: Translating embeddings for modeling multi-relational data. In: Advances in Neural Information Processing Systems, pp. 2787–2795 (2013)
2. Cheng, K., Ahmed, N.K., Sun, Y.: Neural compositional rule learning for knowledge graph reasoning. arXiv preprint arXiv:2303.03581 (2023)
3. Cheng, K., Liu, J., Wang, W., Sun, Y.: RLogic: recursive logical rule learning from knowledge graphs. In: SIGKDD, pp. 179–189 (2022)
4. Cheng, K., Yang, Z., Zhang, M., Sun, Y.: UniKER: a unified framework for combining embedding and definite horn rule reasoning for knowledge graph inference. In: EMNLP, pp. 9753–9771 (2021)
5. Cohen, W.W.: TensorLog: a differentiable deductive database. arXiv preprint arXiv:1605.06523 (2016)

6. Dettmers, T., Minervini, P., Stenetorp, P., Riedel, S.: Convolutional 2D knowledge graph embeddings. In: AAAI, pp. 1811–1818 (2018)
7. Galárraga, L., Teflioudi, C., Hose, K., Suchanek, F.M.: Fast rule mining in ontological knowledge bases with AMIE+. VLDB J. **24**(6), 707–730 (2015)
8. Galárraga, L.A., Teflioudi, C., Hose, K., Suchanek, F.: AMIE: association rule mining under incomplete evidence in ontological knowledge bases. In: WWW, pp. 413–422 (2013)
9. Guo, S., Wang, Q., Wang, L., Wang, B., Guo, L.: Jointly embedding knowledge graphs and logical rules. In: EMNLP, pp. 192–202 (2016)
10. Guo, S., Wang, Q., Wang, L., Wang, B., Guo, L.: Knowledge graph embedding with iterative guidance from soft rules. In: AAAI, pp. 4816–4823 (2018)
11. Ji, G., He, S., Xu, L., Liu, K., Zhao, J.: Knowledge graph embedding via dynamic mapping matrix. In: IJCNLP, pp. 687–696 (2015)
12. Lin, Y., Liu, Z., Sun, M., Liu, Y., Zhu, X.: Learning entity and relation embeddings for knowledge graph completion. In: AAAI, pp. 2181–2187 (2015)
13. Meilicke, C., Chekol, M.W., Ruffinelli, D., Stuckenschmidt, H.: Anytime bottom-up rule learning for knowledge graph completion. In: IJCAI, pp. 3137–3143 (2019)
14. Muggleton, S.: Inductive Logic Programming, No. 38: Morgan Kaufmann (1992)
15. Nickel, M., Tresp, V., Kriegel, H.P., et al.: A three-way model for collective learning on multi-relational data. In: ICML, pp. 809–816, No. 10.5555 (2011)
16. Niu, G., Zhang, Y., Li, B., Cui, P., Liu, S., Li, J., Zhang, X.: Rule-guided compositional representation learning on knowledge graphs. In: AAAI, pp. 2950–2958, No. 03 (2020)
17. Omran, P.G., Wang, K., Wang, Z.: Scalable rule learning via learning representation. In: IJCAI, pp. 2149–2155 (2018)
18. Qian, W., Fu, C., Zhu, Y., Cai, D., He, X.: Translating embeddings for knowledge graph completion with relation attention mechanism. In: IJCAI, pp. 4286–4292 (2018)
19. Qu, M., Chen, J., Xhonneux, L.P., Bengio, Y., Tang, J.: RNNLogic: learning logic rules for reasoning on knowledge graphs. arXiv preprint arXiv:2010.04029 (2020)
20. Sadeghian, A., Armandpour, M., Ding, P., Wang, D.Z.: DRUM: end-to-end differentiable rule mining on knowledge graphs. In: NeurIPS, pp. 15347–15357 (2019)
21. Salvat, E., Mugnier, M.-L.: Sound and complete forward and backward chainings of graph rules. In: Eklund, P.W., Ellis, G., Mann, G. (eds.) ICCS-ConceptStruct 1996. LNCS, vol. 1115, pp. 248–262. Springer, Heidelberg (1996). https://doi.org/10.1007/3-540-61534-2_16
22. Sun, Z., Deng, Z.H., Nie, J.Y., Tang, J.: RotatE: knowledge graph embedding by relational rotation in complex space. arXiv preprint arXiv:1902.10197 (2019)
23. Toutanova, K., Chen, D.: Observed versus latent features for knowledge base and text inference. In: Proceedings of the 3rd workshop on continuous vector space models and their compositionality, pp. 57–66 (2015)
24. Trouillon, T., Welbl, J., Riedel, S., Gaussier, É., Bouchard, G.: Complex embeddings for simple link prediction. In: International Conference on Machine Learning, pp. 2071–2080. PMLR (2016)
25. Wang, Z., Zhang, J., Feng, J., Chen, Z.: Knowledge graph embedding by translating on hyperplanes. In: AAAI, pp. 1112–1119 (2014)
26. Wong, C.M., et al.: Improving conversational recommender system by pretraining billion-scale knowledge graph. In: ICDE, pp. 2607–2612. IEEE (2021)
27. Wu, H., Wang, Z., Wang, K., Shen, Y.D.: Learning typed rules over knowledge graphs. In: KR, pp. 494–503, No. 1 (2022)

28. Xiao, H., Huang, M., Hao, Y., Zhu, X.: TransG: a generative mixture model for knowledge graph embedding. arXiv preprint arXiv:1509.05488 (2015)
29. Yang, B., Yih, W.t., He, X., Gao, J., Deng, L.: Embedding entities and relations for learning and inference in knowledge bases. arXiv preprint arXiv:1412.6575 (2014)
30. Yang, F., Yang, Z., Cohen, W.W.: Differentiable learning of logical rules for knowledge base reasoning. In: Advances in Neural Information Processing Systems, vol. 30 (2017)
31. Yasunaga, M., Ren, H., Bosselut, A., Liang, P., Leskovec, J.: QA-GNN: reasoning with language models and knowledge graphs for question answering. In: NAACL, pp. 535–546 (2021)
32. Zhang, W., et al.: Iteratively learning embeddings and rules for knowledge graph reasoning. In: WWW, pp. 2366–2377 (2019)

Explore Across-Dimensional Feature Correlations for Few-Shot Learning

Tianyuan Li and Wenming Cao(✉)

Shenzhen University, College of Electronics and Information Engineering, Nanhai, Shenzhen 518060, Guangdong, China
litianyuan2021@email.szu.edu.cn, wmcao@szu.edu.cn

Abstract. Few-shot learning (FSL) aims to learn new concepts with only few examples, which is a challenging problem in deep learning. Attention-oriented methods have shown great potential in addressing the FSL problem. However, many of these methods tend to separately focus on different dimensions of the samples, like channel dimension or spatial dimension only, which may lead to limitations in extracting discriminative features. To address this problem, we propose an across-dimensional attention network (ADANet) to explore the feature correlations across channel and spatial dimensions of input samples. The ADANet can capture across-dimensional feature dependencies and produce reliable representations for the similarity metric. In addition, we also design a three-dimensional offset position encoding (TOPE) method that embeds the 3D position information into the across-dimensional attention, enhancing the robustness and generalization ability of the model. Extensive experiments on standard FSL benchmarks have demonstrated that our method can achieve better performances compared to the state-of-the-art approaches.

Keywords: Few-shot learning · Across-dimension · Position encoding · Attention mechanism

1 Introduction

In recent years, deep learning has achieved remarkable accomplishments in a wide range of tasks, such as image classification [2,16], object detection [12,14] and semantic segmentation [7,18]. Many of these achievements can be attributed to the ability of deep learning models to learn complex patterns and representations from large amounts of training data. However, despite the impressive performance, deep learning models encounter significant limitations when faced with situations where the availability of labeled training data is scarce. And this is where few-shot learning (FSL) comes into play, enabling models to learn new concepts or tasks with only few labeled examples [10].

The goal of FSL is to learn an efficient and flexible model that can adapt to new tasks or concepts quickly, without requiring a large amount of labeled data.

Q. Liu et al. (Eds.): PRCV 2023, LNCS 14427, pp. 151–163, 2024.
https://doi.org/10.1007/978-981-99-8435-0_12

To better solve FSL, attention mechanisms have been widely used to identify the most relevant features of the few labeled samples and then utilize them to classify new samples. For example, CAN [6] produces the cross-attention map for each pair of class feature and query sample feature. SaberNet [11] introduces vision self-attention to generate and purify features of samples.

However, many attention-based FSL methods often suffer from the limitation of analyzing single-dimensional attention(like channel or spatial attention) separately or combining them in a simple sequential manner, which can lead to limited capacity to capture complex relationships between spatial and channel dimensions. To address this problem, we propose an across-dimensional attention network (ADANet) that views channel and spatial dimensions as an integral unit and focuses on informative blocks along two dimensions. By jointly considering the features from spatial and channel dimensions, our model can better capture the complicated correlations hidden in the feature maps and generate more reliable representations. Besides, according to the characteristics of across-dimensional attention, we design a three-dimensional offset position encoding (TOPE) approach which allows our model to better utilize the position information of images and improve the generalization ability for FSL tasks. We evaluate our proposed method on standard FSL datasets and achieve high performance. Additionally, we conduct a series of ablation studies and the results demonstrate the effectiveness of our ADANet and TOPE method. The main contributions can be summarized as follows:

- We pay attention to the feature dependencies between spatial and channel dimensions, and propose an across-dimensional attention network to capture the complex feature correlations.
- We design a three-dimensional offset position encoding method for our attention network which can incorporate 3D position information into the across-dimensional attention.
- Experiments on standard FSL benchmarks show our method achieves excellent performance, and ablation studies validate the effectiveness of each component.

2 Related Work

2.1 Few-Shot Learning

Few-shot learning (FSL) aims to learn a model that can recognize new classes with only a few labeled examples [3,22]. Several approaches have been proposed for few-shot learning: **Metric-based methods**, which aim to learn a similarity metric between examples and utilize nearest neighbor-based classification. For example, Prototypical Network [17] learns a prototype representation for each class by taking the mean of the class examples. During testing, the closest prototype to a query example is selected as the predicted class. DeepEMD [21] chooses Earth Mover's Distance (EMD) as a metric to compute the distance between image representations. ReNAP [9] adaptively learns class prototypes and produces more accurate representations for the learned distance metric. **Optimization-based methods**, which aim to learn good initialization of models that can be quickly adapted to new classes with only few samples. For example, MAML [4] learns to optimize for fast adaptation during training and is fine-tuned to the new task with few samples during testing. Reptile [15] works by iteratively updating the model's parameters towards the average gradient direction of tasks in the support set. **Generative-based methods**, which aim to learn a generative model that can produce new samples from few labeled examples. Variational Autoencoder (VAE) and Generative Adversarial Network (GAN) are common-used generative models.

2.2 Attention Mechanisms in Few-Shot Learning

In recent years, attention mechanisms have emerged as a promising approach for improving the performance of FSL algorithms. Hou et al. [6] develop a cross-attention module to generate cross-attention maps between support and query samples. The attention maps help highlight the important regions in images and make the extracted features more discriminative. SEGA [19] utilizes both visual and semantic prior knowledge to guide attention towards discriminative features of a new category, leading to more discriminative embeddings of the novel class even with few samples.

In this work, we explore the feature correlations across different dimensions and propose an across-dimensional attention network (ADANet). Given a feature map, our model extracts local blocks around each pixel across spatial and channel dimensions, and then calculates the across-dimensional attention exerted on every pixel. By cooperating with the three-dimensional offset position encoding (TOPE) method, our model is also able to leverage the position information of pixels. As a result, the feature correlations hidden across dimensions can be fully explored.

3 Methodology

3.1 Preliminary

For few-shot classification, datasets are typically divided into three parts: a train set D_{train} from classes C_{train}, a validation set D_{val} from classes C_{val} and a test set D_{test} from C_{test}, where $C_{train} \cap C_{val} \cap C_{test} = \emptyset$. All three parts consist of multiple episodes, each of which contains a support set $S = \left\{(x_s^l, y_s^l)\right\}_{l=1}^{N\times K}$ with only K image-label pairs for each N classes and a query set $Q = \left\{(x_q, y_q)\right\}_{q=1}^{N\times M}$ with M instances for each N classes. A single episode is called a N-way K-shot task. In the training stage, the FSL model is optimized using randomly sampled tasks from D_{train}. Then the D_{val} is used to select a model with the best performance during training. In the test stage, the selected model utilizes the learned knowledge from D_{train} and the support set S in an episode sampled from D_{test} to classify x_q as one of N classes in the query set Q from the same episode.

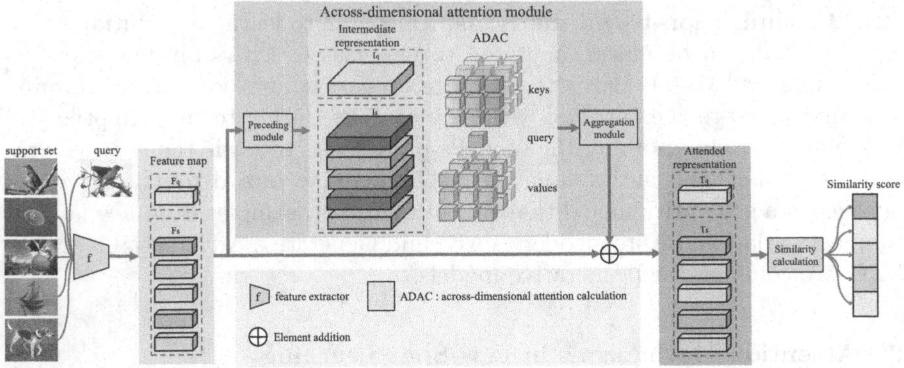

Fig. 1. The overall framework of ADANet. The feature extractor transforms query and support images to base feature maps, F_s and F_q. The preceding module expands the dimensions of feature maps and produces the intermediate representations, I_s and I_q. The across-dimensional attention is computed in intermediate representations and then aggregated with the aggregation module. By combining with base feature maps, the attended representations T_s and T_q are generated for similarity calculation.

3.2 Overall Framework

Figure 1 illustrates the overall architecture of ADANet, mainly consisting of two parts: a feature extractor and an across-dimensional attention (ADA) module. Given a support set and a query image as input, the backbone feature extractor produces feature maps, F_s and $F_q \in \mathbb{R}^{C\times H\times W}$, where C denotes the channel size, $H \times W$ denotes the spatial size. The ADA module first performs the unfold operation along the spatial and channel dimension, transforming the feature maps to intermediate representations I_s and $I_q \in \mathbb{R}^{C\times H\times W\times V}$, where $V = v_c \times v_h \times v_w$ and denotes the size of the across-dimensional block. Then the ADA module calculates the across-dimensional attention using intermediate representation and incorporates the attention with the base feature maps to produce attended representations, T_s and $T_q \in \mathbb{R}^{C\times H\times W}$. We then utilize the cosine similarity to compute the distances between the query and the support set. The query is classified as the class of its nearest support representation.

3.3 Three-Dimensional Offset Position Encoding (TOPE)

Position encoding allows neural networks to process spatial information in a way that is invariant to translation and rotation. The basic idea behind position encoding is to map the position information to a fixed-dimensional vector that captures the relevant patterns and relationships in the samples. There are two popular methods for position encoding: sinusoidal embeddings based on the absolute position and relative positional embeddings. Early experiments suggested that the latter usually has better performance. So following the idea of relative positional embeddings, we design a three-dimensional offset position encoding (TOPE) method using the learned embeddings. Given a pixel

$p \in \mathbb{R}^{1\times1\times1}$ in position (d, i, j) and a block $B_v(d, i, j)$ centered around p, where d denotes the channel position and (i, j) denotes the spatial position, each position $(z, x, y) \in B_v(d, i, j)$ receives three relative distances: a channel offset $z-d$, a spatial row offset $x-i$ and a spatial column offset $y-j$. The three offsets are correlated with embeddings r_{z-d}, r_{x-i} and r_{y-j} respectively. By concatenating the three learned embeddings, we can get the 3D offset embedding $e_{(z,x,y)}$ in position (z, x, y). Figure 2 shows an example of TOPE with block size $V = 3 \times 3 \times 3$.

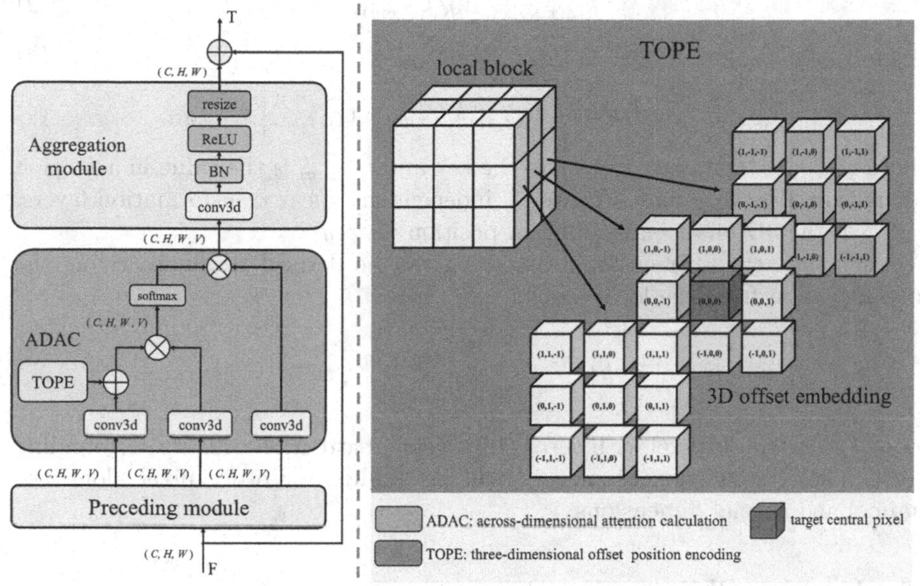

Fig. 2. left: Pipeline of across-dimensional attention module; **right**: an example of three-dimensional offset position encoding with local block size $V = 3 \times 3 \times 3$.

3.4 Across-Dimensional Attention (ADA)

The ADA module takes the base feature map F and transforms it to focus the relevant regions across channel and spatial dimensions in an image, preparing an informative and reliable attended representation for the metric. Figure 2 illustrates the pipeline of the ADA module. The ADA module contains a preceding module, an across-dimensional attention calculation (ADAC) module and an aggregation module. The preceding module expands the dimensions of input feature maps $F \in \mathbb{R}^{C\times H\times W}$ and produces the intermediate representations $I \in \mathbb{R}^{C\times H\times W\times V}$ by extracting the local block $B \in \mathbb{R}^{V}$ around every pixel $p \in \mathbb{R}^{1\times1\times1}$, where $V = v_c \times v_h \times v_w$ denotes the channel extent and $v_h \times v_w$ denotes the spatial extent. Then the positional encoding module inserts the 3D offset embeddings into every block and the ADAC module computes the across-dimensional attention exerted on every pixel. After that, the aggregation module integrates the feature correlations across different dimensions and obtains the attended representations $T \in \mathbb{R}^{C\times H\times W}$ by combining with the base feature maps. The detailed calculation process is as follows.

Given a pixel $f_{(d,i,j)} \in \mathbb{R}^{1\times1\times1}$ from the feature map and its local block of pixels in positions $(z, x, y) \in B_v(d, i, j)$, the attention between the central pixel and the pixel from the local block can be computed following the below process:

$$q_{(d,i,j)} = W_Q f_{(d,i,j)}, \tag{1}$$

$$k_{(z,x,y)} = W_K f_{(z,x,y)}, \tag{2}$$

$$v_{(z,x,y)} = W_V f_{(z,x,y)}, \tag{3}$$

$$a_{(z,x,y)} = softmax\left[q_{(d,i,j)}^{\top}(k_{(z,x,y)} + e_{(z,x,y)})\right] v_{(z,x,y)}, \tag{4}$$

where $q_{(d,i,j)}$ is the query, $k_{(z,x,y)}$ is the key and $v_{(z,x,y)}$ is the value in attention calculation. W_Q, W_K and W_V are all independent linear transformation layers. $e_{(z,x,y)}$ is the 3D offset embedding in position (z, x, y).

Then the across-dimensional attention can be derived by summarizing the pixel attention from the local block:

$$m_{(d,i,j)} = \sum_{(z,x,y)\in B_v(d,i,j)} a_{(z,x,y)}, \tag{5}$$

$m_{(d,i,j)}$ is one single pixel with across-dimensional attention from its surrounding block. The computation of $m_{(d,i,j)}$ will be applied for every pixel along the channel and spatial dimensions.

3.5 Learning Object

We jointly trains the proposed model by combining two losses: the linear loss L_{linear} and the task loss L_{task}. L_{linear} is calculated with a fully-connected classification layer:

$$L_{linear} = -\log \frac{\exp\left(w_l^{\mathrm{T}} GAP\left(T_q\right) + b_l\right)}{\sum_{l'=1}^{C_{train}} \exp\left(w_{l'}^{\mathrm{T}} GAP\left(T_q\right) + b_{l'}\right)}, \tag{6}$$

where T_q is the attended query representation. $\left[w_1^{\top}, w_2^{\top}, \cdots, w_{C_{train}}^{\top}\right]$ are weights in the fully-connected layer, $\left[b_1, b_2, \cdots, b_{C_{train}}\right]$ are biases in the layer. GAP is the global average pooling operation. L_{task} is the loss function for the N-way K-shot task and is calculated by cosine similarity between support prototype and query attended representations. We first average the labeled support attended representations of every category to produce support prototype $\overline{T_s}$:

$$\overline{T_s} = \sum_{l=1}^{K} T_s^l, \tag{7}$$

where K is the number of support samples from every class. Then the L_{task} is computed as:

$$L_{task} = -\log \frac{\exp\left(cos\left(GAP\left(\overline{T_s}\right)^{(n)}, GAP\left(T_q\right)^{(n)}\right)\right)}{\sum_{n'=1}^{N} \exp\left(cos\left(GAP\left(\overline{T_s}\right)^{(n')}, GAP\left(T_q\right)^{(n')}\right)\right)}, \tag{8}$$

where cos is the cosine similarity calculation. N is the number of classes from a N-way K-shot task. n is the nth class from N.

The overall classification loss is defined as:

$$L = \alpha L_{linear} + \beta L_{task}, \tag{9}$$

where hyper-parameters α and β are used to balance the effect of different loss terms. The network can be trained by optimizing L with gradient descent algorithm. The detailed training workflow is shown in Algorithm 1.

Algorithm 1: The workflow of our proposed ADANet.

Input: Train set D_{train}, initial network parameters Φ, number of epochs m_{epoch}
Output: The learned parameters

Randomly initialize Φ.
while Φ has not converged **do**
  for $i = 1, 2, \cdots, m_{epoch}$ **do**
    Sample task $\{S, Q\}$ from D_{train}
    for each task t, $t = 1, 2, \cdots, n$ **do**
      Obtain F_s and F_q from the feature extractor ;
      Get T_s and T_q from the ADA module ;
      Compute the linear loss L_{linear} with Eq. 6 ;
      Compute the task loss L_{task} with Eq. 8 ;
      Calculate the overall classification loss L with Eq. 9 ;
      Calculate the gradients $\nabla_{\Phi} L$ and adopt SGD algorithm to update Φ ;
    end for
  end for
end while

4 Experiments

In this section, we evaluate ADANet on standard FSL benchmarks and compare the results with the recent state-of-the-art methods. We also conduct ablation studies to validate the effect of the major components.

4.1 Experiment Setup

We use standard FSL benchmarks: miniImageNet, tieredImageNet, CUB-200-2011 (CUB) and CIFAR-FS to conduct the extensive experiments. We apply ResNet-12 as our backbone following the recent FSL methods. The spatial size of input images is 84×84. We use stochastic gradient descent (SGD) with weight decay of 0.0005 and momentum of 0.9 to optimize our model. We set 90 training epochs with initial learning rate of 0.1 and use a decay factor of 0.05 at the 60th, 70th, and 80th epochs. During the test, we randomly sample 10000 test episodes with 15 query images for a class in each episode. We calculate and record the average accuracy with 95% confidence intervals over the 10000 episodes. All experiments are conducted with PyTorch on one NVIDIA 3090Ti GPU.

Table 1. Results on miniImageNet and tieredImageNet datasets.

method	backbone	miniImageNet		tieredImageNet	
		5-way 1-shot	5-way 5-shot	5-way 1-shot	5-way 5-shot
PPA	WRN-28-10	59.60 ± 0.41	73.74 ± 0.19	65.65 ± 0.92	83.40 ± 0.65
MTL	ResNet-12	61.20 ± 1.80	75.50 ± 0.80	–	–
LEO	WRN-28-10	61.76 ± 0.08	77.59 ± 0.12	66.33 ± 0.05	81.44 ± 0.09
ProtoNet [17]	ResNet-12	62.39 ± 0.21	80.53 ± 0.14	68.23 ± 0.23	84.03 ± 0.16
MetaOptNet	ResNet-12	62.64 ± 0.82	78.63 ± 0.46	65.99 ± 0.72	81.56 ± 0.53
SimpleShot	ResNet-18	62.85 ± 0.20	80.02 ± 0.14	–	–
S2M2 [13]	ResNet-34	63.74 ± 0.18	79.45 ± 0.12	–	–
CAN [6]	ResNet-12	63.85 ± 0.81	81.57 ± 0.56	69.89 ± 0.51	84.23 ± 0.37
CTM	ResNet-18	64.12 ± 0.82	80.51 ± 0.13	68.41 ± 0.39	84.28 ± 1.73
DeepEMD [21]	ResNet-12	65.91 ± 0.82	82.41 ± 0.56	71.16 ± 0.87	86.03 ± 0.58
RENet [8]	ResNet-12	67.60 ± 0.44	82.58 ± 0.30	71.61 ± 0.51	85.28 ± 0.35
TPMM	ResNet-12	67.64 ± 0.63	83.44 ± 0.43	72.24 ± 0.70	86.55 ± 0.63
SUN	ViT-S	67.80 ± 0.45	83.25 ± 0.30	72.99 ± 0.50	86.74 ± 0.33
FewTRUE [5]	ViT-S	68.02 ± 0.88	84.51 ± 0.53	72.96 ± 0.92	86.43 ± 0.67
SetFeat [1]	ResNet-12	68.32 ± 0.62	82.71 ± 0.46	73.63 ± 0.88	**87.59 ± 0.57**
ADANet(ours)	ResNet-12	**69.56 ± 0.19**	**85.32 ± 0.13**	**74.15 ± 0.20**	87.46 ± 0.16

4.2 Comparison with State-of-the-art

To evaluate the effectiveness of our approach, we conduct comparison experiments with state-of-the-art methods on four FSL datasets. The experimental results are shown in Tables 1 and 2. Under the 5-way 1-shot task setting, our ADANet achieves better performances compared with previous methods on all datasets. Under the 5-way 5-shot setting, our model improves accuracy by 2.61% on miniImageNet, by 0.79% on CUB-200-2011 and by 1.8% on CIFAR-FS, and obtains competitive results with SetFeat on tieredImageNet. Note that we use a smaller backbone compared with some methods (WRN-28-10, ResNet-34, etc.) yet we get an obvious increase in accuracy under both two task settings.

Table 2. Results on CUB and CIFAR-FS datasets.

method	backbone	CUB		CIFAR-FS	
		5-way 1-shot	5-way 5-shot	5-way 1-shot	5-way 5-shot
ProtoNet [17]	ResNet-12	66.09 ± 0.92	82.50 ± 0.58	72.20 ± 0.70	83.50 ± 0.50
RelationNet	ResNet-34	66.20 ± 0.99	82.30 ± 0.58	–	–
MAML [4]	ResNet-34	67.28 ± 1.08	83.47 ± 0.59	–	–
S2M2 [13]	ResNet-34	72.92 ± 0.83	86.55 ± 0.51	62.77 ± 0.23	75.75 ± 0.13
FEAT [20]	ResNet-12	73.27 ± 0.22	85.77 ± 0.14	–	–
RENet [8]	ResNet-12	79.49 ± 0.44	91.11 ± 0.24	74.51 ± 0.46	86.60 ± 0.32
FewTRUE [5]	ViT-S	–	–	76.10 ± 0.88	86.14 ±± 0.64
ADANet(ours)	ResNet-12	**81.82 ± 0.18**	**91.90 ± 0.10**	**76.54 ± 0.20**	**87.94 ± 0.15**

4.3 Ablation Studies

To investigate the effects of modules in ADANet, we conduct ablation and comparison experiments. The experiments are performed on CUB and CIFAR-FS datasets under the 5-way 1-shot setting. We apply the ResNet-12 backbone and cosine similarity metric as the baseline. Table 3 presents the experimental results. The single-dimensional attention (SE and stand-alone) can achieve a slight improvement over the baseline. CBAM tends to perform better than single-dimensional attention. But CBAM just arranges channel and spatial attention in a sequential way which can't capture the complex feature correlations between dimensions. In our work, the feature correlations across different dimensions can be fully explored by adding the ADA module, and the classification accuracy is improved by 2.87% on CUB and by 1.42% on CIFAR-FS. TOPE module is designed to encode the 3D position information into the across-dimensional attention. By incorporating TOPE with ADA, the accuracy can increase by 3.76% on CUB and by 2.57% on CIFAR-FS compared with the baseline.

Table 3. Ablation and comparison experiments on CUB and CIFAR-FS datasets.

method	attention type	position encode	CIFAR-FS	CUB
baseline	none	×	73.97	78.06
(others) SE	channel	×	73.91	78.81
(others) stand-alone	spatial	✓	74.56	78.62
(others) CBAM	channel + spatial	×	74.27	79.33
(ours) baseline+ADA	across-dimension	×	75.39	80.93
(ours) baseline+ADA+TOPE	across-dimension	✓	76.54	81.82

4.4 Convergence Analysis

We record the loss and accuracy during the training and validation stages. Figure 3(a) shows the loss curves of models on miniImageNet under the 5-way 1-shot setting. The orange curve denotes the baseline model and the blue one denotes our model. We adopt multiple tasks to validate the learned model, so only the L_{task} is recorded in validation. During training, for both two models, the loss value drops rapidly in the first ten epochs and then gradually flattens out. At the 60th epoch, with the adjustment of the learning rate, there is a brief and sharp decrease in the value. Then the loss curves remain flattened to the end. Note that, the loss of ADANet drops and flattens faster than the baseline which indicates our model has better optimization performance. During validation, the value of L_{task} only is obviously smaller than the training loss, but the validation loss curves have similar trends as the training phase. Figure 3(b) shows the accuracy trends of models on miniImageNet under the 5-way 1-shot setting. It's clear that ADANet outperforms the baseline in both training and validation.

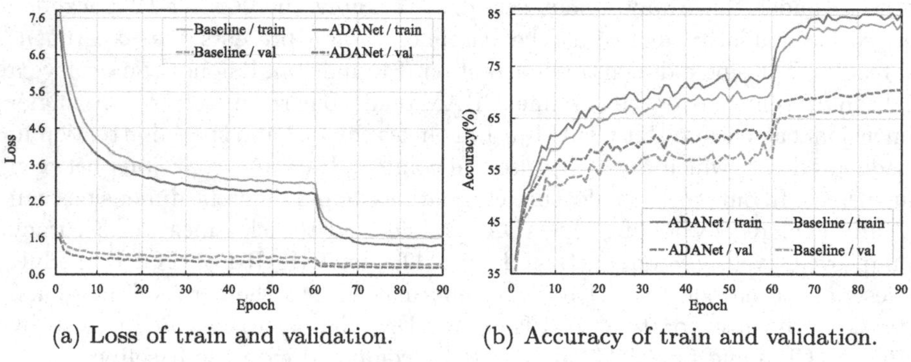

(a) Loss of train and validation. (b) Accuracy of train and validation.

Fig. 3. Loss and accuracy of the baseline model and ADANet on miniImageNet.

4.5 Visualization

To qualitatively analyze the effects of the proposed attention mechanism, we visualize the output of the feature extractor and the across-dimensional attention (ADA) module. We randomly select five images from different categories and another image belonging to one of these categories. These images are all from the miniImageNet dataset and compose a 5-way 1-shot task. Then we feed this task into a well-trained ADANet to produce the Grad-CAM visualization results. As shown in Fig. 4, the feature maps obtained from the feature extractor may contain non-target objects which can affect the accuracy of classification. After adopting the ADA module, the output attended representations significantly improve the focus on target objects and alleviate the distraction of non-target

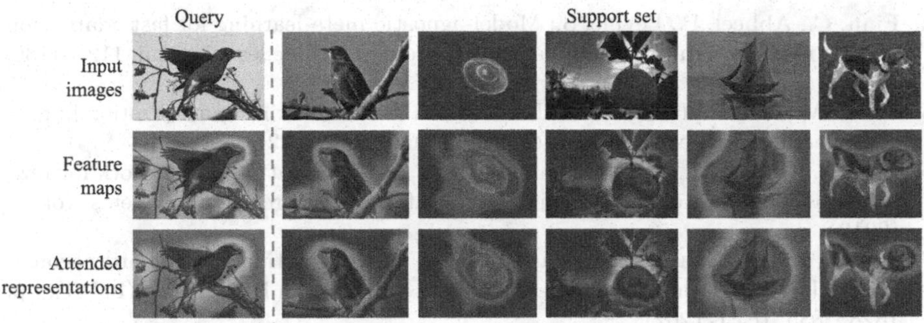

Fig. 4. Grad-CAM visualization of a 5-way 1-shot task on miniImageNet dataset.

objects. The visualization results illustrate that by capturing the feature correlations across different dimensions, ADANet can generate more discriminative representations for few-shot classification.

5 Conclusion

In this work, we propose an across-dimensional attention network (ADANet) to explore the feature correlations between different dimensions. The ADANet extracts local informative blocks and captures across-dimensional feature dependencies for the base feature maps. Additionally, we design a three-dimensional offset position encoding (TOPE) method to encode the 3D offset embedding into the across-dimensional attention, which ensures that the position information can be fully utilized. Extensive experiments show that our method outperforms most recent FSL approaches and both two modules in the ADANet make significant contributions. In future work, we plan to investigate the use of our model in other FSL tasks, such as few-shot object detection and semantic segmentation.

Acknowledgements. This work is supported by National Natural Science Foundation of China, No.61771322, No.61971290, Shenzhen foundation for basic research JCYJ20220531100814033, and the Shenzhen Stability Support General Project (Category A) 20200826104014001.

References

1. Afrasiyabi, A., Larochelle, H., Lalonde, J.F., Gagné, C.: Matching feature sets for few-shot image classification. In: Proceedings of the IEEE/CVF Conference on Computer Vision and Pattern Recognition, pp. 9014–9024 (2022)
2. Chang, D., et al.: The devil is in the channels: mutual-channel loss for fine-grained image classification. IEEE Trans. Image Process. **29**, 4683–4695 (2020)
3. Cheng, Y., et al.: Know you at one glance: a compact vector representation for low-shot learning. In: Proceedings of the IEEE International Conference on Computer Vision Workshops, pp. 1924–1932 (2017)

4. Finn, C., Abbeel, P., Levine, S.: Model-agnostic meta-learning for fast adaptation of deep networks. In: International conference on machine learning, pp. 1126–1135. PMLR (2017)
5. Hiller, M., Ma, R., Harandi, M., Drummond, T.: Rethinking generalization in few-shot classification. arXiv preprint arXiv:2206.07267 (2022)
6. Hou, R., Chang, H., Ma, B., Shan, S., Chen, X.: Cross attention network for few-shot classification. In: Advances in Neural Information Processing Systems, vol. 32 (2019)
7. Hu, H., Cui, J., Zha, H.: Boundary-aware graph convolution for semantic segmentation. In: 2020 25th International Conference on Pattern Recognition (ICPR), pp. 1828–1835. IEEE (2021)
8. Kang, D., Kwon, H., Min, J., Cho, M.: Relational embedding for few-shot classification. In: Proceedings of the IEEE/CVF International Conference on Computer Vision. pp. 8822–8833 (2021)
9. Li, X., et al.: ReNAP: relation network with adaptiveprototypical learning for few-shot classification. Neurocomputing **520**, 356–364 (2023)
10. Li, X., Yang, X., Ma, Z., Xue, J.H.: Deep metric learning for few-shot image classification: a review of recent developments. Pattern Recogn. **138**, 109381 (2023)
11. Li, Z., Hu, Z., Luo, W., Hu, X.: SaberNet: self-attention based effective relation network for few-shot learning. Pattern Recogn. **133**, 109024 (2023)
12. Liang, J., Chen, H., Du, K., Yan, Y., Wang, H.: Learning intra-inter semantic aggregation for video object detection. In: Proceedings of the 2nd ACM International Conference on Multimedia in Asia, pp. 1–7 (2021)
13. Mangla, P., Kumari, N., Sinha, A., Singh, M., Krishnamurthy, B., Balasubramanian, V.N.: Charting the right manifold: manifold mixup for few-shot learning. In: Proceedings of the IEEE/CVF Winter Conference on Applications of Computer Vision, pp. 2218–2227 (2020)
14. Mi, P., et al.: Active teacher for semi-supervised object detection. In: Proceedings of the IEEE/CVF Conference on Computer Vision and Pattern Recognition, pp. 14482–14491 (2022)
15. Nichol, A., Schulman, J.: Reptile: a scalable metalearning algorithm. arXiv preprint arXiv:1803.02999 2(3), 4 (2018)
16. Qian, X., Hang, R., Liu, Q.: ReX: an efficient approach to reducing memory cost in image classification. In: Proceedings of the AAAI Conference on Artificial Intelligence, vol. 36, pp. 2099–2107 (2022)
17. Snell, J., Swersky, K., Zemel, R.: Prototypical networks for few-shot learning. In: Advances in Neural Information Processing Systems, vol. 30 (2017)
18. Xu, M., Zhang, Z., Wei, F., Hu, H., Bai, X.: Side adapter network for open-vocabulary semantic segmentation. In: Proceedings of the IEEE/CVF Conference on Computer Vision and Pattern Recognition, pp. 2945–2954 (2023)
19. Yang, F., Wang, R., Chen, X.: SEGA: semantic guided attention on visual prototype for few-shot learning. In: Proceedings of the IEEE/CVF Winter Conference on Applications of Computer Vision, pp. 1056–1066 (2022)
20. Ye, H.J., Hu, H., Zhan, D.C., Sha, F.: Few-shot learning via embedding adaptation with set-to-set functions. In: Proceedings of the IEEE/CVF conference on computer vision and pattern recognition, pp. 8808–8817 (2020)

21. Zhang, C., Cai, Y., Lin, G., Shen, C.: DeepEMD: few-shot image classification with differentiable earth mover's distance and structured classifiers. In: Proceedings of the IEEE/CVF conference on computer vision and pattern recognition, pp. 12203–12213 (2020)
22. Zhao, F., Zhao, J., Yan, S., Feng, J.: Dynamic conditional networks for few-shot learning. In: Proceedings of the European Conference on Computer Vision (ECCV), pp. 19–35 (2018)

Pairwise-Emotion Data Distribution Smoothing for Emotion Recognition

Hexin Jiang[1], Xuefeng Liang[1,2(✉)], Wenxin Xu[2], and Ying Zhou[1]

[1] School of Artificial Intelligence, Xidian University, Xi'an, China
{hxjiang,yingzhou}@stu.xidian.edu.cn

[2] Guangzhou Institute of Technology, Xidian University, Guangzhou, China
xliang@xidian.edu.cn, wxxv@stu.xidian.edu.cn

Abstract. In speech emotion recognition tasks, models learn emotional representations from datasets. We find the data distribution in the IEMOCAP dataset is very imbalanced, which may harm models to learn a better representation. To address this issue, we propose a novel Pairwise-emotion Data Distribution Smoothing (PDDS) method. PDDS considers that the distribution of emotional data should be smooth in reality, then applies Gaussian smoothing to emotion-pairs for constructing a new training set with a smoother distribution. The required new data are complemented using the mixup augmentation. As PDDS is model and modality agnostic, it is evaluated with three state-of-the-art models on two benchmark datasets. The experimental results show that these models are improved by 0.2% $\sim$ 4.8% and 0.1% $\sim$ 5.9% in terms of weighted accuracy and unweighted accuracy. In addition, an ablation study demonstrates that the key advantage of PDDS is the reasonable data distribution rather than a simple data augmentation.

Keywords: Pairwise-emotion Data Distribution Smoothing · Gaussian Smoothing · Mixup Augmentation · Model-modality Agnostic

1 Introduction

Speech emotion recognition (SER) is of great significance to understanding human communication. SER techniques have been applied to many fields, such as video understanding [13], human-computer interaction [8], mobile services [16] and call centers [14]. Up to now, plenty of deep learning based SER methods have been proposed [20,21,27]. Recently, multimodal emotion recognition attracted more attention [3,9,17,23] because of its richer representation and better performance. Most of these studies assigned one-hot labels to utterances. In fact, experts often have inconsistent cognition of emotional data, thus, the one-hot label is obtained by majority voting. Figure 1(a) shows the emotional data distribution in the IEMOCAP dataset [6].

This work was supported in part by the Guangdong Provincial Key Research and Development Programme under Grant 2021B0101410002.

Q. Liu et al. (Eds.): PRCV 2023, LNCS 14427, pp. 164–175, 2024.
https://doi.org/10.1007/978-981-99-8435-0_13

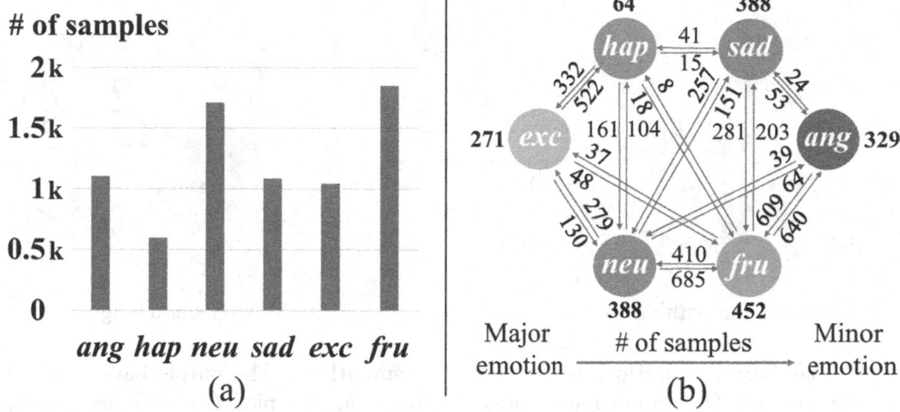

Fig. 1. (a) Data distribution in the IEMOCAP when the label is one-hot (ang, angry; hap, happiness; neu, neutral state; sad, sadness; exc, excited; fru, frustration). (b) The statistics of clear and ambiguous samples in the IEMOCAP. Each vertex represents an emotion category with only clear samples. The bold number is the quantity of clear samples. Each directed edge denotes a set of ambiguous samples, whose tail is the major emotion with most votes and the head is the minor emotion with fewer votes. The quantity of ambiguous samples is on the edge.

Later, some studies argued that the one-hot label might not represent emotions well. They either addressed this problem through multi-task learning [19], soft-label [1,11], multi-label [2] and polarity label [7], or enhanced the model's learning ability of ambiguous emotions by dynamic label correction [12], progressive co-teaching [24] and label reconstruction through interactive learning [26]. These works usually defined the samples with consistent experts' votes as clear samples and those with inconsistent votes as ambiguous samples. The statistics of clear and ambiguous samples in the IEMOCAP dataset are shown in Fig. 1(b). After statistical analysis, we find that the data distribution is imbalanced, especially for ambiguous samples. For example, the quantity of ambiguous samples between *anger* and *frustration* is abundant, while that between *anger* and *sadness* is very few. Meanwhile, the quantity of clear samples of *happiness* counts much less than other clear emotion categories. We consider such distribution is unreasonable that may prevent the model from learning a better emotional representation. The possible reason is that the votes come from few experts, which is a rather sparse sampling of human population.

We think that clear and ambiguous emotional data shall follow a smooth statistical distribution in the real world. Based on this assumption, we propose the Pairwise-emotion Data Distribution Smoothing (PDDS) to address the problem of unreasonable data distribution. PDDS applies Gaussian smoothing on the data distribution between clear emotion-pairs, which augments ambiguous samples up to reasonable quantities, meanwhile balances the quantities of clear samples in all categories to be close to each other. Figure 2 shows the data

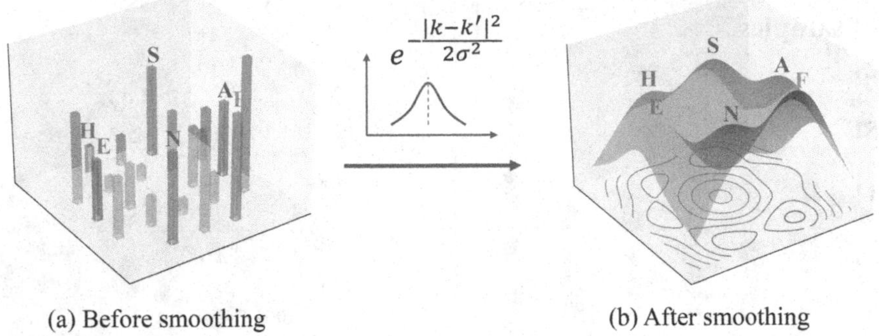

Fig. 2. The data distributions before and after smoothing. The purple bars represent clear samples and the orange bars represent ambiguous samples. (Color figure online)

distributions before and after smoothing. To complement the missing data, we use a feature-level mixup between clear samples to augment the data.

As PDDS is model and modality agnostic, we evaluate it on three state-of-the-art (SOTA) methods, as well as the unimodal and bimodal data on two benchmark datasets. The results show that these models are improved by 0.2% $\sim$ 4.8% and 0.1% $\sim$ 5.9% in terms of weighted accuracy and unweighted accuracy. Our proposed CLTNet achieves the best performance. The ablation study reveals that the nature of superior performance of PDDS is the reasonable data distribution rather than simply increasing the data size.

The contributions of this work are threefold:

1. We study the imbalance of the distribution of clear and ambiguous samples in existing speech emotion datasets, which will harm models to learn a better representation.
2. We propose the Pairwise-emotion Data Distribution Smoothing (PDDS) to construct a more reasonable training set with a more reasonable distribution by applying Gaussian smoothing to emotion-pairs, and complementing required data using a mixup augmentation.
3. Extensive experiments demonstrate that our proposed PDDS is model and modality agnostic, which considerably improves three SOTA methods on two benchmark datasets. We also proposed a model called CLTNet which achieved the best performance.

2 Method

Figure 3 shows that our proposed framework includes three modules: (1) Preprocessing module. It extracts audio features using the pre-trained data2vec [4] and text features using the pre-trained Bert [10]. (2) Pairwise-emotion Data Distribution Smoothing module (PDDS). It smooths the unreasonable data distribution, and is a plug-in module that can be applied to other SER methods. (3) CLTNet: a proposed model for utterance-level multimodal emotion recognition.

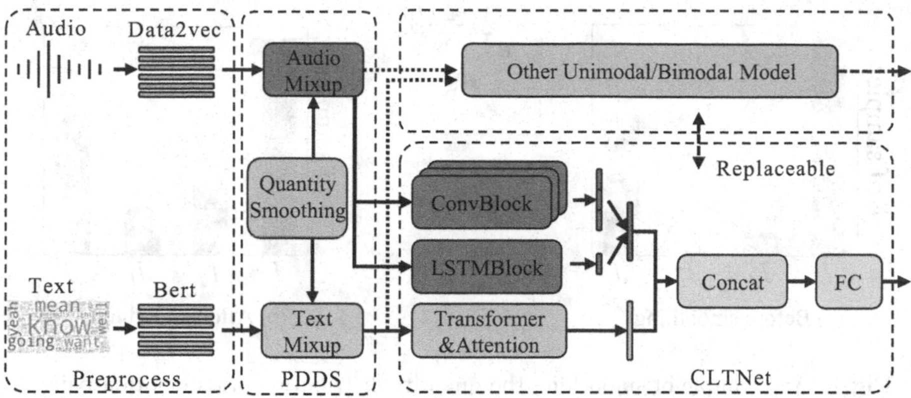

Fig. 3. The framework of our approach. It consists of three modules: (1) Data preprocessing; (2) Pairwise-emotion Data Distribution Smoothing (PDDS); (3) CLTNet.

2.1 Pairwise Data Distribution Smoothing

Quantity Smoothing. Suppose there are c emotions in the dataset. As the label of an ambiguous sample often comes from two distinct emotions, we construct a clear-ambiguous-clear distribution for the population of four types of samples between every two emotions i and j, where $i, j \in \{1, \ldots, c\}, i \neq j$. They are *clear samples* I only containing emotion i, *ambiguous samples* I_J containing major emotion i and minor emotion j, *ambiguous samples* J_I containing major emotion j and minor emotion i, and *clear samples* J only containing emotion j, as shown in Fig. 4.

We think that the quantity distribution of these four types of samples in every emotion-pair should be statistically smooth, so a Gaussian kernel is convolved with the distribution to have a smoothed version,

$$n_k = \sum_{k' \in K} e^{-\frac{|k-k'|^2}{2\sigma^2}} n_{k'} \Bigg/ \sum_{k' \in K} e^{-\frac{|k-k'|^2}{2\sigma^2}}, \tag{1}$$

where $K = \{k-1, k, k+1\}$ denotes the indexes of $\{I, I_J, J_I\}$ or $\{I_J, J_I, J\}$ when k is the index of I_J or J_I. The index of I, I_J, J_I, J can be 0,1,2,3, respectively. $n_{k'}$ is the quantity of samples in type k before smoothing, and n_k is the quantity of samples after smoothing. For instance, there are four types of samples for the emotion-pair of *happiness* and *sadness*: *hap*, hap_{sad}, sad_{hap} and *sad*. We first convert the four types into integer indexes 0, 1, 2 and 3, and their corresponding quantities of samples are $n_{0'}, n_{1'}, n_{2'}$ and $n_{3'}$. We can calculate the number of *ambiguous samples* (hap_{sad} and sad_{hap}) after smoothing according to Eq. 1, when $k = 1$, $K = \{0, 1, 2\}$, and when $k = 2$, $K = \{1, 2, 3\}$. For *clear samples*, if their quantity in an emotion category is too small, they are augmented until the quantity reaches that in other categories. The smoothed quantity distribution of all emotion-pairs is shown in Fig. 2(b).

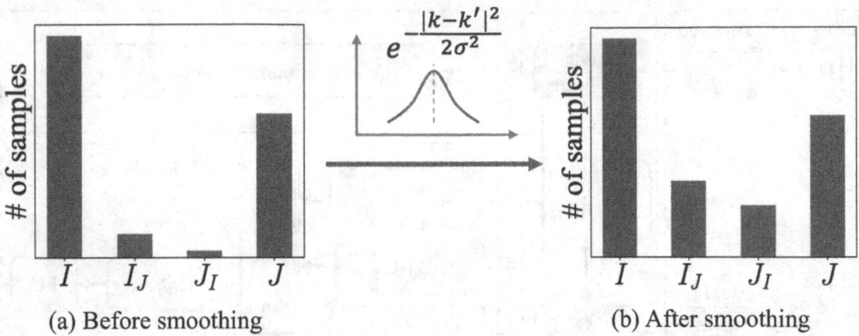

Fig. 4. An example of smoothing the quantity distribution of an emotion-pair.

Mixup Augmentation. After smoothing the data distribution, the quantities of some ambiguous samples in the original dataset are less than the expectation. To complete data, a feature-level data augmentation, Mixup [25], is applied to augment those samples. Mixup can improve the generalization ability of the model by generating new training samples by linearly combining two clear samples and their labels. It follows the rules

$$x_{mix} = px_{\alpha} + (1-p)x_{\beta}, \tag{2}$$

$$y_{mix} = py_{\alpha} + (1-p)y_{\beta}, p \in [0,1], \tag{3}$$

where x_{α} and x_{β} are the features from a clear sample of the major emotion and a clear sample of the minor emotion, respectively. p is a parameter that controls the proportion of each emotion in the augmented sample. x_{mix} is the feature of the new sample. y_{α} and y_{β} are the one-hot labels of x_{α} and x_{β}, and y_{mix} is the label distribution of the new sample. To avoid undersampling, we use the original data when the original quantity of ambiguous samples has met or exceeded the smooth distribution.

2.2 CLTNet

To further verify the effectiveness of PDDS, we design a simple but effective utterance-level multimodal fusion network, named CLTNet, which uses convolutional network(CNN), long short term memory network(LSTM) and Transformer to extract multimodal emotional features as shown in Fig. 3. Firstly, for the audio modality, the acoustic features are fed into three convolutional blocks with different scales to capture richer local semantic patterns. Each of them has a 1D convolutional layer and a global pooling layer. To capture the temporal dependencies in the acoustic sequence, an LSTM layer and a pooling layer are employed. The obtained 4 encoded feature vectors are concatenated and fed into a fully-connected layer to get the audio representations as follows,

$$h_A = Concat(x_{conv_A^1}, x_{conv_A^2}, x_{conv_A^3}, x_{lstm_A})W_A + b_A, \tag{4}$$

$$x_{conv_A^i} = ConvBlock(x_A), \tag{5}$$

$$x_{lstm_A} = LSTMBlock(x_A), \tag{6}$$

where
$ConvBlock(\cdot) = MaxPool(Relu(Conv1D(\cdot))), LSTMBlock(\cdot) = MaxPool(LSTM(\cdot)), x_A \in \mathbb{R}^{t_A \times d_A}$ and $x_{conv_A^i}$, $x_{lstm_A} \in \mathbb{R}^{d_1}$ are input features and output features of convolution blocks and LSTM blocks, respectively, $W_A \in \mathbb{R}^{4d_1 \times d}$ and $b_A \in \mathbb{R}^d$ are trainable parameters. For the text modality, the text features are fed into a transformer encoder of N layers to capture the interactions between each pair of textual words. An attention mechanism [18] is applied to the outputs of the last block to focus on informative words and generate text representations,

$$h_T = (a_{fuse}^T z_T)W_T + b_T, \tag{7}$$

$$z_T = TransformerEncoder(x_T), \tag{8}$$

$$a_{fuse} = softmax(z_T W_z + b_z), \tag{9}$$

where $x_T \in \mathbb{R}^{t_T \times d_T}$ and $z_T \in \mathbb{R}^{t_T \times d_2}$ are input features and output features of the transformer encoder, $W_z \in \mathbb{R}^{d_2 \times 1}$, $b_z \in \mathbb{R}^1$, $W_T \in \mathbb{R}^{d_2 \times d}$ and $b_T \in \mathbb{R}^d$ are trainable parameters. $a_{fuse} \in \mathbb{R}^{t_T \times 1}$ is the attention weight. Finally, the representations of the two modalities are concatenated and fed into three fully-connected layers (FC) with a residual connection, followed by a softmax layer. As using the label distribution to annotate emotional data, we choose the KL loss to optimize the model,

$$Loss_{KL} = \sum_{i=1}^{C} y_i log \frac{y_i}{\hat{y_i}}, \tag{10}$$

$$h = Concat(h_A, h_L)W_u + b_u, \tag{11}$$

$$\hat{y} = Softmax((h + ReLU(hW_h + b_h))W_c + b_c), \tag{12}$$

where $W_u \in \mathbb{R}^{2d \times d}$, $b_u \in \mathbb{R}^d, W_h \in \mathbb{R}^{d \times d}$, $b_h \in \mathbb{R}^d, W_c \in \mathbb{R}^{d \times c}$ and $b_c \in \mathbb{R}^c$ are trainable parameters, $y \in \mathbb{R}^c$ is the true label distribution, $\hat{y} \in \mathbb{R}^c$ is the predicted label distribution.

3 Experiment

3.1 Dataset and Evaluation Metrics

To evaluate PDDS, we conduct the performance test on two benchmark datasets.

1) Interactive Emotional Dyadic Motion Capture (IEMOCAP): IEMOCAP is a commonly used English corpus for SER [6]. There are five sessions in the dataset, where each sentence is annotated by at least three experts. Following previous work [15], we choose six emotions: *anger*, *happiness*, *sadness*, *neutral*, *excited* and *frustration*. Session 1 to 4 are used as the training set and session 5 is used as the testing set, and PDDS is only applied on the training set.

2) FAU-AIBO Emotion Corpus (FAU-AIBO): FAU-AIBO is a corpus of German children communicating with Sony's AIBO pet robot [5], where each word is annotated by five experts and the chunk labels are obtained by a heuristic approach. Following previous work [22], we choose four emotions: **A***nger*, **M***otherese*, **E***mphatic* and **N***eutral*. We find that FAU-AIBO also has an imbalanced data distribution, for example, the quantity of ambiguous samples between *anger* and *emphatic* is abundant, while that between *anger* and *neutral* is very few. Experiments are performed using 3-fold cross validation and PDDS is applied on the training set in each fold.

For both datasets, the weighted accuracy (WA, i.e., the overall accuracy) and unweighted accuracy (UA, i.e., the average accuracy over all emotion categories) are adopted as the evaluation metrics.

3.2 Implementation Details

For audio, 512-dimensional features are extracted from the raw speech signal by the pre-trained data2vec [4]. The frame size and frame shift are set to 25 ms and 20 ms, respectively. For text, the pre-trained Bert [10] on English and German corpora are used for IEMOCAP and FAU-AIBO, respectively, to embed each word into a 768-dimensional vector.

In the original IEMOCAP dataset, the quantity of clear samples belonging to happiness is very small, we then augment the data using mixup in order to balance the quantities of clear samples in all emotion categories. The augmentation settings are $p = 0.5$, $\alpha = \beta = happiness$, and the quantity is increased from 43 to 215. Afterward, the data distribution smoothing is applied to the ambiguous samples of each emotion-pair with Gaussian kernel $\sigma = 1$.

The kernel sizes of three CNN blocks in CLTNet are 3, 4 and 5, and the number of hidden units is 300 in LSTM. The number of layers in the transformer encoder is 5 and each layer is with 5 attention heads. The embedding dimensions d_1, d_2 and d in Eq. (4) and (7) are set to 300, 100 and 100, respectively. CLTNet is trained by the Adam optimizer. The learning rate, weight decay and batch size

Table 1. The comparison of models without and with PDDS on the IEMOCAP dataset.

Model	w/o PDDS		w/ PDDS	
	WA (%)	UA (%)	WA (%)	UA (%)
SpeechText [3]	55.7	50.6	**58.8**	**55.1**
MAT [9]	54.5	51.0	**58.0**	**55.5**
MCSAN [23]	60.0	56.5	**60.2**	**58.0**
CLTNet(audio only)	44.9	39.0	**48.4**	**44.9**
CLTNet(text only)	50.9	45.8	**51.6**	**50.1**
CLTNet	55.9	52.3	**60.7**	**58.2**

Table 2. The comparison of models without and with PDDS on the FAU-AIBO dataset.

Model	w/o PDDS		w/ PDDS	
	WA (%)	UA (%)	WA (%)	UA (%)
SpeechText [3]	66.5	55.2	**68.1**	**57.5**
MAT [9]	67.9	58.9	**68.7**	**59.0**
MCSAN [23]	68.2	57.8	**69.7**	**62.2**
CLTNet(audio only)	63.7	48.0	**66.1**	**51.7**
CLTNet(text only)	64.9	54.9	**65.8**	**59.0**
CLTNet	68.2	57.5	**70.9**	**61.6**

are set to 0.001, 0.00001 and 64, respectively. The training stops if the loss does not decrease for 10 consecutive epochs, or at most 30 epochs. All experiments were repeated 3 times with different random seeds, and the average results were reported in the following sections.

3.3 Validation Experiment

As PDDS is model-modality agnostic, we select SpeechText [3], MAT [9], and MCSAN [23], which aim at utterance-level emotion recognition, to evaluate its effectiveness. Our proposed CLTNet and its unimodal branches are also tested. The experimental results on the IEMOCAP dataset and the FAU-AIBO dataset are shown in Table 1 and Table 2, respectively. On IEMOCAP, we can observe that all models are significantly improved when the training data are processed by PDDS, with 0.2% $\sim$ 4.8% increase on WA and 1.5% $\sim$ 5.9% increase on UA. Among them, our proposed CLTNet achieves the best performance. On FAU-AIBO, we can observe similar results, with 0.8% $\sim$ 2.7% increase on WA and 0.1% $\sim$ 4.4% increase on UA. This result demonstrates that a reasonable data distribution in the training set does help models learn a better emotional representation.

A more detailed analysis, the confusion matrices of CLTNet on the IEMOCAP dataset with and without PDDS, are shown in Fig. 5. One can see that the classification accuracies in most emotion categories are increased. Only two more samples are misclassified in *frustration*. However, the rightmost columns in the confusion matrices illustrate that the model trained on the original dataset inclines to misclassify more samples as *frustration*. By contrast, PDDS considerably alleviates this tendency.

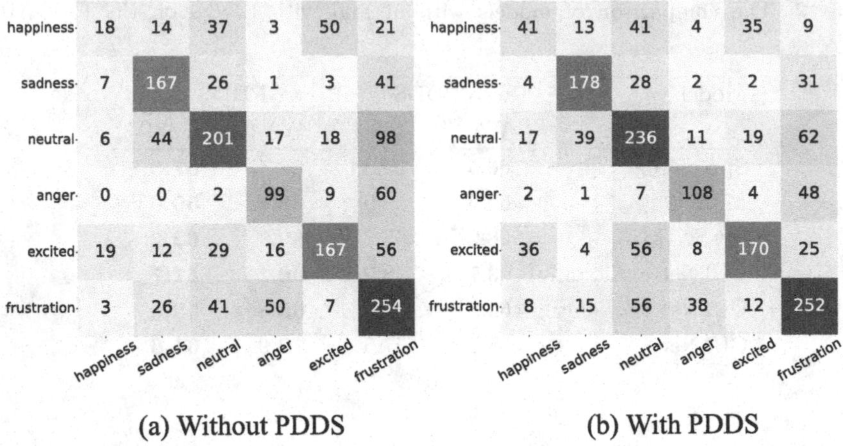

Fig. 5. Confusion matrices of CLTNet on the IEMOCAP dataset without and with PDDS.

3.4 Ablation Study

To verify the rationality of PDDS, two additional experiments on the IEMOCAP dataset are designed and tested on the CLTNet model: (a) Only balancing *happiness*: Mixup augmentation is only applied on clear samples of *happiness* rather than ambiguous samples. (b) Uniform balancing: the clear samples in each category and the ambiguous samples of each emotion-pair are augmented to 400 if their quantity is less than 400 (Most of them are less than 400 in the IEMOCAP dataset).

The results are shown in Table 3. We can observe that all three data augmentations boost the performance of the model. Compared to only balancing the *happiness* category, augmenting both ambiguous and clear samples can help the model perform better. Although Uniform balancing has the largest training dataset, PDDS performs best on WA and UA with 4.8% and 5.9% improvements. This result reveals the nature of advantage of PDDS is the reasonable data distribution instead of simply increasing the data size.

We also investigate different mixup augmentations. Our default setting is mixing data by randomly selecting persons. Considering different persons have different voiceprints, we also test the mixup from the same person, and the results on the IEMOCAP dataset are listed in Table 4. We can observe that the mixup created by the same person performs worse than the one created by random persons. One possible explanation is that mixup is a data augmentation method to improve the model robustness, which has been applied to the speech recognition and sentence classification. Random mixup increases the diversity of the data. When the augmentation is created by the same person, the diversity of the augmented data is reduced. The model's robustness deteriorates as a result.

Table 3. Evaluation of different data augmentation methods.

Data	WA (%)	UA (%)
Original training data	55.9	52.3
Only balancing *happiness*	57.4	54.1
Uniform balancing	58.6	56.0
With PDDS	**60.7**	**58.2**

Table 4. Evaluation of mixing up different persons on the IEMOCAP dataset.

Model	same person mixup		random mixup	
	WA (%)	UA (%)	WA (%)	UA (%)
SpeechText [3]	56.7	53.2	**58.8**	**55.1**
MAT [9]	56.5	54.5	**58.0**	**55.5**
MCSAN [23]	59.7	**58.3**	**60.2**	58.0
CLTNet(audio only)	48.3	43.6	**48.4**	**44.9**
CLTNet(text only)	51.4	49.0	**51.6**	**50.1**
CLTNet	59.3	56.5	**60.7**	**58.2**

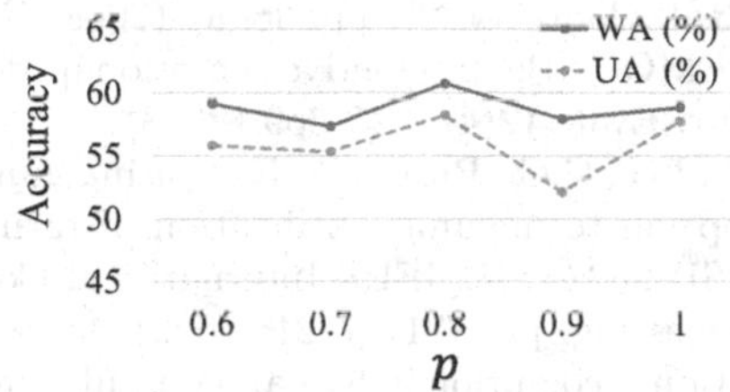

Fig. 6. Effect of different p values of CLTNet on the IEMOCAP dataset.

We further evaluate the effectiveness of p values in the mixup augmentation, and show the result of CLTNet on the IEMOCAP dataset in Fig. 6. One can see the best results are achieved when $p = 0.8$.

4 Conclusion

In this paper, we address the imbalanced data distribution in the existing emotion datasets by proposing the Pairwise-emotion Data Distribution Smoothing (PDDS) method. PDDS constructs a more reasonable training set with smoother distribution by applying Gaussian smoothing to emotion-pairs, and complements required data using a mixup augmentation. Experimental results show that PDDS considerably improves three SOTA methods. Meanwhile, our proposed CLTNet achieves the best result. More importantly, the ablation study verifies that the nature of superiority of PDDS is the reasonable data distribution instead of simply increasing the amount of data. In future work, we will explore a more reasonable data distribution for a better emotional representation learning.

References

1. Ando, A., Kobashikawa, S., Kamiyama, H., Masumura, R., Ijima, Y., Aono, Y.: Soft-target training with ambiguous emotional utterances for DNN-based speech emotion classification. In: 2018 IEEE International Conference on Acoustics, Speech and Signal Processing, pp. 4964–4968 (2018)
2. Ando, A., Masumura, R., Kamiyama, H., Kobashikawa, S., Aono, Y.: Speech emotion recognition based on multi-label emotion existence model. In: Proc. Interspeech 2019, pp. 2818–2822 (2019)
3. Atmaja, B.T., Shirai, K., Akagi, M.: Speech emotion recognition using speech feature and word embedding. In: 2019 Asia-Pacific Signal and Information Processing Association Annual Summit and Conference (APSIPA ASC), pp. 519–523 (2019)
4. Baevski, A., Hsu, W.N., Xu, Q., Babu, A., Gu, J., Auli, M.: data2vec: A general framework for self-supervised learning in speech, vision and language. In: Proceedings of the 39th International Conference on Machine Learning. vol. 162, pp. 1298–1312 (2022)
5. Batliner, A., Steidl, S., Nöth, E.: Releasing a thoroughly annotated and processed spontaneous emotional database: the FAU Aibo emotion corpus. In: Proc. Workshop Lang. Resour. Eval. Conf. vol. 28, pp. 28–31 (2008)
6. Busso, C., et al.: IEMOCAP: Interactive emotional dyadic motion capture database. Lang. Resour. Eval. **42**(4), 335–359 (2008)
7. Chou, H.C., Lin, W.C., Lee, C.C., Busso, C.: Exploiting annotators' typed description of emotion perception to maximize utilization of ratings for speech emotion recognition. In: ICASSP 2022–2022 IEEE International Conference on Acoustics, Speech and Signal Processing, pp. 7717–7721 (2022)
8. Cowie, R., et al.: Emotion recognition in human-computer interaction. IEEE Signal Process. Mag. **18**(1), 32–80 (2001)
9. Delbrouck, J.B., Tits, N., Dupont, S.: Modulated fusion using transformer for linguistic-acoustic emotion recognition. In: Proceedings of the First International Workshop on Natural Language Processing Beyond Text, pp. 1–10 (2020)
10. Devlin, J., Chang, M.W., Lee, K., Toutanova, K.: BERT: Pre-training of deep bidirectional transformers for language understanding. In: Proceedings of the 2019 Conference of the North American Chapter of the Association for Computational Linguistics: Human Language Technologies. vol. 1, pp. 4171–4186 (2019)
11. Fayek, H.M., Lech, M., Cavedon, L.: Modeling subjectiveness in emotion recognition with deep neural networks: Ensembles vs soft labels. In: 2016 International Joint Conference on Neural Networks (IJCNN), pp. 566–570 (2016)
12. Fujioka, T., Homma, T., Nagamatsu, K.: Meta-learning for speech emotion recognition considering ambiguity of emotion labels. In: Proc. Interspeech 2020, pp. 2332–2336 (2020)
13. Gao, X., Zhao, Y., Zhang, J., Cai, L.: Pairwise emotional relationship recognition in drama videos: Dataset and benchmark. In: Proceedings of the 29th ACM International Conference on Multimedia, pp. 3380–3389 (2021)
14. Gupta, P., Rajput, N.: Two-stream emotion recognition for call center monitoring. In: Proc. Interspeech 2007, pp. 2241–2244 (2007)

15. Hazarika, D., Poria, S., Mihalcea, R., Cambria, E., Zimmermann, R.: ICON: Interactive conversational memory network for multimodal emotion detection. In: Proceedings of the 2018 Conference on Empirical Methods in Natural Language Processing, pp. 2594–2604 (2018)
16. Huahu, X., Jue, G., Jian, Y.: Application of speech emotion recognition in intelligent household robot. In: 2010 International Conference on Artificial Intelligence and Computational Intelligence. vol. 1, pp. 537–541 (2010)
17. Lian, Z., Chen, L., Sun, L., Liu, B., Tao, J.: GCNet: Graph completion network for incomplete multimodal learning in conversation. IEEE Trans. Pattern Anal. Mach. Intell. **45**(7), 8419–8432 (2023)
18. Lian, Z., Liu, B., Tao, J.: CTNet: Conversational transformer network for emotion recognition. IEEE/ACM Trans. Audio, Speech, Lang. Process. **29**, 985–1000 (2021)
19. Lotfian, R., Busso, C.: Predicting categorical emotions by jointly learning primary and secondary emotions through multitask learning. In: Proc. Interspeech 2018, pp. 951–955 (2018)
20. Parry, J., Palaz, D., et al: Analysis of Deep Learning Architectures for Cross-Corpus Speech Emotion Recognition. In: Proc. Interspeech 2019, pp. 1656–1660 (2019)
21. Poria, S., Cambria, E., Hazarika, D., Majumder, N., Zadeh, A., Morency, L.P.: Context-dependent sentiment analysis in user-generated videos. In: Proceedings of the 55th Annual Meeting of the Association for Computational Linguistics. vol. 1, pp. 873–883 (2017)
22. Seppi, D., et al: Patterns, prototypes, performance: classifying emotional user states. In: Proc. Interspeech 2008, pp. 601–604 (2008)
23. Sun, L., Liu, B., Tao, J., Lian, Z.: Multimodal cross-and self-attention network for speech emotion recognition. In: ICASSP 2021-2021 IEEE International Conference on Acoustics, Speech and Signal Processing, pp. 4275–4279 (2021)
24. Yin, Y., Gu, Y., Yao, L., Zhou, Y., Liang, X., Zhang, H.: Progressive co-teaching for ambiguous speech emotion recognition. In: ICASSP 2021-2021 IEEE International Conference on Acoustics, Speech and Signal Processing (ICASSP), pp. 6264–6268 (2021)
25. Zhang, H., Cisse, M., Dauphin, Y.N., Lopez-Paz, D.: mixup: Beyond empirical risk minimization. In: International Conference on Learning Representations (2018)
26. Zhou, Y., Liang, X., Gu, Y., Yin, Y., Yao, L.: Multi-classifier interactive learning for ambiguous speech emotion recognition. IEEE/ACM Trans. Audio, Speech, Lang. Process. **30**, 695–705 (2022)
27. Zou, H., Si, Y., Chen, C., Rajan, D., Chng, E.S.: Speech emotion recognition with co-attention based multi-level acoustic information. In: ICASSP 2022-2022 IEEE International Conference on Acoustics, Speech and Signal Processing, pp. 7367–7371 (2022)

SIEFusion: Infrared and Visible Image Fusion via Semantic Information Enhancement

Guohua Lv[1,2,3(✉)], Wenkuo Song[1,2,3], Zhonghe Wei[1,2,3], Jinyong Cheng[1,2,3], and Aimei Dong[1,2,3]

[1] Key Laboratory of Computing Power Network and Information Security, Ministry of Education, Shandong Computer Science Center (National Supercomputer Center in Jinan), Qilu University of Technology (Shandong Academy of Sciences), Jinan, China
guohualv@qlu.edu.cn

[2] Shandong Provincial Key Laboratory of Computer Networks, Shandong Fundamental Research Center for Computer Science, Jinan, China

[3] Faculty of Computer Science and Technology, Qilu University of Technology (Shandong Academy of Sciences), Jinan 250353, China

Abstract. At present, most of existing fusion methods focus on the metrics and visual effects of image fusion, but ignore the requirements of high-level tasks after image fusion, which leads to the unsatisfactory performance of some methods in subsequent high-level tasks such as semantic segmentation and object detection. In order to address this problem and obtain images with rich semantic information for subsequent semantic segmentation tasks, we propose a fusion network for infrared and visible images based on semantic information enhancement named SIEFusion. In our fusion network, we design a cross-modal information sharing module(CISM) and a fine-grained detail feature extraction module(FFEM) to obtain better fused images with more semantic information. Extensive experiments show that our method outperforms the state-of-the-art methods both in qualitative and quantitative comparison, as well as in the subsequent segmentation tasks.

Keywords: Image fusion · Semantic information · Deep learning · Subsequent vision task

1 Introduction

Due to technical and theoretical limitations, a single modality imaging device cannot completely describe an imaging scene [20]. For example, infrared sensors can capture the thermal radiation information emitted by objects, which can effectively highlight prominent targets such as pedestrians and vehicles, but lack of detailed description of the scenes. On the contrary, visible images usually

This work was supported by the Natural Science Foundation of Shandong Province, China (ZR2020MF041 and ZR2022MF237).

Q. Liu et al. (Eds.): PRCV 2023, LNCS 14427, pp. 176–187, 2024.
https://doi.org/10.1007/978-981-99-8435-0_14

contain rich scene information and texture information, but are easily affected by extreme weather conditions and illumination. With the continuous development of imaging device technology, it is found that different types of imaging devices contain a large amount of complementary information, which inspires researchers to integrate these complementary information into a single image. This is also the significance of image fusion technology.

Among numerous types of fusion tasks, the complementary of visible and infrared images encourages researchers to incorporate useful information from multi-modal images into a single fused image to facilitate human perception and subsequent vision applications, such as object detection [1], object tracking [10,11], and semantic segmentation [6]. In the past decades, a number of infrared and visible image fusion methods have been proposed.

In general, these fusion methods can be divided into traditional methods and deep learning-based methods. On the one hand, traditional methods can be divided into five categories, i.e., multi-scale transform (MST)-based methods [2], sparse representation (SR)-based methods [17], subspace-based methods [25], optimization-based methods [18], and hybrid methods [24]. On the other hand, deep learning-based fusion methods can be classified into four categories, i.e., AE-based [12,13], CNN-based [30–32,35], GAN-based [16,23], and Transformer-based methods [21,28]. Meanwhile, with the development of deep learning, more and more fusion methods adopt the technology of deep learning.

However, although the existing fusion methods have achieved satisfactory results, there are still some problems to be solved. On the one hand, most of existing fusion methods only focus on better visual quality and higher evaluation metrics, and rarely consider the related requirements of subsequent high-level task such as semantic segmentation. Some studies have shown that considering only visual quality and evaluation metrics is of little help for subsequent high-level tasks [15,26]. Some scholars have used segmentation masks to guide the image fusion process, but the mask only segments some salient objects [22], and its effect on enhancing semantic information is limited, because other non-salient objects also contain semantic information [19]. On the other hand, there is also work that has made efforts and explored subsequent high-level mission requirements such as SeAFusion [31]. It combines the semantic loss and the image fusion task, and takes into account the semantic requirements of subsequent high-level tasks. However, SeAFusion does not perform particularly well in extreme environments especially in dim scenes, which affects the subsequent segmentation tasks even more.

In order to address the aforementioned problems, we propose a based on semantic information enhancement image fusion network called SIEFusion. In our fusion network, we design a cross-modal information sharing module(CISM) and a fine-grained feature extraction module(FFEM). CISM exploits the design concept of spatial attention to make full use of the complementary information between visible and infrared images. FFEM uses dense connections on the backbone to make full use of the extracted feature information, meanwhile, it uses skip connection and gradient operator on the branch to extract fine-grained information of images. To be concrete, the main contributions of our work are as follows:

(1) We propose an infrared and visible image fusion network that can obtain more semantic information for subsequent semantic segmentation tasks.
(2) We design a cross-modal information sharing module(CISM) to extract more information for image fusion before feature fusion, so that final fused images contain richer semantic information.
(3) We design a fine-grained feature sharing module(FFEM) to obtain more abundant fine-grained features for feature fusion so as to obtain more semantic information for subsequent semantic segmentation.

2 Method

2.1 Problem Formulation

Given a pair of registered infrared image $I_i \in R^{H\times W\times 1}$ and visible image $I_v \in R^{H\times W\times 3}$, they will be fed into our fusion network. Under the constraint of the loss function, after feature extraction, feature fusion, and image reconstruction, the fused image we want is finally generated.

Specifically, we extract features from the input infrared image and visible image through the feature extraction module E_f, and the process of extracting features can be expressed as

$$F_i = E_f(I_i), F_v = E_f(I_v), \tag{1}$$

where F_i and F_v mean infrared feature maps and visible feature maps, respectively. Then, the extracted features will be integrated. The feature fusion process can be expressed as

$$F_f = C(F_i, F_v), \tag{2}$$

where $C(\cdot)$ represents feature maps concatenate in the channel dimension. Finally, the fused image $I_f \in R^{H\times W\times 3}$ is reconstructed from the fused features F_f via the image reconstructor R_i, which is presented as:

$$I_f = R_i(F_f). \tag{3}$$

2.2 Network Architecture

Figure 1 illustrates the overall framework of our proposed method, including the fusion network and the segmentation network. The fusion network contains our proposed cross-modal information sharing modules, fine-grained detail feature extraction modules, and an image reconstructor.

CISM. Cross-modal information sharing module(CISM), inspired by the design concept of dual stream correlation attention module and spatial attention of CBAM attention module, which is different from the channel attention of CMDAF module in PIAFusion, is designed to take advantage of the shared and complementary features. At the same time, it also fully aggregates complementary information with pixel weighting. Thus CISM can be formulated as follows:

$$\hat{F}_v = F_v \oplus (F_v \otimes \sigma(Conv(C(Avg(F_i - F_v) + Max(F_i - F_v))))), \tag{4}$$

$$\hat{F}_i = F_i \oplus (F_i \otimes \sigma(Conv(C(Avg(F_v - F_i) + Max(F_v - F_i))))), \quad (5)$$

where $Max(\cdot)$ and $Avg(\cdot)$ denote max-pooling and average-pooling, and $Conv(\cdot)$ is convolution operation, $\hat{F}_v$ and $\hat{F}_i$ represent feature maps of visible and infrared image after a certain feature extraction, respectively, $\oplus$ and $\otimes$ stand for element-wise summation and channel-wise multiplication, respectively, $\sigma(\cdot)$ means sigmoid function. We pass the obtained single-channel feature map through the sigmoid function to obtain a coefficient, and then multiply the coefficient with the original feature map and add it to the original feature map. Thus, more complementary information are assembled through CISM.

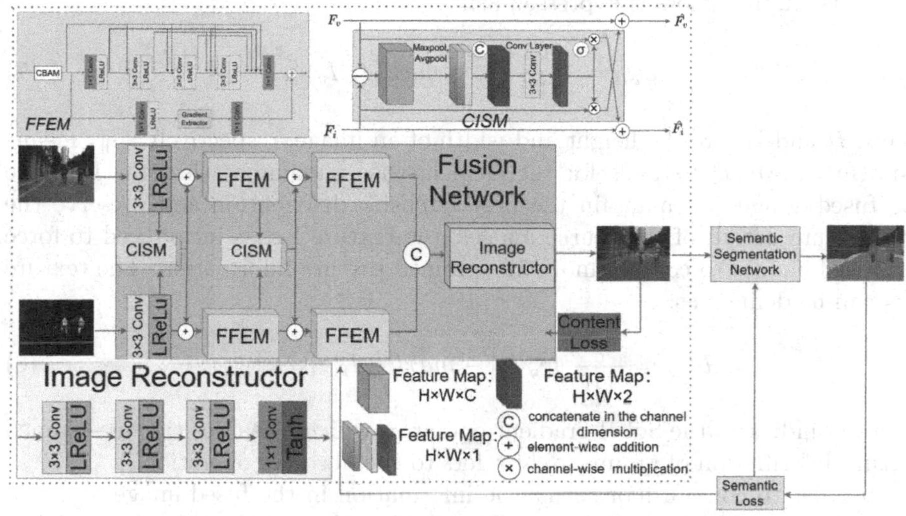

Fig. 1. The overall framework of our proposed method.

FFEM. The fine-grained feature extraction module, inspired by the design concepts of resblock [8] and denseblock [9]. We design two 1×1 convolutional layers, three 3×3 convolutional layers, and a CBAM module(a mixed channel and spatial attention mechanism module, designed to capture more distinctive features) [34] in the backbone. Besides, we set two 1×1 convolutional layers and an edge extraction module in the branch which uses Sobel gradient operator to extract more fine-grained features containing more semantic information.

Image Reconstructor. The image reconstructor consists of three 3×3 convolutional layers and one 1×1 convolutional layer. The 3×3 convolutional layers all use LReLU as the activation function, but the activation function of the 1×1 convolutional layer is Tanh.

Segmentation Network. We introduce a segmentation network to predict the segmentation results of fused images and use it to construct semantic loss. Then, the semantic loss is used to guide the training of the fusion network, and final fused images are forced to contain more semantic information. More details about the segment network can be found in [27].

2.3 Loss Function

In order to get higher quality fused images, content loss is introduced to our model, which includes intensity loss L_{int} and texture loss L_{tex}. The content loss is expressed as follows:

$$L_{con} = L_{int} + \eta L_{tex}, \tag{6}$$

where L_{int} represents the intensity loss, which measures the difference between fused images and source images at the pixel level, L_{tex} means the gradient loss, which is used to incorporate more information from the source images into the fused images, η is the parameter used to control the balance between L_{int} and L_{tex}. The intensity loss is expressed as:

$$L_{int} = \frac{1}{HW} \left\| I_f - max\left(I_i, I_v\right) \right\|_1, \tag{7}$$

where H and W are the height and width of an image, respectively, $\|\cdot\|_1$ means l_1-norm and $max(\cdot)$ stands for the element-wise maximum selection. To make the fused images to maintain the best intensity distribution and preserve the rich texture details of the source images, the texture loss is introduced to force the fused images to contain more fine-grained texture information. The texture loss can be defined as:

$$L_{tex} = \frac{1}{HW} \left\| \left| \nabla I_f \right| - max(\left| \nabla I_i \right|, \left| \nabla I_v \right|) \right\|_1, \tag{8}$$

where ∇ indicates the Sobel gradient operator, which measures the fine-grained texture information of an image, $|\cdot|$ refers to the absolute operation.

In order to contain more semantic information in the fused images, we add semantic loss to the fusion network. The semantic loss can be roughly expressed as the gap between segmentation result I_s and label L_s, which is expressed as:

$$L_{sea} = \phi(I_s, L_s), \tag{9}$$

where ϕ denotes the function that measures the difference between I_s and L_s. To be precise, the segmentation model is used to segment the fused images to obtain the main segmentation result I_s and auxiliary segmentation result I_{sr}. Meanwhile, the main semantic loss L_{main} and auxiliary semantic loss L_{aux} are obtained at the same time, which are expressed as :

$$L_{main} = \frac{-1}{HW} \sum_{h=1}^{H} \sum_{w=1}^{W} \sum_{c=1}^{C} L_{so}^{(h,w,c)} \log(I_s^{(h,w,c)}), \tag{10}$$

$$L_{aux} = \frac{-1}{HW} \sum_{h=1}^{H} \sum_{w=1}^{W} \sum_{c=1}^{C} L_{so}^{(h,w,c)} \log(I_{sr}^{(h,w,c)}), \tag{11}$$

where L_{so} denotes a one-hot vector transformed from the segmentation label L_s. Ultimately, the semantic loss is expressed as:

$$L_{sea} = L_{main} + \lambda L_{aux}, \tag{12}$$

where λ is a constant for balancing the L_{main} and L_{aux}, which is set to 0.1 [27]. In addition, semantic loss is used not only to constrain the fusion model, but also to train the segmentation model.

Finally, the loss used to guide the model training process is expressed as:

$$L_{overall} = L_{con} + \zeta L_{sea}, \tag{13}$$

where ζ is the trade-off parameter which is used for balancing L_{con} and L_{sea}.

3 Experiments

3.1 Experimental Configurations

To evaluate our method, we conduct qualitative and quantitative experiments on MSRS [31] and TNO [33] datasets. Our method is evaluated against seven state-of-the-art algorithms, including one traditional method, i.e., MDLatLRR [14], two AE-based methods, i.e., NestFuse and DenseFuse, one GAN-based method, i.e., FusionGAN, and three CNN-based methods, i.e., PIAFusion, SeAFusion and U2Fusion. The parameters of seven methods are set the same as reported in the original papers, and we selected six metrics for quantitative comparison, including standard deviation (SD), average gradient (AG), entropy(EN) [29], visual information fidelity (VIF) [7], Q_{abf} and spatial frequency (SF) [5].

In the training phase, we train our model on the training set of MSRS including 1083 pairs of images. We iteratively train the fusion and segmentation networks 4 times, with the remaining hyper-parameters set as follows: $\eta = 10$, $\zeta = 0$, 1, 2, 3 in every iteration, respectively [31]. We optimize our fusion model using Adam optimizer with a batch size of 4, initial learning rate of 0.001, weight decay 0.0002 with the guidance of $L_{overall}$. Besides, we utilize mini-batch stochastic gradient descent whose initial learning rate is set as 0.01 [27] to optimize the segmentation network with a batch size of 8, momentum of 0.9, and weight decay of 0.0005. Our method is implemented on the PyTorch platform.

In addition, in order to facilitate our image fusion work, we convert the visible RGB images into YCbCr color space before sending the images to the fusion network, and only use the Y channel of visible images and infrared images for fusion. After completing the above work, the fused images will be obtained, and then the fused images and Cb, Cr channels will be recombined into YCbCr color space. Finally, the images are converted into RGB images.

3.2 Results and Analysis

Qualitative Comparison. As we can see in the Fig. 2, the fused image generated by FusionGAN is relatively blurred and poorly represented in terms of edge and texture details, with only a few salient targets being represented and poorly represented in terms of scene information. The conventional method MDLatLRR and other methods perform well in the representation of scene information relative to FusionGAN, with clear target edges and well-preserved texture information. However, there is a certain amount of overexposure in the image generated by U2Fusion for the scene is too bright, and the influence of noise on the image is serious. Meanwhile, FusionGAN, DenseFuse, MDLatLRR, NestFuse, PIAFusion, and SeAFusion generate images with low brightness. On the contrary, our method is clearer on salient targets while the edges of objects are clearer, texture information is well preserved, and the luminance is more consistent with subjective vision than other methods. Meanwhile, some details are also better represented as the boxes show.

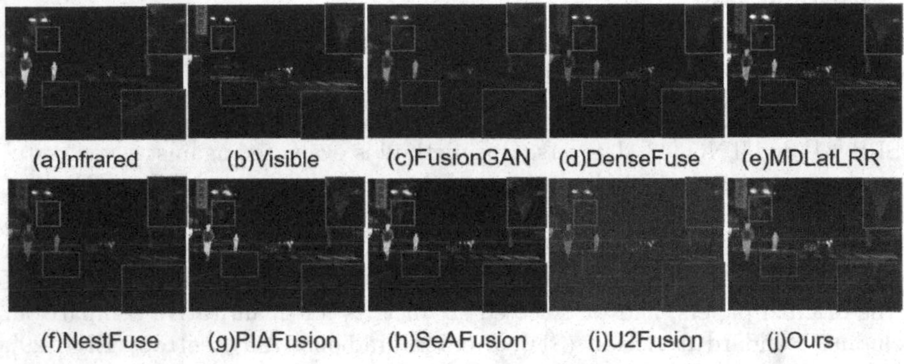

Fig. 2. Visual quality comparison of our method with seven state-of-the-art infrared and visible image fusion methods on 00881N image from MSRS dataset.

Table 1. Quantitative results on MSRS dataset. The best and second best are indicated in bold and underline, respectively.

	NestFuse	MDLatLRR	FusionGAN	DenseFuse	PIAFusion	SeAFusion	U2Fusion	Ours
EN	5.0401	6.0138	5.2417	5.6527	6.0380	<u>6.2061</u>	4.0655	**6.2239**
VIF	0.7456	0.7751	0.7732	0.7746	1.0210	<u>1.0457</u>	0.4290	**1.0831**
AG	2.0501	2.3241	1.3108	1.8080	**2.8460**	2.8204	1.4139	<u>2.8329</u>
SD	31.1480	29.2763	17.4636	24.9197	<u>38.6515</u>	38.3165	16.5399	**38.6664**
Q_{abf}	0.4775	0.5064	0.4976	0.4388	<u>0.6555</u>	0.6537	0.2540	**0.6817**
SF	7.5791	7.4311	4.1852	5.8038	**9.7444**	9.4031	5.4620	<u>9.6592</u>

Quantitative Comparison. We randomly select 50 pairs of images for comparison from testing set of MSRS which consists of 361 pairs of images. As shown in Table 1, we can see that our method achieves good performance especially on EN, VIF, SD, and Q_{abf}, which indicates that our method achieves good results in retaining the source images information and the edge information of images. Meanwhile, it also maintains richer texture details of the images and maintains better visual effects. These can also be reflected in Fig. 2, other methods do not perform well enough in texture details and visual effects such as FusionGAN and U2Fusion, and remaining methods are not particularly satisfactory in terms of details, losing some scene information, and in some places the scene is dark, resulting in some detailed information not being represented.

Generalization Experiments. In order to show the generalization of our method, we directly test it and other comparison methods on the TNO dataset. In Fig. 3, we can see that other methods such as FusionGAN, DenseFuse, NestFuse, and U2Fusion do not perform well enough in the scene brightness, resulting in the loss of some detail information. As the red boxes show, FusionGAN can not even see the target, and the target is almost completely lost. The remaining methods are PIAFusion, SeAFusion, and our method. All of them perform well enough in scene brightness and the target is clear, but the target edge of PIAFusion is not smooth, only SeAFusion and our method achieve good results in scene brightness and target edge. Additionally, from green boxes we can see our method can preserve more information of source images compared to SeAFusion.

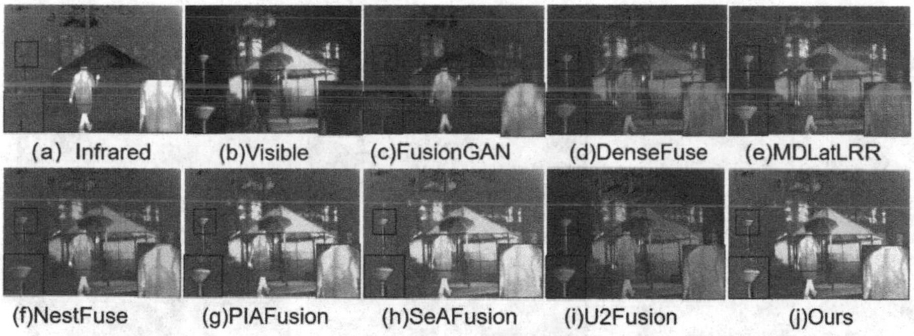

Fig. 3. Visual quality comparison of our method with seven state-of-the-art infrared and visible image fusion methods on "umbrella" from TNO dataset.

As shown in Table 2, our method has also achieved good results in quantitative comparison of generalization experiments especially on VIF, SF, and Q_{abf}, which indicates that our method has enough advantages in terms of preserving richer edge and texture details, more edge information, and more scene information compared to other methods. Overall, the results of generalization experiments effectively demonstrate the generalization of our method.

Table 2. Quantitative results on TNO dataset. The best and second best are indicated in bold and <u>underline</u>, respectively.

	NestFuse	MDLatLRR	FusionGAN	DenseFuse	PIAFusion	SeAFusion	U2Fusion	Ours
EN	7.0466	6.0731	6.5580	6.8192	6.8887	**7.1335**	6.9366	<u>7.1166</u>
VIF	0.7818	0.6263	0.4105	0.6583	<u>0.7438</u>	0.7042	0.6060	**0.7843**
AG	3.8485	2.6918	2.4210	3.5600	4.4005	**4.9802**	4.8969	<u>4.9273</u>
SD	41.8759	27.8343	30.6636	34.8250	39.8827	**44.2436**	36.4150	<u>43.7810</u>
Q_{abf}	<u>0.5270</u>	0.4389	0.2327	0.4463	0.5674	0.4879	0.4267	**0.5712**
SF	10.0471	7.9978	6.2753	8.9853	11.3358	<u>12.2525</u>	11.6061	**12.3076**

3.3 Ablation Study

We propose a cross-modal information sharing module (CISM) to take full advantage of the complementary information from different modalities images. At the same time, a fine-grained feature extraction module(FFEM) is designed to extract fine-grained image features. After removing them, we conduct experiments again. The comparison results are shown in Fig. 4, from red and green boxes we can see that after losing the information sharing module(CISM), the image brightness is significantly reduced and the target edge details are not clear enough, and after losing the fine-grained feature extraction module(FFEM), the detail features of the images are not clear enough, and images are relatively fuzzy, which proves the effectiveness of our designed module.

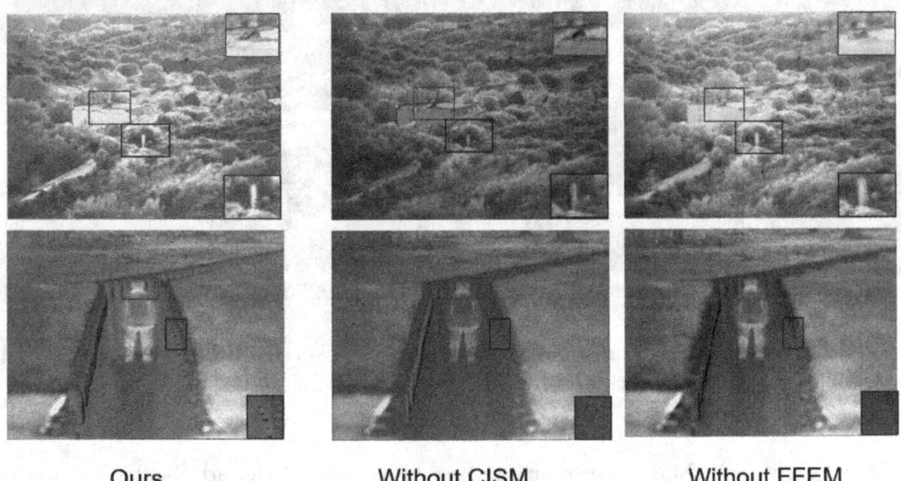

Fig. 4. Results of ablation studies

3.4 Segmentation Performance

We segment the images generated by each fusion method using DeepLabv3+ [4], and Fig. 5 shows final segmentation results. As we can see from the red and green boxes, methods such as FusionGAN, MDLatLRR, and NestFuse cannot segment the semantic targets of the sky and terrain, the rest of methods can split out the sky and terrain, but cannot preserve more semantic information of terrain. On the contrary, our method can preserve as much semantic information about the sky and terrain as possible, thus proving the superiority of our method.

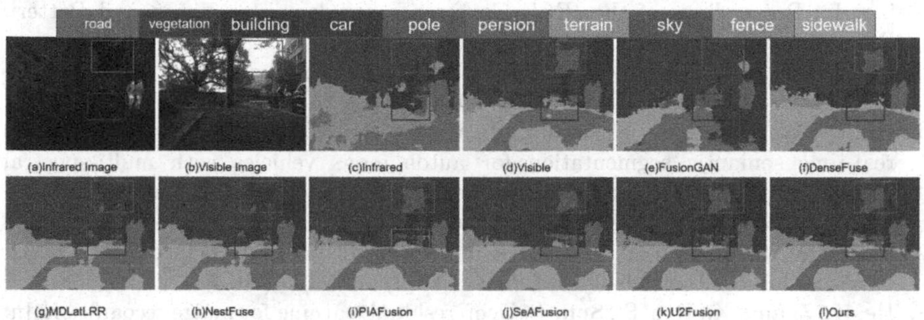

Fig. 5. Segmentation results for infrared, visible and all fused images from the MSRS dataset. The segmentation model is DeepLabv3+, which is pretrained on the Cityscapes dataset [3]. The scene is: 00555D.

4 Conclusion

In this paper, we propose an infrared and visible image fusion method named SIEFusion based on semantic information enhancement. In our method, a cross-modal information sharing module(CISM) is designed to extract the complementary information between images of different modalities, and a fine-grained feature extraction module(FFEM) is designed to extract more fine-grained image features. The final fused images contain as much semantic information as possible to meet the semantic requirements of subsequent semantic segmentation tasks. Through a series of comparative experimental results, our method performs well in terms of visual effects and quantitative metrics, and its effectiveness is verified in subsequent segmentation tasks.

References

1. Cao, Y., Guan, D., Huang, W., Yang, J., Cao, Y., Qiao, Y.: Pedestrian detection with unsupervised multispectral feature learning using deep neural networks. Inf. Fusion **46**, 206–217 (2019)
2. Chen, J., Li, X., Luo, L., Mei, X., Ma, J.: Infrared and visible image fusion based on target-enhanced multiscale transform decomposition. Inf. Sci. **508**, 64–78 (2020)
3. Chen, L.C., Zhu, Y., Papandreou, G., Schroff, F., Adam, H.: Encoder-decoder with atrous separable convolution for semantic image segmentation. In: Proceedings of the European Conference on Computer Vision (ECCV), pp. 801–818 (2018)
4. Cordts, M., et al.: The cityscapes dataset for semantic urban scene understanding. In: Proceedings of the IEEE Conference on Computer Vision and Pattern Recognition, pp. 3213–3223 (2016)
5. Eskicioglu, A.M., Fisher, P.S.: Image quality measures and their performance. IEEE Trans. Commun. **43**(12), 2959–2965 (1995)
6. Ha, Q., Watanabe, K., Karasawa, T., Ushiku, Y., Harada, T.: MFNet: Towards real-time semantic segmentation for autonomous vehicles with multi-spectral scenes. In: 2017 IEEE/RSJ International Conference on Intelligent Robots and Systems (IROS), pp. 5108–5115. IEEE (2017)
7. Han, Y., Cai, Y., Cao, Y., Xu, X.: A new image fusion performance metric based on visual information fidelity. Inf. fusion **14**(2), 127–135 (2013)
8. He, K., Zhang, X., Ren, S., Sun, J.: Deep residual learning for image recognition. In: Proceedings of the IEEE Conference on Computer Vision and Pattern Recognition, pp. 770–778 (2016)
9. Huang, G., Liu, Z., Van Der Maaten, L., Weinberger, K.Q.: Densely connected convolutional networks. In: Proceedings of the IEEE Conference on Computer Vision and Pattern Recognition, pp. 4700–4708 (2017)
10. Jin, L., et al.: Rethinking the person localization for single-stage multi-person pose estimation. IEEE Trans. Multimedia (2023)
11. Li, C., Zhu, C., Huang, Y., Tang, J., Wang, L.: Cross-modal ranking with soft consistency and noisy labels for robust RGB-T tracking. In: Proceedings of the European Conference on Computer Vision (ECCV), pp. 808–823 (2018)
12. Li, H., Wu, X.J.: Densefuse: a fusion approach to infrared and visible images. IEEE Trans. Image Process. **28**(5), 2614–2623 (2018)
13. Li, H., Wu, X.J., Durrani, T.: Nestfuse: an infrared and visible image fusion architecture based on nest connection and spatial/channel attention models. IEEE Trans. Instrum. Meas. **69**(12), 9645–9656 (2020)
14. Li, H., Wu, X.J., Kittler, J.: MDLatLRR: a novel decomposition method for infrared and visible image fusion. IEEE Trans. Image Process. **29**, 4733–4746 (2020)
15. Li, S., et al.: Single image deraining: A comprehensive benchmark analysis. In: Proceedings of the IEEE/CVF Conference on Computer Vision and Pattern Recognition, pp. 3838–3847 (2019)
16. Liu, J., et al.: Target-aware dual adversarial learning and a multi-scenario multi-modality benchmark to fuse infrared and visible for object detection. In: Proceedings of the IEEE/CVF Conference on Computer Vision and Pattern Recognition, pp. 5802–5811 (2022)
17. Liu, Y., Chen, X., Ward, R.K., Wang, Z.J.: Image fusion with convolutional sparse representation. IEEE Signal Process. Lett. **23**(12), 1882–1886 (2016)

18. Ma, J., Chen, C., Li, C., Huang, J.: Infrared and visible image fusion via gradient transfer and total variation minimization. Inf. Fusion **31**, 100–109 (2016)
19. Ma, J., et al.: Infrared and visible image fusion via detail preserving adversarial learning. Inf. Fusion **54**, 85–98 (2020)
20. Ma, J., Ma, Y., Li, C.: Infrared and visible image fusion methods and applications: a survey. Inf. Fusion **45**, 153–178 (2019)
21. Ma, J., Tang, L., Fan, F., Huang, J., Mei, X., Ma, Y.: SwinFusion: cross-domain long-range learning for general image fusion via swin transformer. IEEE/CAA J. Automatica Sinica **9**(7), 1200–1217 (2022)
22. Ma, J., Tang, L., Xu, M., Zhang, H., Xiao, G.: STDFusionnet: an infrared and visible image fusion network based on salient target detection. IEEE Trans. Instrum. Meas. **70**, 1–13 (2021)
23. Ma, J., Yu, W., Liang, P., Li, C., Jiang, J.: FusionGAN: a generative adversarial network for infrared and visible image fusion. Inf. fusion **48**, 11–26 (2019)
24. Ma, J., Zhou, Z., Wang, B., Zong, H.: Infrared and visible image fusion based on visual saliency map and weighted least square optimization. Infrared Phys. Technol. **82**, 8–17 (2017)
25. Mou, J., Gao, W., Song, Z.: Image fusion based on non-negative matrix factorization and infrared feature extraction. In: 2013 6th International Congress on Image and Signal Processing (CISP). vol. 2, pp. 1046–1050. IEEE (2013)
26. Pei, Y., Huang, Y., Zou, Q., Lu, Y., Wang, S.: Does haze removal help CNN-based image classification? In: Proceedings of the European Conference on Computer Vision (ECCV), pp. 682–697 (2018)
27. Peng, C., Tian, T., Chen, C., Guo, X., Ma, J.: Bilateral attention decoder: a lightweight decoder for real-time semantic segmentation. Neural Netw. **137**, 188–199 (2021)
28. Qu, L., Liu, S., Wang, M., Song, Z.: TransMEF: a transformer-based multi-exposure image fusion framework using self-supervised multi-task learning. In: Proceedings of the AAAI Conference on Artificial Intelligence. vol. 36, pp. 2126–2134 (2022)
29. Roberts, J.W., Van Aardt, J.A., Ahmed, F.B.: Assessment of image fusion procedures using entropy, image quality, and multispectral classification. J. Appl. Remote Sens. **2**(1), 023522 (2008)
30. Tang, L., Deng, Y., Ma, Y., Huang, J., Ma, J.: SuperFusion: a versatile image registration and fusion network with semantic awareness. IEEE/CAA J. Automatica Sinica **9**(12), 2121–2137 (2022)
31. Tang, L., Yuan, J., Ma, J.: Image fusion in the loop of high-level vision tasks: a semantic-aware real-time infrared and visible image fusion network. Inf. Fusion **82**, 28–42 (2022)
32. Tang, L., Yuan, J., Zhang, H., Jiang, X., Ma, J.: PIAFusion: a progressive infrared and visible image fusion network based on illumination aware. Inf. Fusion **83**, 79–92 (2022)
33. Toet, A.: The TNO multiband image data collection. Data Brief **15**, 249–251 (2017)
34. Woo, S., Park, J., Lee, J.Y., Kweon, I.S.: CBAM: convolutional block attention module. In: Proceedings of the European Conference on Computer Vision (ECCV), pp. 3–19 (2018)
35. Xu, H., Ma, J., Jiang, J., Guo, X., Ling, H.: U2Fusion: a unified unsupervised image fusion network. IEEE Trans. Pattern Anal. Mach. Intell. **44**(1), 502–518 (2020)

DeepChrom: A Diffusion-Based Framework for Long-Tailed Chromatin State Prediction

Yuhang Liu[1], Zixuan Wang[2], Jiaheng Lv[1], and Yongqing Zhang[1(✉)]

[1] School of Computer Science, Chengdu University of Information Technology, Chengdu 610225, China
zhangyq@cuit.edu.cn

[2] College of Electronics and Information Engineering, Sichuan University, Chengdu 610065, China

Abstract. Chromatin state reflects distinct biological roles of the genome that can systematically characterize regulatory elements and their functional interaction. Despite extensive computational studies, accurate prediction of chromatin state remains a challenge because of the long-tailed class imbalance. Here, we propose a deep-learning framework, DeepChrom, to predict long-tailed chromatin state directly from DNA sequence. The framework includes a diffusion-based model that balances the samples of different classes by generating pseudo-samples and a novel dilated CNN-based model for chromatin state prediction. On top of that, we further develop a novel equalization loss to increase the penalty on generated samples, which alleviates the impact of the bias between ground truth and generated samples. DeepChrom achieves outstanding performance on nine human cell types with our designed paradigm. Specifically, our proposed long-tailed learning strategy surpasses the traditional training method by 0.056 in Acc. To our knowledge, DeepChrom is pioneering in predicting long-tailed chromatin states by the diffusion-based model to achieve sample balance.

Keywords: Chromatin state · Diffusion model · Long-tailed learning · Bioinformatics

1 Introduction

Chromatin state refers to chromatin's different structural and functional states in different cell types [1]. Due to its diverse array of functions, which encompass pivotal roles in gene regulation, cellular differentiation, and disease pathogenesis,

This work is supported by the National Natural Science Foundation of China under Grant No. 62272067; the Sichuan Science and Technology Program under Grant No. 2023NSFSC0499; the Scientific Research Foundation of Sichuan Province under Grant Nos.2022001 and MZGC20230078; the Scientific Research Foundation of Chengdu University of Information Technology under Grant No.KYQN202208.

Q. Liu et al. (Eds.): PRCV 2023, LNCS 14427, pp. 188–199, 2024.
https://doi.org/10.1007/978-981-99-8435-0_15

it has gained increasing attention [2–5]. The epigenetic modifications on DNA sequences are the primary factor determining chromatin state. For example, Ernst et al. defined 15 chromatin states with distinct biological roles by mapping nine chromatin marks [6]. Similarly, Wang et al. defined 18 chromatin states from six histone marks by using ChIP-seq data [7]. These studies have revealed that chromatin states exhibit a long-tail distribution, with certain states being more abundant than others. For example, the number of enhancers is significantly large than insulators [8]. Although genomic assays, such as ChIP-seq, can reveal the chromatin state, it needs more expensive and time-consuming experiments. Therefore, computational methods for long-tailed chromatin state prediction are urgently required.

Many efforts have been made to predict chromatin states through deep-learning algorithms. DeepSEA is a pioneering work that constructs a CNN network to predict 919 chromatin features from DNA sequences [9]. Similarly, Chen et al. proposed a dilated CNN network, Sei, to predict chromatin profiles for discovering the regulatory basis of traits [10]. In addition to the above methods, there are many other chromatin feature predictors, such as DanQ, DeepFormer, and DeepATT, among others [11–14]. However, these methods usually ignore the long-tailed problem between chromatin states. Specifically, some approaches typically achieve sample balancing by shuffling positive samples to generate negative ones, leading to bias in actual situations. Other methods directly predict the long-tailed chromatin states, causing an imbalance between head and tailed classes. Therefore, a computational framework is needed to predict long-tailed chromatin states.

Considering these issues, we propose a deep-learning framework, DeepChrom, to predict long-tailed chromatin states by data augmentation. We leverage the probabilistic diffusion model (DM) [15] to expand long-tailed class samples for data augmentation. Such DM-expanded samples are generated from Gaussian noise through a denoising UNet. On top of that, we further apply a novel equalization loss, which aims to reduce the impact of the bias between ground truth and generated samples [16]. Combinations of these two strategies provide a simple and agnostic solution for real-world applications. In addition, we propose a novel chromatin state prediction model that uses CNN to capture the sequence motif information and dilated CNN to learn the high-order syntax. With this chromatin state predictor, the chromatin state can be accurately predicted. To demonstrate the effectiveness of our approach, we evaluate its performance on nine cell types. The results show that our proposed methods significantly outperform the competing ones.

In summary, the key contribution of this study is as follows.

- We propose a diffusion-based framework that generates pseudo-sample for solving long-tailed problems in chromatin state prediction.
- We propose an equalization loss that increases the penalty on generated samples to mitigate the impact of the bias between ground truth and generated samples.

- We conduct extensive experiments on nine human cell types and achieve superior performance on long-tailed chromatin state prediction.

The rest of this article will be organized as follows: Sect. 2 briefly reviews the related work. Section 3 provides a detailed introduction to our proposed methods. Section 4 presents the experimental results. Finally, we discuss and conclude this study in Sect. 5.

2 Related Work

2.1 Chromatin State Prediction

Chromatin state predictor aims to build a deep-learning model that learns the genome language to accurately identify chromatin states from DNA sequences. Zhou et al. developed DeepSEA, which pioneered the use of neural networks in predicting chromatin states. Following the groundbreaking work of DeepSEA [9], numerous researchers have made valuable explorations and breakthroughs in improving the performance of chromatin state prediction algorithms. The main effects have focused on model architecture. Quang et al. proposed a simple yet effective method that consists of a single CNN layer, a BiLSTM layer, and a dense layer [11]. Specifically, CNN is used to learn motif information, and BiLSTM is used to learn regulatory grammar. Yao et al. proposed a hybrid DNN model DeepFormer, which utilizes CNN and flow attention to achieve accurate chromatin feature prediction under a limited number of parameters [12]. Chen et al. performed better by integrating dilated convolution to expand the perceptual field without degrading the spatial resolution [10]. Inspired by these studies, we propose a novel chromatin state predictor, which integrates CNN, dilated CNN, and attention for better prediction.

2.2 Long-Tailed Learning

Long-tailed learning aims to train a well-performing model from many samples that follow a long-tailed class distribution. However, in real-world applications, trained models are typically biased towards head classes, causing poor performance on tail classes. To tackle this problem, many efforts have been conducted in recent years, making tremendous progress. Specifically, existing long-tailed learning studies can be grouped into three main categories [17]: class re-balancing [18], information augmentation [19], and module improvement [20]. Among these methods, resampling methods for class re-balancing are the most straightforward and practical. However, this method can lead to some drawbacks, such as information loss, overfitting, and underrepresentation. Therefore, several generative models have been proposed to address the class imbalance by generating pseudo-samples to achieve class balance [21]. Recently, probabilistic diffusion models (DMs) [15] have achieved state-of-the-art results in pseudo-sample generating. The generative power of this method stems from the parameterized

Markov chain, which contains forward and backward processes, respectively, corresponding to the process of adding noise and denoising. In this study, we apply DM-based methods to generate high-quality tailed DNA sequence samples and use a novel loss function to reduce the bias between ground truth and generated samples.

3 Methods

3.1 Methodology Overview

This paper proposes an end-to-end chromatin state prediction method based on a diffusion model and dilated CNN. As shown in Fig. 1, the proposed framework includes a diffusion model for generating pseudo-samples and a chromatin state predictor. First, the one-hot encoding strategy is applied to convert the DNA sequence to an $n \times 4$ matrix, where n is the length. Then the encoded data is fed into the diffusion model for training to generate tail class sequences from the Gaussian noise. Specifically, the diffusion model contains forward and backward steps, in which the data is gradually added to Gaussian noise in the forward step until it becomes random noise. Moreover, the backward step is a denoising process that gradually restores the data by denoising UNet. Finally, the generated DNA sequences are fed with the original sequences into a dilated CNN for training. During chromatin state predictor training, the generated sequences receive more discouraging gradients to reduce the impact of the bias caused by generated sequences.

3.2 Pseudo Sequences Generation

We employ a diffusion model for pseudo-sample generation, which consists of forward and backward processes.

Forward Process. It aims to gradually add Gaussian noise to the original data until it becomes pure noise (Fig. 1A). Let $q(x_0)$ represent the real data distribution for the DNA sequence, and x_0 is a real sequence sampled from $q(x_0)$. The forward process can be defined as $q(x_t|x_{t-1})$, which can be formulated as:

$$q(x_t|x_{t-1}) = \mathcal{N}\left(x_t; \sqrt{1-\beta_t}x_{t-1}, \beta_t I\right) \tag{1}$$

where $\mathcal{N}$ represent normal distribution with mean and variance $\sqrt{1-\beta_t}x_{t-1}, \beta_t I$ respectively. Specifically, by sampling the standard normal distribution $\epsilon\,\mathcal{N}(0, I)$, the feature vector at t-th step can be formulated as:

$$x_t = \sqrt{1-\beta_t}x_{t-1} + \sqrt{\beta_t}\epsilon_{t-1} \tag{2}$$

where x_t is the vector of each DNA sequence after the t-th addition of the noise; β_t is a constant that takes values between 0 and 1, and β linear increases with t; ϵ_{t-1} is the base noise obtained from the $(t-1)$-th sampling. Especially, x_0 is the one-hot encoding DNA sequence.

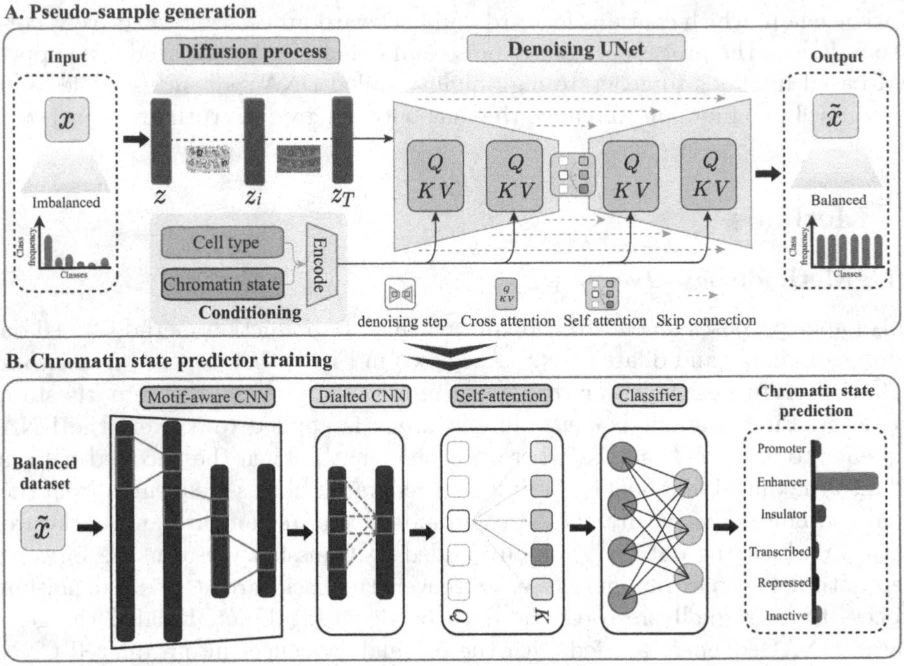

Fig. 1. Overview framework of our proposed method. (A) The diffusion-based model is used to generate long-tailed class sequences. The DNA sequences of long-tailed classes are first converted to noise via diffusion. Then, the denoising UNet is applied to restore the sequence by fusing the cell type and chromatin state information. (B) The illustration of the chromatin state predictor. It uses the generated balanced sample for training.

Backward Process. It aims to learn a data distribution $p(x)$ by gradual denoising from a Gauss distribution, equivalent to learning the inverse process of a Markov chain of length T. We add conditions in the backward process to construct a general model that can generate different chromatin state sequences across different cell types. It can be defined as $p\left(x_{t-1}|x_t, c\right)$. However, this process cannot be solved directly. Therefore, a deep-learning model is applied to fit this distribution, which can be formulated as follows:

$$p_{\theta,\phi}\left(x_{t-1}|x_t\right) = N\left(x_{t-1}; \mu_\theta\left(x_t, t\right), \Sigma_{\theta(x_t,t)}\right) p_\phi(c, x_t) \tag{3}$$

where $\mu_\theta\left(x_t, t\right)$ and $\Sigma_{\theta(x_t,t)}$ represent the mean and variance of the normal distribution $\mathcal{N}$; c represents condition, namely cell type and chromatin state. Specifically, we fix the variance $\Sigma_{\theta(x_t,t)} = \sigma_t^2 I$ and the network learns only the mean. We refer to the most successful diffusion model for noise predictor and use a UNet-based model to restore the sequences from noise [22]. This model relies on a re-weighted variant of the variational lower bound on $p(x)$, which mirrors denoising score-matching. The condition c can be considered by the cross-attention.

Specifically, it can be interpreted as an equally weighted sequence of denoising autoencoders $\epsilon_\theta(x_t, t)$. The objective can be described as follows:

$$L_{DM} = \mathbb{E}_{x,c,\epsilon \in \mathcal{N}(0,1),t}\left[\|\epsilon - \epsilon_\theta(x_t, t, c)\|_2^2\right] \tag{4}$$

where x_t is a noisy version of the input x, c is the condition, and t uniformly sampled from $\{1, \ldots, T\}$.

3.3 Chromatin State Prediction

As shown in Fig. 1B, the chromatin state predictor is a dilated CNN-based network, which takes the DNA sequence as input and predicts the corresponding chromatin state. It is composed of (i) a motif-aware CNN block for extracting sequence patterns (motifs), (ii) a dilated CNN block for learning the syntax, (iii) a self-attention block for capturing the internal correlation of the syntax, and (vi) a classification block for modeling the chromatin state for each sequence. Given the input s, the computational process can be described as follow.

$$x = Conv(s) \tag{5}$$

$$x = Dropout(ReLU(x), 0.2) \tag{6}$$

where $Conv()$ represent the convolution operation. The output of the motif-aware CNN block is fed to the dilated CNN block.

$$x = dConv(x, d) \tag{7}$$

$$x = Dropout(ReLU(x), 0.2) \tag{8}$$

where $dConv()$ represents dilated convolution operation, and d represents the dilation size. After that, self-attention is used to learn long and short dependency.

$$Attention(Q, K, V) = softmax\left(\frac{QK^T}{\sqrt{d_k}}\right)V \tag{9}$$

where Q, K, and V represent queries, keys, and values vectors; $\sqrt{d_k}$ is the dimension of the input key vector. The multi-head is applied to obtain the final values to increase the model capacity. After that, the classification block is used to make the final prediction.

$$y = Activation(MLP(x)) \tag{10}$$

where y is the predicted chromatin state. $Activation()$ is the activation function.

3.4 Equalization Loss

Our equalization loss aims to reduce the impact of the bias between ground truth and generated sequences. We achieve this by re-weighting the conventional softmax cross-entropy loss function, which can be formulated as:

$$L = -w \sum_{j=1}^{C} y_j log(p_j) \tag{11}$$

where C is the number of categories, p_j is probability of class j, which is calculated by $softmax$ funciton. y_j is the true class label, which can be formulated as:

$$y_j = \begin{cases} 1 & \text{if } j = c, \\ 0 & \text{otherwise.} \end{cases} \tag{12}$$

For equalization loss, we introduce a weight term w, which can be formulated as:

$$w = \begin{cases} 1 & \text{if } x \text{ is ground truth sequence,} \\ \mu & \text{otherwise.} \end{cases} \tag{13}$$

where μ is the manually selected practical value.

4 Experiments

4.1 Experimental Settings

Dataset. We collect all nine cellular chromatin states from ChromHMM [8], which defines 15 chromatin states for each cellular. These chromatin states can be divided into six broad classes: enhancer, transcribed, inactive, promoter, insulator, and repressed. The DNA sequences are derived from '.bed' files for dataset processing, which record the genome coordinates, and correspondent chromatin states [23]. The lengths of each sequence are expanded to 1000bp by centering on the coordinate midpoint because the median of each chromatin state is about 1000bp.

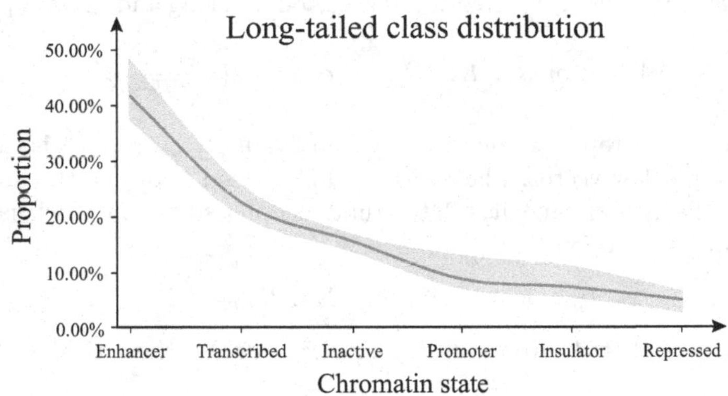

Fig. 2. The proportion of each chromatin state. The blue line represents the average proportion, while the light blue area represents the maximum and minimum proportion of the nine cells. (Color figure online)

In the experiment, we divide the dataset into the training set, validation set, and test set, which account for 80%, 10%, and 10%, respectively. Specifically, the training set contains ground truth and generated sequences, while validation and test sets only contain ground truth sequences. Since DeepChrom requires binary vectors as input, each DNA sequence is converted to one-hot format, namely A=[1, 0, 0, 0], C=[0, 1, 0, 0], G=[0, 0, 1, 0], and T=[0, 0, 0, 1]. After that, each sequence is encoded as a 1000 × 4 matrix. The proportion of each chromatin state is shown in Fig. 2.

Comparison Methods and Evaluation Metrics. To measure the performance of our proposed framework, several well-known methods are used for comparison, including DeepSEA [9], DanQ [11], and Sei [10].

The performance of our proposed framework is evaluated by several metrics, including accuracy (Acc), upper reference accuracy (UA), relative accuracy (RA), and AUROC. Specifically, Acc represents the top-1 accuracy, UA is the maximal value between vanilla accuracy A_v on a balanced training set and balanced accuracy A_b on a balanced training set, defined by $A_u = max\,(A_v, A_b)$, and the relative accuracy is calculated by the empirically upper reference accuracy divided by top-1 accuracy [17], which is defined by $A_r = \frac{A_t}{A_u}$.

Implementation Details. After extensive experiments, we finally determined the hyperparameters. For diffusion model training, we train the network about 15 epochs with a batch size 64. We adopt AdamW as the optimizer with the learning rate of 5×10^{-4} and weight decay as 1×10^{-4}. As for the structure of the diffusion model, we apply a linear β schedule, ranging from 0.0002 to 0.04, and the time step T=400. We train the network about 15 epochs for chromatin state predictor training with a batch size of 128. we also utilize AdamW as the optimizer with the same learning rate and weight decay. The γ is set to 0.9, and μ is set to 1.2. We stack three motif-aware CNN blocks for the model structure with kernel size = 9 and the max pooling size = 3. The dilation is (2, 4, 8) for three dilated CNN blocks, respectively. The number of transformer layers is 2 with 4 attention heads. In the experiment, we adopt Pytorch to implement and train the model on Nvidia 3080.

4.2 Effectiveness of Our Proposed Long-Tailed Learning Methods

Quality of the Generated Sequence. The quality of the generated sequence largely determines the model's performance. Figure 3 shows the performance of the different models on ground truth and generated sequences. The results show that chromatin prediction methods can effectively discriminate the generated sequences. Our chromatin predictor performs better among competing methods, resulting in 0.645 for Acc in generated sequences. This result demonstrates the validity of the generated sequence. However, we notice that the performance of generated sequence is poorer than the ground truth sequence, resulting in a decrease in Acc by 0.031. We hypothesize that the generated sequence may have a bias from the ground truth sequence, so the generated sequences do not fully reflect the chromatin state. Overall, these results demonstrate the effectiveness of the generated sequences but have a bias with the ground truth sequence.

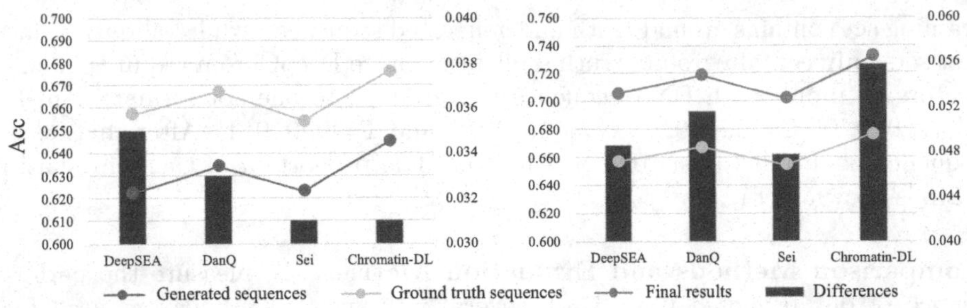

Fig. 3. The performance of the proposed approach. The left figure compares the average accuracy of nine cells when training using generated and ground truth sequences. At the same time, the right figure compares the average accuracy when training using ground truth sequences and balanced samples with equalization loss.

Superior of Our Proposed Framework. We test the performance of data augmentation with equalization loss for long-tail chromatin state prediction as shown in Fig. 3 and 4, our proposed long-tailed learning strategy demonstrates accurate prediction of the long-tailed chromatin state, outperforming traditional training methods. We observe that (i) By utilizing a diffusion model for data augmentation and combining it with equalization loss, we effectively enhance the prediction of long-tailed chromatin state. This approach improves 0.048-0.056 in Acc and 0.040-0.054 in AUROC for different models. (ii) Our chromatin state predictor performs better than competing methods. It surpasses the second-best method by 0.013 in both Acc and AUROC. These results indicate that our approach is practical for long-tailed chromatin state prediction.

4.3 Ablation Study

We conduct extensive ablation studies on nine cell-type datasets to explore the effectiveness of data augmentation and equalization loss in long-tailed chromatin state prediction. One strategy is varied to control computational cost at a time, with all others set to their default setting. As shown in Fig. 3 and Table 1, using DM for tail class sample generation can effectively improve performance and achieve Acc gains of 0.014-0.022 among different methods. Given the bias between the generated sequence and the ground truth sequence, we then test the effectiveness of our proposed equalization loss. As shown in Table 1, the proposed equalization loss can mitigate the impact of the bias, achieving additional improvement of 0.026-0.024 for Acc and 0.027-0.032 for RA in the different models.

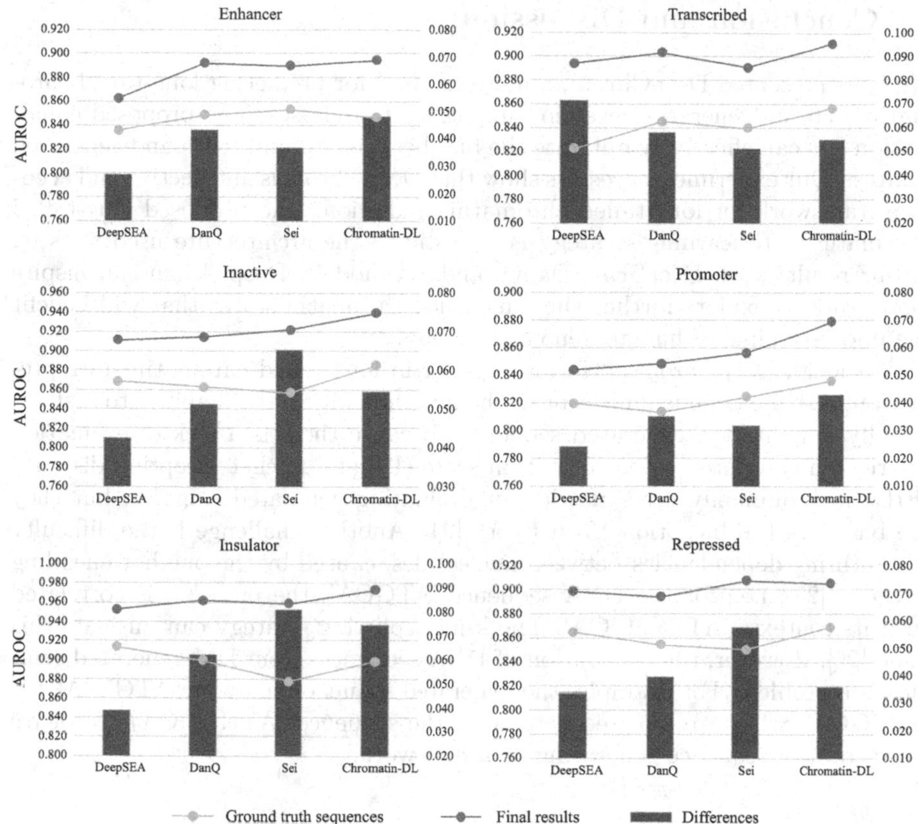

Fig. 4. The comparison of average AUROC performance using ground truth sequences and balanced sequences with equalization loss in six chromatin states.

Table 1. The average results on nine cell types for four methods in terms of accuracy (Acc), upper reference accuracy (UA), and relative accuracy (RA) under equalization loss or without equalization loss.

Method	Without equalization loss			With equalization loss		
	Acc	UA	RA	Acc	UA	RA
DeepSEA	0.671	0.746	0.899	0.706	0.758	0.931
DanQ	0.683	0.758	0.901	0.719	0.772	0.931
Sei	0.676	0.755	0.895	0.702	0.761	0.922
DeepChrom	**0.691**	**0.762**	**0.907**	**0.732**	**0.781**	**0.937**

5 Conclusion and Discussion

We have presented DeepChrom, a novel method for predicting long-tailed chromatin state by generating pseudo-samples for tail classes. The proposed equalization loss can effectively mitigate the bias between ground truth and generated samples. Our experimental results show that DeepChrom is an effective and accurate framework for long-tailed chromatin prediction. The proposed long-tailed chromatin state learning strategy is agnostic to the architecture used, as seen in our results with DeepSEA, DanQ, and Sei models. DeepChrom can inspire more work to explore further the long-tailed chromatin states that will benefit the understanding of human genome function.

Although DeepChrom performs superior in long-tailed chromatin state prediction, it still has some limitations. The first difficulty is the inability to systematically evaluate the generated sequences because there is a lack of evaluation metrics. In computer vision, inception score (IS) and fréchet inception distance (FID) are commonly used metrics for evaluating generated samples, but they are based on the Inception V3 network [24]. Another challenge is the difficulty in capturing dependencies between nucleotides, caused by the one-hot encoding strategy [25], i.e., for the DNA sequence 'ATGCA', the base 'G' is correlated with its contexts 'AT' and 'CA'. The k-mer splitting strategy can mitigate this issue [26]. However, the restoration of DNA sequences from DM-generated samples is a problem. For example, the generated sequences may be 'ATG', 'AGC', and 'GCA', which are not recoverable to the sequence (ATG(A)CA). We leave this aspect of sequence generating as future work.

References

1. Ernst, J., Kellis, M.: Discovery and characterization of chromatin states for systematic annotation of the human genome. Nat. Biotechnol. **28**(8), 817–825 (2010)
2. Orouji, E., Raman, A.T.: Computational methods to explore chromatin state dynamics. Briefings in Bioinformatics 23(6), bbac439 (2022)
3. Dupont, S., Wickström, S.A.: Mechanical regulation of chromatin and transcription. Nat. Rev. Genet. **23**(10), 624–643 (2022)
4. Preissl, S., Gaulton, K.J., Ren, B.: Characterizing cis-regulatory elements using single-cell epigenomics. Nat. Rev. Genet. **24**(1), 21–43 (2023)
5. Chao, W., Quan, Z.: A machine learning method for differentiating and predicting human-infective coronavirus based on physicochemical features and composition of the spike protein. Chin. J. Electron. **30**(5), 815–823 (2021)
6. Ernst, J., et al.: Mapping and analysis of chromatin state dynamics in nine human cell types. Nature **473**(7345), 43–49 (2011)
7. Wang, Z., et al.: Prediction of histone post-translational modification patterns based on nascent transcription data. Nat. Genet. **54**(3), 295–305 (2022)
8. Vu, H., Ernst, J.: Universal annotation of the human genome through integration of over a thousand epigenomic datasets. Genome Biol. **23**, 1–37 (2022)
9. Zhou, J., Troyanskaya, O.G.: Predicting effects of noncoding variants with deep learning-based sequence model. Nat. Methods **12**(10), 931–934 (2015)

10. Chen, K.M., Wong, A.K., Troyanskaya, O.G., Zhou, J.: A sequence-based global map of regulatory activity for deciphering human genetics. Nat. Genet. **54**(7), 940–949 (2022)
11. Quang, D., Xie, X.: DanQ: a hybrid convolutional and recurrent deep neural network for quantifying the function of DNA sequences. Nucleic Acids Res. **44**(11), e107–e107 (2016)
12. Yao, Z., Zhang, W., Song, P., Hu, Y., Liu, J.: DeepFormer: a hybrid network based on convolutional neural network and flow-attention mechanism for identifying the function of DNA sequences. Briefings in Bioinformatics 24(2), bbad095 (2023)
13. Li, J., Pu, Y., Tang, J., Zou, Q., Guo, F.: Deepatt: a hybrid category attention neural network for identifying functional effects of dna sequences. Briefings in bioinformatics 22(3), bbaa159 (2021)
14. Kelley, D.R., Snoek, J., Rinn, J.L.: Basset: learning the regulatory code of the accessible genome with deep convolutional neural networks. Genome Res. **26**(7), 990–999 (2016)
15. Ho, J., Jain, A., Abbeel, P.: Denoising diffusion probabilistic models. Adv. Neural. Inf. Process. Syst. **33**, 6840–6851 (2020)
16. Tan, J., et al.: Equalization loss for long-tailed object recognition. In: Proceedings of the IEEE/CVF Conference on Computer Vision and Pattern Recognition, pp. 11662–11671 (2020)
17. Zhang, Y., Kang, B., Hooi, B., Yan, S., Feng, J.: Deep long-tailed learning: a survey. IEEE Trans. Pattern Anal. Mach. Intell. (2023)
18. Kang, B., et al.: Decoupling representation and classifier for long-tailed recognition. arXiv preprint arXiv:1910.09217 (2019)
19. Li, S., Gong, K., Liu, C.H., Wang, Y., Qiao, F., Cheng, X.: Metasaug: Meta semantic augmentation for long-tailed visual recognition. In: Proceedings of the IEEE/CVF Conference on Computer Vision and Pattern recognition, pp. 5212–5221 (2021)
20. Zhang, S., Li, Z., Yan, S., He, X., Sun, J.: Distribution alignment: a unified framework for long-tail visual recognition. In: Proceedings of the IEEE/CVF Conference on Computer Vision and Pattern Recognition, pp. 2361–2370 (2021)
21. Kingma, D.P., Dhariwal, P.: Glow: Generative flow with invertible 1x1 convolutions. Advances in neural Inf. Process. Syst. 31 (2018)
22. Dhariwal, P., Nichol, A.: Diffusion models beat GANs on image synthesis. Adv. Neural. Inf. Process. Syst. **34**, 8780–8794 (2021)
23. Liu, Y., Wang, Z., Yuan, H., Zhu, G., Zhang, Y.: Heap: a task adaptive-based explainable deep learning framework for enhancer activity prediction. Briefings in Bioinformatics, p. bbad286 (2023)
24. Szegedy, C., Vanhoucke, V., Ioffe, S., Shlens, J., Wojna, Z.: Rethinking the inception architecture for computer vision. In: Proceedings of the IEEE Conference on Computer Vision and Pattern Recognition, pp. 2818–2826 (2016)
25. Zhang, Q., et al.: Base-resolution prediction of transcription factor binding signals by a deep learning framework. PLoS Comput. Biol. **18**(3), e1009941 (2022)
26. Zhang, Y., et al.: Uncovering the relationship between tissue-specific TF-DNA binding and chromatin features through a transformer-based model. Genes **13**(11), 1952 (2022)

Adaptable Conservative *Q*-Learning for Offline Reinforcement Learning

Lyn Qiu[1], Xu Li[2], Lenghan Liang[3], Mingming Sun[2], and Junchi Yan[1(✉)]

[1] MoE Key Lab of Artificial Intelligence, Shanghai Jiao Tong University, Shanghai, China
{lyn_qiu,yanjunchi}@sjtu.edu.cn

[2] Cognitive Computing Lab, Baidu Research, Beijing, China
{lixu13,sunmingming01}@baidu.com

[3] School of Artificial Intelligence, Peking University, Beijing, China
975529547@pku.edu.cn

Abstract. The Out-of-Distribution (OOD) issue presents a considerable obstacle in offline reinforcement learning. Although current approaches strive to conservatively estimate the Q-values of OOD actions, their excessive conservatism under constant constraints may adversely affect model learning throughout the policy learning procedure. Moreover, the diverse task distributions across various environments and behaviors call for tailored solutions. To tackle these challenges, we propose the Adaptable Conservative Q-Learning (ACQ) method, which capitalizes on the Q-value's distribution for each fixed dataset to devise a highly generalizable metric that strikes a balance between the conservative constraint and the training objective. Experimental outcomes reveal that ACQ not only holds its own against a variety of offline RL algorithms but also significantly improves the performance of CQL on most D4RL MuJoCo locomotion tasks in terms of normalized return.

Keywords: Offline Reinforcement Learning · Deep Reinforcement Learning · Out-of-distribution problem · Conservative algorithm

1 Introduction

In recent years, reinforcement learning (RL) has showcased its immense potential across a vast array of domains, spanning from robotic control [14] and strategy games [21,37], to recommendation systems [27,43]. Nevertheless, the substantial costs and safety concerns associated with agent-environment interactions, particularly in real-world scenarios, pose significant challenges [7]. Offline RL (Batch RL) [20,22] offers an enticing alternative, enabling agents to learn from

L. Qiu—Work partly done during internship at Cognitive Computing Lab, Baidu Research.

J. Yan—The SJTU authors were supported by NSFC (61972250, U19B2035), Shanghai Municipal Science and Technology Major Project (2021SHZDZX0102).

Q. Liu et al. (Eds.): PRCV 2023, LNCS 14427, pp. 200–212, 2024.
https://doi.org/10.1007/978-981-99-8435-0_16

extensive, pre-collected datasets without necessitating further interactions. However, since the target value in policy evaluation relies on the learned policy for sampling actions, and the Q-value during training employs the behavior policy for action sampling, this inevitably leads to out-of-distribution (OOD) actions that may impede effective training.

To alleviate the distribution shift issue, most existing methods [24,32,35] employ explicit policy constraints, ensuring that the learned policy remains closely aligned with the behavior policy. Additionally, they utilize learning latent actions and uncertainty estimation to minimize overestimation [2,26,29]. Among these techniques, maintaining conservatism in Q-value estimation [19] is widely supported in offline RL, as it penalizes OOD state-action pairs' values while simultaneously limiting generalization [25]. To strike a balance, a previous method [38] adopts neural networks as flexible weights and incorporates surrogate losses to learn suitable adaptive functions. However, this approach encounters unstable training issues and exhibits considerable fluctuations across different task settings, necessitating unified and appropriate measures for various environments and behaviors.

In this work, we introduce Adaptable Conservative Q-Learning (ACQ), an innovative algorithm designed to tackle the OOD problem. Our main contributions are as follows:

1) Building upon the foundation of Conservative Q-learning (CQL), we develop adaptive conservatism by updating the loss weight for constraints according to the moving average difference between the predicted Q-value and a predefined anchor value.
2) Through a quantitative analysis of the datasets' statistical properties and attempts to employ the median percentile of the anchor value, we find that ACQ can maintain the Q-value under the learned policy close to the expected Q-value through adaptive conservatism.

Experiments on D4RL-v0 and D4RL-v2 benchmarks demonstrate that our approach surpasses CQL in 10 out of 15 tasks, achieving a higher average score and rendering it competitive with recent state-of-the-art offline RL methods.

2 Related Work

Model-Free Offline RL. Numerous model-free offline RL methods strive to maximize returns while ensuring that the learned policy and the behavior policy remain sufficiently close. This is typically achieved by incorporating explicit policy constraints, such as KL-divergence [12], Wasserstein distance [39], or Euclidean distance [9]. Alternatively, implicit constraints can be enforced by leveraging importance sampling [28], learning latent actions [3], and uncertainty quantification [2,4]. However, these methods may produce over-pessimistic Q-value functions and suffer from unstable Q-value volatility, resulting in fluctuating performance in non-expert datasets.

Model-Based Offline RL. In contrast to model-free approaches, model-based offline RL methods [30,41] learn the dynamics model in a supervised manner and utilize the learned dynamics for policy optimization [25]. Despite their potential, these methods may be hindered by issues such as lack of interpretability, weak generalizability, and high computational costs [42].

Conservative Algorithms. The conservative algorithm [25,26,34] aims to provide a conservative estimate of the Q-function when the dataset is not abundant enough, which has demonstrated considerable success in offline RL. For instance, the principle has been incorporated into a modified version of classical value iteration called model-based offline value iteration [5], in which a penalty term is subtracted from the estimated Q-values. This approach has been shown to achieve impressive sample efficiency in recent studies [13,33]. It is worth noting that the model-based approach relies on constructing an empirical transition kernel, which necessitates a specific representation of the environment, as discussed in [1,23].

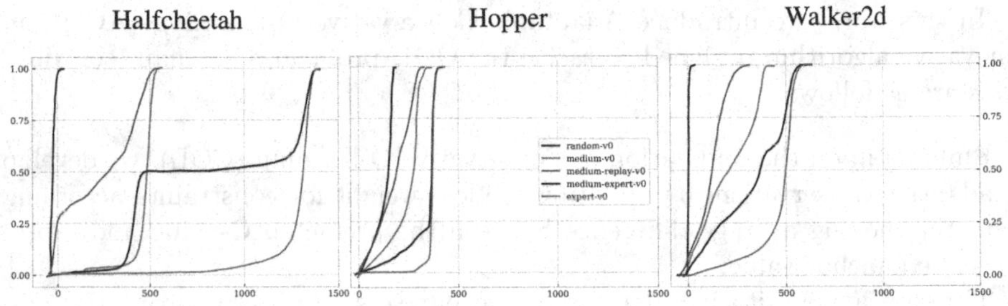

Fig. 1. The cumulative distribution function of D4RL-v0 datasets in all MuJoCo locomotion tasks.

3 Prelinminaries

We represent the environment as an infinite-horizon Markov Decision Process (MDP) defined by a tuple $(\mathcal{S}, \mathcal{A}, \mathcal{P}, r, \rho, \gamma)$, where $\mathcal{S}$ is the state space, $\mathcal{A}$ is the action space, $P(\cdot \mid s, a)$ is the transition distribution,r is the reward function, $\rho_0(s)$ is the initial state distribution, and $\gamma \in (0,1)$ is the discount factor. The goal of RL agent is to learn a policy $\pi(a \mid s)$ that maximizes the expected cumulative discounted reward $\sum_{i=t+1}^{\infty} \gamma^i r\left(s_i, a_i, s_{i+1}\right)$.

Given a replay buffer $\boldsymbol{D}=\left\{\left(s, a, r, s^{\prime}\right)\right\}$, the policy iteration and policy improvement can be represented as follows:

$$\hat{Q}^{k+1} \leftarrow \arg\min_Q \mathbb{E}_{s,a\sim\mathcal{D}}\left[\left(\hat{\mathcal{B}}^\pi \hat{Q}^k(s,a) - Q(s,a)\right)^2\right] \text{ (policy evaluation)} \quad (1)$$

$$\hat{\pi}^{k+1} \leftarrow \arg\max_\pi \mathbb{E}_{s\sim\mathcal{D},a\sim\pi^k(a|s)}\left[\hat{Q}^{k+1}(s,a)\right] \text{ (policy improvement)} \quad (2)$$

where k is the iteration number, $\hat{\mathcal{B}}^{\pi}$ represents the bellman operator on the fixed dataset following the policy $\hat{\pi}(a \mid s)$, and the bellman operator is defined as $\hat{\mathcal{B}}^{\pi}\hat{Q}^{k}(s,a) = \mathbb{E}_{s,a,s' \sim \mathcal{D}}\left[r(s,a) + \gamma \mathbb{E}_{a' \sim \hat{\pi}^{k}(a'|s')}\left[\hat{Q}^{k}\left(s', a'\right)\right]\right]$. The main difference between online RL and offline RL is that dataset $\boldsymbol{D}$ needs to be collected by a behavior policy $\pi_{\beta}(a \mid s)$.

Anchor Value. The Anchor value represents an estimate of the expected cumulative reward that an agent can obtain from a given state or state-action pair, following a specific policy in reinforcement learning. In our approach, we employ the 50th percentile of the Q-values obtained from the collected datasets. This is because, in the case of a relatively stable data distribution (as shown in Fig. 1), the 50th percentile lies in the middle of the data distribution and better reflects the central tendency of the data, being less affected by extreme values. Meanwhile, the anchor value is computed based on the expected discounted reward under the 50 percent best policy present in the dataset.

Algorithm 1. Adaptable Conservative Q-Learning

1: Initialize Offline dataset $\mathcal{D}$, Q-function Q_{θ}, policy π_{ϕ}, Anchor value Q^{Anchor}. Initialize balance weight $\eta^{0} = 0$ and moving average factor w.
2: **For** k =1 to **K do**
3: Sample batch B from D
4: Train the Q-function $Q_{\theta}^{k}(s,a)$ using G_{Q} gradient steps on objective from Eq. (12): $G_{\theta} = \nabla$ ACQ (θ)
5: Improve policy π_{ϕ} via G_{π} gradient steps on objective with SAC-style entropy regularization [11]: $G_{\phi} = \nabla_{\phi}\mathbb{E}_{s \sim \mathcal{B}, a \sim \pi_{\phi}^{k}(\cdot|s)}\left[Q_{\theta}^{k}(s,a) - \log \pi_{\phi}^{k}(a \mid s)\right]$
6: Update the balance weight α^{k} via Equation (5): $\eta^{k+1} \leftarrow \eta^{k} + w \cdot \left(\mathbb{E}_{s,a \sim \mathcal{B}} Q_{\theta}^{k}(s,a) - Q^{\text{Anchor}}\right)$
7: **End for**

4 Methodology

In this section, we first present the instantiation of our proposed algorithm, Adaptable Conservative Q-Learning (ACQ). Following this, we provide a theoretical analysis to verify the convergence of ACQ. Lastly, we offer practical implementation details.

4.1 Adaptable Conservative Q-Learning

We use Conservative Q-Learning [19] as our baseline algorithm, its objective is to minimize the expected Q-value under the dataset distribution of (s, a) pairs. CQL uses a conservative estimate of the Q-function to address the challenges of distributional shift and overestimation bias. Specifically, it adds a lower bound to penalize the expected Q-value at distribution $\mu(a \mid s)$ of OOD state-action pairs and improve the bound with an additional maximization term under the learned data distribution $\hat{\pi}_\beta(a \mid s)$, iteration update can be expressed as:

$$\begin{aligned}\hat{Q}^{k+1} \leftarrow \arg\min_{Q} \frac{1}{2}\mathbb{E}_{s,a,s'\sim\mathcal{D}}\left[\left(Q(s,a) - \hat{\mathcal{B}}^{\pi}\hat{Q}^{k}(s,a)\right)^{2}\right] \\ + \alpha \cdot \left(\mathbb{E}_{s\sim\mathcal{D},a\sim\mu(a|s)}[Q(s,a)] - \mathbb{E}_{s\sim\mathcal{D},a\sim\hat{\pi}_\beta(a|s)}[Q(s,a)]\right)\end{aligned} \tag{3}$$

where the former part indicates the conservatism part, while the latter part corresponds to a standard Bellman error objective.

$$\begin{aligned}\eta^{k+1}\left(\mathbb{E}_{s\sim\mathcal{D},a\sim\mu(a|s)}[Q(s,a)] - \mathbb{E}_{s\sim\mathcal{D},a\sim\hat{\pi}_\beta(a|s)}[Q(s,a)]\right) \\ + \min_{Q} \frac{1}{2}\mathbb{E}_{s,a\sim\mathcal{D}}\left[\left(Q(s,a) - \hat{\mathcal{B}}^{\pi}\hat{Q}^{k}(s,a)\right)^{2}\right]\end{aligned} \tag{4}$$

$$\eta^{k+1} \leftarrow \eta^{k} + w \cdot \left(\mathbb{E}_{s,a\sim\mathcal{D}}Q^{k}(s,a) - Q^{\text{Anchor}}\right) \tag{5}$$

In our algorithm, we set a smoothed moving average factor w [10] for trade-off factor η^{k+1} to facilitate the convergence of the expected Q-value $Q^k(\mathrm{s,a})$ towards the designated anchor value Q^{Anchor}.
In cases where the estimated Q-value exceeds the anchor value, the adjustable factor associated with conservative components is increased to prevent sudden escalations in the Q-value. On the other hand, when the estimated Q-value falls below the anchor value, the adjustable factor is decreased to avoid the estimated Q-value prediction from becoming indistinct, thereby hindering the learning of distinct action advantages.

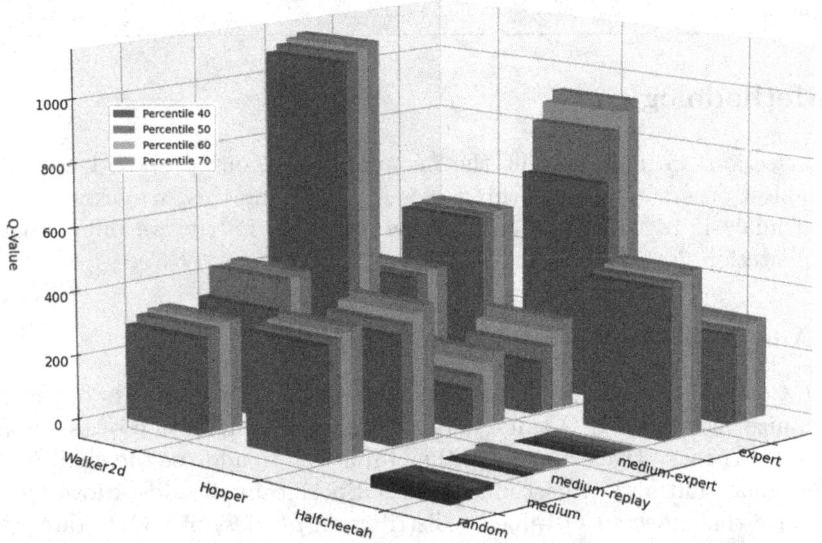

Fig. 2. Visualization of percentile of Q-value distribution for all task settings on D4RL-v0 environment.

For a better understanding of the ACQ, setting the derivative of Eq. (4) to 0, we can represent the Q^{k+1} iterative expression as:

$$\forall s, a \in \mathcal{D}, k, \quad \hat{Q}^{k+1}(s,a) = \hat{\mathcal{B}}^{\pi}\hat{Q}^{k}(s,a) - \eta^{k+1}\frac{\mu(a \mid s)}{\hat{\pi}_{\beta}(a \mid s)} \tag{6}$$

Proposition 1. Let $\tau = w \cdot \frac{\mu(\mathrm{a}|\mathrm{s})}{\hat{\pi}_{\beta}(\mathrm{a}|\mathrm{s})}$ represent a coupled indicating factor between the adjustable penalty weight and the importance sampling factor $\frac{\mu(\mathrm{a}|\mathrm{s})}{\hat{\pi}_{\beta}(\mathrm{a}|\mathrm{s})}$. If $\tau < (1-\sqrt{\gamma})^2$, The sequence of Q-value $Q^k(s,a)$ under the policy $\mu(a \mid \mathrm{s})$ of ACQ will converge to the given anchor value Q^{Anchor}.

Proof. Using the Bellman iteration with the balance weight, we can express Q^{k+1} as:

$$Q^{k+1}(s,a) = \mathbf{r} + \gamma Q^{k}(s,a) - \tau \cdot \sum_{i=0}^{k} Q_i \tag{7}$$

For simplicity, let $Q^k = Q^{k+1}(s,a) - Q^{\mathrm{Anchor}}$, then we could get:

$$Q^{k+1} = \gamma Q^{k} - \tau \cdot \sum_{i=0}^{k} Q_i \tag{8}$$

For ease of analysis, let $J_k = \sum_{i=0}^{k} Q_i$. The recursive expression can be formulated as follows:

$$\begin{bmatrix} J_{k+1} \\ J_k \end{bmatrix} = \begin{bmatrix} 1+\gamma-\tau & -\gamma \\ 1 & 0 \end{bmatrix} \begin{bmatrix} J_k \\ J_{k-1} \end{bmatrix} \tag{9}$$

Let λ_i denote an eigenvalue of $\begin{bmatrix} 1+\gamma-\tau & -\gamma \\ 1 & 0 \end{bmatrix}$, then for all $\lambda_i \in \{\lambda_1, \ldots, \lambda_n\}$, $|\lambda_i| < 1$, where $0 \leq \tau < (1-\sqrt{\gamma})^2$. The convergence properties of Q-function iterating holds after Q-function iterations.

4.2 Variants and Practical Object

To shorten the computation cost of (3), CQL($\mathcal{R}$) [19] use $\mu(a \mid s)$ to approximate the policy for maximizing the current Q-value integrate with a regularizer $\mathcal{R}(\mu)$:

$$\begin{aligned} &\min_{Q}\max_{\mu}\frac{1}{2}\mathbb{E}_{s,a,s'\sim\mathcal{D}}\left[\left(Q(s,a) - \hat{\mathcal{B}}^{\pi_k}\hat{Q}^{k}(s,a)\right)^2\right] \\ &\quad + \alpha\cdot\left(\mathbb{E}_{s\sim\mathcal{D},a\sim\mu(a|s)}[Q(s,a)] - \mathbb{E}_{s\sim\mathcal{D},a\sim\hat{\pi}_{\beta}(a|s)}[Q(s,a)]\right) + \mathcal{R}(\mu) \end{aligned} \tag{10}$$

During actual practice, it's noticed that setting the regularizer as a KL divergence against a uniform prior distribution, $\mathcal{R}(\mu) = -\bar{D}_{\mathrm{KL}}(\mu, \rho)$ and $\rho = \mathrm{Unif}(a)$, could be more steady with high dimensional actions spaces as CQL($\mathcal{H}$):

$$\min_{Q} \frac{1}{2}\mathbb{E}_{s,a,s'\sim\mathcal{D}}\left[\left(Q - \widehat{\mathcal{B}}^{\pi_k}\hat{Q}^k\right)^2\right] \\ +\alpha \cdot \mathbb{E}_{s\sim\mathcal{D}}\left[\log\sum_a \exp(Q(s,a)) - \mathbb{E}_{a\sim\widehat{\pi}_\beta(a|s)}[Q(s,a)]\right] \tag{11}$$

The first term in the second part corresponds to a soft maximum of the Q-values for any state s. In practice, we refer to Eq. (12) and present a variant of Eq. (4) as follows:

$$\begin{aligned} \mathrm{ACQ}(\theta) =& \frac{1}{2}\mathbb{E}_{s,a,s'\sim\mathcal{D}}\left[\left(Q_\theta - \widehat{\mathcal{B}}^{\pi_k}\hat{Q}^k_\theta\right)^2\right] \\ &+ \eta^{k+1} \cdot \mathbb{E}_{s\sim\mathcal{D}}\left[w\log\sum_a \exp(Q_\theta(s,a)) - \mathbb{E}_{a\sim\widehat{\pi}_\beta(a|s)}[Q_\theta(s,a)]\right] \end{aligned} \tag{12}$$

Among them, the anchor value is linearly from 0 to the percentile of the Q-value distribution during the training process to guarantee the stability of resilient conservatism. The complete pseudo-code is shown in Algorithm 1.

4.3 Implementation Settings

Q-Value. We compute the Q-value for each state-action pair (s, a) by employing the Monte Carlo return, which is the sum of discounted rewards from the state and action pair until the termination of the episode. Although the Gym Mujoco environments are intrinsically infinite-horizon non-episodic continuing tasks, the maximum episode length is pragmatically set to 1000 time steps. As a result, the Q-value might not precisely approximate the infinite-horizon return for state-action pairs occurring near the end of an episode due to this constraint. Nonetheless, for state-action pairs appearing near the beginning of an episode, the Q-value can provide a close approximation of the infinite-horizon return.

Lower Bound Guarantee. Meanwhile, since we adapt our ACQ based on the CQL framework, we need to guarantee a lower bound of η(which is α in CQL) [19]:

$$\eta \geq \max_{s,a} \frac{C_{r,T,\delta} R_{\max}}{(1-\gamma)\sqrt{|\mathcal{D}(s,a)|}} \cdot \max_{s,a}\left[\frac{\mu(a \mid s)}{\hat{\pi}_\beta(a \mid s)}\right]^{-1} \tag{13}$$

The size of state-action pair counts is represented by $\mathcal{D}(s, a)$, and $C_{r,T,\delta}$ is a constant that depends on the properties of $r(s, a)$ and $T(s' \mid s, a)$. Thus, to mitigate the overestimation introduced by the OOD actions, we need to set the α properly.

5 Experiments

In this section, we showcase the capabilities of ACQ across a diverse range of scenarios. We begin by analyzing the Q-value distribution for each problem setting in D4RL datasets [8]. Following this, we conduct experiments to illustrate

Table 1. Average normalized return of all algorithms in the 'v2' dataset of D4RL [8] over 4 random seeds. We highlighted the key comparison between ACQ and CQL. We use r as random, m as medium, m-r as medium-replay, m-e as medium-expert, and e as expert.

	BRAC-v	MOReL	DT	MCQ	UWAC	TD3 + BC	IQL	CQL	**ACQ(ours)**
halfcheetah-r	31.2	25.6	-0	28.5	2.3 ± 0.0	11.0 ± 1.1	13.1	17.5 ± 1.5	17.8 ± 1.1
hopper-r	12.2	53.6	–	31.8	2.7 ± 0.3	8.5 ± 0.6	7.9	7.9 ± 0.4	10.2 ± 0.5
walker2d-r	1.9	37.3	–	17	2.0 ± 0.4	1.6 ± 1.7	5.4	5.1 ± 1.3	13.7 ± 2.3
halfcheetah-m	46.3	42.1	42.6	64.3	42.2 ± 0.4	48.3 ± 0.3	47.4	47.0 ± 0.5	51.4 ± 0.9
hopper-m	31.1	95.4	67.6	78.4	50.9 ± 4.4	59.3 ± 4.2	66.2	53.0 ± 28.5	76.6 ± 7.6
walker2d-m	81.1	77.8	74	91	75.4 ± 3.0	83.7 ± 2.1	78.3	73.3 ± 17.7	67.5 ± 0.2
halfcheetah-m-r	47.7	40.2	36.6	56.8	35.9 ± 3.7	44.6 ± 0.5	44.2	45.5 ± 0.7	81.2 ± 0.1
hopper-m-r	0.6	93.6	82.7	101.6	25.3 ± 1.7	60.9 ± 18.8	94.7	88.7 ± 12.9	48.8 ± 1.5
walker2d-m-r	0.9	49.8	66.6	91.3	23.6 ± 6.9	81.8 ± 5.5	73.8	81.8 ± 2.7	97.6 ± 6.5
halfcheetah-m-e	41.9	53.3	86.8	87.5	42.7 ± 0.3	90.7 ± 4.3	86.7	75.6 ± 25.7	78.1 ± 18.6
hopper-m-e	0.8	108.7	107.6	111.2	44.9 ± 8.1	98.0 ± 9.4	91.5	105.6 ± 12.9	94.3 ± 16.1
walker2d-m-e	81.6	95.6	108.1	114.2	96.5 ± 9.1	110.1 ± 0.5	109.6	107.9 ± 1.6	106.8 ± 0.6
Average Above	31.4	64.4	–	72.8	37 ± 3.19	58.2 ± 4.1	59.9	59.1 ± 8.9	62.0 ± 4.7
halfcheetah-e	–	–	–	96.2	92.9 ± 0.6	96.7 ± 1.1	95	96.3 ± 1.3	90.6 ± 1.9
hopper-e	–	–	–	111.4	110.5 ± 0.5	107.8 ± 7	109.4	96.5 ± 28.0	103.2 ± 3.7
walker2d-e	–	–	–	107.2	108.4 ± 0.4	110.2 ± 0.3	109.9	108.5 ± 0.5	107.4 ± 0.8
Total Average	–	–	–	79.2	50.41 ± 2.7	67.55 ± 3.8	68.9	67.35 ± 9.1	69.68 ± 4.2

ACQ's performance against various offline RL algorithms, encompassing both model-based and model-free techniques. Moreover, we perform ablation studies to evaluate the stability of the trade-off factor η in comparison to CQL and examine the percentile sensitivity in our method.

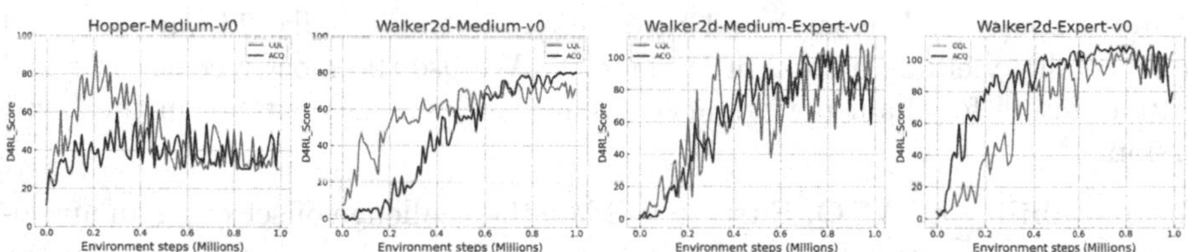

Fig. 3. Comparison of learning curves between ACQ and CQL on the v0 version of D4RL datasets. Curves are averaged over the 4 seeds.

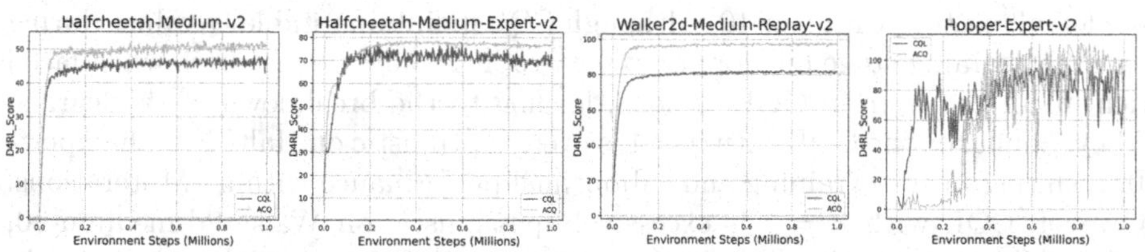

Fig. 4. Comparison of learning curves between ACQ and CQL on the v2 version of D4RL datasets. Curves are averaged over the 4 seeds.

5.1 Experimental Details

We conducted all our experiments using a single NVIDIA RTX 3090 GPU, Python 3.7, and the PyTorch 1.8.0 framework [31]. The experiments were performed in the MuJoCo 2.10 environment [36] and Mujoco-py 2.1.2.14. We employed the AdamW optimizer [16] with a batch size of 256, a discount factor of 0.99, critic learning rate and actor learning rate are both $3 \times e^{-4}$, and a smoothed moving average parameter of $2 \times e^{-4}$.

5.2 Q-Value Distribution and Effect of the Percentile

Figure 1 displays the Q-value distribution across various problem settings in the D4RL-v2 dataset. Within the same environment, the variance, mean, and nature of the Q-value distribution exhibit significant differences due to distinct behavior policies. As a result, identifying a universally suitable penalty weight for all environments and policies is challenging. We conduct an ablation study on the percentile of the Q-value in different problem settings, as illustrated in Fig. 2, demonstrating the robustness of the Q-value percentile depending on the problem settings. The results underline the need to adjust penalty weights based on the learned Q-value and the Q-value distribution of the environments and behavior policies.

5.3 Deep Offline RL Benchmarks

We conduct experiments in MuJoCo locomotion tasks of the D4RL dataset, which are made up of five types of datasets (random, medium, medium-replay, medium-expert, and expert) and three environments (Walker2d, Hopper, and Halfcheetah), yielding a total of 15 datasets. We use the most recently released "-v2" datasets for main performance evaluation, and also extend on the "-v0" version.

The instability of ACQ. First, to validate the challenge of selecting an appropriate constant η for each problem set, we perform a parameter study on the more demanding D4RL-v0 dataset. We observe that CQL is sensitive to the trade-off factor η, and its performance on certain problem settings declines as η leans towards pessimistic ($\eta = 14$) or optimistic ($\eta = 2$). Specifically, we employ the default setting with $\eta = 10$. Although CQL maintains stable Q-values during the early training stage on Hopper-expert-v0, it rapidly deteriorates when faced with minor errors from OOD actions, leading to the breakdown of the learned policy. Similarly, the CQL with $\eta = 4$ is overly optimistic on HalfCheetah-expert-v0, resulting in slow training and suboptimal performance within 1M iterations. However, CQL with $\eta = 4$ is excessively pessimistic on Walker2d-medium-v0, yielding Q-values significantly lower than the true Q-values and achieving suboptimal performance within 1M iterations.

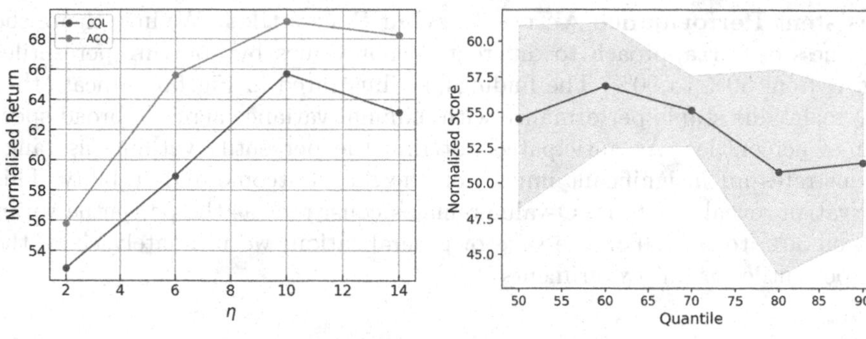

(a) The performance of ACQ against CQL in the D4RL-v2 datasets over the tradeoff factor α.

(b) The performance of an ablation over percentile.

Fig. 5. xxxx

Main Results. We compare the ACQ against several strong baseline methods and choose their implementation results from reliable official sources. Specifically, we obtain results for MOReL [15], MCQ [25], UWAC [40], CQL [19], TD3BC [9] and IQL [17] from the authors' implementation, and the numbers of BRAC-v [39] and DT [6] are taken from the D4RL [8] paper. Besides, the results of CQL are obtained by compiling the official codebase [18]

As depicted in Table 1, our method's performance remarkably surpasses prior work on most problem settings in the D4RL-v2 dataset. We find that CQL and TD3-BC perform the best among all baselines, and ACQ outperforms these two methods in most of the tasks. Notably, most baseline algorithms exhibit inconsistent performance across the problem settings due to the various transition distributions, achieving good performance with high variance. In contrast, our ACQ consistently performs well across almost all tasks, benefiting from the resilient conservatism that depends on the moving average difference between the expectation of the learned Q-value and the anchor value. Figure 3 and 4 further demonstrate that ACQ exhibits enhanced stability compared to CQL, attributable to the flexible conservatism.

5.4 Ablation Study

Robust Performance Across Various η Values. Recall that the performance of CQL is sensitive to the choice of η 4.3. In this section, we conduct extensive experiments on the D4RL-v2 dataset to assess this sensitivity. As illustrated in Fig. 5a, CQL's performance exhibits substantial fluctuations under different η values, raising concerns about its reliability in real-world applications. By incorporating ACQ, we empower CQL to adaptively adjust the trade-off between the conservative term and the RL objective, thereby achieving consistent performance improvements and maintaining stability across a range of values.

Consistent Performance Across Varying Percentiles. We investigate the robustness of our approach to different anchor values by applying percentiles ranging from 50% to 90%. The findings, as illustrated in Fig. 5b, indicate that ACQ maintains stable performance with minimal variance across a broad spectrum of percentiles. As anticipated, altering the percentile within this range demonstrates an insignificant impact on the final outcome and stability. This observation reveals that the Q-value remains consistent as the percentile varies between 50% to 90%. For the sake of generalization, we ultimately chose the 50th percentile for our experiments.

6 Conclusion

Advanced offline RL methods predominantly emphasize constraining the learned policy to stay near the behavior policy present in the dataset, aiming to mitigate overestimation due to out-of-distribution actions. However, they often overlook the importance of striking a balance between conservative properties and the RL objective, which leads to the adoption of a constant weight for the penalty term or manual hyperparameter selection for different problem settings. Real-world applications, on the other hand, demand robust and stable algorithms that minimize the need for exploration and evaluation within the environment. In this paper, we find that the 50th percentile of the Q-value distribution serves as a qualified metric that adapts to robustness based on the dataset's distribution and problem settings. Leveraging this statistic, we propose the Adaptable Conservative Q-Learning towards Anchor Value via Adjustable Weight (ACQ), an approach that relies on adaptive conservatism towards the medium percentile of the Q-value distribution. ACQ seamlessly integrates the optimal technique with the classical method CQL, consequently enhancing the algorithm's robustness.

References

1. Agarwal, A., Kakade, S., Yang, L.F.: Model-based reinforcement learning with a generative model is minimax optimal. In: Conference on Learning Theory, pp. 67–83. PMLR (2020)
2. Agarwal, R., Schuurmans, D., Norouzi, M.: An optimistic perspective on offline reinforcement learning. In: ICML, pp. 104–114. PMLR (2020)
3. Ajay, A., Kumar, A., Agrawal, P., Levine, S., Nachum, O.: Opal: Offline primitive discovery for accelerating offline reinforcement learning. arXiv preprint arXiv:2010.13611 (2020)
4. An, G., Moon, S., Kim, J.H., Song, H.O.: Uncertainty-based offline reinforcement learning with diversified q-ensemble. Adv. Neural. Inf. Process. Syst. **34**, 7436–7447 (2021)
5. Azar, M.G., Osband, I., Munos, R.: Minimax regret bounds for reinforcement learning. In: International Conference on Machine Learning, pp. 263–272. PMLR (2017)
6. Chen, L., et al.: Decision transformer: reinforcement learning via sequence modeling. Adv. Neural. Inf. Process. Syst. **34**, 15084–15097 (2021)

7. Dulac-Arnold, G., et al.: Challenges of real-world reinforcement learning: definitions, benchmarks and analysis. Mach. Learn. **110**(9), 2419–2468 (2021)
8. Fu, J., Kumar, A., Nachum, O., Tucker, G., Levine, S.: D4rl: datasets for deep data-driven reinforcement learning. arXiv preprint arXiv:2004.07219 (2020)
9. Fujimoto, S., Gu, S.S.: A minimalist approach to offline reinforcement learning. Adv. Neural. Inf. Process. Syst. **34**, 20132–20145 (2021)
10. Guiñón, J.L., Ortega, E., García-Antón, J., Pérez-Herranz, V.: Moving average and savitzki-golay smoothing filters using mathcad. Papers ICEE **2007**, 1–4 (2007)
11. Haarnoja, T., Zhou, A., Abbeel, P., Levine, S.: Soft actor-critic: off-policy maximum entropy deep reinforcement learning with a stochastic actor. In: ICML, pp. 1861–1870. PMLR (2018)
12. Jaques, N., et al.: Way off-policy batch deep reinforcement learning of implicit human preferences in dialog. arXiv preprint arXiv:1907.00456 (2019)
13. Jin, Y., Yang, Z., Wang, Z.: Is pessimism provably efficient for offline rl? In: International Conference on Machine Learning, pp. 5084–5096. PMLR (2021)
14. Kendall, A., et al.: Learning to drive in a day. In: ICRA, pp. 8248–8254. IEEE (2019)
15. Kidambi, R., Rajeswaran, A., Netrapalli, P., Joachims, T.: Morel: Model-based offline reinforcement learning. Adv. Neural. Inf. Process. Syst. **33**, 21810–21823 (2020)
16. Kingma, D.P., Ba, J.: Adam: a method for stochastic optimization. arXiv preprint arXiv:1412.6980 (2014)
17. Kostrikov, I., Nair, A., Levine, S.: Offline reinforcement learning with implicit q-learning. arXiv preprint arXiv:2110.06169 (2021)
18. Kumar, A., Zhou, A., Tucker, G., Levine, S.: Conservative q-learning for offline reinforcement learning https://arxiv.org/abs/2006.04779
19. Kumar, A., Zhou, A., Tucker, G., Levine, S.: Conservative q-learning for offline reinforcement learning. Adv. Neural. Inf. Process. Syst. **33**, 1179–1191 (2020)
20. Lange, S., Gabel, T., Riedmiller, M.: Batch reinforcement learning. Reinforcement learning: State-of-the-art, pp. 45–73 (2012)
21. Leibo, J.Z., Zambaldi, V., Lanctot, M., Marecki, J., Graepel, T.: Multi-agent reinforcement learning in sequential social dilemmas. arXiv preprint arXiv:1702.03037 (2017)
22. Levine, S., Kumar, A., Tucker, G., Fu, J.: Offline reinforcement learning: tutorial, review, and perspectives on open problems. arXiv preprint arXiv:2005.01643 (2020)
23. Li, G., Wei, Y., Chi, Y., Gu, Y., Chen, Y.: Breaking the sample size barrier in model-based reinforcement learning with a generative model. Adv. Neural. Inf. Process. Syst. **33**, 12861–12872 (2020)
24. Liu, Y., Swaminathan, A., Agarwal, A., Brunskill, E.: Off-policy policy gradient with state distribution correction. arXiv preprint arXiv:1904.08473 (2019)
25. Lyu, J., Ma, X., Li, X., Lu, Z.: Mildly conservative q-learning for offline reinforcement learning. arXiv preprint arXiv:2206.04745 (2022)
26. Ma, Y., Jayaraman, D., Bastani, O.: Conservative offline distributional reinforcement learning. Adv. Neural. Inf. Process. Syst. **34**, 19235–19247 (2021)
27. Munemasa, I., Tomomatsu, Y., Hayashi, K., Takagi, T.: Deep reinforcement learning for recommender systems. In: ICOIACT, pp. 226–233. IEEE (2018)
28. Nachum, O., Dai, B., Kostrikov, I., Chow, Y., Li, L., Schuurmans, D.: Algaedice: policy gradient from arbitrary experience. arXiv preprint arXiv:1912.02074 (2019)
29. O'Donoghue, B., Osband, I., Munos, R., Mnih, V.: The uncertainty bellman equation and exploration. In: International Conference on Machine Learning, pp. 3836–3845 (2018)

30. Ovadia, Y., et al.: Can you trust your model's uncertainty? evaluating predictive uncertainty under dataset shift. In: Advances in Neural Information Processing Systems 32 (2019)
31. Paszke, Aet al.: Pytorch: an imperative style, high-performance deep learning library. In: Advances in Neural Information Processing Systems 32 (2019)
32. Precup, D., Sutton, R.S., Dasgupta, S.: Off-policy temporal-difference learning with function approximation. In: ICML, pp. 417–424 (2001)
33. Rashidinejad, P., Zhu, B., Ma, C., Jiao, J., Russell, S.: Bridging offline reinforcement learning and imitation learning: a tale of pessimism. Adv. Neural. Inf. Process. Syst. **34**, 11702–11716 (2021)
34. Sinha, S., Mandlekar, A., Garg, A.: S4rl: surprisingly simple self-supervision for offline reinforcement learning in robotics. In: Conference on Robot Learning, pp. 907–917. PMLR (2022)
35. Sutton, R.S., Mahmood, A.R., White, M.: An emphatic approach to the problem of off-policy temporal-difference learning. J. Mach. Learn. Res. **17**(1), 2603–2631 (2016)
36. Todorov, E., Erez, T., Tassa, Y.: Mujoco: a physics engine for model-based control. In: 2012 IEEE/RSJ International Conference on Intelligent Robots and Systems, pp. 5026–5033. IEEE (2012)
37. Vinyals, O., et al.: Starcraft ii: A new challenge for reinforcement learning. arXiv preprint arXiv:1708.04782 (2017)
38. Wu, K., et al.: Acql: an adaptive conservative q-learning framework for offline reinforcement learning (2022)
39. Wu, Y., Tucker, G., Nachum, O.: Behavior regularized offline reinforcement learning. arXiv preprint arXiv:1911.11361 (2019)
40. Wu, Y., et al.: Uncertainty weighted actor-critic for offline reinforcement learning. arXiv preprint arXiv:2105.08140 (2021)
41. Yu, T., Kumar, A., Rafailov, R., Rajeswaran, A., Levine, S., Finn, C.: Combo: Conservative offline model-based policy optimization. Adv. Neural. Inf. Process. Syst. **34**, 28954–28967 (2021)
42. Yu, T., et al.: Mopo: model-based offline policy optimization. Adv. Neural. Inf. Process. Syst. **33**, 14129–14142 (2020)
43. Zou, L., Xia, L., Ding, Z., Song, J., Liu, W., Yin, D.: Reinforcement learning to optimize long-term user engagement in recommender systems. In: SIGKDD, pp. 2810–2818 (2019)

Boosting Out-of-Distribution Detection with Sample Weighting

Ao Ke, Wenlong Chen, Chuanwen Feng, and Xike Xie(✉)

Data Darkness Lab, MIRACLE Center, Suzhou Institute for Advanced Research, University of Science and Technology of China, Hefei, China
{sa21225249,wlchency,chuanwen}@mail.ustc.edu.cn, xkxie@ustc.edu.cn

Abstract. To enhance the reliability of machine learning models in the open world, there is considerable interest in detecting out-of-distribution (OOD) inputs that lie far away from the training distribution. Existing distance-based methods have shown promising performance in OOD detection. These methods generally follow a common approach of capturing the distances between training samples and each test sample in the feature space. This can be understood as encoding the distance information to assign sample weights, which are then used to calculate a weighted distance score to determine if the input is OOD. However, these methods often adopt a coarse-grained weighting approach, where only a small fraction of the training samples are considered and given weights. Consequently, they fail to fully leverage the complete distance information of the training data, leading to occasional difficulties in effectively distinguishing OOD samples. In this paper, we propose a novel approach to encode the complete distance information of the training data by assigning a weight to each sample based on its distance from the test sample, with the weights decaying as the distance increases. Furthermore, we introduce a weighted distance-based method for OOD detection. We demonstrate the superiority of our method over most existing supervised OOD detectors. Particularly, on a hard OOD detection task involving CIFAR-100 vs. CIFAR-10, our method achieves a reduction of 1.82% in the average FPR95 compared to the current best method KNN.

Keywords: Out-of-distribution detection · Contrastive training

1 Introduction

The prevailing assumption underlying the availability of machine learning models is that test inputs adhere to a distribution similar to the training data. However, this assumption is easily violated when these models are deployed in the open world [18,22]. The presence of out-of-distribution (OOD) inputs, which deviate from the distribution of the training data referred to as in-distribution (ID), can lead to incorrect predictions by the trained models [5]. To prevent the model from making unreliable or unsafe predictions, it is crucial to detect OOD inputs and distinguish them from ID inputs.

Q. Liu et al. (Eds.): PRCV 2023, LNCS 14427, pp. 213–223, 2024.
https://doi.org/10.1007/978-981-99-8435-0_17

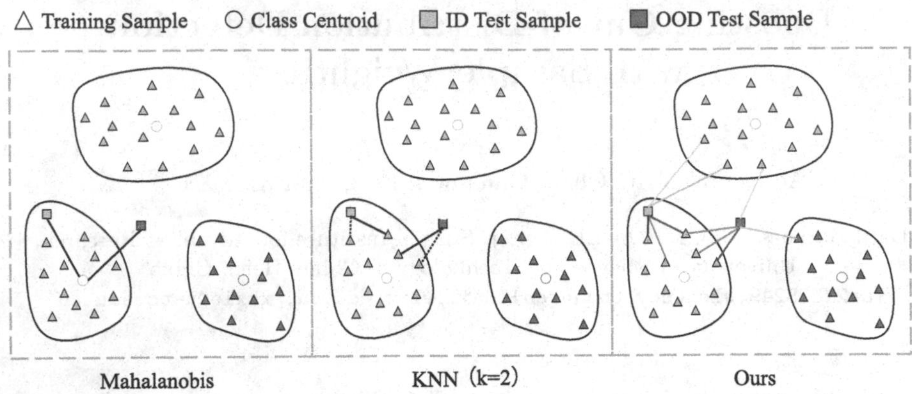

Fig. 1. Illustration of sample weighting in three methods. The solid lines between training and test samples represent the sample weighting, where darker colors indicate higher weights. In our method, solid lines connect the test sample to all training samples, but most connections are omitted for clarity. The score functions for ID and OOD test samples are similar in the first two methods, whereas in our method, the score functions for ID and OOD test samples show distinct differences.

Among plenty of proposed OOD detection algorithms [5,12,29], distance-based methods [11,21,23,24] caught much research attention due to their promising performance. Under the assumption that OOD samples remain relatively far away from the ID samples in the feature space, these methods leverage the distance-based metrics calculated from the extracted feature embeddings for detection. For example, `Mahalanobis` [11] utilizes the minimum Mahalanobis distance between a test sample and all class centroids as a detection metric under the additional assumption that the trained features can be fitted well by multivariate Gaussian distributions yet not always hold in the real world. Another state-of-the-art distance-based method `KNN` [23] computes the Euclidean distance between a test sample and its k-th closest training sample. Although no distributional assumptions are required, this method suffers from high sensitivity to the parameter k, as it only considers a single pairwise distance.

A distinctive characteristic of distance-based methods is their utilization of distance information in a dual manner. They encode the distance information to assign sample weights and calculate weighted distances as detection scores. For example, `Mahalanobis` assigns a weight of one to the class centroid closest to the test sample while assigning weights of zero to the remaining class centroids. Similarly, `KNN` assigns a weight of one to the k-th nearest neighbor and weights of zero to other training samples. However, these methods, called coarse-grained weighting schemes, neglect most of the sample distances. As a result, they occasionally fail to effectively differentiate between ID and OOD samples, as illustrated in Fig. 1, where the score functions for ID and OOD test samples are similar. Particularly in *near OOD* settings [27], where the ID and OOD distributions exhibit meaningful similarity, OOD detection becomes a challenging task that requires careful consideration. This observation has motivated us to investigate the potential benefits of different distance information encoding approaches for OOD detection.

In this paper, we propose a novel OOD detection method that encodes the complete distance information of the training data and assigns weights to each sample based on their distance from the test sample. Our approach is motivated by an alternative interpretation of the aforementioned common assumption on distance-based methods. We argue that compared to OOD samples, ID samples are relatively closer to the ID data since they originate from the same distribution. Therefore, the distance information of the training samples (not the centroids) closer to the test sample plays a critical role in discriminating between OOD and ID samples. To capture this intuition, we assign higher weights to samples closer to the test sample and lower weights to samples farther away. As depicted in Fig. 1, our method increases the distinction between the score functions for ID and OOD samples by leveraging the complete distance information, allowing for more effective OOD detection. By utilizing pairwise Euclidean distance between the test and training samples, our method is liberated from the Gaussian distribution assumption about the underlying feature space.

To summarize, the key contributions of the paper are as follows:

- By exploring an approach to encode distance information between samples, we propose an effective OOD detection method based on sample weighting. Our method offers a simple and flexible perspective for OOD detection and is easy to use, requiring no data distribution assumptions or additional OOD training data.
- On various OOD detection benchmarks, our method establishes state-of-the-art results. Compared to the best method `KNN`, our method reduces the average FPR95 by 1.82%. Notably, our method demonstrates excellent performance on challenging hard tasks in OOD detection.

2 Related Work

There have been many works for out-of-distribution detection [4–6, 11, 16, 20, 32, 34]. A straightforward approach is to utilize the output of the trained neural network as some estimation for test input [4, 5, 12, 26]. For example, [5] proposed a baseline that adopts the predicted softmax probability as an indicator for detecting OOD samples, which is based on the empirical observation that the softmax probability of OOD samples tends to be lower than that of the ID samples [31]. However, the results are inconsistent since the trained neural network can also give a high confidence for some OOD samples. To improve the performance of OOD detection, [12] applies temperature scaling and small perturbation to inputs. Following up this road, [8] proposed to decompose the confidence of predicted class probabilities and modify the input pre-processing.

The other line [1, 6, 13, 14, 17] is to leverage auxiliary OOD data during the training stage to enhance the discrepancy between OOD and ID samples. For instance, [1] proposed a theoretically motivated method that mines the auxiliary OOD data for improving the performance of OOD detection. To improve the efficiency of using OOD samples, [14] proposed a posterior sampling-based method, selecting the most informative OOD samples from the large outlier set, which also potentially refined the decision boundary of ID and OOD samples. However, due to the diversity of OOD sample space, it is difficult to collect the representative OOD auxiliary data in the real world.

More recently, distance-based methods [2,11,23,25] have shown a prominent performance in OOD detection. It adopts some distance metrics between test inputs and ID data in the feature space to differentiate OOD samples from test inputs under the assumption that OOD samples lie far from the ID data. For instance, [11,19] employed the Mahalanobis distance metric to distinguish OOD samples from test inputs, accompanied by an implicitly distributional assumption. To relieve distributional assumption, [23] proposed calculating the distance to the k-th nearest neighbor as a metric from a non-parametric standpoint. In this case, other ID samples in training data have the same zero weights for test input.

3 Methodology

In this section, we describe our weighted distance-based method for OOD Detection. In Sect. 3.1, we introduce the setup for the OOD detection problem. In Sect. 3.2, we explain how distance information is encoded to derive weights and model the score function for OOD detection. In Sect. 3.3, we explore the use of supervised contrastive training to obtain richer feature representation, aiming for more informative OOD detection capabilities.

3.1 Problem Setting

In the context of multi-class classification, we define the input space as $\mathcal{X}$ and the label space as $\mathcal{Y} = \{1, 2, \ldots, C\}$. The training dataset is denoted as $\mathcal{D}_{tr}^{in} = \{(\boldsymbol{x}_i, y_i)\}_{i=1}^{n}$, comprising in-distribution (ID) data drawn from the joint distribution $\mathcal{P}_{\mathcal{X}\mathcal{Y}}^{in}$. Additionally, we represent the marginal distribution over $\mathcal{X}$ as $\mathcal{P}_{\mathcal{X}}$. The feature extractor $f : \mathcal{X} \rightarrow \mathbb{R}^e$, typically a parameterized deep neural network and often referred to as the penultimate layer, is employed to map the input $\boldsymbol{x}$ to a high-dimensional feature embedding $f(\boldsymbol{x})$. Leveraging the data labels available in supervised training, we incorporate a neural network g, commonly a linear classifier, and employ $g \circ f : \mathcal{X} \rightarrow \mathbb{R}^{|\mathcal{Y}|}$ to generate a logit vector for predicting the label of an input sample.

Out-of-Distribution Detection. In the real world, apart from achieving reliable predictive performance on in-distribution samples, a trustworthy machine learning model should be capable of identifying out-of-distribution instances that originate from a different distribution than the training data. Specifically, given a test set of samples from $\mathcal{P}_{\mathcal{X}}^{in} \times \mathcal{P}_{\mathcal{X}}^{ood}$, OOD detection aims to perform a binary classification task by determining whether each sample $\boldsymbol{x} \in \mathcal{X}$ belongs to the ID ($\mathcal{P}_{\mathcal{X}}^{in}$) or the OOD ($\mathcal{P}_{\mathcal{X}}^{ood}$).

3.2 Weighted Distance as Score Function

In this section, we present our method for OOD detection based on weighted distance calculation, which leverages the dual utilization of distance information: encoding the distance information to derive sample weights and further calculating the weighted distances as detection scores. To begin, we obtain the embedding sets of training samples $\mathcal{D}_{tr} = \{\boldsymbol{x}_i\}_{i=1}^n$, denoted as $\mathbb{Z} = \{\mathbf{z}_i\}_{i=1}^n$ where $\mathbf{z}_i = f(\boldsymbol{x}_i)/||f(\boldsymbol{x}_i)||_2$ is the normalized penultimate feature. During testing, for a given test sample $\hat{\boldsymbol{x}}$, we derive its normalized penultimate feature $\hat{\mathbf{z}}$ and calculate the Euclidean distances $||\hat{\mathbf{z}} - \mathbf{z}_i||_2$ with respect to each training sample $\boldsymbol{x}_i \in \mathcal{D}_{tr}$. Next, we encode the distance information using an exponential decay function to generate weights w_i for each training sample $\boldsymbol{x}_i$:

$$w_i = \frac{exp(-||\hat{\mathbf{z}} - \mathbf{z}_i||_2)}{\sum_{k=1}^n exp(-||\hat{\mathbf{z}} - \mathbf{z}_k||_2)}. \tag{1}$$

Subsequently, to evaluate whether a given input sample $\hat{\boldsymbol{x}}$ is OOD, we calculate the weighted Euclidean distance to each training sample as the score $\mathcal{S}_{\hat{\boldsymbol{x}}}$, given by:

$$\mathcal{S}_{\hat{\boldsymbol{x}}} = \sum_{i=1}^n w_i ||\hat{\mathbf{z}} - \mathbf{z}_i||_2. \tag{2}$$

A low score $\mathcal{S}_{\hat{\boldsymbol{x}}}$ indicates that the test sample lies close to some training samples in the feature space. Conversely, a high score suggests that the test sample is significantly distant from all the training samples, indicating a higher likelihood of it being an OOD sample.

3.3 Contrastive Training for OOD Detection

Distance-based OOD detection methods rely on the quality of the feature space defined by the feature extractor, and learning a good feature representation is one of the critical steps in OOD detection frameworks. This is because a rich understanding of the key semantics of ID data, which can be obtained through training, allows OOD data lacking related semantics to be far away in the feature space [21]. Even though existing supervised detectors have achieved notable performance [8,11–13], the semantic representations generated by supervised learning only enable distinguishing between classes labeled in the dataset, which is insufficient for OOD detection [27]. Current state-of-the-art OOD detection methods provide additional means to further enrich the feature representation space, such as accessing extra OOD data [6,15] or adding rotation-based self-supervised loss [7]. However, these approaches are limited by the need for sophisticated hyperparameter tuning, the practical challenges of collecting OOD data, or the problematic assumption that irrelevant auxiliary tasks can produce favorable representations. Exempt from these limitations, contrastive learning techniques can learn richer semantic features from the ID dataset than supervised learning, making them more effective for OOD detection [27].

We utilize supervised contrastive training (SupCon) [9] to learn feature representations, which aims to train a feature extractor f that maps samples belonging to the same class closer in the feature space and those from different classes further apart. Given a training set $\{\boldsymbol{x}_i, \boldsymbol{y}_i\}_{i=1}^{N}$ where $\boldsymbol{x}_i$ is a sample and $\boldsymbol{y}_i$ is its corresponding label, we create $2N$ training pairs $\{\tilde{\boldsymbol{x}}_i, \tilde{\boldsymbol{y}}_i\}_{i=1}^{2N}$ by applying two random augmentations, where $\tilde{\boldsymbol{x}}_{2i-1}$ and $\tilde{\boldsymbol{x}}_{2i}$ are two augmentations of $\boldsymbol{x}_i$ and $\tilde{\boldsymbol{y}}_{2i-1} = \tilde{\boldsymbol{y}}_{2i} = \tilde{\boldsymbol{y}}_i$. We treat all other samples of the same class as positives and minimize the following loss:

$$\mathcal{L}oss = \frac{1}{2N}\sum_{i=1}^{2N} -log\frac{\frac{1}{|P(i)|}\sum_{j\in P(i)} e^{\mathbf{u}_i^T\mathbf{u}_j/\tau}}{\sum_{k=1,k\neq i}^{2N} e^{\mathbf{u}_i^T\mathbf{u}_k/\tau}}; \quad \mathbf{u}_i = \frac{h(f(\boldsymbol{x}_i))}{||h(f(\boldsymbol{x}_i))||_2}, \tag{3}$$

where $h(.)$ is a projection header, τ is the temperature, $P(i)$ is the set of indices of all positives distinct from i, and $|P(i)|$ is its cardinality.

4 Experiment

4.1 Common Setup

Datasets. Following the common benchmarks used in literature, we utilize CIFAR-10 and CIFAR-100 [10] as the in-distribution datasets and evaluate our method on iSUN [28], LSUN [30], Textures [3] and Places365 [33] as the OOD datasets in our primary experiments. For the large dataset Places365, we randomly select 10,000 test images. All images have a size of 32×32.

Experiment Details. We use ResNet18 as the backbone in our major experiments and train each network for 500 epochs using both the conventional cross-entropy (SupCE) loss and SupCon loss. The batch size is 512, and we utilize stochastic gradient descent with momentum 0.9 and weight decay 10^{-4}. Following the settings of `KNN+` [23] with the SupCon loss, we initialize the learning rate at 0.5 and employ the cosine annealing. The temperature τ is 0.1. The dimension of the penultimate feature is 512, and the dimension of the projection head is 128.

Baselines. We compare our method with the non-distance-based baselines in the `CIDER` [15] and refer to the experimental results based on CIFAR-10. To demonstrate the effectiveness of our method, we replicate the state-of-the-art supervised distance-based methods, `KNN` and `Mahalanobis`.

Evaluation Metrics. We evaluate the performance of our OOD detectors using the following three metrics, which are widely adopted in the field: (1) the area under the receiver operating characteristic curve (AUROC), (2) the area under the precision-recall curve (AUPR), (3) the false positive rate (FPR95) of OOD samples when the true positive rate of ID samples is equal to 95%. AUROC and AUPR are both threshold independent metrics.

Table 1. Comparison with different OOD detection methods on CIFAR-10 with SupCE loss. ↑ indicates larger values are better and vice versa. Bold numbers indicate superior results.

Method	OOD Dataset									
	iSUN		LSUN		Textures		Places365		Average	
	AUROC	FPR	AUROC	FPR	AUROC	FPR	AUROC	FPR	AUROC	FPR
	↑	↓	↑	↓	↑	↓	↑	↓	↑	↓
MSP [5]	90.39	59.77	85.35	69.71	78.97	69.33	64.56	66.79	79.81	66.40
Energy [13]	83.21	73.53	91.22	56.42	86.62	64.35	87.22	65.91	87.06	65.05
ODIN [12]	94.09	37.35	94.93	31.45	85.52	70.51	90.38	48.49	91.23	46.95
GODIN [8]	**94.94**	**34.03**	95.08	32.43	89.69	46.91	87.31	62.63	91.75	44.00
Mahalanobis [11]	80.06	71.98	78.94	77.15	93.36	**28.32**	65.99	89.28	79.58	66.68
KNN [23]	93.72	39.02	94.81	35.37	93.59	41.01	90.63	48.52	93.18	40.98
Ours	93.93	37.56	**95.17**	**32.20**	**93.86**	39.43	**90.78**	**47.47**	**93.43**	**39.16**

Table 2. Evaluation on **hard OOD detection** tasks. The model is trained using ResNet18 and SupCE loss.

Task	Method	AUROC↑	AUPR↑	FPR↓
CIFAR-100 vs. CIFAR-10	KNN	76.94	72.98	81.78
	Ours	**77.76**	**73.47**	**81.35**
CIFAR-10 vs. CIFAR-100	KNN	89.99	88.32	52.40
	Ours	**90.11**	**88.54**	**51.69**

4.2 Main Results

Evaluations on Common Benchmarks. We present results in Table 1, which includes a comprehensive comparison with state-of-the-art OOD detection methods. Our method demonstrates favorable performance across various evaluation metrics. All models are trained using ResNet18 with the SupCE loss, utilizing CIFAR-10 as the in-distribution dataset. Compared to the current best method, `KNN`, our method reduces the average FPR95 by 1.82%, which is a relatively 4.44% reduction in error. Moreover, unlike `Mahalanobis`, which relies on the assumption that embeddings of each class can be fitted well by multivariate Gaussian distribution in feature space, our method, without making distribution assumptions, exhibits significant improvements in the average AUROC by 13.85% and the average FPR95 by 27.52%. These findings further support the effectiveness of our proposed approach.

Evaluations on Hard OOD Tasks. We evaluate our method on hard OOD tasks where the ID and OOD samples are semantically similar, as discussed in the literature [27]. Specifically, we consider the CIFAR-100 vs. CIFAR-10 task and vice versa, which pose significant challenges for OOD detection. As shown in Table 2, our method consistently outperforms `KNN` across all evaluation metrics. Notably, in the more challenging task where CIFAR-100 is used as the

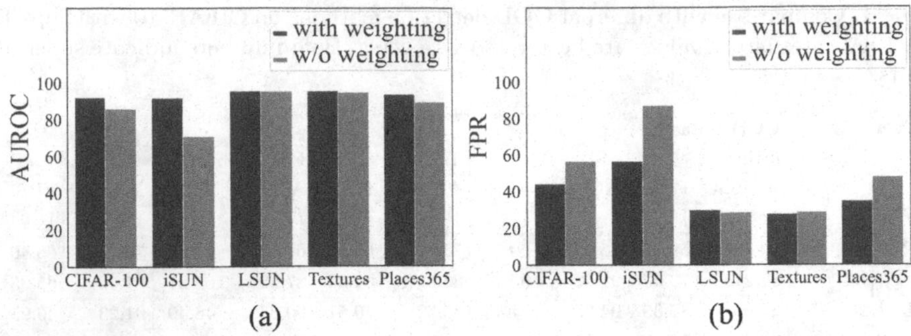

Fig. 2. Ablation study on weighting scheme. Results are based on CIFAR-10 (ID) in terms of (a) AUROC and (b) FPR95. The model is trained using ResNet18 and SupCon loss.

ID dataset, our method achieves a 0.82% improvement in AUROC, a 0.49% improvement in AUPR, and a 0.43% reduction in FPR95 compared to KNN. These results highlight the effectiveness of our proposed approach in encoding distance information with a decaying strategy, thereby enhancing performance in challenging OOD detection tasks.

Impact of Supervised Contrastive Learning. We investigate the impact of utilizing supervised contrastive loss in our method. As depicted in Table 3, when using CIFAR-10 as the ID dataset, our method trained with SupCon loss outperforms the variant trained with SupCE loss in most test datasets. Specifically, our approach with SupCon loss exhibits a 2.76% increase in AUROC and a significant 13.01% reduction in FPR95 based on the evaluation using the Places365 test set.

Table 3. Evaluation using supervised contrastive learning loss. Results are based on CIFAR-10 trained with ResNet18.

Training Loss	OOD Dataset									
	iSUN		**LSUN**		**Textures**		**Places365**		Average	
	AUROC↑	FPR↓	AUROC↑	FPR↓	AUROC↑	FPR↓	AUROC↑	FPR↓	AUROC↑	FPR↓
SupCE	**93.93**	**37.56**	95.17	32.20	93.86	39.43	90.78	47.47	93.43	39.16
SupCon	91.80	56.41	**95.53**	**29.72**	**95.43**	**27.53**	**93.54**	**34.46**	**94.07**	**37.03**

4.3 Ablation Studies

In this section, we conduct ablations to explore the influence of different factors on the performance of our method. To ensure consistency, we conduct all ablation studies using CIFAR-10 as the ID dataset and models trained with SupCon loss, as described in Sect. 4.1.

Ablation on the Weighting Scheme. The key idea of our method lies in the encoding of distance information, where closer training samples receive higher weights. In this ablation study, we compare the performance of our method with and without sample weighting. The version without sample weighting assigns the same weight value $w = 1/n$ to all training samples. As illustrated in Fig. 2, employing sample weighting improves the AUROC by 20.93% and reduces the FPR95 by 30.70% on the iSUN dataset, compared to the counterpart without sample weighting.

Table 4. Ablation study on distance metric. Results are presented for hard OOD detection tasks based on ResNet18 and SupCon loss.

Task	Distance	AUROC↑	AUPR↑	FPR↓
CIFAR-100 vs. CIFAR-10	Mahalanobis	71.13	65.98	87.70
	Cosine	59.82	59.94	88.11
	Euclidean	**74.21**	**69.36**	**85.22**
CIFAR-10 vs. CIFAR-100	Mahalanobis	91.72	**90.51**	44.29
	Cosine	89.62	88.57	48.38
	Euclidean	**92.04**	90.38	**44.04**

Ablation on Distance Metric. In our work, we employ the calculation of Euclidean distance between embeddings from the penultimate layer. Furthermore, we investigate the advantages of using the Euclidean distance by comparing it with other commonly used distance metrics, namely the Mahalanobis distance and cosine distance, as presented in Table 4. The results indicate that employing the Euclidean distance yields superior performance compared to the Mahalanobis distance and cosine distance in the context of hard OOD detection tasks.

5 Conclusion

In the paper, we introduce a novel method for OOD detection based on sample weighting. Unlike existing distance-based methods that only consider partial sample distances, our proposed method assigns weights to each training sample based on its distance from the test sample, with the weights decaying as the distance increases. By incorporating the complete distance information, our method achieves superior performance on diverse OOD detection benchmarks, including hard OOD tasks. We believe that our work provides valuable insights into the encoding of distance information and can inspire further research of exploring alternative strategies to sample weighting for OOD detection.

Acknowledgements. This work is supported by NSFC (No.61772492, 62072428) and the CAS Pioneer Hundred Talents Program.

References

1. Chen, J., Li, Y., Wu, X., Liang, Y., Jha, S.: ATOM: robustifying out-of-distribution detection using outlier mining. In: Oliver, N., Pérez-Cruz, F., Kramer, S., Read, J., Lozano, J.A. (eds.) ECML PKDD 2021. LNCS (LNAI), vol. 12977, pp. 430–445. Springer, Cham (2021). https://doi.org/10.1007/978-3-030-86523-8_26
2. Chen, X., Lan, X., Sun, F., Zheng, N.: A boundary based out-of-distribution classifier for generalized zero-shot learning. In: Vedaldi, A., Bischof, H., Brox, T., Frahm, J.-M. (eds.) ECCV 2020. LNCS, vol. 12369, pp. 572–588. Springer, Cham (2020). https://doi.org/10.1007/978-3-030-58586-0_34
3. Cimpoi, M., Maji, S., Kokkinos, I., Mohamed, S., Vedaldi, A.: Describing textures in the wild. In: Proceedings of the IEEE Conference on Computer Vision and Pattern Recognition, pp. 3606–3613 (2014)
4. DeVries, T., Taylor, G.W.: Learning confidence for out-of-distribution detection in neural networks. arXiv preprint arXiv:1802.04865 (2018)
5. Hendrycks, D., Gimpel, K.: A baseline for detecting misclassified and out-of-distribution examples in neural networks. In: Proceedings of International Conference on Learning Representations (2017)
6. Hendrycks, D., Mazeika, M., Dietterich, T.G.: Deep anomaly detection with outlier exposure. In: 7th International Conference on Learning Representations, ICLR 2019, New Orleans, LA, USA, 6-9 May (2019)
7. Hendrycks, D., Mazeika, M., Kadavath, S., Song, D.X.: Using self-supervised learning can improve model robustness and uncertainty. In: Neural Information Processing Systems (2019)
8. Hsu, Y.C., Shen, Y., Jin, H., Kira, Z.: Generalized odin: detecting out-of-distribution image without learning from out-of-distribution data. In: Proceedings of the IEEE/CVF Conference on Computer Vision and Pattern Recognition, pp. 10951–10960 (2020)
9. Khosla, P., et al.: Supervised contrastive learning. Adv. Neural. Inf. Process. Syst. **33**, 18661–18673 (2020)
10. Krizhevsky, A., Hinton, G., et al.: Learning multiple layers of features from tiny images (2009)
11. Lee, K., Lee, K., Lee, H., Shin, J.: A simple unified framework for detecting out-of-distribution samples and adversarial attacks. In: Advances in Neural Information Processing Systems 31 (2018)
12. Liang, S., Li, Y., Srikant, R.: Enhancing the reliability of out-of-distribution image detection in neural networks. In: International Conference on Learning Representations (2018)
13. Liu, W., Wang, X., Owens, J., Li, Y.: Energy-based out-of-distribution detection. Adv. Neural. Inf. Process. Syst. **33**, 21464–21475 (2020)
14. Ming, Y., Fan, Y., Li, Y.: Poem: out-of-distribution detection with posterior sampling. In: International Conference on Machine Learning (2022)
15. Ming, Y., Sun, Y., Dia, O., Li, Y.: How to exploit hyperspherical embeddings for out-of-distribution detection? In: Proceedings of the International Conference on Learning Representations (2023)
16. Osawa, K., et al.: Practical deep learning with bayesian principles. In: Advances in Neural Information Processing Systems 32 (2019)
17. Papadopoulos, A.A., Rajati, M.R., Shaikh, N., Wang, J.: Outlier exposure with confidence control for out-of-distribution detection. Neurocomputing **441**, 138–150 (2021)

18. Parmar, J., Chouhan, S., Raychoudhury, V., Rathore, S.: Open-world machine learning: applications, challenges, and opportunities. ACM Comput. Surv. **55**(10), 1–37 (2023)
19. Ren, J., Fort, S., Liu, J., Roy, A.G., Padhy, S., Lakshminarayanan, B.: A simple fix to mahalanobis distance for improving near-ood detection. arXiv preprint arXiv:2106.09022 (2021)
20. Ren, J., et al.: Likelihood ratios for out-of-distribution detection. Advances in neural information processing systems **32** (2019)
21. Sehwag, V., Chiang, M., Mittal, P.: Ssd: a unified framework for self-supervised outlier detection. In: International Conference on Learning Representations (2021)
22. Sun, Y., Li, Y.: Opencon: open-world contrastive learning. Trans. Mach. Learn. Res
23. Sun, Y., Ming, Y., Zhu, X., Li, Y.: Out-of-distribution detection with deep nearest neighbors. In: International Conference on Machine Learning (2022)
24. Tack, J., Mo, S., Jeong, J., Shin, J.: Csi: novelty detection via contrastive learning on distributionally shifted instances. Adv. Neural. Inf. Process. Syst. **33**, 11839–11852 (2020)
25. Techapanurak, E., Suganuma, M., Okatani, T.: Hyperparameter-free out-of-distribution detection using cosine similarity. In: Proceedings of the Asian Conference on Computer Vision (2020)
26. Thulasidasan, S., Chennupati, G., Bilmes, J.A., Bhattacharya, T., Michalak, S.: On mixup training: improved calibration and predictive uncertainty for deep neural networks. In: Advances in Neural Information Processing Systems 32 (2019)
27. Winkens, J., et al.: Contrastive training for improved out-of-distribution detection. arXiv preprint arXiv:2007.05566 (2020)
28. Xu, P., Ehinger, K.A., Zhang, Y., Finkelstein, A., Kulkarni, S.R., Xiao, J.: Turkergaze: crowdsourcing saliency with webcam based eye tracking. arXiv preprint arXiv:1504.06755 (2015)
29. Yang, J., Zhou, K., Li, Y., Liu, Z.: Generalized out-of-distribution detection: a survey. arXiv preprint arXiv:2110.11334 (2021)
30. Yu, F., Seff, A., Zhang, Y., Song, S., Funkhouser, T., Xiao, J.: Lsun: construction of a large-scale image dataset using deep learning with humans in the loop. arXiv preprint arXiv:1506.03365 (2015)
31. Yu, Q., Aizawa, K.: Unsupervised out-of-distribution detection by maximum classifier discrepancy. In: Proceedings of the IEEE/CVF International Conference on Computer Vision, pp. 9518–9526 (2019)
32. Zaeemzadeh, A., Bisagno, N., Sambugaro, Z., Conci, N., Rahnavard, N., Shah, M.: Out-of-distribution detection using union of 1-dimensional subspaces. In: Proceedings of the IEEE/CVF Conference on Computer Vision and Pattern Recognition, pp. 9452–9461 (2021)
33. Zhou, B., Lapedriza, A., Khosla, A., Oliva, A., Torralba, A.: Places: a 10 million image database for scene recognition. IEEE Trans. Pattern Anal. Mach. Intell. **40**(6), 1452–1464 (2017)
34. Zhou, Y.: Rethinking reconstruction autoencoder-based out-of-distribution detection. In: Proceedings of the IEEE/CVF Conference on Computer Vision and Pattern Recognition, pp. 7379–7387 (2022)

Causal Discovery via the Subsample Based Reward and Punishment Mechanism

Jing Yang[1,2](✉), Ting Lu[1,2], and Fan Kuai[1,2]

[1] Intelligent Interconnected Systems Laboratory of Anhui Province, Hefei, People's Republic of China
[2] School of Computer Science and Information Engineering, Hefei University of Technology, Hefei 230009, Anhui, People's Republic of China
jsjyj0801@163.com

Abstract. The discovery of causal relationships between variables from a large number of nonlinear observations is an important research direction in the field of data mining. Extensive studies have highlighted the challenge of constructing accurate causal networks using existing algorithms using large-scale, nonlinear, and high-dimensional data. To address the challenge of this, we propose a general method for causal discovery algorithms applicable to handling large sample data, namely the subsample based reward and punishment mechanism (SRPM) method, which can handle nonlinear large sample data more effectively. We mainly made three contributions. First, we determine the size of the sample split based on experiments to obtain better learning results. Secondly, we developed a reward and punishment mechanism where each sub-sample dynamically changes its own weight in the process of learning the network skeleton, and proposed the SRPM, a subsample method based on the reward and punishment mechanism. Finally, we combine SRPM with three different additive noise model structure learning algorithms applicable to non-Gaussian nonlinear data respectively, and demonstrate the effectiveness of the method through experiments on data generated by a variety of nonlinear function dependencies. Compared with the existing algorithms, the causal network construction algorithm based on SRPM method has a great improvement in accuracy and time performance, and the effectiveness of the method is also verified in the real power plant data.

Keywords: causal discovery · nonlinear · high-dimensional · large sample

---

This work was supported by the National Natural Science Foundation of China (No. 62176082 and 61902068), Guangdong Basic and Applied Basic Research Foundation (No. 2020A1515011499).

Q. Liu et al. (Eds.): PRCV 2023, LNCS 14427, pp. 224–238, 2024.
https://doi.org/10.1007/978-981-99-8435-0_18

1 Introduction

Exploring and discovering causal relationships between variables is a burgeoning field in data mining, with significant applications in statistics and machine learning. In the past few decades, causality discovery methods based on observational data has made great progress in basic theory [1], algorithm design [2] and practical application [3,4], attracting the attention of scholars in related fields. The current classical approaches to causal discovery fall into three broad categories. The first method is a causal discovery method based on dependency analysis, which consists of two stages: causal skeleton learning to causal direction determination. Firstly, learning the causal skeleton between variables (SK) using conditional independence tests based on the causal Markov assumption, and then determining the causal direction based on conditional independence and collision identification methods. Wermuth et al [5] proposed a theoretical algorithm for building directed graphs based on the idea of dependency analysis, which laid the foundation for research based on dependency analysis methods. The second approach is a causal discovery method based on scoring search, which consists of two parts: a scoring function [6] and a search algorithm. The third method is a hybrid approach, incorporating methods based on dependency analysis and scoring search. These methods first obtain the adjacent nodes of a node based on dependency analysis to reduce the search space and reduce the spatial complexity of the search, and then search for the best network based on the scoring function, which is a mainstream method for exploring the causal discovery problem at present.

A plethora of causal discovery algorithms have already been proposed, such as the KMI (kernel mutual information) method proposed by Gretton A et al. [7] and the classical Granger causality (GC) framework reformulated by Andrea et al. [8] based on the Echo State Network (ESN) model for multivariate signals generated by arbitrarily complex networks can detect nonlinear causality very well. However, these algorithms typically have average accuracy and high algorithmic complexity and must be better suited for high-dimensional scenarios. FI method [9] and HDDM model [10] break the curse of high-dimensional data and can effectively infer the causal order of nonlinear multi-variable data. Nevertheless, due to limited sample sizes, these methods cannot meet the demands of the current big data society. Therefore, the challenge of discovering causal relationships from large sample, nonlinear, and high-dimensional data remains an active area of research.

We developed a novel subsample-based method with a reward and punishment mechanism to overcome these challenges, drawing inspiration from the Adaboost algorithm [11]. This approach effectively handles large-scale, nonlinear, and high-dimensional data and significantly contributes to the field of causal discovery. The critical contributions of our work are as follows:

(1) The relationship between the correctness of learning causal networks with additive noise models for nonlinear sample data and the number of samples

is explored, whereby the large sample data is divided into multiple data streams of appropriate size.

(2) A reward and punishment mechanism is developed to give the skeleton different weights and fed back to learn the causal network skeleton. A new method SRPM is proposed that can effectively construct causal relationships for large sample data with non-linearity and high dimensionality.
(3) In separate experiments on simulated and real data, we combined the SRPM method with three different structural learning algorithms for additive noise models applicable to nonlinear non-Gaussian data.

The remainder of this article is organized as follows. In the second part, we review related work in the field of causal discovery. In the third part, we propose improved algorithms for learning nonlinear and non-Gaussian causality using the SRPM method. In the fourth part, we offer experimental results and analysis, comparing our approach to existing methods on both synthetic and real-world datasets. Finally, in the last part, we summarize our work.

2 Related Work

Bayesian networks are a classical causal model based on directed acyclic graphs (DAGs). Traditional models require many independence tests when performing dependency analysis. To reduce the computational complexity, Yang proposed the PCS (partial correlation statistics) algorithm [12]. While achieving higher accuracy, the algorithm's complexity is also significantly reduced. However, the partial correlation method used in this algorithm cannot handle nonlinear data. Yang et al. further successively proposed three algorithms that can effectively handle nonlinear non-Gaussian data to address the complexity posed by data generated by nonlinear structural equation models. First, Yang et al. proposed an RCS algorithm based on rank correlation [13]. Rank correlation calculates correlations based on the relative positions of variable values, regardless of whether the relationship between the variables is linear or nonlinear. The RCS algorithm can uncover nonlinear relationships hidden in the data. However, rank correlation does not eliminate the influence of other variables, which may lead to some bias. Therefore, Yang et al. proposed a PRCS algorithm based on partial rank correlation [14]. Partial rank correlation can eliminate the influence of other variables and reveal the true correlation between variables. Both of these algorithms measure the correlation of variables based on the relative positions of variable values, ignoring the influence of the size information of variable values, which may lead to the loss of important information in the learning network process. Therefore, Yang et al. simplified the HISC coefficient formula by using the mapping relation of the function, proposed the coefficients SC (Spatial Coordinates) and CSC (Conditional Spatial Coordinates), and designed a feature selection algorithm based on these two coefficients. A nonlinear causal structure learning algorithm (SCB) is proposed, significantly reducing the computational complexity of independence analysis and improving judgment accuracy.

3 Introduction to the Algorithmic Framework

This section will introduce the SRPM approach and its use in causal network construction. Like traditional structure learning algorithms, algorithms based on the SRPM method have two stages: skeleton discovery and DAG search. Since the DAG search stage still applies the original MDL scoring function [15] and climbing search algorithm [16], it will not be described in detail here. Next, we analyze the process and feasibility of the skeleton discovery of the algorithm.

3.1 Introduction to SRPM Method

By studying three additive noise model causal learning algorithm applicable to nonlinear non-Gaussian data, we find an interesting phenomenon, that is, when the number of samples reaches a certain number, the number of network learning structure errors will increase with the increase of the number of samples. Why does this happen? Back to the RCS, PRCS and SCB algorithms themselves, the model is unable to learn all the characteristic information of the data quantity, nor could it effectively remove the interference information. The degree of fitting of the function curve showed a trend of first rising and then falling. Too little data led to over-fitting, and too much data led to under-fitting. Thus these algorithms are outwardly manifested by a larger number of structural errors with large sample data, and inwardly by more redundant edges in the skeleton search phase. Can we control the input sample size to an appropriate level to optimize the effect of causality learning? Here, we thought of data segmentation. According to the data of each subsample after segmentation, the corresponding skeleton structure is learned separately. We consider that as long as two variables are found to be neighboring nodes in half of all skeleton structures, an edge exists between these two variables in the average network skeleton (the skeleton obtained from each subsample data is synthesized in equal proportion to the final skeleton used in the search phase).

To further optimize the accuracy of causal discovery, we developed a reward and punishment mechanism to assign different weights to each skeleton of the subsample data. It compares the results between any two variables in the skeleton of the subsample data with the results of the average network skeleton described above, and the count is increased by one if the results are consistent (a consistent result means that the correct edge was found or no non-existent edge was found in the skeleton search phase), and decreased by one if the opposite is true. The initial probability of each skeleton is obtained according to equation (1).

$$P = \frac{\text{count}}{n * (n-1)} \tag{1}$$

Where n represents the number of variables in the original network. The count is a counter, from which the reward and punishment mechanism is applied. In order to eliminate the influence of magnitude between indicators, data standardization is required to address the comparability between data indicators. The above P-values were further processed according to equation (2).

$$W_i = \frac{P_i}{\sum_{j=1}^{m} p_j} \tag{2}$$

The probability of each skeleton is divided by the sum of their probabilities, and the resulting W_i is the final weight of the skeleton corresponding to the ith copy of the data. Where, m denotes the number of copies divided by large samples.

3.2 Correlation Measures and Hypothesis Testing

The correlation coefficient is a statistical analysis index reflecting the degree of correlation. Using different correlation coefficients will have different effects on the skeleton discovery stage of causal construction. Here we use the rank correlation coefficient as an example to illustrate the correlation principle of the correlation coefficient. "Rank" can be understood as a kind of order or sort, so it is solved according to the sorting position of the original data.

For the complete data set $\mathbf{D} = \{x_1, x_2, \ldots, x_m\}$, where m represents the total number of samples, and the corresponding variable set is $\mathbf{V}$. The data set is sorted to obtain a new data set $\mathbf{D}' = \{x'_1, x'_2, \ldots, x'_m\}$. For any two variables $X_i, X_j \subset \mathbf{V}$, Spearman correlation coefficient between variables X_i and X_j is calculated as follows:

$$r_{ij} = 1 - \frac{6\sum_{k=1}^{m}\left(d_{ij}^{k}\right)^2}{m\left(m^2-1\right)} \tag{3}$$

where d_{ij} denotes the order difference between variables X_i and X_j, and d_{ij}^k represents the order difference between variables X_i and X_j in the kth sample.

Theorem 1 ([13]). *For the data generated by the additive noise model, the disturbances conform to arbitrary distribution and are not correlated with each other.* $\mathbf{V}$ *is the set of variables, m is the number of samples, and m is large enough. For any variables X_i and X_j, r_{ij} represents the Spearman correlation coefficient between two variables X_i and X_j, and the test statistic t_0 follows Student's t distribution, and the degree of freedom is m-2.*

$$t_0 = \frac{r_{ij}}{\sqrt{\left(1-r_{ij}^2\right)/(m-2)}} \tag{4}$$

We construct the T-statistic using the statistic t_0 and then introduce the hypothesis test from the paper [17] to define the strong and weak correlations between nodes X_i and X_j in the causal network model.

Definition 1. *Strong correlation: X_i and X_j are strongly correlated when and only when* $\mathrm{p-value}\left(X_i, X_j\right) \leq K_\alpha$.

Definition 2. *Weak correlation: X_i and X_j are weakly correlated when and only when* $\mathrm{p-value}\left(X_i, X_j\right) \geq K_\alpha$.

3.3 Skeleton Discovery Algorithm Based on SRPM Method (SRPM-SK)

After data splitting, we find the candidate neighbor nodes for each node through some method., which is similar to the constraint phase of the SC algorithm [18], the MMHC algorithm [19] and the L1MB algorithm [20]. This requires a "strong correlation" between two variables so that they are "adjacent" to each other in the network. Correlation between variables is measured by correlation. The algorithmic framework for this stage is shown in Fig. 1.

```
Algorithm SRPM-SK
Input: dataset D={x_1, x_2, ..., x_n},threshold K_α, rate
Output: Skeleton network SK
1. Split D={D_1, D_2, ..., D_m}(D_k = {x_1, x_2, ..., x_n})
2. Set the neighbor sets NV(X_i)=∅(i = 1, ..., n)
3. Let the elements of SK_k(i, j), SK(i, j)(i, j =1,...,n) (i, j =1,...,n) to 0
4. for each D_k
     for i =1 to n
       for j =1to i, i≠j
         Calculate r(X_i, X_j)
         Calculate p-value(X_i, X_j)
         if p-value(X_i, X_j)< K_α then
           NV_k(X_i) ∪ X_j, SK_k(i, j) = 1
5. Calculate weight of SK_k
6. for i =1 to n
     for j =1to n
       flag = 0
       for each SK_k
         if SK_k(i, j) == 1
           Calculate flag += weight
         if flag > rate
           SK(i, j) = 1
   return SK
```

Fig. 1. SRPM-SK algorithm framework.

The main task of the first phase of the SRPM-SK algorithm is to reduce the complexity of the causal network by segmenting the data and defining strong and weak correlations. The input data set for the algorithm is $\mathbf{D} = \{x_1, x_2, \ldots, x_n\}$, hypothesis testing threshold is K_α and the skeleton discovery threshold is rate, the total number of samples is n, and the output is the skeleton SK of the causal network. We have a fixed hypothesis test threshold of 0.005 and a rate value of 0.5. The specific implementation process is as follows:

Step 1 The original data set **D** was divided into m parts. Each subsample data $\mathbf{D_k}$ contained all variables in **D**.

Step 2 Initialize all neighbor nodes of X_i as null.

Step 3 Initialize the adjacency matrix SK_k of the subsample skeleton and the adjacency matrix SK of the final combined skeleton. All the elements in the two matrices are 0.

Step 4 Take dataset $\mathbf{D_k}$ as input, obtain coefficient matrix r by correlation coefficient calculation, perform hypothesis testing, calculate p-value value, distinguish strong and weak correlation by Definition 1 and Definition 2, update the causal network skeleton and the set of neighbor nodes of the current variable.

Step 5 Calculate the weight of subsample skeleton SK_k according to SRPM method introduced in Sect. 3.1.

Step 6 Construct the final skeleton according to the weight. To be specific: first, determine whether there is an edge between two variables by the value of a position in the adjacency matrix of the subsample skeleton, determine the probability of having an edge by a flag value flag calculated from the weights, compare the flag with the skeleton discovery threshold rate, and generally, we consider that as long as the probability of having an edge is more than half, then determine the final skeleton with an edge between these two variables.

4 Experimental Results and Analysis

This section uses the SRPM-SK algorithm to discover the causal skeleton and then learn the causal structure. We compare it with the three algorithms separately to complete the experiments on simulated data on low and high-dimensional networks and real data. On the low-dimensional network, the algorithm based on our proposed SRPM method performs better than the original algorithm in terms of accuracy and time performance. Still, this paper does not show experimental results on the low-dimensional network due to space limitations.

4.1 Benchmark Network and Data Sets

For the simulation of low-dimensional large-sample data, the benchmark network we used is sourced from the Bayesian Network Repository (BNR). These networks are obtained from real decision systems in various fields (medical, biological, financial, etc.) and can be used to evaluate the accuracy of structural

Table 1. Network information.

Network	Number of nodes	Number of edges
alarm10	370	570
Child10	200	257
gene	801	972
hailfinder10	560	1017
Insurance10	270	556

learning algorithms. For the simulation of high-dimensional large-sample data, the benchmark network we used is a high-dimensional network synthesized from 10 identical low-dimensional networks (excluding the gene network). The following table shows the basic information of the selected benchmark networks (Table 1).

The data sets in the experiments were generated by ANMs (additive noise models). We used four different nonlinear functions to generate the simulated data, as follows:
$\langle 1\rangle$ $x_i = \mathbf{w}_{X_i}^T \sin(\mathbf{pa}(X_i)) + rand(0,1)$
$\langle 2\rangle$ $x_i = \mathbf{w}_{X_i}^T e^{(-\mathbf{pa}(X_i))} + rand(0,1)$
$\langle 3\rangle$ $x_i = \mathbf{w}_{X_i}^T cos(\mathbf{pa}(X_i)) + rand(0,1)$
$\langle 4\rangle$ $x_i = \mathbf{w}_{X_i}^T (\forall(cos(\mathbf{pa}(X_i)) + e^{(-\mathbf{pa}(X_i))} + \sin(\mathbf{pa}(X_i)) + \mathbf{pa}(X_i)^2 + \mathbf{pa}(X_i))) + rand(0,1)$
Where $\mathbf{w}_{X_i}^T$ is randomly generated, i.e. $\mathbf{w}_{X_i}^T = rand(0,1)$, $\mathbf{pa}(X_i)$ denotes the parent node of X_i. Each X_i can be obtained by superimposing multiple parent nodes through their respective functional relationships. The values of the node variables for nodes without parents are generated by the distribution function $rand(0,1)$. The values of the other node variables are generated by the formulas specified by ANMs, i.e., linear combinations of the values of each parent node variable after transforming the nonlinear function, with weights uniformly distributed in the range (0,1).

4.2 High Dimensional Network Analysis

The data faced by the current society not only has the characteristics of a large number of samples, but also has the characteristics of high dimension. We designed the following experiments: using four structural equation models to generate experimental data for five high-dimensional networks with sample sizes set to (100000,200000,500000,1000000) respectively. The algorithm based on the SRPM method is compared with its corresponding original algorithm to evaluate the performance of the SRPM method under high-dimensional large sample data. Due to the limited space, only some experimental results are shown in this paper.

Figures 2 and 3 show the number of structural errors generated by the learning network of the algorithm using the SRPM method and its corresponding point original algorithm, compared to the benchmark network of ANM$\langle 2\rangle$ and ANM$\langle 4\rangle$, respectively. All the dotted lines in the figure are above the corresponding real lines. Visually, there is a large gap between the two comparison algorithms. That is, the algorithm based on the SRPM method is much better than its original counterpart in terms of the accuracy of the learning network and is particularly prominent in the PRCS algorithm and the SCB algorithm, where the number of structural errors can be reduced to at most 1/8 of the original one.

The original algorithm obtains the interrelationships between variables through a large number of calculations. As the number of samples increases,

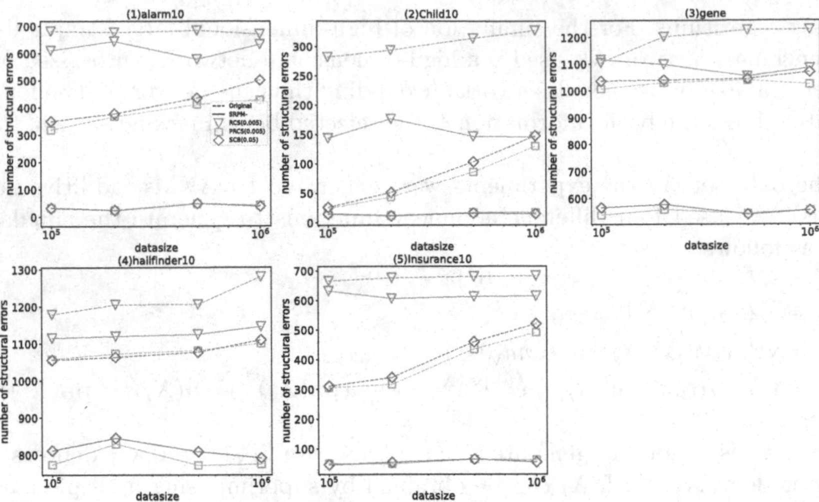

Fig. 2. For ANM⟨2⟩, the number of structural errors in the high-dimensional network.

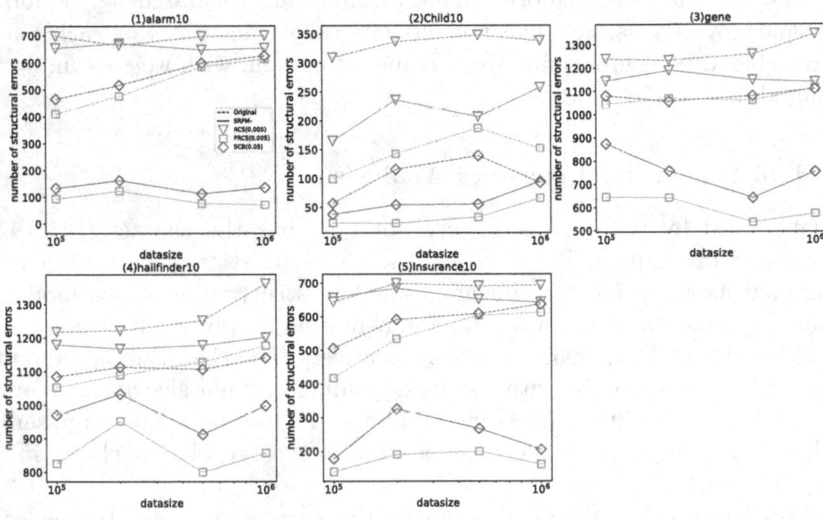

Fig. 3. For ANM⟨4⟩, the number of structural errors in the high-dimensional network.

it may cause a certain direct or indirect effect on two variables that had no connection initially and lead to the generation of redundant edges, manifested in the increase in structural errors. However, the algorithm based on the SRPM method controls the calculation of correlation within the appropriate range through data division, which greatly reduces the complexity of the independence test and the incidence of false positives. In addition, the reward and punishment mechanism defined by us assigns different weights to each subsample. It selects whether there is a mutual relationship between two or more variables according to the sample weights, which optimizes the phenomenon of excess edges in the learning network. In addition, the RCS algorithm only considers the simple correlation between two variables, while the PRCS algorithm and SCB algorithm consider partial correlation. To some extent, there is a correlation between the variables. Still, other related variables influence this correlation, and the simple correlation approach does not consider this influence, so it cannot show the true correlation between the two variables. The true correlation can only be genuinely obtained after removing the effects of other relevant variables. Partial correlation can remove the influence of other variables and obtain the true correlation between variables. Therefore, under the large sample data of a high-dimensional network, the accuracy of the SRPM-PRCS algorithm and SRPM-SCB algorithm is more prominent than the original algorithm.

Table 2. Execution times under high-dimensional networks (one decimal place reserved, unit: s).

models	networks	algorithms	datasets size			
			100000	200000	500000	1000000
ANM<2>	alarm10	RCS/SRPM-RCS	25.8/**9.6**	61.9/**25.1**	167.6/**58.9**	821.6/**120.7**
		PRCS/SRPM-PRCS	**16.1**/29.6	**32.2**/41.2	**88.6**/152.3	**225.4**/302.7
		SCB/SRPM-SCB	**69.1**/163.3	472.5/**282.0**	1377.5/**1155.0**	**3931.6**/3990.0
	Child10	RCS/SRPM-RCS	12.1/**8.4**	30.1/**21.4**	64.1/**49.5**	317.5/**100.3**
		PRCS/SRPM-PRCS	**12.6**/12.7	26.8/**22.2**	66.0/**65.0**	208.3/**151.5**
		SCB/SRPM-SCB	**22.9**/49.6	152.0/**92.4**	442.1/**369.9**	1326.9/**1283.6**
	gene	RCS/SRPM-RCS	72.8/**31.7**	170.9/**63.1**	541.7/**152.4**	3261.8/**337.3**
		PRCS/SRPM-PRCS	**30.7**/109.9	**65.7**/122.9	**158.8**/536.2	**460.4**/1067.6
		SCB/SRPM-SCB	**309.0**/792.4	2146.6/**1262.8**	6219.6/**5193.6**	**18079.0**/18272.0
	hailfinder10	RCS/SRPM-RCS	70.2/**24.9**	149.7/**54.7**	569.5/**158.9**	2334.4/**269.4**
		PRCS/SRPM-PRCS	**22.0**/55.9	**52.2**/67.0	**135.4**/290.9	**374.5**/563.6
		SCB/SRPM-SCB	**152.6**/380.5	1067.0/**617.4**	3055.7/**2564.0**	8948.9/**8761.3**
	Insurance10	RCS/SRPM-RCS	26.1/**9.4**	60.0/**23.4**	143.6/**49.5**	710.1/**110.9**
		PRCS/SRPM-PRCS	**14.6**/19.3	28.5/**28.1**	**85.8**/104.4	**219.6**/223.0
		SCB/SRPM-SCB	**39.4**/92.7	265.4/**149.4**	750.2/**645.4**	**2178.5**/2236.5
ANM<4>	alarm10	RCS/SRPM-RCS	37.4/**13.4**	71.4/**29.9**	275.5/**72.3**	1137.1/**164.0**
		PRCS/SRPM-PRCS	**17.1**/30.5	**32.8**/40.7	**98.5**/159.6	**228.2**/321.1
		SCB/SRPM-SCB	**68.8**/168.1	471.1/**275.7**	1369.5/**1179.8**	**2661.8**/2758.1
	Child10	RCS/SRPM-RCS	12.5/**7.9**	30.9/**20.4**	65.5/**49.6**	387.4/**100.9**
		PRCS/SRPM-PRCS	14.8/**12.9**	36.4/**21.9**	98.2/**67.7**	267.7/**185.4**
		SCB/SRPM-SCB	**24.4**/52.4	154.3/**90.6**	450.9/**371.2**	880.5/**867.9**
	gene	RCS/SRPM-RCS	150.6/**50.7**	265.0/**100.1**	741.6/**224.4**	3705.0/**544.2**
		PRCS/SRPM-PRCS	**38.0**/110.0	**78.7**/125.1	**194.5**/540.4	**590.0**/1052.0
		SCB/SRPM-SCB	**308.7**/762.0	2119.0/**1215.0**	5562.2/**4648.0**	13353.0/**11922.0**
	hailfinder10	RCS/SRPM-RCS	102.0/**35.3**	272.3/**95.1**	1359.6/**274.4**	3877.3/**1008.9**
		PRCS/SRPM-PRCS	**24.9**/56.3	**65.2**/67.1	**175.0**/281.6	**1349.8**/6476.3
		SCB/SRPM-SCB	**154.7**/370.8	1038.1/**613.3**	2836.5/**2293.3**	**7332.2**/10760.0
	Insurance10	RCS/SRPM-RCS	33.5/**13.2**	103.8/**45.6**	391.3/**96.1**	1132.8/**173.9**
		PRCS/SRPM-PRCS	**15.5**/19.5	31.0/**29.2**	**90.5**/110.0	**226.5**/233.1
		SCB/SRPM-SCB	**39.8**/90.4	267.0/**160.1**	673.3/**579.7**	1816.7/**1567.6**

Table 2 shows the performance of the execution time of each algorithm under each model in the high-dimensional network. In this table, we highlight (bold) each experiment's results that take less time. As shown in it, the execution time of the SRPM-RCS algorithm is better than that of the RCS algorithm at any time. This is determined by the correlation coefficient calculation method used in the original algorithm. The RCS algorithm calls the Spearman correlation coefficient to directly calculate and return the correlation coefficient between two variable nodes. Therefore, the time performance of the SRPM-RCS algorithm is consistent with its performance in low-dimensional networks. Because of data segmentation, each part is processed faster, and the larger the number of samples, its time performance is better than the original algorithm. The SRPM-SCB algorithm has less execution time than the SCB algorithm for sample numbers 200000 and 500000, with a maximum difference of nearly 1000 s, but does not perform as well as the SCB algorithm for sample numbers 100000 and 1000000. This is because the SCB algorithm processes the data by cycling the number of samples and finally relies on matrix numerical multiplication operations to find its spatial coordinate (SC) coefficients. On the one hand, the SRPM-SCB algorithm reduces the number of cycles and, on the other hand, increases the number of high-dimensional matrix multiplication operations. And the good or bad time performance it exhibits depends on the trade-off between these two counts. According to the experimental results, it is reasonable to believe that when the sample number is 200000 and 500000, the reduced number of cycles has a greater impact on the time performance of the SRPM-SCB algorithm than the increased number of matrix multiplication. At 100,000 samples, the opposite is true. When the number of samples is 1000000, the two effects are similar; whether the performance is improved in a specific time depends on the ANM used. The execution time of the SRPM-PRCS algorithm is almost always inferior to that of the PRCS algorithm. This is because the PRCS algorithm computes the conditional correlation coefficient of the entire matrix. In high-dimensional networks, the algorithm will be challenged by the dimensionality of the matrix. Therefore, the corresponding algorithm based on SRPM is not outstanding in time performance on other networks except that the execution time of the Child10 network (the network has the smallest dimension among the five high-dimensional benchmark networks we used and the network information given in Table 2 is referred to) is improved compared with the original algorithm. In summary, the original algorithm has a specific impact on the execution time of the algorithm based on the SRPM method, but the difference in time is generally slight. Compared with the performance improved by the number of structural errors, the increase in execution time is entirely acceptable.

Through the above experimental verification, we can conclude that the SRPM method suits the complex situation of high-dimensional and large-sample networks with nonlinear data and can achieve good results in various indexes.

4.3 Real Data Analysis

The actual data in the experiment came from the sensor data recorded by a power plant in 2019. The data recorded the operation status of 412 detection points between 2019-1-22 and 2019-3-19. The equipment collected data every 30 s, generating 60,480 data every week on average. The data from the natural environment has the characteristics of high dimensionality and a large amount of data. So we construct the causal network structure diagram (as shown in Fig. 4, due to the excessive number of measurement points, only some of the causality diagrams between measurement points are shown here) of this data using the SRPM method to find out the potential causal relationship between the measurement points, and then the prediction model of the target measurement points was trained to verify the accuracy and validity of the causal analysis. The experiment uses the long short-term memory network LSTM as the prediction model. Select the data from January 24 to January 30 as the training data and from January 31 as the test data. As shown in Fig. 5, the model prediction trend plots for measurement points 2,9,14 and 21 are shown respectively. Figure(a) represents the trend plot of all measurement points without causality screening features for model prediction. Figure(b) illustrates the trend of model prediction after feature selection using the SRPM method to determine causality.

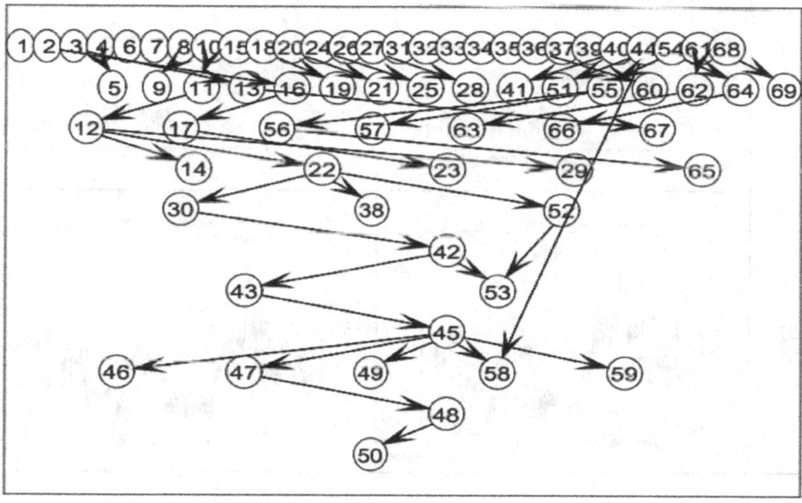

Fig. 4. Causality diagram between some measurement points.

Figure 5 shows a complex nonlinear function relationship between the raw data of each measurement point of the power plant. Suppose all the measurement points are used to predict the trend of the current measurement point without feature selection. In that case, the results will be somewhat different from the actual data, i.e., the results will be disturbed by many irrelevant measurement

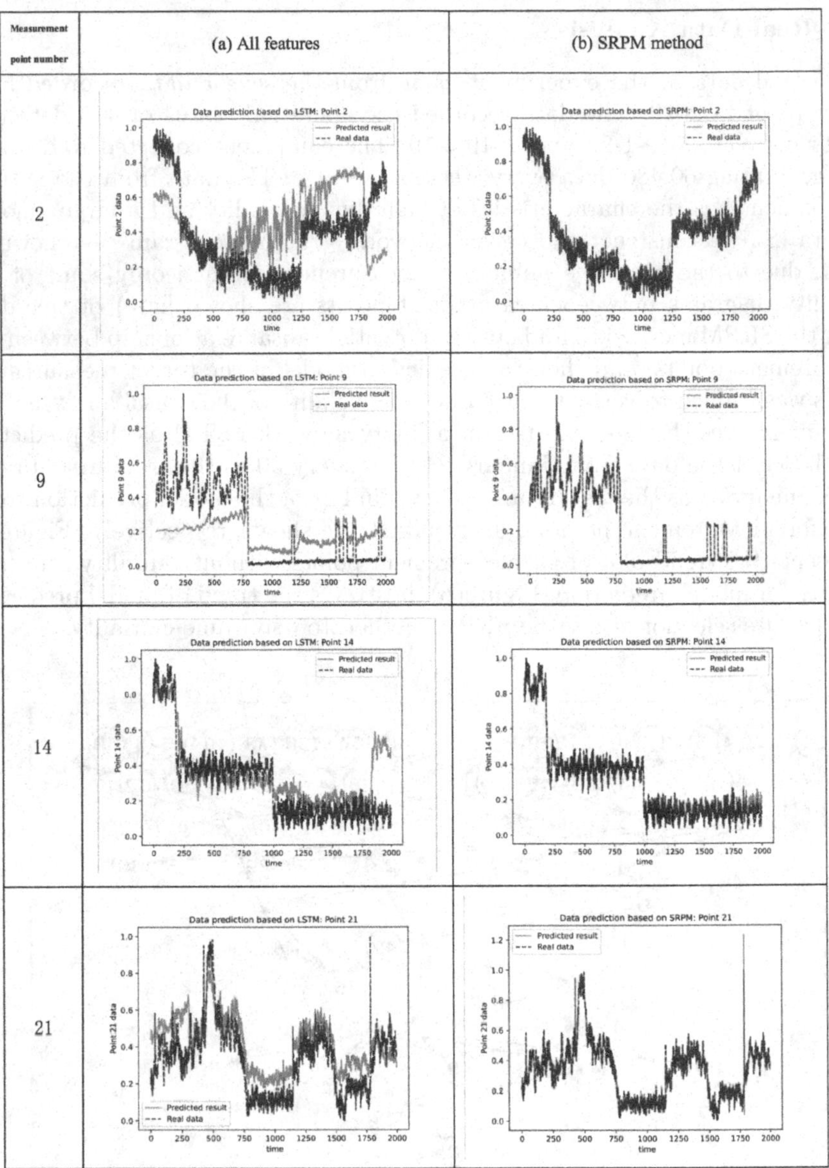

Fig. 5. Experimental results of prediction of some measurement points in power plants.

point data. If the causal network relationship constructed by the SRPM method is used as the basis, and after feature selection, predictions are made based on these selected relevant measurement points. The obtained predicted trends match the real trend data of the current measurement points. The model prediction results for the above four measurement points (2,9,14,21) show that the

SRPM method can effectively remove the influence of irrelevant measurement points on the observed measurement points and has good causality detection performance for realistic nonlinear large sample data.

5 Conclusion and Outlook

Facing the challenges of high-dimensional and nonlinear causal discovery on large sample datasets, a method SRPM is proposed for complex nonlinearities by studying causal structure learning on additive noise network models. And with the help of extensive experiments and results analysis, it is concluded that the method SRPM proposed in this paper can deal with the problem of learning causal network structures in a variety of nonlinear environments in a general and effective way, can cope with large sample data sets efficiently, and also solves the difficulties of high time complexity and curse of dimensionality brought by high dimensionality. In addition to its good performance on simulated data, the algorithm also performs well in real recorded power plant data, which is a good aid for fault analysis and early warning in power plants and has a wide range of application prospects.

References

1. Vowels, M.J., Camgoz, N.C., Bowden, R.: D'ya like dags? a survey on structure learning and causal discovery. ACM Comput. Surv. **55**(4), 1–36 (2022)
2. Shen, X., Ma, S., Vemuri, P., Simon, G.: Challenges and opportunities with causal discovery algorithms: application to alzheimer's pathophysiology. Sci. Rep. **10**(1), 2975 (2020)
3. Zhu, J.Y., Sun, C., Li, V.O.: An extended spatio-temporal granger causality model for air quality estimation with heterogeneous urban big data. IEEE Trans. Big Data **3**(3), 307–319 (2017)
4. Assaad, C.K., Devijver, E., Gaussier, E.: Survey and evaluation of causal discovery methods for time series. J. Artifi. Intell. Res. **73**, 767–819 (2022)
5. Wermuth, N., Lauritzen, S.L.: R: Graphical and Recursive Models for Contingency Tables. Eksp. Aalborg Centerboghandel (1982)
6. Cooper, G.F., Herskovits, E.: A bayesian method for constructing bayesian belief networks from databases. In: Uncertainty Proceedings 1991, pp. 86–94. Elsevier (1991)
7. Gretton, A., Herbrich, R., Smola, A,J.: The kernel mutual information. In: 2003 IEEE International Conference on Acoustics, Speech, and Signal Processing, 2003. Proceedings (ICASSP 2003), vol. 4, pp. IV-880. IEEE (2003)
8. Duggento, A., Guerrisi, M., Toschi, N.: Recurrent neural networks for reconstructing complex directed brain connectivity. In: 2019 41st Annual International Conference of the IEEE Engineering in Medicine and Biology Society (EMBC), pp. 6418–6421. IEEE (2019)
9. Liu, F., Chan, L.-W.: Causal inference on multidimensional data using free probability theory. IEEE Trans. Neural Netw. Learn. Syst. **29**(7), 3188–3198 (2017)
10. Zeng, Y., Hao, Z., Cai, R., Xie, F., Huang, L., Shimizu, S.: Nonlinear causal discovery for high-dimensional deterministic data. IEEE Trans. Neural Netw. Learn. Syst. (2021)

11. Freund, Y., Schapire, R.E.: A decision-theoretic generalization of on-line learning and an application to boosting. J. Comput. Syst. Sci. **55**(1), 119–139 (1997)
12. Yang, J., An, N., Alterovitz, G.: A partial correlation statistic structure learning algorithm under linear structural equation models. IEEE Trans. Knowl. Data Eng. **28**(10), 2552–2565 (2016)
13. Yang, J., Fan, G., Xie, K., Chen, Q., Wang, A.: Additive noise model structure learning based on rank correlation. Inf. Sci. **571**, 499–526 (2021)
14. Yang, J., Jiang, L., Xie, K., Chen, Q., Wang, A.: Causal structure learning algorithm based on partial rank correlation under additive noise model. Appl. Artif. Intell. **36**(1), 2023390 (2022)
15. Rissanen, J.: Modeling by shortest data description. Automatica **14**(5), 465–471 (1978)
16. Teyssier, M., Koller, D.: Ordering-based search: A simple and effective algorithm for learning bayesian networks. arXiv preprint arXiv:1207.1429 (2012)
17. Zar, J.H.: Significance testing of the spearman rank correlation coefficient. Jo. Am. Stat. Assoc. **67**(339), 578–580 (1972)
18. Friedman, N., Nachman, I., Pe'er, D.: Learning bayesian network structure from massive datasets: The" sparse candidate" algorithm. arXiv preprint arXiv:1301.6696 (2013)
19. Tsamardinos, I., Brown, L.E., Aliferis, C.F.: The max-min hill-climbing bayesian network structure learning algorithm. Mach. Learn. **65**, 31–78 (2006)
20. Schmidt, M., Niculescu-Mizil, A., Murphy, K., et al.: Learning graphical model structure using l1-regularization paths. AAAI **7**, 1278–1283 (2007)

Local Neighbor Propagation Embedding

Wenduo Ma[1], Hengzhi Yu[1], Shenglan Liu[2](✉), Yang Yu[3], and Yunheng Li[1]

[1] Faculty of Electronic Information and Electrical Engineering, Dalian University of Technology, Dalian 116024, Liaoning, China

[2] School of Innovation and Entrepreneurship, Dalian University of Technology, Dalian 116024, Liaoning, China

liusl@dlut.edu.cn

[3] The Alibaba Group, Hangzhou 310051, Zhejiang, China

Abstract. Manifold learning occupies a vital role in the field of nonlinear dimensionality reduction and serves many relevant methods. Typical linear embedding methods have achieved remarkable results. However, these methods perform poorly on sparse data. To address this issue, we introduce neighbor propagation into Locally Linear Embedding (LLE) and propose a new method named Local Neighbor Propagation Embedding (LNPE). LNPE enhances the local connections between neighborhoods by extending 1-hop neighbors into n-hop neighbors. The experimental results show that LNPE can obtain faithful embeddings with better topological and geometrical properties.

Keywords: Local Neighbor Propagation · Manifold Learning · Geometrical Properties · Topological

1 Introduction

Over the past few decades, manifold learning has already attracted extensive attention and applied in video prediction [2], video content identification [14], spectral reconstruction [10], image super-resolution [3], etc. Manifold learning research can be divided into local methods such as Locally Linear Embedding [18], Local Tangent Space Alignment (LTSA) [25] and some global methods such as Isometric Mapping (ISOMAP) [20] and Maximum Variance Unfolding (MVU) [22]. Besides, compared with nonlinear manifold learning methods, some linear embedding methods are more effective for practical applications such as Locality Preserving Projection (LPP) [7], Neighborhood Preserving Embedding (NPE) [6], Neighborhood Preserving Projection (NPP) [15].

Locally linear methods in manifold learning such as LLE have been widely studied and applied. Based on LLE, many improved methods are proposed, such as Hessian Locally Linear Embedding (HLLE) [4], Modified Locally Linear Embedding (MLLE) [24] and Improved Locally Linear Embedding (ILLE) [23]. Among the LLE-based methods, HLLE has high computational complexity and MLLE does not work on weak-connected data. Actually, the computational complexity and robustness of algorithms are issues to be considered for

Q. Liu et al. (Eds.): PRCV 2023, LNCS 14427, pp. 239–249, 2024.
https://doi.org/10.1007/978-981-99-8435-0_19

many LLE-based improved methods. Inspired by Graph Convolutional Networks (GCNs), we propose a simple and unambiguous LLE-based improved method, named Local Neighbor Propagation Embedding (LNPE). In contrast to previous approaches, LNPE extends the neighborhood size through neighbor propagation layer by layer. This architecture enhances the topological connections within each neighborhood. From the view of global mapping, neighborhood interactions increase with neighbor propagation to improve the global geometrical properties. Experimental results verify the effectiveness and robustness of LNPE.

The remainder of this paper is organized as follows. We first introduce related work in Sect. 2. Section 3 presents the main body of this paper, which includes the motivation of LNPE, the mathematical background, the framework of LNPE and the analysis of computational complexity. The experimental results are presented in Sect. 4 to verify the effectiveness and robustness of LNPE and Sect. 5 summarizes our work.

2 Related Work

The essence of manifold learning is how to maintain the relationship corresponding to the intrinsic structure between samples in two different spaces. Researchers have done lots of work to measure the relationship from different aspects.

PCA maximizes the global variance to reduce dimensionality, while Multidimensional Scaling (MDS) [9] considers the low-dimensional distance between samples which is consistent with high-dimensional data. Based on MDS, ISOMAP utilize the shortest path algorithm to realize global mapping. Besides, MVU realizes an unfolding manifold through positive semi-definite and kernel technique and RML [11] obtains the intrinsic structure of the manifold with the Riemannian methods instead of Euclidean distance. Compared with ISOMAP, LLE represents local manifold learning, which obtains neighbor weights with locally linear reconstruction. Furthermore, Locally Linear Coordination (LLC) [17] constructs a local model and makes a global alignment and contributes to both LLE and LTSA. LTSA describes local curvature through the tangent space of samples and takes the curvature as the weight of tangent space to realize global alignment.

Another type of manifold structure preserving is graph-based embedding. Classical methods are still representative such as LE [1] and its linear version LPP. More related methods include NPE, Orthogonal Neighborhood Preserving Projections (ONPP) [8], etc. The graph-based method produces a far-reaching influence on machine learning and related fields. For instance, the graph is introduced into semi-supervised learning in LE, and LLE is also a kind of neighbor graph. Moreover, L-1 graph-based methods such as Sparsity preserving Projections (SPP) [16] and its supervised extension [5] are proposed with the wide application of sparse methods.

3 Local Neighbor Propagation Embedding

In this section, the idea of neighbor propagation is introduced gradually. First, we present the motivation for LNPE and describe the origin of LNPE. Then the basic algorithm prototype is briefly introduced. Finally, the LNPE framework is introduced.

3.1 Motivation

LLE, which is a classical method, should be representative of methods based on local information in manifold learning. In the case of simple data distribution, LLE tends to get satisfactory results. But once the data distribution becomes sparse, it is difficult for LLE to maintain topological and geometrical properties. The following items show the reasons.

1. The neighborhood size is hard to determine for sparse data. Inappropriate neighbors will be selected with a larger neighborhood size.
2. LLE focuses more on each single neighborhood, but is weak in the interaction between different neighborhoods. Thus, it is difficult to obtain the ideal effect in geometrical structure preservation.

There is a natural contradiction between Item 1 and Item 2. More specifically, the interaction of neighborhood information will be weakened by small neighborhoods inevitably. To improve the capability and robustness of LLE, we introduce neighbor propagation into LLE and propose Local Neighbor Propagation Embedding (LNPE), inspired by GCNs. The 1-hop neighborhoods are extended to n-hop neighborhoods through neighbor propagation, which will enhance the connections of points in different neighborhoods. In this way, LNPE avoids short circuits (Item 1) by setting a small neighborhood size k. Meanwhile, i-hop neighborhoods expand neighborhood size to depict local parts adequately by enhancing the topological connections. Furthermore, correlations between different neighborhoods (Item 2) are generated in neighbor propagation to produce more overlapping information, which is conducive to preserving the geometrical structure.

3.2 Mathematical Background

Based on LLE, LNPE introduces neighbor propagation to improve its applicability. Before starting the LNPE framework, we first review the original LLE.

Suppose $\mathbf{X} = \{\mathbf{x}_1, \ldots, \mathbf{x}_n\} \subset \mathrm{R}^D$ indicates a high-dimensional dataset that lies on a smooth D-dimensional manifold, LLE tends to embed the intrinsic manifold from high-dimensional space into lower-dimensional subspace with preserving geometrical and topological structures. Based on the assumption of local linearity, LLE firstly reconstructs each high-dimensional data point $\mathbf{x}_i$ through linear combination within each neighborhood $\mathbf{N}_i$, where $\mathbf{N}_i$ indicates the k-nearest neighbors of $\mathbf{x}_i$ and $\mathbf{N}_i = \{\mathbf{x}_{i_1}, \ldots, \mathbf{x}_{i_k}\}$. Then the reconstruction weights matrix $\mathbf{W}$ in high-dimensional space can be determined by minimizing the total reconstruction error ε_1 of all data points.

$$\varepsilon_1(\mathbf{W}) = \|\mathbf{X}\mathbf{W} - \mathbf{X}\|_F^2 \tag{1}$$

where $\mathbf{W} = [\vec{\mathbf{w}}_1, \vec{\mathbf{w}}_2, \cdots, \vec{\mathbf{w}}_n] \in \mathrm{R}^{n\times n}$, the i-th column vector $\vec{\mathbf{w}}_i$ indicates the reconstruction weights of data point $\mathbf{x}_i$ and $\|\cdot\|_F$ is the Frobenius norm. To remove the influence of transformations including translation, scaling, and rotation, a sum-to-one constraint $\vec{\mathbf{w}}_i^T\vec{1} = 1$ is enforced for each neighborhood.

Let $\mathbf{Y} = \{\mathbf{y}_1, \ldots, \mathbf{y}_n\} \subset \mathrm{R}^d$ be the corresponding dataset in low dimensions. The purpose of LLE is to preserve the same local structures reconstructed in high-dimensional space. Then in the low-dimensional space, LLE chooses to utilize the same weights $\mathbf{W}$ to reproduce the local properties. The objective is to minimize the total cost function

$$\varepsilon_2(\mathbf{Y}) = \|\mathbf{Y}\mathbf{W} - \mathbf{Y}\|_F^2 \tag{2}$$

under the constraint $\mathbf{Y}\mathbf{Y}^T = \mathbf{I}$. Thus, the high-dimensional coordinates are finally mapped into lower-dimensional observation space.

3.3 Local Neighbor Propagation Framework

A faithful embedding is root in more sufficient within-neighborhood and between-neighborhood information of local and global structure. For LLE, the interactive relationship in single neighborhood is not enough to reproduce detailed data distribution, especially when the neighborhood size k is not large enough. Neighborhood propagation is introduced into LLE to intensify the topological connections within neighborhoods and interactions between neighborhoods.

High-Dimensional Reconstruction. Based on the single reconstruction in each neighborhood, LNPE propagates neighborhoods and determines propagating weight matrix. Similarly, we define the high-dimensional and low-dimensional data as $\mathbf{X} = \{\mathbf{x}_1, \ldots, \mathbf{x}_n\} \subset \mathrm{R}^D$ and $\mathbf{Y} = \{\mathbf{y}_1, \ldots, \mathbf{y}_n\} \subset \mathrm{R}^d$, respectively. Suppose that we have finished the single reconstruction with LLE (Eq. 1) and got the weight matrix $\mathbf{W}_1$. In the meantime, the first reconstructed data $\mathbf{X}^{(1)} = \mathbf{X}\mathbf{W}_1$ are obtained from the first reconstruction with LLE. LNPE is to reuse the reconstructed data such as $\mathbf{X}_1$ to reconstruct the original data points in $\mathbf{X}$ again with the reconstructed data previously. Then the first neighborhood propagation is to utilize $\mathbf{X}_1$ to reconstruct original data $\mathbf{X}$ through

$$\varepsilon(\mathbf{W}_2) = \|\mathbf{X}\mathbf{W}_1\mathbf{W}_2 - \mathbf{X}\|_F^2 \tag{3}$$

where $\mathbf{W}_2$ corresponds to the weight matrix in the first neighbor propagation. Thus, we can simply combine the weight matrices $\mathbf{W}_1$ and $\mathbf{W}_2$ as $\mathbf{W} = \mathbf{W}_1\mathbf{W}_2$. Compared with $\mathbf{W}_1$ in LLE, $\mathbf{W}$ extends the 1-hop neighbors to 2-hop neighbors, which expands the neighborhood size through neighbor propagation. One of the most important advantages is that neighbor propagation can enhance topological connections while avoiding short circuits. Besides, a truth worth noting is that the multi-hop neighbors hold lower weights in the process of propagation. Then after $i-1$ propagations, each data point to be reconstructed establishes relations with its i-hop neighbors. The $(i-1)$-th neighbor propagation can be formulated as

$$\varepsilon(\mathbf{W}_i) = \|\mathbf{X}\mathbf{W}_1 \cdots \mathbf{W}_{i-1}\mathbf{W}_i - \mathbf{X}\|_F^2 \tag{4}$$

where $\mathbf{W}_i$ indicates the weight matrix in the $(i-1)$-th neighbor propagation. From the perspective of weight solution and optimization, the i-th neighbor propagation depends on all the first $i-1$ weight matrices and each weight matrix in the propagation must be determined in turn.

Global Low-Dimensional Mapping. Similar to LLE, the low-dimensional embedding in LNPE is to reproduce the high-dimensional properties determined in the reconstruction. LNPE aims to preserve all the topological connections from 1-hop neighborhoods to n-hop neighborhoods with n weight matrices. We define that the matrix product $\mathbf{P}_i = \mathbf{W}_1\mathbf{W}_2 \cdots \mathbf{W}_i$ is the product of weight matrices in the first $i-1$ neighbor propagation. And then the low-dimensional total optimization function can be expressed as

$$\varepsilon(\mathbf{Y}) = \sum_{i=1}^{t+1} \|\mathbf{Y}\mathbf{P}_i - \mathbf{Y}\|_F^2 \tag{5}$$

where the parameter $t+1$ denotes the total hop in the high-dimensional reconstruction with t neighbor propagations. According to the properties of F-norm, the low-dimensional objective Eq. 6 can be written as

$$\varepsilon(\mathbf{Y}) = \mathbf{Y}\left(\sum_{i=1}^{t+1}(\mathbf{P}_i - \mathbf{I})(\mathbf{P}_i - \mathbf{I})^T\right)\mathbf{Y}^T \tag{6}$$

where $\mathbf{I} \in \mathrm{R}^{n\times n}$ indicates the identity matrix. Under the constraint $\mathbf{YY} = I$, the low-dimensional coordinates can be easily obtained by decomposing the target matrix. The detailed algorithm is shown as Algorithm 1.

Equation 5 and Eq. 6 show that LNPE aims to preserve all the learned properties in the high-dimensional reconstruction. Specifically speaking, for the sequence $i = 1, 2, \cdots, t$, a smaller i aims at maintaining the topological connections within each neighborhood, while a larger i pays more attention to expanding the neighborhood interactions between different neighborhoods.

3.4 Computational Complexity

The computational complexity of LNPE follows LLE. Calculating the k nearest neighbors scales as $O(Dn^2)$. In some special data distributions, the computational complexity can be reduced to $O(n \log n)$ with K-D trees [19]. Computing the weight matrix in $t+1$ reconstructions scales as $O((t+1)nk^3)$. Besides,

Algorithm 1. *LNPE Algorithm*

```
Require:
    high-dimensional data X ⊂ R^D;
    neighborhood size k;
    target dimensionality d;
    the neighbor propagation times t.
Ensure:
    low-dimensional coordinates Y ⊂ R^d;
  Find the k-nearest neighbors N_i = {x_{i_1}, ··· , x_{i_k}} for each data point.
  Compute the weight matrix W_1 with LLE.
  Initialize a zero matrix M.
  for e = 1 : t do
      Compute the matrix product P_e = W_1 W_2 ··· W_e
      Compute M = M + (P_i − I)(P_i − I)^T
      Compute the reconstruction data X^(e) = X P_e
      for each sample x_i, i = 1, ··· , n do
          Compute w_i^(e+1) in W_{e+1} with X and X^(e) through minimizing ||x_i − X^(e) w_i^(e+1)||_2^2
      end for
  end for
  Compute M = M + (P_{t+1} − I)(P_{t+1} − I)^T
  Solve Y = arg min_Y Tr(Y M Y^T)
```

computing matrices $\mathbf{P}$ and $\mathbf{M}$ scales as $O(p_1 n^3)$ with sparse matrix and the final calculation of eigenvectors of a sparse matrix has computational complexity $O(p_2 n^2)$, where p_1 and p_2 is parameters related to the ratio of nonzero elements in sparse matrix [21].

4 Experimental Results

In this section, we verify LNPE on both the synthetic and real-world datasets. To conduct a fair comparison, we set the same configuration for the comparison methods and LNPE.

4.1 Synthetic Datasets

On sparse data, curvature-based dimensionality reduction methods, such as HLLE [4] and LTSA [25], have performance degradation. This is because the instability of local relationship among sparse data increases, resulting in less smoothness of the local curvature. As shown in Fig. 1, HLLE [4] is unable to accomplish dimensionality reduction on the S-curve and the Swiss, while the dimensionality reduction performance of LTSA [25] is unstable. When the neighborhood size k is increased, the dimensionality reduction performance of LTSA improves. However, it is not a good tendency to obtain better dimensionality reduction performance by increasing the neighborhood size k, because too large

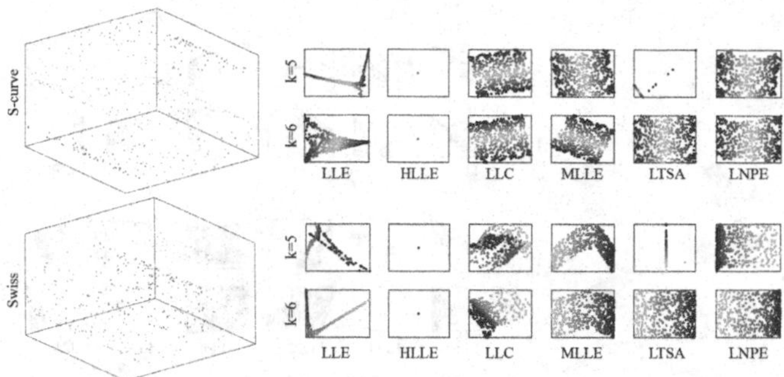

Fig. 1. The S-Curve with 500 samples and its embedding results and the Swiss with 500 samples and its embedding results.

k will lead to short circuits. For LLC [17], each point is represented as a linear combination of other points, where the weights used are controlled by a local constraint matrix. When the data become sparse, the constraint matrix is easily disturbed by noise and outliers, which leads to unstable and inaccurate weights. Unlike these methods, LNPE can better capture the relationship between data through local neighborhood propagation. This is because LNPE can not only better model the linear relationship in the neighborhood, but also strengthen the relationship between the center point and the points outside the neighborhood. LNPE can learn the topological and geometrical structure of the data and embed it in the low-dimensional space faithfully.

To further verify that LNPE has outstanding dimensionality reduction performance on data with complex topology, we select the Swiss-Hole and the Changing-Swiss for our experiments, as shown in Fig. 2. For complex data, the relationship between local points and global neighbors is not easy to maintain. With LLE, the connection of a single neighborhood is not sufficient to fully describe the original data distribution. Especially, LLE performs poorly when the neighborhood size k is not large enough, and too large k can cause short circuits. HLLE [4] has a tendency to depend on the neighborhood size k for the Swiss Hole and cannot accomplish the dimensionality reduction on the Changing-Swiss. Although LTSA [25] performs well on the Swiss Hole, it does not achieve satisfactory results on the Changing-Swiss. MLLE and LLC tend to project the high-dimensional data directly into the low-dimensional space, which leads to the inability to restore topological and geometrical structures well. LNPE is able to completely unfold the topological structure of the original data on the low-dimensional space.

4.2 Real-World Datasets

To evaluate LNPE on real-world applications, we conduct experiments on pose estimation tasks including the Statue face dataset and Teapot dataset. The Statue face dataset, which consists of 698 images with $64 \times 64 = 4096$ pix-

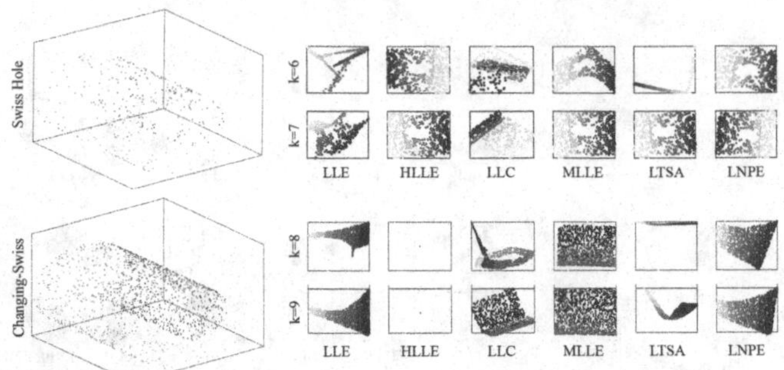

Fig. 2. The Swiss Hole with 500 samples and its embedding results and the Changing-Swiss with 1500 samples and its embedding results.

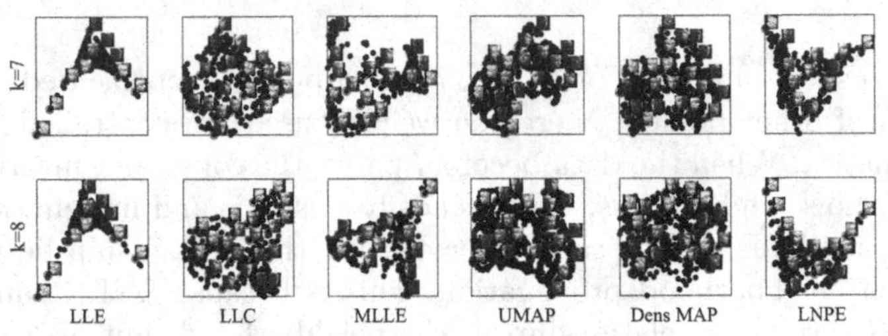

Fig. 3. The Statue Face dataset and its embedding results.

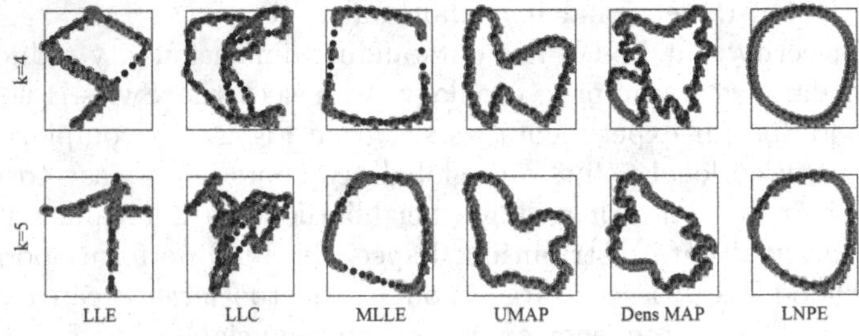

Fig. 4. The teapot dataset and its embedding results.

els, is first used in ISOMAP [20]. The Teapot dataset is a set of 400 teapot images in total [22]. Each image in the Teapot dataset can be seen as a high-dimensional vector consisting of 76 × 101 × 3 RGB pixels. Samples in these two datasets lie on smooth underlying manifolds. All images in the two datasets are tensed into their respective data spaces by bit-pose variations, and all data points obey a streamlined data distribution in this high-dimensional data space. To test the performance of LNPE on sparse data, we downsampled the number of samples on both datasets to half of the original which is feasible due to uniform distribution.

Figure 3 shows the low-dimensional embedding of LNPE and the comparison algorithms on the Statue face dataset, where the experimental results are compared under different choices of nearest neighbor ranges. Specifically, the LLC [17], UMAP [12] and DensMAP [13] perform poorly, and the topology and geometric structure of their embedding results are obviously destroyed. The LLE and MLLE can maintain good topology, but their geometric structure remains weak. In contrast, since MLLE employs a multi-weight structure, it has an advantage in geometric structure retention. The LNPE proposed in this paper is able to recover well the geometric structure at k=7, while it is better in geometric structure maintenance at k=8. Meanwhile, the dimensionality reduction results of LNPE do not exhibit significant topological damage, which indicates that the LNPE is more robust in the neighborhood range. To validate the efficacy of LNPE, we also compare our LNPE with other methods on the teapot dataset are shown in Fig. 4. The previous dimensionality reduction methods perform poorly when $k = 4$ and 5. In the same setup of neighbor, LNPE achieves better dimensionality reduction and obtains good geometric and topological properties.

Overall, compared with previous dimensionality reduction methods, LNPE can achieve competitive dimensionality reduction results with less dependence on the neighborhood size k. This is because it captures the global and local relationship of the data well and has good robustness to sparse data.

5 Conclusion

LLE-based methods usually fail in addressing sparse data where the local connections and neighborhood interactions are difficult to obtain. In this paper, we introduce neighbor propagation in LLE and propose Local Neighbor Propagation Embedding. LNPE expands 1-hop neighborhoods to n-hop neighborhoods, and can be solved iteratively based on the solution of LLE. Although computational complexity increases, LNPE enhances topological connections within neighborhoods and interactions between neighborhoods. Experimental results on both synthetic and real-world datasets show that LNPE improves embedding performance and is more robust than previous methods.

Author contributions. Wenduo Ma and Hengzhi Yu–These authors contributed equally to this work.

References

1. Belkin, M., Niyogi, P.: Laplacian eigenmaps for dimensionality reduction and data representation. Neural Comput. **15**(6), 1373–1396 (2003)
2. Cai, Y., Mohan, S., Niranjan, A., Jain, N., Cloninger, A., Das, S.: A manifold learning based video prediction approach for deep motion transfer. In: 2021 IEEE/CVF International Conference on Computer Vision Workshops (ICCVW) pp. 4214–4221 (2021). https://doi.org/10.1109/ICCVW54120.2021.00470

3. Dang, C., Aghagolzadeh, M., Radha, H.: Image super-resolution via local self-learning manifold approximation. IEEE Signal Process. Lett. **21**(10), 1245–1249 (2014). https://doi.org/10.1109/LSP.2014.2332118
4. Donoho, D.L., Grimes, C.: Hessian eigenmaps: locally linear embedding techniques for high-dimensional data. Proc. Natl. Acad. Sci. **100**(10), 5591–5596 (2003)
5. Gui, J., Sun, Z., Jia, W., Hu, R., Lei, Y., Ji, S.: Discriminant sparse neighborhood preserving embedding for face recognition. Pattern Recogn. **45**(8), 2884–2893 (2012)
6. He, X., Cai, D., Yan, S., Zhang, H.J.: Neighborhood preserving embedding. In: Tenth IEEE International Conference on Computer Vision (ICCV 2005) Volume 1, vol. 2, pp. 1208–1213. IEEE (2005)
7. He, X., Niyogi, P.: Locality preserving projections. In: Advances in Neural Information Processing Systems 16 (2003)
8. Kokiopoulou, E., Saad, Y.: Orthogonal neighborhood preserving projections. In: Fifth IEEE International Conference on Data Mining (ICDM 2005), pp. 8–pp. IEEE (2005)
9. Kruskal, J.B., Wish, M.: Multidimensional scaling, vol. 11. Sage (1978)
10. Li, Y., Wang, C., Zhao, J.: Locally linear embedded sparse coding for spectral reconstruction from rgb images. IEEE Signal Process. Lett. **25**(3), 363–367 (2018). https://doi.org/10.1109/LSP.2017.2776167
11. Lin, T., Zha, H.: Riemannian manifold learning. IEEE Trans. Pattern Anal. Mach. Intell. **30**(5), 796–809 (2008)
12. McInnes, L., Healy, J., Melville, J.: Umap: uniform manifold approximation and projection for dimension reduction. arXiv preprint arXiv:1802.03426 (2018)
13. Narayan, A., Berger, B., Cho, H.: Assessing single-cell transcriptomic variability through density-preserving data visualization. Nat. Biotechnol. (2021). https://doi.org/10.1038/s41587-020-00801-7
14. Nie, X., Liu, J., Sun, J., Liu, W.: Robust video hashing based on double-layer embedding. IEEE Signal Process. Lett. **18**(5), 307–310 (2011). https://doi.org/10.1109/LSP.2011.2126020
15. Pang, Y., Zhang, L., Liu, Z., Yu, N., Li, H.: Neighborhood preserving projections (NPP): a novel linear dimension reduction method. In: Huang, D.-S., Zhang, X.-P., Huang, G.-B. (eds.) ICIC 2005. LNCS, vol. 3644, pp. 117–125. Springer, Heidelberg (2005). https://doi.org/10.1007/11538059_13
16. Qiao, L., Chen, S., Tan, X.: Sparsity preserving projections with applications to face recognition. Pattern Recogn. **43**(1), 331–341 (2010)
17. Roweis, S., Saul, L., Hinton, G.E.: Global coordination of local linear models. In: Advances in Neural Information Processing Systems 14 (2001)
18. Roweis, S.T., Saul, L.K.: Nonlinear dimensionality reduction by locally linear embedding. Science **290**(5500), 2323–2326 (2000)
19. Saul, L.K., Roweis, S.T.: An introduction to locally linear embedding. unpublished. http://www.cs.toronto.edu/~roweis/lle/publications.html (2000)
20. Tenenbaum, J.B., Silva, V.d., Langford, J.C.: A global geometric framework for nonlinear dimensionality reduction. Science **290**(5500), 2319–2323 (2000)
21. Van Der Maaten, L., Postma, E., Van den Herik, J., et al.: Dimensionality reduction: a comparative. J. Mach. Learn. Res. **10**(66–71), 13 (2009)
22. Weinberger, K.Q., Saul, L.K.: An introduction to nonlinear dimensionality reduction by maximum variance unfolding. In: AAAI, vol. 6, pp. 1683–1686 (2006)
23. Xiang, S., Nie, F., Pan, C., Zhang, C.: Regression reformulations of lle and ltsa with locally linear transformation. IEEE Trans. Syst. Man Cybern. Part B (Cybernetics) **41**(5), 1250–1262 (2011)

24. Zhang, Z., Wang, J.: Mlle: Modified locally linear embedding using multiple weights. In: Advances in Neural Information Processing Systems 19 (2006)
25. Zhang, Z., Zha, H.: Nonlinear dimension reduction via local tangent space alignment. In: Liu, J., Cheung, Y., Yin, H. (eds.) IDEAL 2003. LNCS, vol. 2690, pp. 477–481. Springer, Heidelberg (2003). https://doi.org/10.1007/978-3-540-45080-1_66

Inter-class Sparsity Based Non-negative Transition Sub-space Learning

Miaojun Li, Shuping Zhao, Jigang Wu(✉), and Siyuan Ma

School of Computer Science and Technology, Guangdong University of Technology, Guangzhou 510006, Guangdong, China
asjgwucn@outlook.com

Abstract. Least squares regression has shown promising performance in the supervised classification. However, conventional least squares regression commonly faces two limitations that severely restrict their effectiveness. Firstly, the strict zero-one label matrix utilized in least squares regression provides limited freedom for classification. Secondly, the modeling process does not fully consider the correlations among samples from the same class. To address the above issues, this paper proposes the inter-class sparsity-based non-negative transition sub-space learning (ICSN-TSL) method. Our approach exploits a transition sub-space to bridge the raw image space and the label space. By learning two distinct transformation matrices, we obtain a low-dimensional representation of the data while ensuring model flexibility. Additionally, an inter-class sparsity term is introduced to learn a more discriminative projection matrix. Experimental results on image databases demonstrate the superiority of ICSN-TSL over existing methods in terms of recognition rate. The proposed ICSN-TSL achieves a recognition rate of up to 98% in normal cases. Notably, it also achieves a classification accuracy of over 87% even on artificially corrupted images.

Keywords: Least squares regression · inter-class sparsity · transition sub-space learning

1 Introduction

Least squares regression (LSR) is a widely used approach in statistics for data analysis. It aims to find a projection matrix by minimizing the sum of squared errors [11]. Several variations of LSR-based models have been proposed, such as LASSO regression [12], Ridge regression [7], elastic-net regression [17], and partial LSR [14]. These methods flexibly introduce the regularization terms in

This work was supported in part by the Natural Science Foundation of China under Grant No. 62106052 and Grant No. 62072118, in part by the Guangdong Basic and Applied Basic Research Foundation under Grant No. 2021B1515120010, in part by the Huangpu International Sci&Tech Cooperation Fundation of Guangzhou, China, under Grant No.2021GH12.

Q. Liu et al. (Eds.): PRCV 2023, LNCS 14427, pp. 250–262, 2024.
https://doi.org/10.1007/978-981-99-8435-0_20

the LSR framework. But the classification performance of the LSR framework still suffers from some limitations.

In the classification task, maximizing the distance between different classes is essential. When the label matrix is strictly binary, the regression target for each class is constrained to be a fixed value, typically zero or one. Consequently, the Euclidean distance between any two samples from different classes becomes invariant, as the regression targets are held constant. To address this limitation, researchers have proposed techniques to relax the strict zero-one label matrix and learn a more discriminative regression target. The most representative method is discriminative LSR (DLSR) [15]. DLSR introduces ε-draggings to drag the regression targets of different classes in opposite directions. However, this technique may lead to an increase in the regression distances between samples from the same class. Fisher DLSR (FDLSR) [2] is a technique that combines the Fisher criterion and ε-draggings to simultaneously achieve small intra-class distances and large inter-class distances. FDLSR aims to improve the discriminative capability of the learned projection matrix. Discriminant and sparsity based least squares regression with l_1 regularization (DS_LSR) [19] introduces an orthogonal relaxed term to relax the label matrix. This approach incorporates sparsity-based regularization with l_1-norm to promote a more sparse and discriminative regression target. Inter-class sparsity based discriminative least square regression (ICS_DLSR) [13] relaxes the strict regression target using a sparsity error term with $l_{2,1}$-norm. However, one drawback of this method is that the learned projection matrix is sensitive to noise, which can significantly impact the classification performance when dealing with "unclear" data samples. To address this issue, low-rank inter-class sparsity based semi-flexible target least squares regression (LIS_StLSR) [18] introduces a semi-flexible regression matrix and a low-rank inter-class sparsity constraint. This approach enhances the model's robustness to noise, making it more reliable in challenging scenarios. By incorporating different regularization terms and relaxation strategies, these methods aim to enhance the discriminative capability and robustness of the learned projection matrix.

In the above methods, there is a common limitation of using a single transformation matrix, which can restrict the margin of the learned model. To address this limitation, double relaxed regression (DRR) [6] proposes the idea of using more transformation matrices to provide more freedom. DRR uses two transformation matrices and introduces the class compactness graph to solve the overfitting problem. Sparse non-negative transition subspace learning (SN-TSL) [1] extends the idea by incorporating a transition sub-space in the learning process. However, SN-TSL with l_1-norm sparsity constraint may not effectively discover the intrinsic group structure of data, and it is time-consuming.

In this study, we propose a novel approach called inter-class sparsity based non-negative transition sub-space learning (ICSN-TSL). The flow chart of the proposed model is shown in Fig. 1. The learning process of ICSN-TSL consists of two steps. In the first step, the original samples are transformed into a transition sub-space using the projection matrix W. This step aims to capture the essential information and underlying structure of the data. In the second step, the transition sub-space is further projected into a binary label space using the projection matrix Q. This transformation facilitates the classification process by mapping the samples into distinct classes. The proposed ICSN-TSL model offers several contributions.

- By incorporating the inter-class sparsity term, our method promotes the learning of a more discriminative projection matrix. This enables better separation and classification of different classes.
- The introduction of a non-negative transition sub-space helps bridge the gap between the original sample space and the label space. It facilitates the extraction of meaningful features and improves the interpretability of the learned projection matrix.

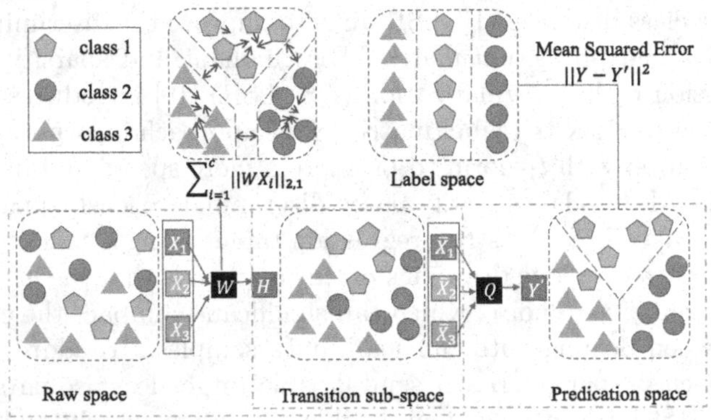

Fig. 1. The flow chart of the proposed method is illustrated. X_i is the raw data of the ith class samples, H represents the projection matrix of the transition sub-space, c is the number of class.

The rest of the paper is organized as follows. Section 2 presents some notations and related works. Section 3 describes the proposed method and its optimal solution, and analyzes the algorithm. Then, we perform some experiments in Sect. 4 to demonstrate the effectiveness of the proposed method. The conclusions are given in Sect. 5.

2 Related Work

2.1 Notations

In this subsection, to ensure the completeness and readability of this work, we first introduce the notations used throughout this paper. The matrix is denoted using capital letters, e.g., X, and the element in the i-th row and j-th column of the matrix X is denoted as $X_{i,j}$. The vectors are denoted using lowercase letters, e.g., x. For a matrix X, its l_F-norm and $l_{2,1}$-norm are calculated as

$$\|X\|_F^2 = \sum_{i=1}^{d}\sum_{j=1}^{n} X_{i,j}^2 = Tr(X^T X) \tag{1}$$

$$\|X\|_{2,1} = \sum_{i=1}^{d} \|X_{i.}\|_2 = \sum_{i=1}^{d}\sqrt{\sum_{j=1}^{n} X_{i,j}^2} \tag{2}$$

where X^T denotes the transposed matrix of X. $Tr(X)$ is the trace operator.

2.2 StLSR

Given a training set $X = [x_1, x_2, ..., x_n] \in R^{d\times n}$ where contains n training samples with d dimensions, and a label matrix $Y = [y_1, y_2, ..., y_n] \in R^{c\times n}$ whose each element $y_i = [0, 0, ..., 0, 1, 0, ..., 0]$ (only the index of the correct class is 1) is a zero-one vector to indicate the label of sample x_i from c classes. We first give the formula of the standard LSR model:

$$\min_{P} \|Y - PX\|_F^2 + \lambda \|P\|_F^2 \tag{3}$$

where $P \in R^{c\times d}$ is the transformation matrix, λ is a positive regularization parameter. The problem has a closed solution as $P = YX^T(XX^T + \lambda I)^{-1}$.

2.3 ICS_DLSR

The label matrix used by the standard LSR is not suitable for classification. ICS_DLSR uses label relaxation terms with $l_{2,1}$-norm to fit the transformed data into a rigid binary label matrix. Moreover, ICS_DLSR introduces inter class sparsity constraints to utilize the correlation between samples within the same class. This constraint aims to induce a common sparse structure among samples of the same class after the transformation, leading to a reduction in intra-class boundaries and an expansion of inter-class boundaries. The model of ICS_DLSR is formulated as follows:

$$\min_{P,E} \frac{1}{2}\|Y + E - PX\|_F^2 + \frac{\lambda_1}{2}\|P\|_F^2 + \lambda_2 \|E\|_{2,1} + \lambda_3 \sum_{i=1}^{c} \|PX_i\|_{2,1} \tag{4}$$

where E is a sparsity error term, in order to relax the zero-one label matrix. $\sum_{i=1}^{c} \|PX_i\|_{2,1}$ is the inter-class sparsity constraint item.

2.4 SN-TSL

In contrast to the methods DLSR and ICS_DLSR of relaxing the binary label matrix, SN-TSL utilizes a sparse non-negative transition sub-space, instead of the ε-draggings technique and inter-class sparsity terms. The objective function of SN-TSL is formulated as follows:

$$\min_{W,Q,H} \frac{1}{2}\|WX - H\|_F^2 + \alpha \|H\|_1 + \frac{\beta}{2}\|QH - Y\|_F^2 + \frac{\lambda}{2}\left\{\|W\|_F^2 + \|Q\|_F^2\right\} \quad s.t.\ H \geq 0 \tag{5}$$

where $W \in R^{r\times d}$ represents the projection matrix from the raw space to transition sub-space, $Q \in R^{c\times r}$ represents the projection matrix from the transition sub-space to predication space, $H \in R^{r\times n}$ is the transition sub-space, r is its dimensionality and SN-TSL peaks when r is approximately equal to c or $2c$. Here, we set r to $2c$ for all datasets.

3 The Proposed Method

3.1 Problem Formulation and Learning Model

Based on Eq. (3), we introduce a transition subspace in the transformation process to find two projection matrices W and Q. The original samples are first transformed into a transition sub-space, and then mapped to the binary label space. Our objective is to optimize the transition matrix by minimizing the classification error. Based on the previous discussion, this problem can be formulated as follows:

$$\min_{W,Q,H} \frac{1}{2}\|QH - Y\|_F^2 + \frac{\beta}{2}\|WX - H\|_F^2 + \frac{\lambda_1}{2}\left\{\|W\|_F^2 + \|Q\|_F^2\right\} \tag{6}$$

We add sparsity and non-negative constraints to the transition matrix H to ensure that the final regression objective is both sparse and non-negative. To avoid the l_1-norm disadvantage, we utilize the $l_{2,1}$-norm instead. The $l_{2,1}$-norm is effective in reducing the negative impact of noise and outliers while promoting sparsity. Thus, we propose to impose the following constraint on the objective function:

$$\begin{aligned} n \min_{W,Q,H} & \frac{1}{2}\|QH - Y\|_F^2 + \frac{\beta}{2}\|WX - H\|_F^2 + \frac{\lambda_1}{2}\left\{\|W\|_F^2 + \|Q\|_F^2\right\} \\ & + \alpha\|H\|_{2,1} \quad s.t.\ H \geq 0 \end{aligned} \tag{7}$$

Classification aims to assign each record to its corresponding class, so we want the distance between samples from different classes to be large enough during training, and samples between the same class to be more compact. To effectively capture the relationships between samples, we introduce an inter-class sparsity constraint. Therefore, the objective function of our method can be defined as follows:

$$\begin{aligned} \min_{W,Q,H} & \frac{1}{2}\|QH - Y\|_F^2 + \frac{\beta}{2}\|WX - H\|_F^2 + \frac{\lambda_1}{2}\left\{\|W\|_F^2 + \|Q\|_F^2\right\} \\ & + \alpha\|H\|_{2,1} + \lambda_2\textstyle\sum_{i=1}^{c}\|WX_i\|_{2,1} \quad s.t.\ H \geq 0 \end{aligned} \tag{8}$$

where λ_1, λ_2, α, and β are positive regularization parameters, the $l_{2,1}$-norm based $\|H\|_{2,1}$ and $H \geq 0$ ensures the learnt H is sparse and non-negative in rows, which is robust to noise in the data.

3.2 Solution to ICSN-TSL

The objection function in (8) cannot be directly optimized, we use the alternating direction method of multipliers (ADMM) to optimize of the learning model [2, 15]. The variables D and F are first introduced to make the objective function (8). It can be separated and optimized as follows:

$$\begin{aligned}\min_{W,Q,H,D,F} \frac{1}{2}\|QH-Y\|_F^2 + \frac{\beta}{2}\|WX-H\|_F^2 + \frac{\lambda_1}{2}\left\{\|W\|_F^2 + \|Q\|_F^2\right\} \\ + \alpha\|F\|_{2,1} + \lambda_2\textstyle\sum_{i=1}^{c}\|D_i\|_{2,1}\ s.t.\ D = WX, F = H, H \geq 0\end{aligned} \tag{9}$$

Then convert (9) into the following augmented Lagrange function:

$$\begin{aligned}L(W,Q,H,D,F) = \min_{W,Q,H,D,F} & \frac{1}{2}\|QH-Y\|_F^2 + \frac{\beta}{2}\|WX-H\|_F^2 + \alpha\|F\|_{2,1} \\ & + \frac{\lambda_1}{2}\left\{\|W\|_F^2 + \|Q\|_F^2\right\} + \lambda_2\textstyle\sum_{i=1}^{c}\|D_i\|_{2,1} \\ & + \frac{\mu}{2}\left\{\left\|D - WX + \frac{C_1}{\mu}\right\|_F^2 + \left\|F - H + \frac{C_2}{\mu}\right\|_F^2\right\}\end{aligned} \tag{10}$$

where C_1 and C_2 denote the Lagrange multiplier and $\mu > 0$ is a penalty parameter. We will obtain the solutions for all variables by solving the problem (10) alternately. The details are as follows:

$$L(W) = \min_W \frac{\beta}{2}\|WX-H\|_F^2 + \frac{\lambda_1}{2}\|W\|_F^2 + \frac{\mu}{2}\left\|D - WX + \frac{C_1}{\mu}\right\|_F^2 \tag{11}$$

$$L(Q) = \min_Q \frac{1}{2}\|QH-Y\|_F^2 + \frac{\lambda_1}{2}\|Q\|_F^2 \tag{12}$$

$$L(D) = \min_D \lambda_2\sum_{i=1}^{c}\|D_i\|_{2,1} + \frac{\mu}{2}\left\|D - WX + \frac{C_1}{\mu}\right\|_F^2 \tag{13}$$

$$\begin{aligned}L(H) = \min_H \frac{1}{2}\|QH-Y\|_F^2 + \frac{\beta}{2}\|WX-H\|_F^2 + \frac{\mu}{2}\left\|F - H + \frac{C_2}{\mu}\right\|_F^2 \\ s.t. H \geq 0\end{aligned} \tag{14}$$

$$L(F) = \min_F \alpha\|F\|_{2,1} + \frac{\mu}{2}\left\|F - H + \frac{C_2}{\mu}\right\|_F^2 \tag{15}$$

By setting the derivation of (11) with respect to W, we can achieve $W = (\beta H + \mu D + C_1)X^T((\beta+\mu)XX^T + \lambda_1 I_d)^{-1}$. The optimal Q can be obtained by setting the derivation of (12) to zero, where Q can be calculated $Q = YH^T(HH^T + \lambda_1 I_r)^{-1}$. By setting the derivation of (14) to zero and considering the non-negativity of H, H can be obtained as $H = \max(\hat{H}, 0)$, where $\hat{H} = \left[(\mu+\beta)I_r + Q^TQ\right]^{-1}(\beta WX + Q^TY + \mu F + C_2)$. To update D, the $l_{2,1}$-norm is utilized for (13) and D can be obtained as

$$[D_i]_{j,:} = \begin{cases} \frac{\left\|[T_i]_{j,:}\right\|_2 - \lambda_2\mu}{\left\|[T_i]_{j,:}\right\|_2}[T_i]_{j,:}, & if \quad \left\|[T_i]_{j,:}\right\|_2 > \lambda_2\mu \\ 0, & otherwise \end{cases}$$

where $T = WX - \frac{C_1}{\mu}$, and D_i and T_i are i-th subset of D and T corresponding to samples of the i-th class, respectively.

The solution of F in (15) can be solved as

$$F_{j,:} = \begin{cases} \frac{\|U_{j,:}\|_2 - \alpha\mu}{\|U_{j,:}\|}U_{j,:}, & if \quad \|U_{j,:}\|_2 > \alpha\mu \\ 0, & otherwise \end{cases}$$

where $U = H - \frac{C_2}{\mu}$, and $F_{j,:}$ and $U_{j,:}$ are the j-th row vector of F and U, respectively.

3.3 Classification

The detailed procedures of the proposed method are outlined in Algorithm 1, after which the two learned projection matrices W and Q are used to obtain a regression vector QWy of a new test sample y, which is then classified using the nearest-neighbor (NN) classifier.

3.4 Computational Time Complexity

In the ICSN-TSL learning, the variables W, Q, H, D, and F are updated alternatively. The most expensive computation is the matrix inverse, and other computational costs are negligible. We can obtain that the computational cost of updating the variable W, Q, and H are bounded by $O(d^3)$, $O(r^3)$, and $O(r^3)$ respectively. Therefore, the main computational complexity of Algorithm 1 is $O(\tau(d^3 + n^3 + 2r^3))$, where τ is the number of iterations.

Algorithm 1. Solving ICSN-TSL by ADMM

Input: Training samples matrix $X \in R^{d \times n}$, label matrix $Y \in R^{c \times n}$, parameters $\lambda_1, \lambda_2, \alpha, \beta$.

Output: W and Q.

Initialization: Random and normalized W and Q; $H = 0$; $D = WX$; $F = H$; $C_1 =$; $C_2 = 0$; $\mu = 10^{-5}$; $\mu_{\max} = 10^8$; $\rho = 1.1$.

while not converged **do**

1. Update the variables of W, Q, H, D and F as (11)-(15), respectively.
2. Update C_1, C_2, and μ by using

$$\begin{cases} C_1 &= C_1 + \mu(D - WX) \\ C_2 &= C_2 + \mu(F - H) \\ \mu &= \min(\rho\mu, \mu_{\max}) \end{cases}$$

where ρ and $\mu_{\max}$ are positive constants.

end while

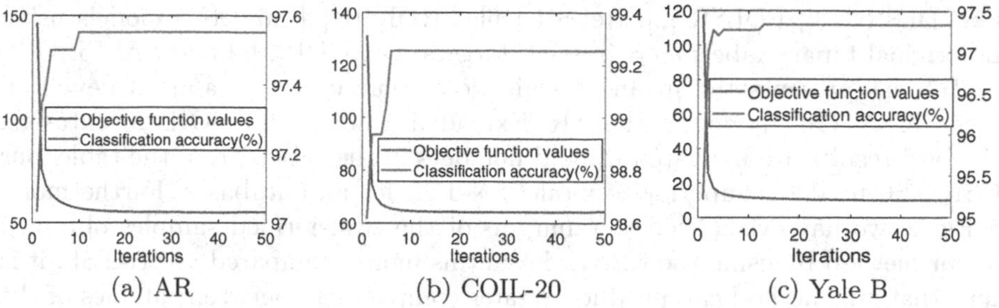

(a) AR (b) COIL-20 (c) Yale B

Fig. 2. The convergence property of the proposed method on (a) AR, (b) COIL-20,and (c) Yale B.

3.5 Convergence Analysis

The convergence of the proposed algorithm is demonstrated through experiments conducted on datasets such as AR, COIL-20, and Yale B. In Fig. 2, we observe that the value of the objective function decreases consistently with an increasing number of iterations. Eventually, the algorithm converges to a smaller value after a finite number of iterations. This convergence behavior indicates that the algorithm is effectively optimizing the objective function and improving the performance of the model.

4 Experiments and Analysis

In this section, we evaluate the performance of the proposed method on public databases.

4.1 Data Sets

We will validate the classification performance of the ICSN-TSL model on several publicly available datasets, such as AR [8], Extended Yale B [5], and COIL-20 [10]. The detailed information of the datasets are introduced as follows: 1) The AR face dataset has over 4,000 color face images, we select a subset of 3,120 images containing 120 people and cropped each image to a size of 50×40. 2) The Extended Yale B dataset contains 2,414 images of 38 humans under different lighting conditions, we adjust the images in advance to 32×32 pixels. 3) The COIL-20 object has total of 1,440 grey-scale images corresponding to 20 classes. In the experiments, each image is resized to 32×32.

4.2 Experimental Results and Analysis

In this sub-section, we compare the proposed ICSN-TSL model with six classification methods based on the LSR model, including SN-TSL [1], DLSR [15],

ICS_DLSR [13], FDLSR [2], ReLSR [16], LRDLSR [3], and two models using the original binary labels as regression targets, i.e., CDPL [9] and SALPL [4].

Tables 1 presents the mean classification accuracy and standard deviation of the ICSN-TSL model on the AR, Extended Yale B, and COIL-20 datasets. The best results are indicated in bold numbers in the table. From the table, our ICSN-TSL model is superior to model SN-TSL for all databases. Furthermore, in Fig. 3, we have visualized the margins of the transformed samples obtained by our method by using the t-SNE. From this figure, compared to SN-TSL, it is clear that our method can produce greater compactness between samples of the same class, and it can also produce greater distances between different classes.

Table 1. Classification accuracies (mean ± std %) of different methods on these three databases.

Database	No.	LRDLSR	FDLSR	SALPL	CPDL	DLSR	ReLSR	ICS_DLSR	SN-TSL	ICSN-TSL
AR	4	91.58 ± 1.05	78.94 ± 1.51	73.24 ± 1.61	71.14 ± 2.75	89.13 ± 0.56	84.77 ± 1.25	91.64 ± 0.70	79.79 ± 0.93	**91.90 ± 0.56**
	6	94.89 ± 0.69	88.18 ± 0.95	84.89 ± 1.26	80.14 ± 3.28	91.06 ± 0.97	92.25 ± 1.26	95.36 ± 0.52	90.15 ± 0.78	**95.90 ± 0.43**
	8	96.32 ± 0.35	93.35 ± 0.64	91.49 ± 1.02	87.81 ± 2.19	93.95 ± 0.59	95.05 ± 0.99	97.08 ± 0.69	94.50 ± 0.71	**97.30 ± 0.59**
	10	97.75 ± 0.35	95.12 ± 0.73	94.98 ± 0.84	91.60 ± 1.13	94.48 ± 0.76	96.34 ± 0.56	98.07 ± 0.46	96.74 ± 0.41	**98.18 ± 0.43**
Extended Yale B	10	83.30 ± 1.45	89.78 ± 0.98	73.85 ± 1.50	74.65 ± 4.11	85.99 ± 1.66	83.77 ± 2.34	87.75 ± 1.12	81.19 ± 1.72	**89.43 ± 1.03**
	15	88.34 ± 1.09	92.76 ± 0.97	84.86 ± 1.34	82.10 ± 3.60	89.77 ± 0.65	88.74 ± 1.07	91.98 ± 1.06	88.97 ± 1.00	**93.31 ± 0.94**
	20	91.77 ± 1.05	94.38 ± 0.59	89.70 ± 0.95	85.78 ± 3.87	92.22 ± 0.45	91.01 ± 1.21	94.82 ± 0.94	93.04 ± 0.82	**95.88 ± 0.94**
	25	93.52 ± 0.70	95.74 ± 0.65	91.72 ± 1.06	87.58 ± 2.86	94.58 ± 0.74	93.32 ± 1.06	96.39 ± 0.48	95.12 ± 0.69	**97.20 ± 0.71**
COIL-20	10	91.50 ± 1.11	92.53 ± 1.43	91.34 ± 1.20	90.42 ± 1.23	92.95 ± 0.96	89.12 ± 1.42	93.30 ± 1.13	90.34 ± 1.60	**93.86 ± 2.06**
	15	93.92 ± 1.17	95.67 ± 0.73	95.82 ± 0.97	94.63 ± 0.88	94.77 ± 0.85	93.25 ± 1.58	96.46 ± 0.95	92.59 ± 1.64	**96.74 ± 0.93**
	20	96.06 ± 0.85	97.14 ± 0.73	97.32 ± 1.23	95.90 ± 0.94	96.53 ± 1.17	95.59 ± 1.16	97.83 ± 0.72	94.51 ± 1.36	**98.22 ± 0.42**
	25	97.69 ± 0.69	97.84 ± 0.84	98.07 ± 0.76	97.06 ± 0.88	97.33 ± 0.61	96.63 ± 0.64	98.48 ± 0.72	96.41 ± 0.69	**99.00 ± 0.50**

In addition, to evaluate the robustness of each method, we conducted experiments on the AR face and COIL-20 object datasets with mixed artificial noise. In the dataset AR, 10 images from each class were randomly selected as training samples, while the remaining images were used as the test set. For the COIL-20 dataset, 30 samples from each class were randomly chosen for training, and the remaining images were used for testing. A mixture of artificial noise, i.e., "salt & pepper", and randomly generated black squares are added to the two test sets (see Fig. 4), with the degree of mixed noise expressed as $\{0.1, 0.2, 0.3, 0.4, 0.5\}$, where the values indicate the proportion of noise and occlusion in the image. It can be seen that the performance of the proposed method is significantly better than that of SN-TSL and other methods.

4.3 Parameter Sensitivity Analysis

The proposed ICSN-TSL has four parameters to choose, i.e., α, β, λ_1, and λ_2. We define a candidate set $\{10^{-5}, 10^{-4}, 10^{-3}, 10^{-2}, 10^{-1}, 10^{0}, 10^{1}, 10^{2}\}$. Generally, the larger the parameter value, the greater the importance or influence of the corresponding item. The ICSN-TSL recognition rate varies with the model parameters on the dataset AR in Fig. 6 and Fig. 5 show their relationship. It can be observed that the model is very robust to the values of λ_2 and α over the range of candidate sets. Furthermore, the best performance of ICSN-TSL is obtained when $\beta \in [1, 10^2]$ and $\lambda_1 \in [10^{-5}, 10^{-1}]$.

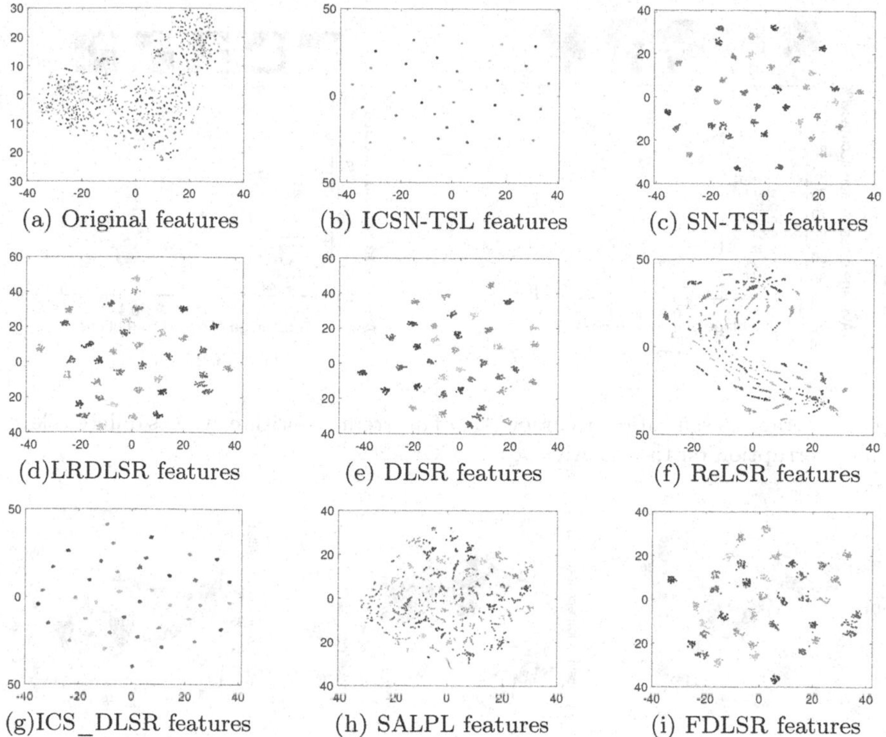

Fig. 3. The t-SNE visualization results of the features extracted by different algorithms on the Extended Yale B.

4.4 Ablation Study

To assess the impact of some constraint terms of the proposed method, an ablation study of the model was conducted. Experimental comparisons are conducted for the following four cases.

(1) ICSN-TSL(λ_1): the case of that $\lambda_1 = 0$ in the ICSN-TSL model, i.e., the model is not prevented from over-fitting.
(2) ICSN-TSL(λ_2): the case of that $\lambda_2 = 0$ in the ICSN-TSL model, i.e., the model is without inter-class sparsity term.
(3) ICSN-TSL(α): the case of that $\alpha = 0$ in the ICSN-TSL model, i.e., the absence of sparsity constraints on the transition sub-space.
(4) ICSN-TSL(β): the case of that $\beta = 0$ in the ICSN-TSL model, i.e., the model is without the regression term.

The experimental results on three datasets are showed in Fig. 7. It can be seen that the performance of ICSN-TSL(β) is extremely poor. This indicates that the performance of the model benefits greatly from the regression term. Therefore, the regression term can explicitly guide the feature learning process, and it is an

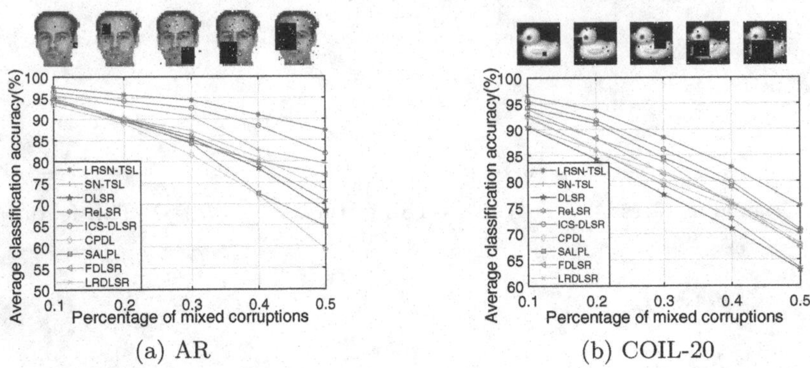

(a) AR (b) COIL-20

Fig. 4. Average classification accuracy (%) of different algorithms versus different levels of data corruption on the (1) AR and (2) COIL-20.

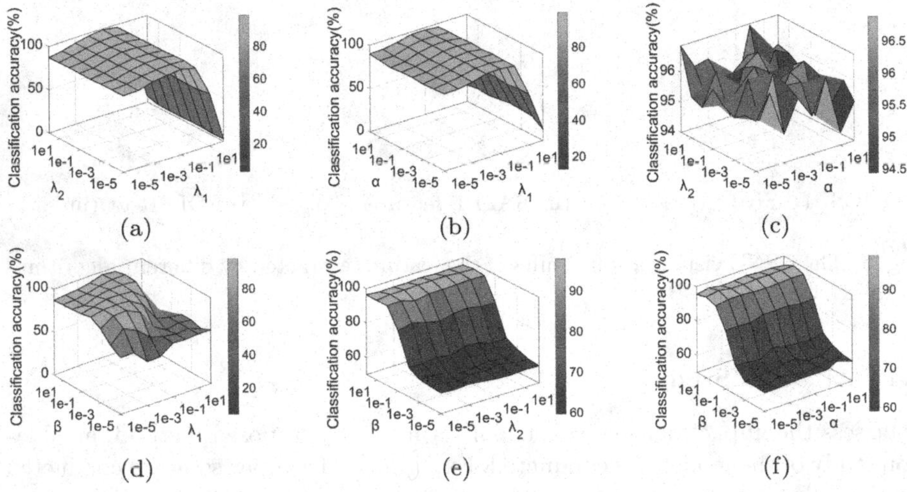

(a) (b) (c)

(d) (e) (f)

Fig. 5. The classification accuracy (%) of ICSN-TSL versus the variations of the parameters (a) λ_1 and λ_2, (b) α and λ_1, (c) α and λ_2, (d) β and λ_1, (e) β and λ_2, (f) β and α on the AR.

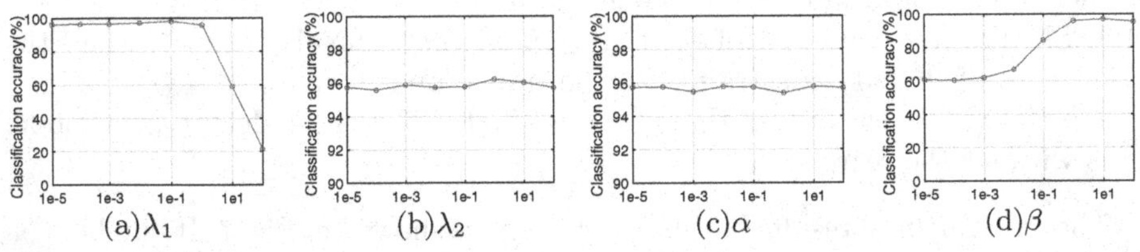

(a)λ_1 (b)λ_2 (c)α (d)β

Fig. 6. The classification accuracy (%) of ICSN-TSL versus the variations of the parameter on the AR.

important role in the final classification. Furthermore, the performance of the model decreases significantly when $\lambda_1 = 0$, i.e., when the over-fitting constraint is missing on the projection matrix. In contrast, the added sparsity constraints and inter-class sparsity can improve the accuracy of the model.

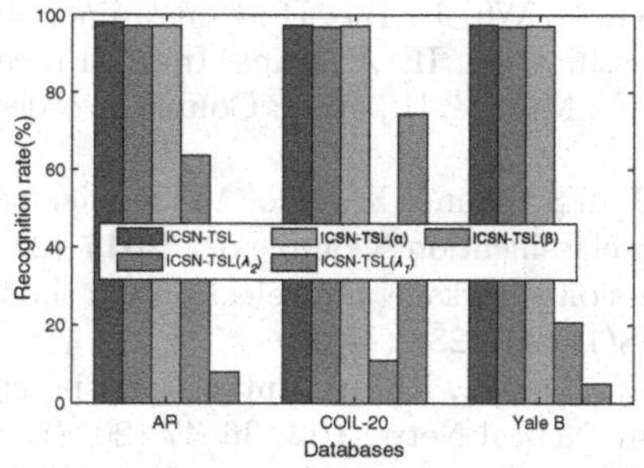

Fig. 7. Ablation study of the proposed method on four datasets.

5 Conclusion

This paper proposes an efficient learning method to overcome the limitations of strict label matrices in classification tasks. The approach introduces a transition sub-space to relax the binary matrix constraint and alleviate the restrictions imposed by a single transformation matrix. The proposed model not only enables the learning of more flexible projection and regression targets but also exhibits robustness to outliers and noise commonly encountered in real-world data. Experimental results demonstrate the effectiveness of the proposed inter-class sparsity-based non-negative transition sub-space learning (ICSN-TSL) model. Especially, ICSN-TSL achieves high classification accuracy, even on artificially corrupted images. For future work, the combination of low-rank constraints and sparse constraints is considered to enhance feature selection and extraction.

References

1. Chen, Z., Wu, X.J., Cai, Y.H., Kittler, J.: Sparse non-negative transition subspace learning for image classification. Signal Process. **183**, 107988 (2021)
2. Chen, Z., Wu, X.J., Kittler, J.: Fisher discriminative least squares regression for image classification. arXiv preprint arXiv:1903.07833 (2019)
3. Chen, Z., Wu, X.J., Kittler, J.: Low-rank discriminative least squares regression for image classification. Signal Process. **173**, 107485 (2020)
4. Fang, X., et al.: Approximate low-rank projection learning for feature extraction. IEEE Trans. Neural Netw. Learn. Syst. **29**(11), 5228–5241 (2018)
5. Georghiades, A.S., Belhumeur, P.N.: Illumination cone models for faces recognition under variable lighting. In: Proceedings of CVPR 1998 (1998)

6. Han, N., et al.: Double relaxed regression for image classification. IEEE Trans. Circuits Syst. Video Technol. **30**(2), 307–319 (2019)
7. Hoerl, A.E., Kennard, R.W.: Ridge regression: biased estimation for nonorthogonal problems. Technometrics **12**(1), 55–67 (1970)
8. Martinez, A., Benavente, R.: The AR face database: CVC technical report, 24 (1998)
9. Meng, M., Lan, M., Yu, J., Wu, J., Tao, D.: Constrained discriminative projection learning for image classification. IEEE Trans. Image Process. **29**, 186–198 (2019)
10. Nene, S.A., Nayar, S.K., Murase, H., et al.: Columbia object image library (COIL-20) (1996)
11. Peng, Y., Zhang, L., Liu, S., Wang, X., Guo, M.: Kernel negative ε dragging linear regression for pattern classification. Complexity **2017** (2017)
12. Tibshirani, R.: Regression shrinkage and selection via the lasso. J. Roy. Stat. Soc.: Ser. B (Methodol.) **58**(1), 267–288 (1996)
13. Wen, J., Xu, Y., Li, Z., Ma, Z., Xu, Y.: Inter-class sparsity based discriminative least square regression. Neural Netw. **102**, 36–47 (2018)
14. Wold, S., Ruhe, A., Wold, H., Dunn, Iii, W.: The collinearity problem in linear regression. The partial least squares (PLS) approach to generalized inverses. SIAM J. Sci. Stat. Comput. **5**(3), 735–743 (1984)
15. Xiang, S., Nie, F., Meng, G., Pan, C., Zhang, C.: Discriminative least squares regression for multiclass classification and feature selection. IEEE Trans. Neural Netw. Learn. Syst. **23**(11), 1738–1754 (2012)
16. Zhang, X.Y., Wang, L., Xiang, S., Liu, C.L.: Retargeted least squares regression algorithm. IEEE Trans Neural Netw. Learn. Syst. **26**(9), 2206–2213 (2014)
17. Zhang, Z., Lai, Z., Xu, Y., Shao, L., Wu, J., Xie, G.S.: Discriminative elastic-net regularized linear regression. IEEE Trans. Image Process. **26**(3), 1466–1481 (2017)
18. Zhao, S., Wu, J., Zhang, B., Fei, L.: Low-rank inter-class sparsity based semi-flexible target least squares regression for feature representation. Pattern Recogn. **123**, 108346 (2022)
19. Zhao, S., Zhang, B., Li, S.: Discriminant and sparsity based least squares regression with L1 regularization for feature representation. In: ICASSP 2020-2020 IEEE International Conference on Acoustics, Speech and Signal Processing (ICASSP), pp. 1504–1508. IEEE (2020)

Incremental Learning Based on Dual-Branch Network

Mingda Dong[1,2], Zhizhong Zhang[2], and Yuan Xie[2(✉)]

[1] Shanghai Institute of AI for Education, East China Normal University, Shanghai, China

[2] School of Computer Science and Technology, East China Normal University, Shanghai, China

{zzzhang,yxie}@cs.ecnu.edu.cn

Abstract. Incremental learning aims to overcome catastrophic forgetting. When the model learns multiple tasks sequentially, due to the imbalance of new and old classes numbers, the knowledge of old classes stored in the model is destroyed by large number of new classes. The existing single-backbone model is difficult to avoid catastrophic forgetting. In this paper, we proposes to use the dual-branch network model to learn new tasks to alleviate catastrophic forgetting. Different from previous dual-branch models that learn tasks in parallel, we propose to use dual-branch network to learn tasks serially. The model creates a new backbone for learning the remaining tasks, and freezes the previous backbone. In this way, the model can reduce damage to the previous backbone parameters used to learn old tasks. The model uses knowledge distillation to preserve the information of old tasks when the model learns new tasks. We also analyze different distillation methods for the dual-branch network model. In this paper we mainly focuses on the more challenging class incremental learning. We use common incremental learning setting on the ImageNet-100 dataset. The experimental results show that the accuracy can be improved by using the dual-branch network.

Keywords: incremental learning · knowledge distillation · catastrophic forgetting

1 Introduction

Deep learning has achieved remarkable success. For example, in image classification tasks, the classification accuracy of the model has exceeded the accuracy of humans. However, humans have the ability to accumulate knowledge when learning multiple tasks. In contrast, catastrophic forgetting occurs when models learn multiple tasks sequentially. For example, after a student learns to distinguish cats from dogs, and then learns to distinguish birds from airplanes, he will be able to distinguish four categories at the same time. After the model has learned these four categories in the same order, the model has poor performance on the first task. When learning the first task, the model parameters are adjusted

Q. Liu et al. (Eds.): PRCV 2023, LNCS 14427, pp. 263–272, 2024.
https://doi.org/10.1007/978-981-99-8435-0_21

to learn the first task. When learning the second task, the first task has less data compared to the second task, and training the model with the current data and memory will cause the model to be more inclined to learn the second task. So the performance on the first task dropped significantly. The problem of catastrophic forgetting has occurred. The research of incremental learning is mainly to solve the problem of catastrophic forgetting. When a new task comes, only a small portion of old tasks data is saved in memory. The model is expected to get high performance on both old and new tasks. This scenario is very practical, such as face recognition security system and recommendation system.

AANets [12] use two types of residual blocks, a stable block and a plastic block, the 2 block learn task in parallel. The model alleviates the forgetting of old tasks by slowly changing the stable block for old tasks and learning to mix with plastic block for new task. We propose a dual-branch model that learns tasks in a sequential manner. The model can better preserve old task information by freezing old branches. In this way, the modification of model parameters that are important to the old task can be less modified, so that the old branch can better preserve the information of the old task. Since only a small number of old task samples are saved, these small samples and a large number of current task samples are used to train the model, which leads to the bias problem of the classifier. [20] propose the the last fully connected layer is biased, they use a valid set to solve the problem. [25] proposes to solve the problem that the gap between the old tasks and new task classifiers' weight are too large. [12] builds a balanced dataset, each class in the balanced dataset has same sample numbers. They train the weight parameters during upper-level in the balanced dataset. [21,24] prose to train classifiers with temperature for softmax function. Following these methods, we propose the baseline method, in the single-branch model, we use softmax function with temperature when training the classifier in the incremental process. We improve the baseline method by changing the model to a two-branch model. Knowledge distillation [8] is often used in incremental learning. We also use knowledge distillation to preserve the knowledge of old tasks. We analyze how to perform distillation in the dual-branch model.

2 Related Work

Incremental learning methods can be divided into regularization methods, replay-based methods, architecture-based methods.

In the regularization-based method, EWC [10] adds constraints to parameters by calculating the importance of each parameter, thereby reducing the modification of important parameters for old tasks. SI [23] finds parameters whose updates contribute greatly to the reduction of the loss function of the previous tasks, and then adds constraints to reduce the modification of these parameters in the future tasks. Adam-NSCL [18] uses Null Space to solve the task incremental learning. For the gradient calculation process, each gradient update should be mapped to the null space of the features of each layer of the old tasks, so as to avoid the increase of loss on the old tasks. It has an elegant mathematical proof, but when actually finding the null space, the conditions are too strict. It is mainly used in task incremental learning rather than class incremental learning. [15] propose to mimicking the oracle, they first analyze oracle which training with all tasks' examples. They found the distribution of each sample by using the oracle is indeed relatively uniform. By mimicking the feature distribution of oracle after each incremental learning, the model will get higher accuracy.

LwF [11] performs distillation on the old and new models, LwF does not save the old samples, only uses the samples of the new task to train the model. LwF optimizes the models by minimizing the KL divergence of the probability distributions from the old and new models. This method is mainly used in the task incremental learning, since the old class samples are not used, the performance in class incremental learning tasks is poor. Many subsequent works are inspired by this work. iCaRl [14] can be seen as an improvement of LwF [11], unlike LwF, iCaRl preserves old samples to train the new model. LwF uses old and new class samples for distillation, and uses new and old classes to calculate the cross-entropy loss. For the problem of classifier deviation, iCaRl proposes to calculate the prototype of each class, and when the model tests, LwF select the class whose prototype has the nearest distance to the sample as the predicted classes. iCaRl also proposes to use herding strategy to keep the old samples. EEIL [3] uses a variety of data augmentation, which is used to increase the number of old samples, and then performing fine-tuning the model to improve performance. PODNet [4] uses the distillation on intermediate layers, instead of only used for output of classifier. SDC [22] finds the drift of old tasks' features when learning the new tasks, SDC calculates the feature drift of each old class, and proposes to compensate the old classes' drift without using the old class examples.

Since only a small number of samples of the old classes can be saved in memory, the number of new classes samples is larger than the old classes samples, which causes the classifier to be more biased towards the new class during the training process. UCIR [9] found that the weight of the new classes is larger than the old classes, and the model tends to predict the samples to new classes. UCIR proposes to use a cosine classifier to avoid the bias of the classifier. BiC [20] further analyzes the problem of classifier bias problem, BiC proposed to train the classifier by using a valid set to reduce the bias problem of the classifier. WA [25] also further studied the bias problem of the classifier. WA adjusts the weights of the old and new classes to alleviate the bias problem of the classifier, the scaled weights are more balanced and improve the accuracy.

In the architecture-based methods, DualNet [13] builds two branches to learn features in parallel, namely the fast learner and slow learner branch. The slow branch uses the self-supervised method to learn knowledge which is task-agnostic. The fast learner to learn knowledge for specific tasks, thereby reducing forgetting. DER [21] use dynamic network expansion to avoid catastrophic forgetting by using a new network for each task.

For samples that cannot be saved due to privacy and other issues, some works proposed to generate samples or features through GAN [6]. CLDGR [16] generates old samples by training GAN. [17] uses a large number of unlabeled data as negative samples to fully learn features, and uses GAN to generate features instead of samples. But these methods need additional training and storing GAN.

OCM [7] analyzes the incremental learning problem from the perspective of mutual information, OCM uses the loss by maximizing the mutual information between x and f(x), f(x) and y, which mainly used in online continual setting. XDER [1] analyzes the blind spot problem occurred in DER [2] and DER++ [2], XDER implant the part of the current logits value to corresponding logits of the memory. Dytox [5] proposes to use transformer architecture in incremental learning, they add and learn task tokens for each task. L2P [19] uses a pre-trained model with promts for incremental learning, the prompts are small learnable parameters which are maintained in pool. L2P learns to select prompts from pool. However L2P need to use a pre-trained model.

3 Method

3.1 Problem Description

Incremental learning can be divided into task incremental learning, domain incremental learning, and class incremental learning. Task and class incremental learning will learn new class in each task. The samples of the new and old tasks do not overlap. Task incremental learning knows the task id of each test sample, so as to select the classifier of the corresponding task for classification, which is easy to solve, thus having higher accuracy. But this setting is relatively rare in practical applications. The more common situation is that the number of the task cannot be obtained during the test. This is the more challenging class incremental problem. For domain incremental learning, the number of categories does not increase with the number of tasks. Instead the domain of the class in the new task is different from the old task. In this paper, we mainly focus on class incremental learning.

Firstly, we define the class incremental learning. The incremental learning has n learning tasks $\{Task_i\}_{i=1}^{n}$, each task $Task_i$ contains the corresponding data $D_i = \{(x_i, y_{x_i}\}_{i=1}^{n_c}$, the categories between different task do not overlap, $\emptyset = y_{x_i} \cap y_{x_j}$. When finish training the task, we select some samples used to train in future tasks. We use memory to store the old task samples, the size of memory is fixed, and the selected samples will be updated during the learning process. The data of the batch in each training process comes from D_i and memory. Memory M stores a subset of samples from previous tasks. During the training process, the feature extractor is recorded as f_θ , and the classifier is recorded as O. The corresponding logits value is recorded as $logits = O(f_\theta(x))$.

3.2 Baseline Method

We propose the baseline method first. When learning multiple tasks sequentially, memory can only keep a small number of old class samples. And samples of new tasks are larger, so we need to adjust the classifier. After each task finish training, model is frozen except for the classifier. Then we build a balanced train dataset which the number of samples of each class is same to train the classifier. We only train the classifier with the cross-entropy loss function, as shown in Eq. 1, t_1 is the temperature coefficient [24], the experiment shows that this method used to fine-tune the single classifier has higher accuracy than other methods that specifically deal with classifier bias problem such as the weight align method. As shown in Fig. 4, the baseline method has high accuracy than other WA. In this paper, we use the model obtained by this way as the baseline methods.

$$\mathcal{L}_{CE}(x_i, y_i) = CE(softmax(O(f_\theta(x)), \frac{y_i}{t_1})) \quad (1)$$

3.3 Model Extension

We analyze the baseline method proposed in 3.2. In the initial stage, the number of saved samples from old tasks in memory is relatively sufficient, which is helpful for the model to distinguish between old and new classes. We found that in the process of learning multiple tasks, the accuracy will decrease quickly. As shown in Fig. 4, accuracy drops significantly as new tasks are learned, the information of the previous task stored by the model f_θ is destroyed by the training of the new tasks. We found it is difficult to effectively save it only by methods such as distillation. DualNet [13] builds two branches to learn tasks. Unlike DualNet, which learns two branch in parallel. The two branch are used for fast learning and slow learning respectively. We think that as the task increases, it will change the two branch parameters, causing catastrophic forgetting. Here we use a two-branch network to learn tasks serially. The two branches are responsible to learn feature representation of certain tasks respectively. Therefore, we proposes to freeze the previous branch of the model as f_{θ_1}, and then create a new branch f_{θ_2} to learn the feature representation of the following tasks. The model concatenates two features obtained by the two branches as Eq. 2, and the concatenated features are sent to the classifier. The structure of the whole process is shown in Fig. 1.

$$feature = concatenate(f_{\theta_1}(x), f_{\theta_2}(x)) \quad (2)$$

3.4 Two Distillation

In the process of learning new tasks, in order to retain the knowledge of old tasks, knowledge distillation is used to retain the information of old classes. This process can be divided into 2 ways to distillation. The first method is to use only the part of the logits value of the task corresponding to the branch for distillation, and only use the corresponding feature to get the logits value, as shown in Fig. 2. The second method is to use the logits values obtained by the two branches to perform distillation, as shown in Fig. 3.

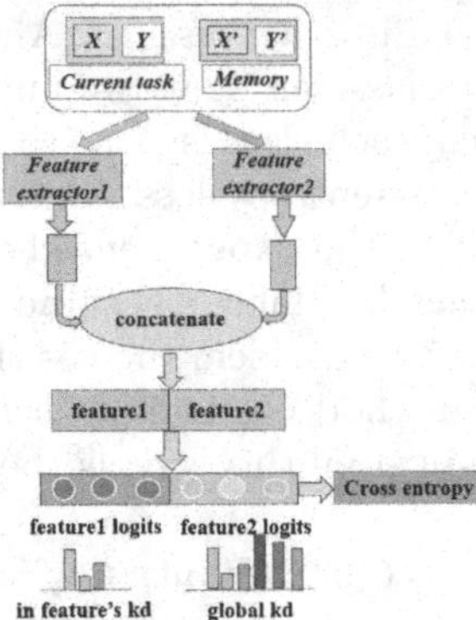

Fig. 1. Dual-branch network model for incremental learning. The model learns two branches sequentially, and the two branches are responsible for the feature representation learning of certain tasks. After learning the task corresponding to the first branch, the first branch will be frozen, and then use the new branch to learn new task, for the same branch used to learn multiple tasks, distillation will be used to retain the old task knowledge.

When there is only one branch, after finish training the corresponding task, the branch is $f_{\theta_{1,t-1}}$, and the loss is calculated as shown in Eq. 3, t is the number of current task, N_c is the number of class in each task.

$$\mathcal{L}_1 = D_{KL}(O(f_{\theta_{1,(t-1)}}(x))_{[1,(t-1)*N_c]}, O(f_{\theta_{1,t}}(x))_{[1,(t-1)*N_c]}) \tag{3}$$

After adding the second branch f_{θ_2}, distillation can be performed by 2 ways. Distillation on the entire logits value $logit_{(t-1)*N_c}$ from two branches, as shown in Eq. 4, the concatenated function cat concatenates the features from the two branches. Distillation on the the model only performed on the logits of the old classes from the corresponding branch, as shown in Eq. 5.

$$\mathcal{L}_2 = D_{KL}(O(cat(f_{\theta_1}(x), f_{\theta_{2,(t-1)}}(x)))_{[1,(t-1)*N_c]}, O(cat(f_{\theta_1(x)}, f_{\theta_{2,t}}(x)))_{[1,(t-1)*N_c]}) \tag{4}$$

$$\mathcal{L}_3 = D_{KL}(O(f_{\theta_{2,(t-1)}}(x))_{[1,(t-1)*N_c]}, O(f_{\theta_{2,t}}(x))_{[1,(t-1)*N_c]}) \tag{5}$$

3.5 Two Stage Training

We use a two-stage training approach for training. At the first stage, the model is trained by using cross-entropy loss and distillation loss. The features are concatenated from two branches and sent to the classifier. The model then uses cross-entropy loss to train the classifier on balanced dataset. The distillation loss uses the methods proposed in Sect. 3.4. Total loss is shown as Eq. 6.

$$\mathcal{L}_{total} = \mathcal{L}_{CE} + \mathcal{L}_{2|5} \tag{6}$$

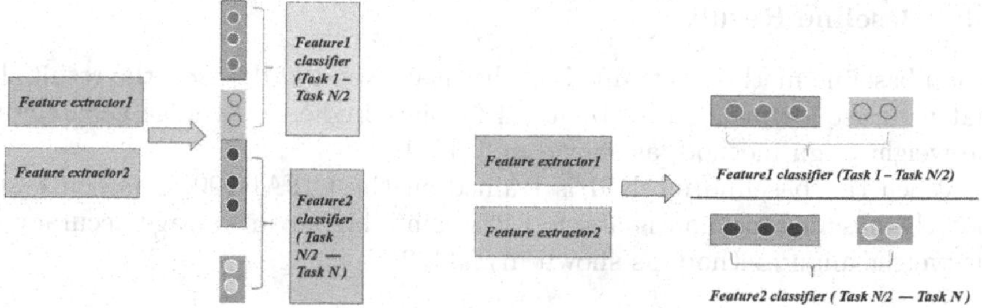

Fig. 2. The distillation method only in the current branch, when the model extends to two branches, each branch is responsible for some task, when the model perform distillation, only the logits of the class corresponding to the current branch is used.

Fig. 3. The method of global distillation. When the model extends a new branch, the model use the information of all the logits corresponding to the splicing of feature obtained by the two branches for distillation.

After the first stage, we get two branches. In order to solve the classifier bias problem, we freeze the two branches and only train the classifier. Specifically, we use the loss as Eq. 1 to retrain the classifier on the balanced datasets.

3.6 Sample Update Policy

The sample update strategy uses the herding algorithm used in iCaRl [14]. The model calculates the mean value of the feature for each class to get prototype. The model will choose the samples with the closest distance to the prototype to save in memory M. When update the M, samples in M which has the largest distance from the prototype will be discarded.

Since there will be two branches, for each task, when there is only one branch, the features calculated by the branch f_{θ_1} are used to calculate the prototype and select samples. When a new branch is added , the feature of each sample from new task will only use the new branch f_{θ_2} to calculate prototype and select samples. And then use the herding algorithm to update the memory $\mathcal{M}$.

4 Experience

The baseline method is trained on ImageNet-100 and the CIFAR100 dataset. Under the setting of 10 classes in the first task, 10 classes are added for each new task, 10 steps in total.

CIFAR100 consists 50000 images for traing, it has 100 classes, each class has 500 images for training, and 100 images for testing. The size of each image is 32 × 32. ImageNet-100 is a subset of ImageNet which is a large-scale dataset with 1,000 classes. ImageNet-100 chooses 100 classes from ImageNet for training and testing.

4.1 Baseline Result

When baseline method is trained on the ImageNet-100 dataset, the result shows that the baseline method is about 1.4% points higher in average accuracy than the weight align method, as shown in Table 1.

When the baseline method is trained on the CIFAR100, experience shows that the baseline method is about 1.2% points higher in average accuracy than the weight align method, as shown in Table 2.

4.2 Result on Imagenet100

As table 1 shows, DualBranch uses two branches on ImageNet-100, and the accuracy of the last task is 63.5%, which is about 7% higher than the baseline method,

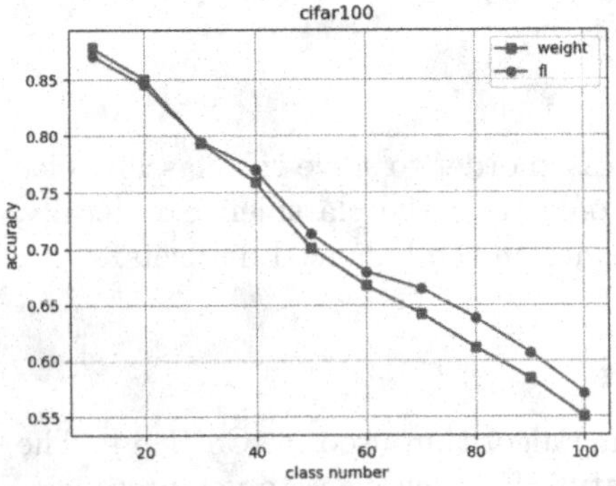

Fig. 4. Test accuracy of the baseline method on the CIFAR100 dataset, the orange line is the weight algin method, and the green line is the baseline method. (Color figure online)

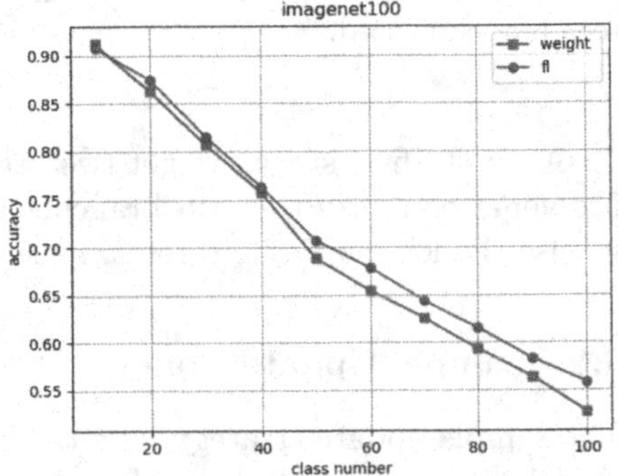

Fig. 5. Test accuracy of the baseline method on ImageNet-100 dataset, the orange line is the weight algin method, and the green line is the baseline method. (Color figure online)

Table 1. Accuracy(%) of different method on ImageNet-100. The first task has 10 classes, and each of the rest 9 tasks adds 10 classes for per task.

Task id	WA	Baseline	DualBranch
Task 5	68.84	70.72	73.28
Task 6	65.40	67.80	70.83
Task 7	62.54	64.37	69.6
Task 8	59.33	61.52	69.0
Task 9	56.36	58.29	65.02
Task 10	52.64	55.78	63.5
Avg	69.90	71.4	74.83

Table 2. Accuracy(%) of baseline method on CIFAR100. The first task has 10 class, and each of the rest 9 tasks adds 10 class per task.

Task id	WeightAlign	Baseline
Task 5	70.06	71.32
Task 6	66.70	67.93
Task 7	64.19	66.43
Task 8	61.14	63.79
Task 9	58.41	60.70
Task 10	54.97	57.10
Avg	70.35	71.53

and the average accuracy is about 3.4% higher than the baseline method. We can see the accuracy in table 1, with the help of the second branch, a significant advantage is achieved as the task increases (Fig. 5).

5 Conclusion

In this paper, we first construct a baseline method, which can more effectively solve the problem of classifier bias. We build a model which has two branches, after creating the second branch, the first branch is frozen, thus preserving the class information of the previous tasks. The new task is then learned on the new branch. Unlike previous models that use two branches to learn tasks in parallel, this serial learning method can better preserve the information of old tasks.

Although the accuracy is improved by adding new branch, but the new branch needs larger space, so we consider reduce the model's parameters while maintaining the accuracy in the future.

References

1. Boschini, M., Bonicelli, L., Buzzega, P., Porrello, A., Calderara, S.: Class-incremental continual learning into the extended der-verse. IEEE Trans. Pattern Anal. Mach. Intell. **45**, 5497–5512 (2022)
2. Buzzega, P., Boschini, M., Porrello, A., Abati, D., Calderara, S.: Dark experience for general continual learning: a strong, simple baseline. In: Advances in Neural Information Processing Systems, vol. 33, pp. 15920–15930 (2020)
3. Castro, F.M., Marín-Jiménez, M.J., Guil, N., Schmid, C., Alahari, K.: End-to-end incremental learning. In: Ferrari, V., Hebert, M., Sminchisescu, C., Weiss, Y. (eds.) ECCV 2018. LNCS, vol. 11216, pp. 241–257. Springer, Cham (2018). https://doi.org/10.1007/978-3-030-01258-8_15
4. Douillard, A., Cord, M., Ollion, C., Robert, T., Valle, E.: PODNet: pooled outputs distillation for small-tasks incremental learning. In: Vedaldi, A., Bischof, H., Brox, T., Frahm, J.-M. (eds.) ECCV 2020. LNCS, vol. 12365, pp. 86–102. Springer, Cham (2020). https://doi.org/10.1007/978-3-030-58565-5_6
5. Douillard, A., Ramé, A., Couairon, G., Cord, M.: DyTox: transformers for continual learning with dynamic token expansion. In: Proceedings of the IEEE/CVF Conference on Computer Vision and Pattern Recognition, pp. 9285–9295 (2022)
6. Goodfellow, I., et al.: Generative adversarial networks. Commun. ACM **63**(11), 139–144 (2020)
7. Guo, Y., Liu, B., Zhao, D.: Online continual learning through mutual information maximization. In: International Conference on Machine Learning, pp. 8109–8126. PMLR (2022)
8. Hinton, G., Vinyals, O., Dean, J.: Distilling the knowledge in a neural network. arXiv preprint: arXiv:1503.02531 (2015)
9. Hou, S., Pan, X., Loy, C.C., Wang, Z., Lin, D.: Learning a unified classifier incrementally via rebalancing. In: Proceedings of the IEEE/CVF Conference on Computer Vision and Pattern Recognition, pp. 831–839 (2019)
10. Kirkpatrick, J., et al.: Overcoming catastrophic forgetting in neural networks. Proc. Natl. Acad. Sci. **114**(13), 3521–3526 (2017)

11. Li, Z., Hoiem, D.: Learning without forgetting. IEEE Trans. Pattern Anal. Mach. Intell. **40**(12), 2935–2947 (2017)
12. Liu, Y., Schiele, B., Sun, Q.: Adaptive aggregation networks for class-incremental learning. In: Proceedings of the IEEE/CVF Conference on Computer Vision and Pattern Recognition, pp. 2544–2553 (2021)
13. Pham, Q., Liu, C., Hoi, S.: DualNet: continual learning, fast and slow. In: Advances in Neural Information Processing Systems, vol. 34, 16131–16144 (2021)
14. Rebuffi, S.A., Kolesnikov, A., Sperl, G., Lampert, C.H.: iCaRl: incremental classifier and representation learning. In: Proceedings of the IEEE Conference on Computer Vision and Pattern Recognition, pp. 2001–2010 (2017)
15. Shi, Y., et al.: Mimicking the oracle: an initial phase decorrelation approach for class incremental learning. In: Proceedings of the IEEE/CVF Conference on Computer Vision and Pattern Recognition, pp. 16722–16731 (2022)
16. Shin, H., Lee, J.K., Kim, J., Kim, J.: Continual learning with deep generative replay. In: Advances in Neural Information Processing Systems, vol. 30 (2017)
17. Tang, Y.M., Peng, Y.X., Zheng, W.S.: Learning to imagine: diversify memory for incremental learning using unlabeled data. In: Proceedings of the IEEE/CVF Conference on Computer Vision and Pattern Recognition, pp. 9549–9558 (2022)
18. Wang, S., Li, X., Sun, J., Xu, Z.: Training networks in null space of feature covariance for continual learning. In: Proceedings of the IEEE/CVF Conference on Computer Vision And Pattern Recognition, pp. 184–193 (2021)
19. Wang, Z., et al.: Learning to prompt for continual learning. In: Proceedings of the IEEE/CVF Conference on Computer Vision and Pattern Recognition, pp. 139–149 (2022)
20. Wu, Y., et al.: Large scale incremental learning. In: Proceedings of the IEEE/CVF Conference on Computer Vision and Pattern Recognition, pp. 374–382 (2019)
21. Yan, S., Xie, J., He, X.: DER: dynamically expandable representation for class incremental learning. In: Proceedings of the IEEE/CVF Conference on Computer Vision and Pattern Recognition, pp. 3014–3023 (2021)
22. Yu, L., et al.: Semantic drift compensation for class-incremental learning. In: Proceedings of the IEEE/CVF Conference on Computer Vision and Pattern Recognition, pp. 6982–6991 (2020)
23. Zenke, F., Poole, B., Ganguli, S.: Continual learning through synaptic intelligence. In: International Conference on Machine Learning, pp. 3987–3995. PMLR (2017)
24. Zhang, X., Yu, F.X., Karaman, S., Zhang, W., Chang, S.F.: Heated-up softmax embedding. arXiv preprint: arXiv:1809.04157 (2018)
25. Zhao, B., Xiao, X., Gan, G., Zhang, B., Xia, S.T.: Maintaining discrimination and fairness in class incremental learning. In: Proceedings of the IEEE/CVF Conference on Computer Vision and Pattern Recognition, pp. 13208–13217 (2020)

Inter-image Discrepancy Knowledge Distillation for Semantic Segmentation

Kaijie Chen[1], Jianping Gou[2(✉)], and Lin Li[1]

[1] School of Computer Science and Communication Engineering, Jiangsu University, Zhenjiang, China

[2] College of Computer and Information Science, College of Software, Southwest University, Chongqing 400715, China

cherish.gjp@gmail.com

Abstract. As a typical dense prediction task, semantic segmentation remains challenging in industrial automation, since it is non-trivial to achieve a good tradeoff between the performance and the efficiency. Meanwhile, knowledge distillation has been applied to reduce the computational cost in semantic segmentation task. However, existing knowledge distillation methods for semantic segmentation mainly mimic the teachers' behaviour using the well-designed knowledge variants from a single image, failing to explore discrepancy knowledge between different images. Considering that the large pre-trained teacher network usually tends to form a more robust discrepancy space than the small student, we propose a new inter-image discrepancy knowledge distillation method (IIDKD) for semantic segmentation. Extensive experiments are conducted on two popular semantic segmentation datasets, where the experimental results show the efficiency and effectiveness of distilling inter-image discrepancy knowledge.

Keywords: Knowledge Distillation · Semantic Segmentation · Discrepancy Distillation

1 Introduction

As a fundamental task in computer vision, semantic segmentation aims to assign each pixel of the input scene image with a category label, which has been widely adopted in different tasks, such as autonomous driving [13], robot navigation and anomaly detection [3]. With the great success of deep learning, fully convolutional networks have far surpassed traditional methods in terms of accuracy for semantic segmentation. However, these semantic segmentation algorithms also rely on complicated deep models with enormous parameters and huge computations, making them very difficult to be deployed on those real-time scenarios or resource-limited embedded devices. Therefore, it is always of great importance to achieve a good balance between computational cost and accuracy in real-world applications. Recently, a lot of efforts have been made in this area.

Q. Liu et al. (Eds.): PRCV 2023, LNCS 14427, pp. 273–284, 2024.
https://doi.org/10.1007/978-981-99-8435-0_22

For example, several lightweight deep neural network architectures have been devised to achieve faster inference, such as ENet [14] and ESPNet [11]. However, there is still a clear performance gap between the above-mentioned compact networks and those cumbersome deep models [7]. Besides redesigning new deep neural network architectures, another promising way is to utilize general model compression strategies, which can be roughly divided into three categories: quantization, pruning, and knowledge distillation [6].

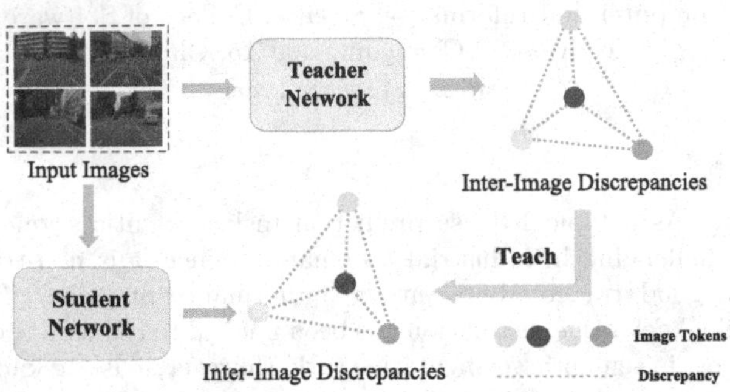

Fig. 1. An illustration of the idea to guide the small student network through mimicking the inter-image discrepancy knowledge from the large teacher network.

Without the required modifications on the network architecture, knowledge distillation can indirectly simplify complex models by transferring the knowledge from a large pre-trained teacher network to a compact student network. Recently, the conventional knowledge distillation methods [1,5,12,15,17,20] have achieved impressive performances in image classification. Compared with image classification, semantic segmentation involves dense prediction of a structured output, where each pixel in an image is required to be assigned with a corresponding label. Previous studies have shown that directly applying KD methods for image classification to semantic segmentation is sub-optimal [9], possibly because that strictly aligning the logits map or feature maps between the teacher and student on a per-pixel basis may ignore the structured dependencies between pixels, leading to a degradation of performance. In the past few years, several knowledge distillation approaches have been introduced for semantic segmentations. For instance, SKD [10] suggests transferring the pairwise similarity between spatial locations, while IFVD [19] requires the student network to mimic the set of similarity between the feature on each pixel and its corresponding class-wise prototype. CWD [16] aims to distill the significant activation regions in each channel. However, the above-mentioned methods strive to extract various sophisticated knowledge from the individual image but ignoring the rich cross-image information between different images. To explore cross-image similarities, CIRKD [21] built pixel dependencies across global images, but ignoring the discrepancy knowledge between diverse images. Given multiple input images, there are valuable discrepancy knowledge in terms of feature or logits maps, which is ignored by previous works.

To address the above-mentioned issues, we propose a novel Inter-Image Discrepancy Knowledge Distillation (IIDKD) for semantic segmentation (Fig. 1), to transfer rich discrepancy knowledge from the teacher network to the student network. Different from the relational knowledge distillation that establishes structured relations between pixels and regions across the samples, the proposed IIDKD mainly focuses on transferring pixel-wise and category-wise discrepancy knowledge between diverse images. Specifically, we first introduce an Attention Discrepancy Module (ADM) that explores the spatial attention discrepancy maps between any two images from the same mini-batch to identify the pixel-wise discrepancy knowledge. We then propose a Soft Probability Discrepancy Module (SDM) that focuses on the most divergent regions of each channel between diverse samples, which pays more attention to the category-wise discrepancy information. Therefore, in the distillation process, the teacher can transfer more noteworthy and discriminative knowledge derived from the critical regions that represents the discrepancies between images to the student. To the best of our knowledge, this is the first time to propose the inter-image discrepancy knowledge distillation (IIDKD) for semantic segmentation that explores the difference information among image samples.

In this paper, the main contributions can be summarized as follows: We propose a novel inter-image discrepancy knowledge distillation for semantic segmentation to enable the student model to acquire more noteworthy and discriminative inter-image discrepancy knowledge during the distillation process; We devise the attention discrepancy module (ADM) and the soft probability discrepancy module (SDM) to fully extract pixel-wise and category-wise discrepancy knowledge from different layers; We conduct extensive experiments on two popular semantic segmentation datasets, where the experimental results demonstrate the effectiveness of distilling inter-image discrepancy knowledge.

2 Method

To deliver effective knowledge distillation by capturing the inter-image discrepancy knowledge, we devise two new modules, the attention discrepancy module and the soft probability discrepancy module, to capture robust and informative discrepancy information between different samples. A combination of the proposed two modules can acquire complementary discrepancy knowledge in terms of both the pixel and the category. We introduce the details of two discrepancy-capturing modules and the overall training procedure in the following sections.

2.1 Notations

Semantic segmentation usually adopts a FCN-like architecture, including a backbone network and a head classifier. Given a mini-batch of n training samples, $X = \{x_1, x_2, \cdots, x_n\}$, the ground-truth labels can be denoted as

$Y = \{y_1, y_2, \cdots, y_n\}$. For the input image X, the backbone network first extracts its feature and aggregates high-level information to generate a dense feature map $\mathbf{F} \in R^{H \times W \times C}$. The head classifier then decodes the feature map into the categorical logits map $\mathbf{Z} \in R^{H \times W \times c}$, where H, W, C, and c denote height, weight, the number of channels, and the number of categories, respectively. Given a teacher network T and a student network S, let Z^t and Z^s denote the logits maps of the teacher model and the student model, respectively. Let F^t denote the feature map from the backbone network of the teacher, and F^s denote the corresponding feature map from the student network.

2.2 Overview

Semantic segmentation is a position-related task, where spatial dependencies play a crucial role. Previous KD works for semantic segmentation strive to design various knowledge variants or explore cross-image relations, which fails to model rich discrepancy knowledge hidden between different images and address the most informative regions. Inspired by the attention mechanism [18] and cross-image contrast [21], we devise an pixel-wise attention discrepancy module to distinguish the important regions of inter-image discrepancy for knowledge distillation.

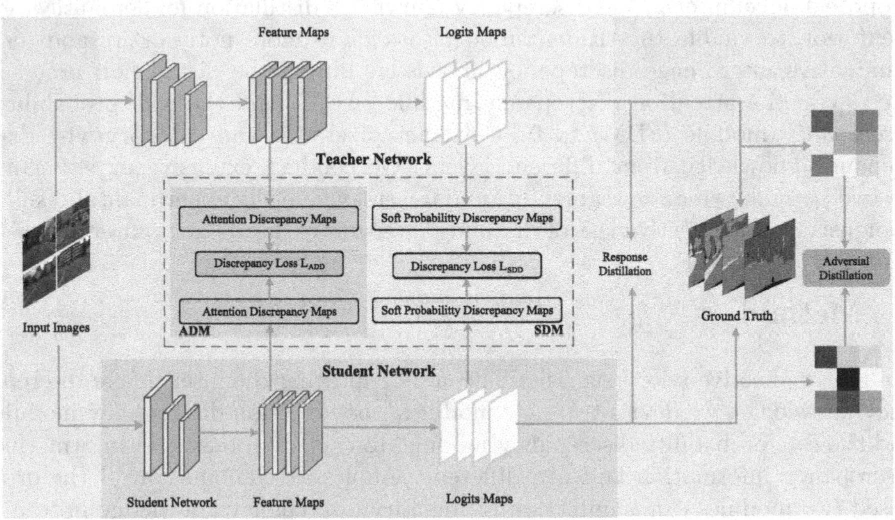

Fig. 2. The overall framework of the proposed inter-image discrepancy knowledge distillation.

Vanilla knowledge distillation aims to match the logits map pixel-by-pixel between the teacher and student. The limitation of the approach lies in that not all pixels in the logits map contribute equally to knowledge distillation.

Strictly aligning each pixel in the probability map may lead to sub-optimal performance due to redundant noise. Motivated by channel knowledge [16], soft probability discrepancy distillation is proposed to capture category-wise difference knowledge as a complement to pixel-wise attention discrepancy distillation. The soft probability discrepancy module can emphasize the most different regions between diverse images in the category dimension through channel regularization and distribution subtraction.

The main IIDKD framework for semantic segmentation is shown in Fig. 2. In our method, the feature maps generated from the backbone and the logits maps generated from the classifier are fed into attention discrepancy distillation module (ADM) and soft probability discrepancy distillation module (SDM), respectively. These two modules calculate the attention discrepancy maps and soft probability maps and then transfer these two types of discrepancy knowledge from teacher to student by the designed loss function. The process is indicated in the diagram by a red dashed line. Besides, we also add the conventional KD loss and the adversarial distillation loss for more robust optimization.

2.3 Attention Discrepancy Distillation

Given a mini-batch of different images, there exists rich discrepancy information in the focus of their feature representation. Considering that the activation value of each pixel in the attention map is indicative of the significance of its corresponding spatial location, we selected the activation-based attention mechanism to construct inter-image attention discrepancies in this paper. Let $\mathcal{F} = \{F_1, F_2, \ldots, F_n\}$ represents n dense feature maps of n training images from one batch. First, given a particular training images x_i, the activation-based attention maps AM_i can be formulated as follows:

$$\mathrm{AM}_i = \frac{vec(\sum_C |F_i|^2)}{\|vec(\sum_C |F_i|^2)\|_2}, \tag{1}$$

where C indicates the number of channels and vec means a vector operation. For a specific training image pair of (x_i, x_j), the inter-image attention discrepancy map can be denoted as:

$$\mathrm{ADM}_{(i,j)} = \|\mathrm{AM}_i - \mathrm{AM}_j\|, \tag{2}$$

where $\mathrm{ADM}_{(i,j)}$ denotes the inter-image attention discrepancy map between i-th and j-th training images. Based on the above equations, we can calculate the inter-image attention discrepancy maps from the teacher and the student, denoted as $\mathrm{ADM}^t_{(i,j)}$ and $\mathrm{ADM}^s_{(i,j)}$. The attention discrepancy distillation loss function for transferring the pixel-wise inter-image attention discrepancy knowledge between any two samples $(x_i, x_j) \in X$ in the teacher model to the student model can be formulated as:

$$L_{add} = \sum_{i=1}^{n}\sum_{j=1}^{n} \mathrm{MSE}\left(\mathrm{ADM}^t_{(i,j)}, \mathrm{ADM}^s_{(i,j)}\right), \tag{3}$$

where $\mathrm{MSE}(\cdot)$ indicates the mean squared loss and $i \neq j$.

2.4 Soft Probability Distillation

As the activation values of each channel tend to encode the saliency of specific categories, we propose the category-wise soft probability discrepancy distillation to highlight the regions with the most significant differences between images in terms of category. Similarly, let $\mathcal{Z} = \{Z_1, Z_2, ..., Z_n\}$ represents n logits maps of n training images from a mini-batch. First, given a specific image pair of (x_i, x_j), we compute the inter-image logits discrepancy map $\text{LDM}_{(i,j)}$ by distribution subtraction operation, which can be denoted as:

$$\text{LDM}_{(i,j)} = \|Z_i - Z_j\| \ . \tag{4}$$

Then, we apply the softmax operation with high temperature to each channel of the logits discrepancy maps, which converts activation values into a soft probability distribution to remove the effect of the magnitude scale. The soft probability discrepancy maps $\text{SDM}_{(i,j)}$ tend to encode the saliency of inter-image semantic discrepancies in the category dimension, which can be computed as:

$$\text{SDM}_{(i,j)} = \frac{exp\left(\frac{\text{LDM}_{(i,j)}}{T}\right)}{\sum_{H\times W} exp\left(\frac{\text{LDM}(i,j)}{T}\right)}, \tag{5}$$

where T presents the hyper-parameter temperature.

Through the above operation, the inter-image soft probability discrepancy maps from the teacher and student can be acquired, which can be denoted as $\text{SDM}^t_{(i,j)}$ and $\text{SDM}^s_{(i,j)}$. We devise the soft probability discrepancy distillation loss function to softly align the category-wise discrepancy knowledge between teacher and student model, which is expressed as:

$$L_{sdd} = \frac{T^2}{c}\sum_{i=1}^{n}\sum_{j=1}^{n}\sum_{k=1}^{c}\text{KL}\left(\text{SDM}^t_{(i,j),c}||\text{SDM}^s_{(i,j),c}\right), \tag{6}$$

where KL indicates the Kullback-Leibler divergence.

2.5 Optimization

Besides the proposed discrepancy distillation loss function terms, we also align each pixel of the logits maps between teacher and student using the conventional knowledge distillation loss for semantic segmentation, as follows:

$$L_{kd} = \frac{1}{H\times W}\sum_{n=1}^{H\cdot W}\varphi\left(\boldsymbol{Z}_n^S\right)\cdot\log\left[\frac{\varphi\left(\boldsymbol{Z}_n^s\right)}{\varphi\left(\boldsymbol{Z}_n^t\right)}\right], \tag{7}$$

where n indicates the n-th pixel in the logits maps and $\varphi(\cdot)$ is the softmax function. The adversarial distillation loss is also applied for a more stable optimization:

$$L_{adv} = E_{S\sim p(S)}[D(S \mid I)] \ , \tag{8}$$

where $E(\cdot)$ is the expectation operator, $D(\cdot)$ is the discriminator, I and S indicates the input image and its corresponding segmentation map, respectively. The overall loss of the proposed IIDKD for knowledge distillation is then formulated as follows:

$$L_{total} = L_{kd} + \alpha L_{add} + \beta L_{sdd} - \gamma L_{adv} \ , \tag{9}$$

where α, β, and γ are the hyper-parameters.

3 Experiments

3.1 Datasets and Setups

Cityscapes [2] consists of 5000 finely annotated images from urban street scene images. It has 2975 training images, 500 validation images, and 1525 testing images respectively.

Pascal VOC 2012 [4] contains 20 foreground object categories and one background category. The resulting dataset is split into three subsets: 10582 images for training, 1449 images for validation and 1456 images for testing.

Experimental Details. During the training process, we crop the input images as 512×512 and train the student model for 80000 iterations with the batch size of 8. We set α, β, γ as 1, 1000, and 0.01, respectively. The temperature hyper-parameter T in soft probability discrepancy distillation is set as 4 to achieve best segmentation performance. We adopt the polynomial learning rate strategy and set the base learning rate as 0.01. The learning rate is calculated as $base\_lr \cdot (1 - \frac{iter}{total\_iter})^{0.9}$. For the data augmentation, we apply random scaling and random horizontal flipping in the range of $[0.5, 2]$. For the optimization, we use the Stochastic Gradient Descent (SGD) as the optimizer, where the momentum and the weight decay are set as 0.9 and 5e-4, respectively.

3.2 Comparisons with Recent Methods

We compare our proposed IIDKD method with the recent state-of-the-art methods, including SKD [10], IFVD [19], CWD [16], and CIRKD [8]. All teacher-student network pairs are built on the same network with identical parameters.

Results on Cityscapes. In the experiments, we use PSPNet with ResNet-101 as the teacher network. PSPNet and DeepLabV3 with different backbones are used to construct four teacher-student architectures to conduct the comparative experiments. Table 1 shows the performance of our IIDKD and four compared methods on Cityscapes dataset, where the proposed IIDKD achieves the best segmentation performance across various student networks with similar or different architectures: 1) our IIDKD improves the PSPNet built on ResNet18 and MobileNetV2 by 5.56% and 4.70% on validation set, respectively; 2) cross-image distillation methods such as CIRKD perform better than most of single-image distillation methods such as SKD and IFVD. Additionally, we can see that the proposed IIDKD outperforms both single-image KD methods and other cross-image KD methods. When replacing the student network with DeepLabV3, the performance gains on ResNet18 and MobileNetV2 reach to 2.57% and 4.63%, respectively, demonstrating the effectiveness and robustness of IIDKD for different network architectures.

Table 1. Comparison of mIoU (%) of different KD methods on Cityscapes dataset.

Method	baseline	SKD [10]	IFVD [19]	CWD [16]	CIRKD [21]	IIDKD (ours)
T: PSPNet-ResNet101	78.50	-	-	-	-	-
S: PSPNet-ResNet18	70.48	72.45	73.80	75.23	74.83	**76.04**
S: DeepLabV3-ResNet18	73.62	74.76	74.19	75.43	75.2	**76.37**
S: PSPNet-MobileNetV2	65.50	67.36	67.65	69.40	68.54	**70.20**
S: DeepLabV3-MobileNetV2	65.80	68.13	68.66	69.72	69.28	**70.43**

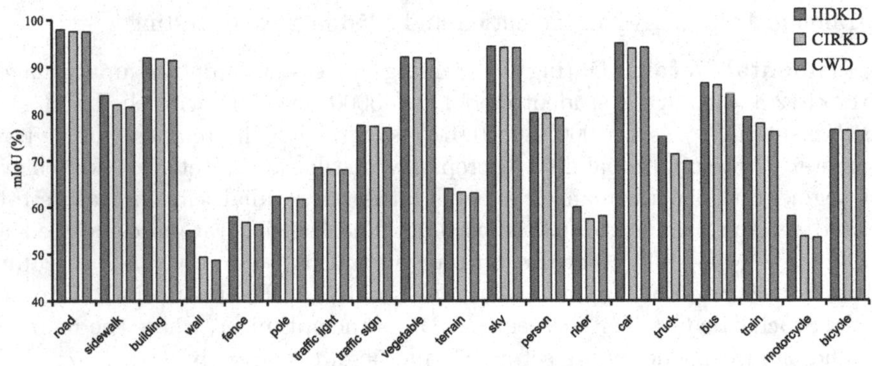

Fig. 3. The individual class IoU scores of our proposed IIDKD and the state-of-the-art CWD and CIRKD on the Cityscapes validation dataset.

In Fig. 3, we also show the performance in terms of IoU scores per category over the student network compared with the state-of-the-art methods. We can observe that our IIDKD performs better on class IoU scores than the most competitive methods, such as CIRKD and CWD. Besides, we further visualize the segmentation results in Fig. 4. It can be seen that our IIDKD generates more consistent semantic labels with the ground truth. The possible reason is that our IIDKD can distill comprehensive inter-image discrepancy knowledge in both pixel and category dimensions for producing more accurate segmentation particulars and facilitating the visual recognition.

Results on Pascal VOC. We use PSPNet with backbone of ResNet101 as the teacher, PSPNet and DeepLabV3 with ResNet18 as the student. Table 2

Table 2. Comparison of mIoU (%) of different KD methods on VOC 2012 dataset.

Method	baseline	SKD [10]	IFVD [19]	CWD [16]	CIRKD [21]	IIDKD (ours)
T: PSPNet-ResNet101	76.60	-	-	-	-	-
S: PSPNet-ResNet18	71.30	72.51	72.68	73.90	74.02	**74.65**
S: DeepLabV3-ResNet18	71.70	72.82	72.54	74.07	74.16	**74.90**

shows the comparisons with state-of-the-art methods on Pascal VOC 2012 validation dataset. It can be clearly seen that the KD methods for transferring cross-image knowledge such as CIRKD perform better than the KD methods for only transferring intra-image pixel affinity such as SKD, IFVD, and CWD. Besides, our IIDKD that considers inter-image discrepancy knowledge has significantly achieved the best results on among all the compared methods. For example, when the student is DeepLabV3 built on ResNet18, the improvement of our IIDKD over CIRKD with the second highest accuracy and over the student baseline is 0.74% and 3.20%. Compared with the cross-image relational knowledge distillation, our IIDKD with better performance implies that the devised attention discrepancy maps and soft probability discrepancy maps as the inter-image discrepancy knowledge are very favorable to improve the performance of knowledge distillation.

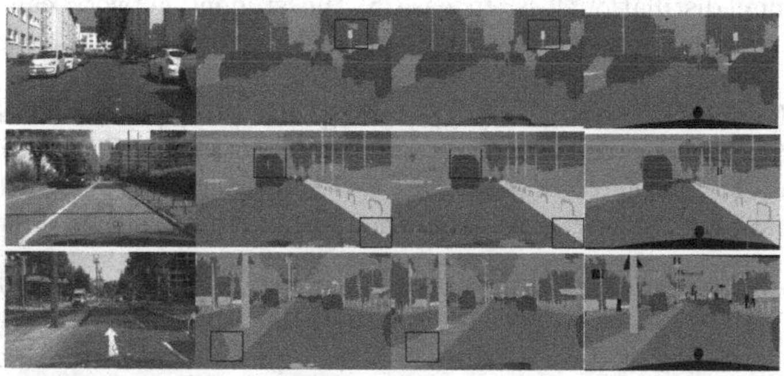

(a) images (b) W/o distillation (c) ours (d) ground truth

Fig. 4. Visualization results on the Cityscapes dataset. Our method generates more accurate and detailed segmentation masks, which are circled by red lines.

3.3 Ablation Studies

To further the contribution of each distillation loss item in overall loss function (i.e., Eq. (9)), we conduct the detailed ablation experiments on the Cityscapes validation dataset. We choose PSPNet with ResNet101 as the teacher network and PSPNet with ResNet18 as the student network. To easily understand, we record the experimental results into five cases according to the addition of the distillation loss:

- Case 1: The student network without any knowledge distillation, which only learns from the ground truth label.
- Case 2: $L_{total} = \lambda L_{kd}$, i.e., the Killback-Leibler divergence of the class probability maps between the teacher and student. The student network learns the response-based logits from the teacher network to update the parameters, but ignore more holistic knowledge.
- Case 3: $L_{total} = L_{kd} - \gamma L_{adv}$, i.e., the adversarial distillation loss. In case 3, the teacher network transfers the high-order relations, while the rich discrepancy knowledge between diverse images is ignored.
- Case 4: $L_{total} = L_{kd} + \alpha L_{add} - \gamma L_{adv}$, where L_{add} is the inter-image attention discrepancy distillation loss, aiming to align the pixel-wise difference between different images. In case 4, the teacher network further transfers the inter-image discrepancy knowledge from the feature layer.
- Case 5: $L_{total} = L_{kd} + \alpha L_{sdd} - \gamma L_{adv}$. L_{sdd} is the inter-image soft probability discrepancy distillation loss. In case 5, the student network can learn the category-wise inter-image discrepancy knowledge of the logits layer from the teacher.
- Case 6: $L_{total} = L_{kd} + \alpha L_{add} + \beta L_{sdd} - \gamma L_{adv}$. In case 6, we combine the holistic adversarial distillation loss and two types of inter-image discrepancy distillation loss to transfer comprehensive and complementary knowledge, which help the student to fully learn from the teacher from multiple dimensions.

In Table 3, compared with the raw student network without any distillation with the mIoU of 70.48% in case 1, the conventional response-based knowledge distillation loss L_{kd} in case 2 brings 1.64% improvement. After applying the adversarial distillation loss L_{adv} in case 3, the performance of the student model can be improved by 1.14% over the model only using the L_{kd}. Then we further add two kinds of discrepancy distillation loss successively to verify their effects respectively. Concretely, on the basis of the student network added L_{kd} and L_{adv}, the pixel-wise attention discrepancy distillation loss L_{add} in case 4 boosts the improvement to 74.57% and the category-wise soft probability distillation loss L_{sdd} in case 5 brings the improvement to 75.40%. Finally, after combining all the aforementioned loss items, the mIoU in case 6 is promoted to 76.04% with a great increase of 5.56%, reducing the gap between student and teacher from 9.46% to 2.49%.

Through the above ablation results, we can conclude that: 1) our proposed IIDKD framework can distill comprehensive and complementary inter-image discrepancy knowledge, which can successfully help the student network attain a sig-

Table 3. Ablation results of distillation loss on Cityscapes.

Method	L_{kd}	L_{adv}	L_{add}	L_{sdd}	mIoU(%)
ResNet101 (T)					78.56
Case 1: ResNet18 (S)	✘	✘	✘	✘	70.48
Case 2: ResNet18 (S)	✔	✘	✘	✘	72.12
Case 3: ResNet18 (S)	✔	✔	✘	✘	73.26
Case 4: ResNet18 (S)	✔	✔	✔	✘	74.47
Case 5: ResNet18 (S)	✔	✔	✘	✔	75.40
Case 6: ResNet18 (S)	✔	✔	✔	✔	76.04

nificant improvement; 2) both pixel-wise discrepancy distillation and category-wise soft probability distillation are effective for improving the performance of the student model; and 3) soft probability distillation is more informative than the attention discrepancy distillation by capturing more noteworthy category-wise knowledge.

4 Conclusion

In this paper, we propose a novel knowledge distillation framework for semantic segmentation, namely inter-image discrepancy knowledge distillation (IIDKD). Specifically, we devise an attention discrepancy distillation module and a soft probability discrepancy distillation module, which aim to well-capture the most important discrepancy regions between diverse images in both pixel and category dimensions. We conduct extensive experiments on the popular segmentation datasets and demonstrate the generality and effectiveness of the proposed method. In the future, as the effective model compression with the informative inter-image discrepancy knowledge transfer, we will consider applying the proposed method into real-time industrial automation scenarios, such as autonomous driving and robot navigation.

Acknowledgement. This work was supported in part by National Natural Science Foundation of China (Grant Nos. 61976107 and 61502208).

References

1. Beyer, L., Zhai, X., Royer, A., Markeeva, L., Anil, R., Kolesnikov, A.: Knowledge distillation: a good teacher is patient and consistent. In: CVPR, pp. 10925–10934 (2022)
2. Cordts, M., et al.: The cityscapes dataset for semantic urban scene understanding. In: CVPR, pp. 3213–3223 (2016)
3. Deng, H., Li, X.: Anomaly detection via reverse distillation from one-class embedding. In: CVPR, pp. 9737–9746 (2022)

4. Everingham, M., Van Gool, L., Williams, C.K., Winn, J., Zisserman, A.: The pascal visual object classes (VOC) challenge. Int. J. Comput. Vision **88**(2), 303–338 (2010)
5. He, R., Sun, S., Yang, J., Bai, S., Qi, X.: Knowledge distillation as efficient pre-training: faster convergence, higher data-efficiency, and better transferability. In: CVPR, pp. 9161–9171 (2022)
6. Hinton, G., Vinyals, O., Dean, J., et al.: Distilling the knowledge in a neural network. arXiv preprint arXiv:1503.02531, vol. 2, no. 7 (2015)
7. Kirillov, A., et al.: Segment anything (2023)
8. Li, M., et al.: Cross-domain and cross-modal knowledge distillation in domain adaptation for 3D semantic segmentation. In: Proceedings of the 30th ACM International Conference on Multimedia, pp. 3829–3837 (2022)
9. Li, Q., Jin, S., Yan, J.: Mimicking very efficient network for object detection. In: CVPR, pp. 6356–6364 (2017)
10. Liu, Y., Shu, C., Wang, J., Shen, C.: Structured knowledge distillation for dense prediction. TPAMI (2020)
11. Mehta, S., Rastegari, M., Caspi, A., Shapiro, L., Hajishirzi, H.: ESPNet: efficient spatial pyramid of dilated convolutions for semantic segmentation. In: ECCV, pp. 552–568 (2018)
12. Mirzadeh, S.I., Farajtabar, M., Li, A., Levine, N., Matsukawa, A., Ghasemzadeh, H.: Improved knowledge distillation via teacher assistant. In: AAAI, vol. 34, pp. 5191–5198 (2020)
13. Pan, H., Chang, X., Sun, W.: Multitask knowledge distillation guides end-to-end lane detection. IEEE Trans. Ind. Inform. (2023)
14. Paszke, A., Chaurasia, A., Kim, S., Culurciello, E.: ENet: a deep neural network architecture for real-time semantic segmentation. arXiv preprint arXiv:1606.02147 (2016)
15. Peng, B., et al.: Correlation congruence for knowledge distillation. In: ICCV, pp. 5007–5016 (2019)
16. Shu, C., Liu, Y., Gao, J., Yan, Z., Shen, C.: Channel-wise knowledge distillation for dense prediction. In: ICCV, pp. 5311–5320 (2021)
17. Tung, F., Mori, G.: Similarity-preserving knowledge distillation. In: ICCV, pp. 1365–1374 (2019)
18. Vaswani, A., et al.: Attention is all you need. In: NeuralPS, vol. 30 (2017)
19. Wang, Y., Zhou, W., Jiang, T., Bai, X., Xu, Y.: Intra-class feature variation distillation for semantic segmentation. In: Vedaldi, A., Bischof, H., Brox, T., Frahm, J.-M. (eds.) ECCV 2020. LNCS, vol. 12352, pp. 346–362. Springer, Cham (2020). https://doi.org/10.1007/978-3-030-58571-6_21
20. Xu, C., Gao, W., Li, T., Bai, N., Li, G., Zhang, Y.: Teacher-student collaborative knowledge distillation for image classification. Appl. Intell. **53**(2), 1997–2009 (2023)
21. Yang, C., Zhou, H., An, Z., Jiang, X., Xu, Y., Zhang, Q.: Cross-image relational knowledge distillation for semantic segmentation. In: CVPR, pp. 12319–12328 (2022)

Vision Problems in Robotics, Autonomous Driving

Cascaded Bilinear Mapping Collaborative Hybrid Attention Modality Fusion Model

Jiayu Zhao[1,2] and Kuizhi Mei[1,2](✉)

[1] College of Artificial Intelligence, Xi'an Jiaotong University, Xi'an, China
meikuizhi@xjtu.edu.cn
[2] National Key Laboratory of Human-Machine Hybrid Augmented Intelligence, National Engineering Research Center of Visual Information and Applications, Institute of Artificial Intelligence and Robotics, Xi'an Jiaotong University, Xi'an, China

Abstract. LiDAR and Camera feature fusion detection is widely used in the field of 3D object detection. A fusion method that integrates the two modalities into a unified representation has been proposed and proven effective. However, there is still a challenge in fusing these two modalities in the same space. To address this issue, we propose a fusion module that leverages bilinear mapping technology and a hybrid attention mechanism. Our method preserves the original independent branches for LiDAR and Camera feature extraction. We then employ a cascade bilinear mapping hybrid attention fusion module to combine the features of these two modalities in the Bird's Eye View (BEV) space. We utilize a hybrid structure of Convolutional Neural Networks (CNN) and Efficient-Attention to process the fused features. Furthermore, Multilayer Perceptron (MLP) is used to extract information from the features and achieve an asymmetric deep fusion of the two modalities in BEV space through the cascaded bilinear mapping method. This approach helps to complement the information provided by both features and achieve improved detection results. On the nuScenes validation set, our model improves the accuracy by 1.2% compared to the current Sota method of feature fusion in BEV space. It is important to note that this improvement is achieved without any data augmentation or Test-Time Augmentation (TTA) techniques.

Keywords: 3D object Detection · Bilinear Mapping · Attention

1 Introduction

3D object detection [2] plays an important role in autonomous driving systems, and its main functions are to detect whether an object exists in the 3D world coordinate system through the data provided by the sensors and to detect its location and category [9]. In recent years, 3D object detection technology has received extensive attention from academia and industry. At present, the mainstream 3D object detection systems mainly include single-modal detectors based

Supported by National Natural Science Foundation of China (Grant No. 62076193).

Q. Liu et al. (Eds.): PRCV 2023, LNCS 14427, pp. 287–298, 2024.
https://doi.org/10.1007/978-981-99-8435-0_23

on single-sensor information and multi-modal detectors based on multi-sensor data fusion. Usually, the sensors in an autonomous driving system include cameras, LiDAR, and radar. Among them, camera sensors capture information in perspective and can provide semantically related information of objects such as color and texture [9], while LiDAR point clouds can provide depth and structure-related geometric information [22]. Different sensors show different performance characteristics according to different scenarios. Making full use of different sensors helps to reduce uncertainty [4] and make accurate and robust predictions.

Currently, there are two primary types of advanced multi-modal detectors: (1)Constructing a unified space, such as Bird's Eye View(BEV) or Voxel, to characterize the two modal data into the same space and fuse their features in this space [12,14,16]; (2)Bottom-up fusion methods based on DETR [21] use token mechanisms to facilitate the transfer of transformer queries across modalities for feature fusion and subsequent prediction.

In this paper, our objective is to improve the fusion effect of two modal features in BEV space by building upon the process proposed by BEVFusion [16]. We aim to achieve deep fusion of LiDAR and Camera features for improved detection accuracy of multi-modal detectors. We propose cascaded bilinear mapping [24,25] to fuse the features of the two modalities in BEV space and information projection to the fusion model using cross-attention-based [10] technique with LiDAR features. Currently, the fusion of multi-modal features in BEV space is achieved through simple concat and CNN module. However, this approach's linear combination of modal features may not sufficiently capture the complex correlation between semantic Camera features and geometric LiDAR features due to significant differences in their distribution. Compared to unilinear mapping, bilinear mapping [6] has been widely adopted in integrating diverse CNN features and Visual Question Answering (VQA) [1] due to its ability to capture distinct modal feature characteristics. However, the high dimensionality of output features and a large number of model parameters may significantly restrict the applicability of bilinear mapping.

This paper employed the bilinear mapping fusion method inspired by Natural Language Processing (NLP), which yields compact output features and powerful expressive capabilities [24]. Additionally, we have devised a CNN-Attention hybrid module to accelerate the calculation while simultaneously achieving comprehensive feature extraction [11]. As well as cross-modal interaction between LiDAR features and fused features following bilinear fusion. Specifically, a CNN combined with efficient cross-attention [7,20] is utilized to encode two kinds of features and project LiDAR features onto the fusion space. Then, the fusion features and the extracted features are fused by bilinear mapping to realize achieve asymmetric deep fusion of multi-modal features in BEV space.

To validate the efficacy of our design, we conducted a comprehensive evaluation of our model's performance on the nuScenes [3] validation set. Our results demonstrate an improvement in accuracy by approximately 1.2% compared to the baseline BEVFusion without incorporating additional techniques such as data augmentation and TTA.

Overall, our research provides three main contributions:

- We proposed a cascaded bilinear mapping hybrid attention fusion module. This module deconstructs and reorganizes the bilinear mapping and attention mechanism, achieving asymmetric deep fusion of features from two modalities while effectively retaining their expressive ability and structural characteristics.
- We have developed and implemented a hybrid CNN-Attention module, which incorporates a lightweight multi-head attention mechanism to reduce cache usage and improve computational efficiency.
- Based on the BEVFusion, we have enhanced the detection performance of multi-modal detectors in the realm of unified representation technology.

2 Related Work

Traditional Classification divides fusion methods into Point-level ,Proposal-level and Object-level. Among them,Point-level [5,18] fusion methods usually draw image semantic features onto foreground LiDAR points and perform LiDAR-based detection on the decorated point cloud input. Proposal-level [2,8] methods is to first project LiDAR points into the feature space or generate proposals, query the relevant Camera features, and then concatenate back into the feature space. Object-level methods mean that each sensor performs deep learning model reasoning separately for the target object, so as to output the results with their own attributes and fuse them at the decision level [26]. Object-level approach always disregards the intrinsic connection between the two modes of representation. Currently, some methods break the boundaries of levels and attempt to use end-to-end approaches for multimodal fusion.

Unified Representation. Some methods opt for building a unified expression that fuses the features of both modalities in the same space. It effectively preserves the information from both modalities. [14,16]Encoding Camera and LiDAR features into the same BEV separately for fusion. [12,13]build the voxel space by sampling features from the image plane according to the predicted depth scores and geometric constraints. The unified representation method can well apply the complementarity of data from different modalities. [4]proposed a method of interactive feature fusion in dual domains, which involves the transformation of camera domain features and voxel domain features into each other's domains and fuses them simultaneously in each domain. The unified representation method can effectively preserve the characteristics of multimodal data, while also addressing the challenge of effectively fusing diverse data.

3 Method

3.1 Overall Architecture

The overall architecture is illustrated in Fig. 1. Building upon BEVFusion [16], this paper proposed a bilinear mapping hybrid attention-based LiDAR-

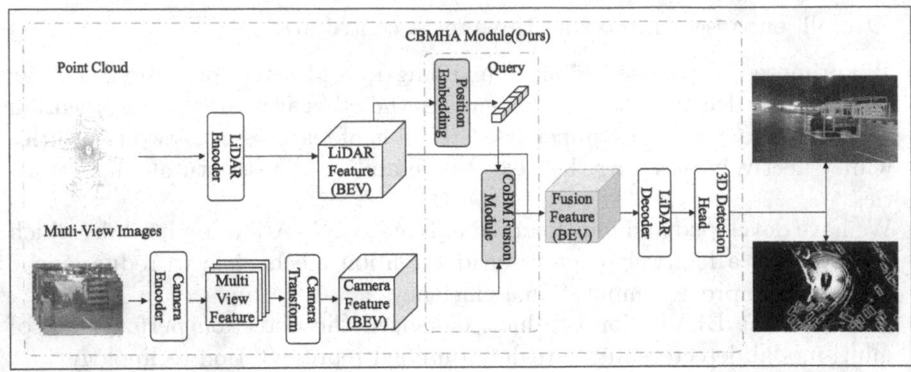

Fig. 1. Overall architecture of the model. The input point cloud and image data are uniformly represented in BEV space through two independent feature extraction branches. The LiDAR features and Camera features are then deeply fused asymmetrically through a cascade bilinear mapping hybrid attention (CBMHA) fusion block. Finally, the fused features are transmitted to the decoder for decoding, and subsequently fed into the 3D detection head for object detection.

Camera fusion detector that treats LiDAR and Camera streams as two independent branches, and uniformly projects them onto the BEV space for bilinear fusion.By inputting LiDAR point cloud and multi-view Camera image, two branches are utilized to extract features of the respective modes. The image input $X_c \in R^{W_c \times H_c \times 3}$ is encoded by Camera Backbone. H_c and H_c represent the height and width data of the image. The LiDAR point cloud data $X_L \in R^{N \times 3}$ is encoded by a voxel-based backbone, where N represents the number of LiDAR points. We propose a cascaded bilinear mapping hybrid attention fusion module to align and fuse the features of two modalities in BEV space. Firstly, the bilinear fusion is executed, followed by utilizing the dense LiDAR point cloud data as Query and the fusion data as Key and Value respectively to facilitate efficient multi-head interactive cross-attention extraction. Bilinear mapping is utilized to perform deep feature fusion between the extracted and fused features. The resulting features are then fed into the network decoder structure for decoding, and subsequently passed onto the target detection head for obtaining accurate 3D detection results.

3.2 Proposed Cascade Bilinear Mapping Hybrid Attention Fusion

Cascade Bilinear Mapping Hybrid Attention Fusion Module. By designing two modules: bilinear fusion and CNN-Attention hybrid, we combine them to construct the structure of feature fusion of two modalities in BEV. We have named this structure as CBMHA Fusion. The architecture is illustrated in Fig. 2 and can be expressed as:

$$
\begin{aligned}
F^m &= BMF(F^{LiDAR}, F^{Camera}), \\
F^f &= BMF(F^m, \phi(F^{LiDAR}, F^m))
\end{aligned}
\tag{1}
$$

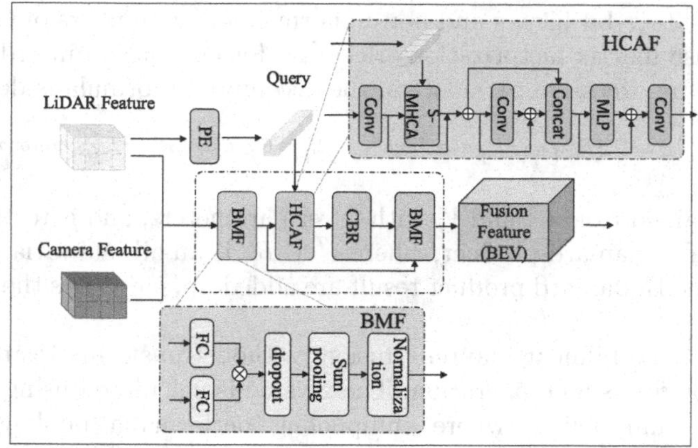

Fig. 2. The overall structural diagram of the cascaded bilinear mapping hybrid attention fusion module is presented, where BMF denotes the fusion module of bilinear mapping, HCAF denotes the hybrid CNN-Attention fusion module, and CBR refers to the module composed of Convolution, BatchNorm and Relu. The query is obtained through Patch Embedding (PE) block for LiDAR features, which are then passed into E-MHCA to enable interactive attention mechanisms. Subsequently, feature convolution is performed to achieve multi-level concatenation before MLP is employed for feature extraction and output generation.

Where F^{LiDAR} and F^{Camera} denote the features from the LiDAR stream and Camera stream, respectively. F^m, F^f denote BMF output. $BMF(\cdot)$ denotes bilinear mapping fusion block, ϕ represents hybrid CNN-Attention fusion block.

Bilinear Mapping Fusion. The inputs of the two modalities are uniformly characterized into BEV space through the feature extraction module, and the Camera features F^{Camera} and LiDAR features F^{LiDAR} are obtained. Due to the different information carried by the two modal features, in order to better fuse the two modal features, considering that the bilinear mapping mechanism can fuse features with different characteristics, the improved bilinear fusion method is used to fuse the two modal data.

Given the LiDAR features $F^{LiDAR} \in R^{h \times w \times m}$ and the Camera features $F^{Camera} \in R^{h \times w \times n}$ in the input BEV space, a basic bilinear model can be obtained as follows:

$$z_i = F^{LiDAR^T} W_i F^{Camera} \tag{2}$$

Where $\mathrm{W}_i \in \mathrm{R}^{m \times n}$ is the projection matrix, $z_i \in \mathrm{R}$ is the output of the bilinear mapping. To get the output z, we need to learn $\mathrm{W} = [\mathrm{W}_i, \ldots, \mathrm{W}_o] \in R^{m \times n \times o}$. However, there is a problem with this method. When m, n is large, the number of parameters increases rapidly, resulting in an increase in the amount of calculation and a certain risk of overfitting. In fact, when the desired output feature channel is the number of o, it is necessary to learn the number of o W_i, which equals to o z_i.

Due to the four-bit arrays and non-uniform channel numbers of Camera and LiDAR, we use matrix factorization tricks [24] for calculation in order to avoid issues caused by excessive parameters, the decomposed formula is described as:

$$z_i = F^{LiDAR^T} U_i V_i^T F^{Camera} = 1^T (U_i^T F^{LiDAR^\circ} V_i^T F^{Camera}) \tag{3}$$

z_i can be obtained from U_i and V_i with fewer parameters, and is further written in the form of Hadamard product, where $1^T \in R^k$ is an all-one variable, and the elements of the Hadamard product result are added up, $\circ$ denotes the Hadamard product.

In general, the bilinear mapping fusion method transforms U and V into a 3D form through reshape operation. The z value is calculated using Hadamard product and sumpooling, where sumpooling performs a pooling of kernel size = stride = k on the result. Finally, we obtain the output by adding dropout to avoid overfitting and normalization to prevent getting stuck in a local minimum. This can be expressed as follows:

$$\begin{aligned} Z_f &= BMF_f(F^{LiDAR}, F^{Camera}) = Dropout(U^T F^{LiDAR} V^T F^{Camera}), \\ Z &= BMF_{sq}(F^{LiDAR}, F^{Camera}) = Normlization(SumPooling(Z_f)) \end{aligned} \tag{4}$$

$BMF_f(\cdot)$, $BMF_{sq}(\cdot)$ denote the expansion and aggregation components of the bilinear fusion module, respectively.

We employ a cascade of multiple BMF blocks, followed by concatenation of the resulting features. The cascaded BMF block is able to fuse multi-modal features more effectively can be described as, and its structure is described as:

$$Z = BMF^p = [z^1, z^2, ..., z^n] \tag{5}$$

Where z^i denotes the output of the i-th block and p represents the number of block cascades.

Hybrid CNN-Attention Fusion Module. In order to address the issues of alignment and global feature extraction, incorporating an attention mechanism into the model can enhance its ability to effectively learn relevant image regions, thereby facilitating more efficient information capture and improving the fusion effect.Inspired by the fusion of image and question in VQA technology, we further use asymmetric deep fusion in bilinear fusion, considering that LiDAR data and Camera data are not completely symmetric and represent images from different viewpoints. Specifically, we apply a cooperative multi-frequency transformer attention mechanism to perform cross-attention on the fused features obtained through bilinear fusion in BEV space.

As shown in Fig. 2, firstly, patch embedding is performed on the LiDAR features to obtain the query. Subsequently, the fused features obtained from the preceding module are subjected to further processing in order to derive the Key and Value components required for attention mechanism. The improved effective multi-head cross-attention module is used to project the LiDAR features on the fusion features, and the fusion feature Z_f and the LiDAR feature F^{LiDAR} are input. The efficient multi-head cross-attention operator can be expressed as follows:

$$\begin{aligned} E-MHCA(Z, F^{LiDAR}) =& Concat(CA_1(Q_{l1}^{(i,j,k)}), K_1^{(i,j,k)}, V_1^{(i,j,k)}), .., \\ & CA_h(Q_{lh}^{(i,j,k)}, K_h^{(i,j,k)}, V_h^{(i,j,k)}))W^P \end{aligned} \tag{6}$$

W^Prepresents the weight matrix, $Q_{li}^{(i,j,k)}$, $K_i^{(i,j,k)}$, $V_i^{(i,j,k)}$ denote the query, key, and value of the i-th head after multi-head division in the channel dimension. The query is derived from the lidar features, while the key and value are extracted from the fused features Z. These are used as inputs to the cross-attention operator $CA(\cdot)$, which is expressed as follows:

$$CA(Q, K, V) = Softmax\left(\frac{(QW^q)\, P_s\, (KW^k)^T}{\sqrt{d}}\right) P_s\left(\, VW^V\right) \tag{7}$$

d denotes scaling factor, QKV denotes the query, key, and value obtained above, and P_s refers to the average pooling operation with step size s. This operation is utilized for downsampling the spatial dimension before performing attention operations in order to reduce computational costs.

The attention block may partially degrade high-frequency information, such as local texture information [19]. Therefore, we have incorporated a CNN module into the collaborative fusion module to work in conjunction with the E-MHCA module for capturing multi-frequency signals. Subsequently, the output features of CNN-Attention are aggregated and further refined by MLP layers to extract more important and unique features before bilinear fusion of the features obtained by projection with the fusion. This method alleviates the memory consumption of transformer to a certain extent, while facilitating more effective integration of the two modalities at a deep level from a global perspective. The formula for the design of our CNN-Attention combination module is as follows:

$$\begin{aligned} &\dot{z}^l, Q_{LiDAR} = Proj(F^m), PE(F^{LiDAR}), \\ &\hat{z}^l = Proj(E-MHCA(\dot{z}^l, Q_{LiDAR}) + \dot{z}^l), \\ &\tilde{z}^l = CBR(\hat{z}^l) + \hat{z}^l, \\ &z^l = MLP(concat(\tilde{z}^l, \hat{z}^l)) + concat(\tilde{z}^l, \hat{z}^l) \end{aligned} \tag{8}$$

$Proj(\cdot)$ represents the projection function, $\hat{z}^l$,$\tilde{z}^l$ and z^l denote $E-MHCA$, CBR, and MLP output respectively. They are utilized in the channel projection point convolution layer with $PE(\cdot)$ denotes patch embedding block and Q_{LiDAR} denotes the query extracted from the lidar features. $E-MHCA(\cdot)$ is an efficient multi-head cross-attention module, which downsamples the spatial dimension before performing attention operations to reduce computational cost. $CBR(\cdot)$ is a Convolution module composed of Convolution, BatchNorm, and Relu.

4 Experiments

4.1 Implementation Details

Our network is constructed based on the bevfusion. The Camera and LiDAR branches adopt swin-T, VoxelNet [27], and FPN [15] as the Backbone to fuse multi-scale Camera features. The Decoder acts as the second stage and TransFusion is used as the detection header at the end. We downsample Camera images to 256×704 and voxelize the LiDAR point cloud with a resolution of 0.075m [16]. In addition, We use the trained LiDAR object detection model as the pre-trained model. It is worth mentioning that we do not use any additional tricks such as data augmentation or TTA in our model. The Pytorch deep learning framework is employed and the programming is implemented with Python. To assess the efficacy of model refinement, in line with BEVFusion, the approach is evaluated on the nuScenes dataset. It is widely used in the field of autonomous driving. It contains 1000 different driving scenes extracted from a spinning LIDAR placed on the roof, five long range RADAR sensors and six cameras with different viewpoints. Among them, training scenario, validation scenario and testing scenario account for 75%, 15% and 15% respectively [3]. We used nuScenes Detection Score (NDS) and mean Average Precision(mAP) to evaluate the model. Since the BEVFusion baseline model used for comparison differs from our experimental environment, we have adjusted the parameters. Based on the above experimental environment, we use the AdamW [17] optimizer with an initial learning rate

Table 1. Comparison with existing methods on nuScenes validation set.

Method	Modality	LiDAR Backbone	Camera Backbone	mAP	NDS
MVP [23]	C+L	CenterPoint	ResNet-50	66.4	70.5
Transfusion [2]	C+L	VoxelNet	ResNet-50	67.5	71.3
AutoAlignV2 [5]	C+L	VoxelNet	CSPNet	67.1	71.2
UVTR [12]	C+L	VoxelNet	ResNet-101	65.4	70.2
FUTR3D [4]	C+L	VoxelNet	ResNet-101	64.5	68.3
CMT [22]	C+L	VoxelNet	ResNet-50	67.9	70.8
3D dual fusion [9]	C+L	VoxelNet	ResNet-50	69.2	72.2
MSMDFusion [8]	C+L	VoxelNet	ResNet-50	69.3	72.0
BEVFusion [14]	C+L	VoxelNet	Dual-Swin-T	67.9	71.0
BEVFusion-bench [16]	C+L	VoxelNet	Swin-T	68.5	71.4
CoBMFusion(ours)	C+L	VoxelNet	Swin-T	**69.7**	**72.2**

of 2.0e-5 and weight decay of 0.01. The model was trained for 12 epochs on a single Tesla A100 GPU using a batch size of 2.

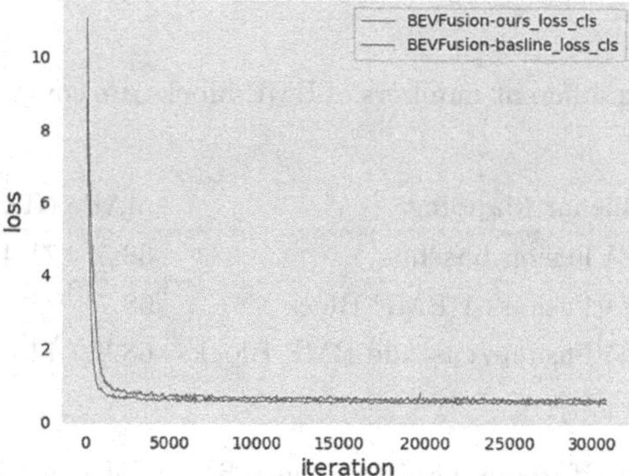

Fig. 3. The training loss of the CoBMFusion model and the BEVFusion model in the same environment in the first epoch.

4.2 State-of-the-Art Comparison

In Table 1, we present the experimental results obtained on the nuScenes validation set in comparison with existing techniques. Without utilizing any data augmentation, TTA or other tricks, our model has achieved exceptional performance with mAP and NDS evaluation criteria of 69.7% and 72.2%, respectively. In comparison to the baseline BEVFusion model used, we have obtained an increase of +1.2% in mAP and +0.8% in NDS through a series of modifications.

Our results outperform the benchmark in the majority category. Among them, truck category increased by 2%, bus category increased by 2.3%, P.E. is increased by 1.2%, motor category increased by 0.6%, bicycle category increased by 2.5%, traffic increases by 1.5%. While our model may not be the Sota model for the nuScenes validation set, it is currently the Sota LiDAR-Camera fusion model in BEV space. Our model retains a simple and effective framework and weakens the dependence of current methods on LiDAR-Camera fusion [16], ensuring independence between modalities while minimizing data-related issues and improving accuracy with minimal complexity. It can be seen that our model maintaining a high degree of competitiveness in the field of fusion in the unified modal representation space.

4.3 Performance Analysis

Training Loss. To more explicitly demonstrate the performance enhancement brought by our improvement, we compare the training loss of the CoBMFusion model with BEVFusion baseline model. By comparing their respective loss functions on the first epoch of 30895 iterations of training on the nuScenes validation dataset. Figure 3 illustrates that our model yields a smaller loss than the benchmark, thereby further validating the effectiveness of our proposed approach.

Table 2. Results using different numbers of BMF blocks are compared on the nuScenes validation set

Billienar Mapping	mAP	NDS
BEVFusion baseline	68.5	71.4
BEVFusion+1 BMF Block	68.7	71.2
BEVFusion+cascade BMF Blocks	68.9	71.5

Ablation Study. To further validate the efficacy of the enhanced model, we have conducted ablation experiments on the fusion module. Table 2 presents the outcomes of ablation experiments, in which we compare the model with added single-layer or cascaded bilinear fusion to the BEVFusion baseline model. Our results indicate that utilizing BMF for multi-modal feature fusion can enhance the accuracy of 3D detection models. Compared to the common concatenation fusion method, our design utilizing single-layer bilinear mapping can enhance accuracy by 0.2%, while cascaded bilinear mapping can further improve it by 0.4%. These results demonstrate the effectiveness of our approach.

Running Time. We also compare fusion components size and evaluate their inference speed using Frames Per Second(FPS) as a performance metric on NVIDIA A100 GPU. As shown in Table 3(a), our approach outperforms the BEVFusion method in terms of performance, while exhibiting faster processing times than other models utilizing attention mechanisms. It validates excellent trade-off between detection performance and inference speed by our method. Table 3(b) presents the number of parameters and floating-point operations per second(FLOPs) inference, where CBR and dynamic fusion are simple fusion modules of two bevfusion model, sparse&decoder belongs to FUTR3D, and CBMHA is our proposed fusion module.

Table 3. Results of different models on the nuScenes validation set. The inference speed and size are measured on a single NVIDIA A100 GPU.

Model	Input Resolution	mAP	NDS	FPS
FUTR3D [4]	1600 × 900	64.5	68.3	2.5
BEVFusion [14]	1600 × 900	67.9	71.0	0.7
BEVFusion [16]	704 × 256	68.5	71.5	**4.2**
CoBMFusion	704 × 256	**69.7**	**72.2**	3.2

(a) Inference Speed Comparison

Model	Fusion Block	Params	FLOPs
BEVFusion [14]	Dynamic fusion	0.9M	25G
BEVFusion [16]	CBR	0.8M	25G
FUTR3D [4]	sampler & decoder	7.5M	2G
CoBMFusion	CBMHA(Ours)	4.7M	63G

(b) Fusion Component Size Comparison

5 Conclusion

We proposed a new structure for fusing LiDAR and camera features in BEV space. Building upon the BEVFusion baseline, our approach leverages cascaded bilinear mapping and a CNN-Attention hybrid architecture to integrate the two modalities. Furthermore, we introduced an asymmetric deep fusion technique using global projection, leveraging the property of bilinear mapping to capture inter-modality feature differences and enabling more effective and comprehensive fusion of modalities. Our experiments on the nuScenes validation set demonstrate significant improvement over the original BEVFusion benchmark results, establishing our designed fusion structure as highly competitive in multi-modal detection models. However, due to hardware limitations such as video memory and so on, there is still room for improvement in terms of feature channel numbers allocated to the network fusion module. These parameters are subject to future tuning. In the future, we will optimize the model's operators to achieve a lightweight design, enhance running speed, and reduce its size.

References

1. Antol, S., et al.: VQA: visual question answering. In: Proceedings of the IEEE International Conference on Computer Vision, pp. 2425–2433 (2015)
2. Bai, X., et al.: Transfusion: robust lidar-camera fusion for 3D object detection with transformers. In: Proceedings of the IEEE/CVF Conference on Computer Vision and Pattern Recognition, pp. 1090–1099 (2022)
3. Caesar, H., et al.: nuScenes: a multimodal dataset for autonomous driving. In: Proceedings of the IEEE/CVF Conference on Computer Vision and Pattern Recognition, pp. 11621–11631 (2020)
4. Chen, X., Zhang, T., Wang, Y., Wang, Y., Zhao, H.: FUTR3D: a unified sensor fusion framework for 3D detection. arXiv preprint arXiv:2203.10642 (2022)
5. Chen, Z., Li, Z., Zhang, S., Fang, L., Jiang, Q., Zhao, F.: Autoalignv2: deformable feature aggregation for dynamic multi-modal 3D object detection. arXiv preprint arXiv:2207.10316 (2022)
6. Cortes, C., Lawarence, N., Lee, D., Sugiyama, M., Garnett, R.: Advances in neural information processing systems 28. In: Proceedings of the 29th Annual Conference on Neural Information Processing Systems (2015)
7. Guo, J., et al.: CMT: convolutional neural networks meet vision transformers. In: Proceedings of the IEEE/CVF Conference on Computer Vision and Pattern Recognition, pp. 12175–12185 (2022)
8. Jiao, Y., Jie, Z., Chen, S., Chen, J., Ma, L., Jiang, Y.G.: Msmdfusion: fusing lidar and camera at multiple scales with multi-depth seeds for 3D object detection. In: Proceedings of the IEEE/CVF Conference on Computer Vision and Pattern Recognition, pp. 21643–21652 (2023)
9. Kim, Y., Park, K., Kim, M., Kum, D., Choi, J.W.: 3D dual-fusion: dual-domain dual-query camera-lidar fusion for 3D object detection. arXiv preprint arXiv:2211.13529 (2022)
10. Li, H., et al.: Delving into the devils of bird's-eye-view perception: a review, evaluation and recipe. arXiv preprint arXiv:2209.05324 (2022)

11. Li, J., et al.: Next-ViT: next generation vision transformer for efficient deployment in realistic industrial scenarios. arXiv preprint arXiv:2207.05501 (2022)
12. Li, Y., Chen, Y., Qi, X., Li, Z., Sun, J., Jia, J.: Unifying voxel-based representation with transformer for 3D object detection. arXiv preprint arXiv:2206.00630 (2022)
13. Li, Y., et al.: Voxel field fusion for 3D object detection. In: Proceedings of the IEEE/CVF Conference on Computer Vision and Pattern Recognition, pp. 1120–1129 (2022)
14. Liang, T., et al.: Bevfusion: a simple and robust lidar-camera fusion framework. arXiv preprint arXiv:2205.13790 (2022)
15. Lin, T.Y., Dollár, P., Girshick, R., He, K., Hariharan, B., Belongie, S.: Feature pyramid networks for object detection. In: Proceedings of the IEEE Conference on Computer Vision and Pattern Recognition, pp. 2117–2125 (2017)
16. Liu, Z., et al.: Bevfusion: multi-task multi-sensor fusion with unified bird's-eye view representation. arXiv preprint arXiv:2205.13542 (2022)
17. Loshchilov, I., Hutter, F.: Decoupled weight decay regularization. arXiv preprint arXiv:1711.05101 (2017)
18. Nabati, R., Qi, H.: Centerfusion: center-based radar and camera fusion for 3D object detection. In: Proceedings of the IEEE/CVF Winter Conference on Applications of Computer Vision, pp. 1527–1536 (2021)
19. Park, N., Kim, S.: How do vision transformers work? arXiv preprint arXiv:2202.06709 (2022)
20. Vaswani, A., et al.: Attention is all you need. In: Advances in Neural Information Processing Systems, vol. 30 (2017)
21. Wang, Y., Guizilini, V.C., Zhang, T., Wang, Y., Zhao, H., Solomon, J.: DETR3D: 3D object detection from multi-view images via 3D-to-2D queries. In: Conference on Robot Learning, pp. 180–191. PMLR (2022)
22. Yan, J., et al.: Cross modal transformer via coordinates encoding for 3D object dectection. arXiv preprint arXiv:2301.01283 (2023)
23. Yin, T., Zhou, X., Krähenbühl, P.: Multimodal virtual point 3D detection (2021)
24. Yu, Z., Yu, J., Fan, J., Tao, D.: Multi-modal factorized bilinear pooling with co-attention learning for visual question answering. In: Proceedings of the IEEE International Conference on Computer Vision, pp. 1821–1830 (2017)
25. Yu, Z., Yu, J., Xiang, C., Fan, J., Tao, D.: Beyond bilinear: generalized multimodal factorized high-order pooling for visual question answering. IEEE Trans. Neural Netw. Learn. Syst. **29**(12), 5947–5959 (2018)
26. Zhang, Y., Zhang, Q., Hou, J., Yuan, Y., Xing, G.: Bidirectional propagation for cross-modal 3D object detection. arXiv preprint arXiv:2301.09077 (2023)
27. Zhou, Y., Tuzel, O.: Voxelnet: end-to-end learning for point cloud based 3D object detection. In: Proceedings of the IEEE Conference on Computer Vision and Pattern Recognition, pp. 4490–4499 (2018)

CasFormer: Cascaded Transformer Based on Dynamic Voxel Pyramid for 3D Object Detection from Point Clouds

Xinglong Li and Xiaowei Zhang(✉)

School of Computer Science and Technology, Qingdao University, Qingdao, China
xiaowei19870119@sina.com

Abstract. Recently, Transformers have been widely applied in 3-D object detection to model global contextual relationships in point cloud collections or for proposal refinement. However, the structural information in 3-D point clouds, especially to the distant and small objects is often incomplete, leading to difficulties in accurate detection using these methods. To address this issue, we propose a Cascaded Transformer based on Dynamic Voxel Pyramid (called CasFormer) for 3-D object detection from LiDAR point clouds. Specifically, we dynamically spread relevant features from the voxel pyramid based on the sparsity of each region of interest (RoI), capturing more rich semantic information for structurally incomplete objects. Furthermore, a cross-stage attention mechanism is employed to cascade the refined results of the Transformer in stage by stage, as well as to improve the training convergence of transformer. Extensive experiments demonstrate that our CasFormer achieves progressive performance in KITTI Dataset and Waymo Open Dataset. Compared to CT3D, our method outperforms it by 1.12% and 1.27% in the moderate and hard levels of car detection, respectively, on the KITTI online 3-D object detection leaderboard.

Keywords: 3-D object detection · Point clouds · Cascaded network

1 Introduction

In the fields of autonomous driving, intelligent transportation, and robot navigation, accurate perception and understanding of the 3-D objects in the surrounding environment are crucial. However, achieving precise, efficient, and robust 3-D object detection remains challenging due to the complexity of the three-dimensional space and the diversity of objects.

In previous studies, researchers have proposed various 3-D object detection methods based on laser point clouds. These methods can be categorized into single-stage detectors [2,18,21] and two-stage detectors [5–7,9,16] based on the steps of generating bounding boxes. In the two-stage detectors, RCNN is used to generate coarse candidate boxes, which are then refined to generate the final detection results with higher accuracy. Recently, with the success of Transformer [3] in various domains, many works have incorporated Transformers into 3-D object detection to model global contextual relationships or refine proposals in point cloud collections.

Q. Liu et al. (Eds.): PRCV 2023, LNCS 14427, pp. 299–311, 2024.
https://doi.org/10.1007/978-981-99-8435-0_24

However, In LiDAR point clouds, distant and small objects are characterized by a very limited number of points, resulting in highly incomplete appearance information. Therefore, performing only single-stage proposal refinement is insufficient to accurately detect them. In 2-D object detection, some networks have addressed these issues by adopting a cascaded structure, which has demonstrated excellent performance. Cascade R-CNN [22] ensures the quality of subsequent proposals by gradually increasing the IoU threshold at each stage. However, in 3-D scenes, the predicted bounding boxes for these objects have very low IoU values. If the cascaded structure with a progressive increase of the threshold is directly applied, the detector would gradually neglect these objects.

Based on the above analysis, we propose a Cascaded Transformer based on Dynamic Voxel Pyramid (CasFormer). CasFormer consists of a Region Proposal Network (RPN) and a Cascaded Transformer Refinement (CTR) network. Specifically, unlike the refinement networks of most existing two-stage detectors, our CTR adopts a cascaded structure. We introduce more connections between stages to progressively improve and complement the proposals, instead of solely using the output of one stage to guide the refinement in the next stage. Additionally, we design a cross-stage attention module to aggregate features from different stages, enhancing the feature representation capability of each stage and enabling comprehensive refinement of region proposals.

To validate the performance of our method across different scenarios and object categories, we conducted a series of experiments and evaluations, comparing it with existing state-of-the-art approaches. Our contributions can be summarized as follows:

- We designed a dynamic voxel pyramid that dynamically extracts features from multiple 3D convolutional layers based on the sparsity of non-empty voxels within the RoI and extends these RoIs to capture more information.
- We propose a cascaded Transformer refinement network that utilizes cross-stage attention mechanisms to progressively aggregate the refinement results from each cascaded stage, aiming to achieve high-quality predictions. Moreover, it achieves better performance with fewer training iterations.
- On the KITTI online 3-D object detection leaderboard, we achieved high detection performance with an average precision (AP) of 82.89% in the moderate car category.

2 Related Work

2.1 CNN-Based 3-D Object Detection

Many existing 3-D object detection methods based on LiDAR point clouds rely on CNNs to extract features for object detection. These methods can be categorized into single-stage detectors and two-stage detectors based on the process of generating bounding boxes, and the two-stage detector usually has higher accuracy. In these detector, the first stage utilizes a RPN to extract region proposals of interest from the input data. The second stage performs refined classification and position adjustment on these region proposals. Typically, these modules operate on the feature regions extracted around the region proposals. Point R-CNN [17] employs a bottom-up approach in the first stage to generate coarse region proposals, followed by region pooling to expand the proposal scale and aggregate RoI features. Voxel R-CNN [9] represents 3-D volumes as voxel centers and extracts RoI features from the voxels using voxel RoI pooling. PV-RCNN [16] introduces an encoding scheme from voxels to keypoints and a multi-scale RoI feature abstraction layer, aggregating RoI features within grid points with multi-scale features based on voxels and points.

These methods aim to improve the accuracy and robustness of 3-D object detection by incorporating two-stage frameworks and utilizing various techniques such as feature aggregation, region pooling, and transformer modules to refine proposals and enhance the representation capabilities of the models.

2.2 Transformer-Based 3-D Object Detection

Transformer [3] have demonstrated outstanding performance in various computer vision tasks, and significant efforts have been made to adapt transformer to 3-D object detection. Pointformer [8] introduces transformer modules to the point backbone to learn context-aware regional features. VoTr [11] replaces the convolutional voxel backbone with transformer modules, proposing sparse voxel and submanifold voxel attention and applying them to voxels. CT3D [5] proposes a channel-wise attention-based transformer detection head for proposal refinement. BEVFormer [23] generates dense queries from the Bird's Eye View (BEV) grid. SST [12] presents a sparse transformer that groups voxels in local regions into patches and applies sparse region attention to the voxels within each patch. SWFormer [14] improves upon SST by enhancing multi-scale feature fusion and voxel diffusion.

These methods leverage transformer to capture contextual information, refine proposals, and enhance feature representation, leading to improved performance in accurately detecting and recognizing 3-D objects in computer vision tasks.

3 Methods

3.1 Overview

Our CasFormer is a two-stage detector, as shown in Fig. 1. We employ a sparse convolution similar to SECOND [21] as the 3D backbone network of our network, generating voxel feature maps with downsampling factors of 2×, 4×, and 8×, represented as $f^{(2)}$, $f^{(3)}$, and $f^{(4)}$, respectively. Following recent research, we compress feature map $f^{(4)}$ along the height dimension to form a BEV map and apply a 2D backbone on it to generate anchors. Then, for each anchor on the feature map, we predict the corresponding class, orientation, and initial 3-D box. Specifically, we propose a Cascaded Transformer Refinement (CTR) network for box refinement in the second stage. Our CTR is a cascaded structure, where each stage first utilizes a Dynamic Voxel Pyramid module (DVP) to dynamically select the required feature levels for each RoI and aggregates the features of each RoI through a pooling operation. The obtained features are fed into a Transformer for initial refinement, which consists of three encoder layers and one decoder layer. In particular, at each stage, we collect the results of preliminary refinement by Transformer from all previous stages and aggregate them using a Cross-Stage Attention (CSA) to guide the generation of high-quality proposals in the current stage. Finally, we gather the predicted boxes from all stages, perform non-maximum suppression (NMS) to obtain the final prediction results. The specific components will be introduced in Sects. 3.2 and 3.3.

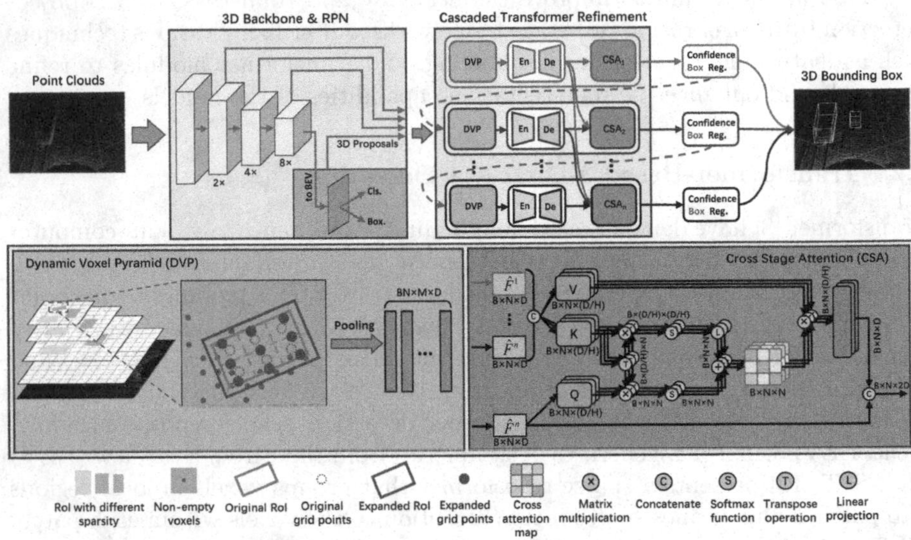

Fig. 1. Our network structure for CasFormer. It generates proposals of 3-D bounding boxes using a voxel-based RPN, which consists of a 3-D backbone and a 2-D detection head. The proposals are progressively refined by a CTR, in which the features from different stages are aggregated by CSAs.

3.2 Cascaded Transformer Refinement

Dynamic Voxel Pyramid. In DVP, we first map each RoI back to the original voxel volume and calculate the non-empty voxel density ρ_{roi}^i within each RoI. Then, based on the density, we determine from which voxel feature layers to extract the features for this RoI. The density ρ_{roi}^i is calculated using the following formula:

$$\rho_{roi}^i = \frac{N_{non}^i}{N_{all}^i}, i = 1, ..., N \tag{1}$$

Where N is the total number of RoIs, N_{non}^i represents the number of non-empty voxels within the i-th RoI, and N_{all}^i is the total number of voxels within the i-th RoI. Next, we set two thresholds, θ_1 and θ_2. If $\rho_{roi}^i \geq \theta_1$, we only extract this RoI feature from $f^{(4)}$. If $\theta_2 \leq \rho_{roi}^i < \theta_1$, we only extract this RoI feature from $f^{(3)}$ and $f^{(4)}$. If $\rho_{roi}^i < \theta_2$, we extract this RoI feature from the last three layers of the 3D Backbone. In this paper, we set θ_1 to 0.7 and θ_2 to 0.4.

We partition the RoI into a regular grid of size G × G × G. For a grid point p_{grid}^{i}, its coordinates are represented as $\{v_p = (x_p, y_p, z_p)\}$ and its features are represented as $\{\varphi_p\}$. Among them, v_p represents the 3-D coordinates of the voxel where p_{grid}^{i} is located, and φ_p represents the features corresponding to that voxel. Specifically, we refer to Pyramid R-CNN [10] and introduce a set of expansion factors $\lambda = [\lambda_1, \lambda_2, \lambda_3]$. This expansion factor is used to enlarge the scope of each RoI, aiding in capturing rich semantic information beyond the RoI. The coordinates of the grid points within the expanded RoI are calculated using the following formula:

$$p_{grid}^{i} = (x_c, y_c, z_c) + ((w, l, h) + 0.5) \cdot \left(\frac{\lambda W}{N_w} \frac{\lambda L}{N_l}, \frac{H}{N_h}\right) \tag{2}$$

Where (W, L, H) represents the width, length, and height of the RoI, respectively. (x_c, y_c, z_c) denotes the coordinates of the bottom-left corner of the RoI. (w, l, h) represents the indices of the grid point along the width, length, and height directions when the bottom-left corner is taken as the origin. (N_w, N_l, N_h) represents the grid size.

Finally, we adopt Voxel RoI Pooling, inspired by Voxel R-CNN [9], to aggregate and align RoI features across different feature layers. After this operation, the feature representation is denoted as $f_{roi} = [f_{roi}^{1}, ..., f_{roi}^{N}] \in \mathbb{R}^{BN \times M \times D}$. Where B represents the batch size, N represents the number of RoIs in a batch, M represents the number of grid points for each RoI, and D represents the feature dimension of the grid points. Next, we input f_{roi} as tokens into a Transformer to refine our proposals.

Cross-Stage Attention. After passing through the decoder in the Transformer, the output feature is denoted as $\hat{F} \in \mathbb{R}^{B \times N \times D}$, which can be directly connected to two Feed-Forward Networks (FFNs) for classification and bounding box regression. However, the shape of distant small objects is highly incomplete, so it is challenging to achieve accurate detection with just one refinement. The analysis results can be found in the ablation experiment results Table 6.

Based on the analysis mentioned in the introduction, we designed a novel cascaded structure to address this issue. In our structure, the proposal refinement results from each stage are not only used to guide the refinement in the next stage but also directly involved in the selection of the final high-quality boxes. In addition, we introduce a cross-stage attention module to aggregate features from different stages in order to more accurately detect such objects.

In the j-th stage of the cascaded structure, we represent the feature vector obtained after decoding in the Transformer as $\hat{F}^{j} \in \mathbb{R}^{B \times N \times D}$. Specifically, we collect the decoding vectors from the current stage and all previous stages, denoted as $\mathbf{F}^{j} = \left[\hat{F}^{1}, \hat{F}^{2}, ..., \hat{F}^{j}\right] \in \mathbb{R}^{BJ \times N \times D}$, where J represents the current stage number. Then, we project the shape of $\mathbf{F}^{j}$ through a linear layer into $B \times N \times D$ and feed it into the subsequent attention module. Then we have $\mathbf{Q}^{j} = \hat{F}^{j}\mathbf{W}_{q}^{j}$, $\mathbf{K}^{j} = \mathbf{F}^{j}\mathbf{W}_{k}^{j}$, and $\mathbf{V}^{j} = \mathbf{F}^{j}\mathbf{W}_{v}^{j}$, where $\mathbf{W}_{q}^{j}$, $\mathbf{W}_{k}^{j}$, and $\mathbf{W}_{v}^{j}$ are linear projections. $\mathbf{Q}^{j}$, $\mathbf{K}^{j}$, and $\mathbf{V}^{j}$ represent query embedding, key embedding, and value embedding, respectively. In the case of H-head attention, the embedding of the h-th head are denoted as $\mathbf{Q}_h^j$, $\mathbf{K}_h^j$, and $\mathbf{V}_h^j$. Specifically, we first compute the matrix multiplication of query embeddings and key embeddings. Then, we calculate the correlations between each channel of the feature dimension, and finally, sum them up to obtain our weights. The attention value for a single head is computed according to the following formula:

$$\hat{F}_h^j = \left[\sigma \left(\frac{Q_h^j \left(K_h^j\right)^T}{\sqrt{D}} \right) \oplus \Lambda \left(\sigma \left(\frac{\left(K_h^j\right)^T \left(K_h^j\right)}{\sqrt{D}} \right) \right) \right] \cdot V_h^j, h = 1, ..., H \quad (3)$$

where $\sigma(\cdot)$ is the softmax function, $\Lambda(\cdot)$ is a linear projection that maps weights of dimensions D/H to dimension N, h represents the attention head index in multi-head attention and $\oplus$ denotes the summation operation.

We concatenate the feature $\hat{F}^j$ with the outputs of all attention heads to obtain the feature vector $F^j = Concat\left(\hat{F}^j, \hat{\mathbf{F}}_1^j, \hat{\mathbf{F}}_2^j, ..., \hat{\mathbf{F}}_H^j\right) \in \mathbb{R}^{B \times N \times 2D}$. Then, we use a linear layer to project the shape of F^j into $N \times 2D$, and connect two FFNs separately for box regression and confidence prediction in this stage. Finally, we gather the predicted boxes from all stages and apply NMS to remove redundant boxes, resulting in more reliable detection outputs.

3.3 Training Losses

We employ an end-to-end training strategy to train our detector, where the overall training loss is the sum of the RPN loss $\mathcal{L}_{RPN}$ and the CTR loss $\mathcal{L}_{CTR}$. We follow SECOND [21] and design the loss of the RPN as a combination of classification loss $\mathcal{L}_{cls}$ and bounding box regression loss $\mathcal{L}_{reg}$, as shown in the following equation:

$$\mathcal{L}_{RPN} = \frac{1}{N_p} \left[\sum_i \mathcal{L}_{cls}(c_i, \hat{c}_i) + \mathbb{I}(IoU_i > u) \sum_i \mathcal{L}_{reg}\left(\delta_i, \hat{\delta}_i\right) \right] \quad (4)$$

Where N_p represents the number of foreground boxes, c_i and $\hat{c}_i$ represent the outputs of the classification branch and the classification labels, respectively. δ_i and $\hat{\delta}_i$ represent the outputs of the box regression branch and the regression targets, respectively. $\mathbb{I}(IoU_i > u)$ represents that only the object proposals with $IoU_i > u$ generate regression loss. $\mathcal{L}_{reg}$ and $\mathcal{L}_{cls}$ are smooth L1 loss and binary cross-entropy loss, respectively.

Our CTR loss $\mathcal{L}_{CTR}$ is the sum of multiple refinement losses in multiple stages. In each refinement stage, we utilize box regression loss $\mathcal{L}_{reg}$ and classification loss $\mathcal{L}_{cls}$, as in [9,16]. For the i-th proposal in the j-th refinement stage, we denote the outputs of the classification branch, the classification labels, the outputs of the box regression branch, and the regression targets as c_i^j, $\hat{c}_i^j$, δ_i^j, and $\hat{\delta}_i^j$, respectively.The loss of CTR is defined as:

$$\mathcal{L}_{CTR} = \frac{1}{N_{all}} \left[\sum_i \sum_j \mathcal{L}_{cls}\left(c_i^j, \hat{c}_i^j\right) + \mathbb{I}\left(IoU_i^j > u^j\right) \sum_i \sum_j \mathcal{L}_{reg}\left(\delta_i^j, \hat{\delta}_i^j\right) \right] \quad (5)$$

Where N_{all} represents the sum of the number of foreground frames in all stages, and $\mathbb{I}\left(IoU_i^j > u^j\right)$ represents that only the object proposals with $IoU_i^j > u^j$ generate regression loss.

4 Experiments

4.1 Dataset

KITTI Dataset. The KITTI dataset [1] consists of 7481 frames in the training set and 7518 frames in the test set. Following previous works, we divide the training data into 3712 frames for training and 3769 frames for validation. We report the results on the validation set using 3D AP at 11 recall thresholds and at a recall threshold of 40. The IoU thresholds used for cars, pedestrians, and cyclists are 0.7, 0.5, and 0.5 respectively.

Waymo Open Dataset. The Waymo Open Dataset [13] consists of 1000 sequences, including 798 training sequences and 202 validation sequences. The official evaluation metric for 3D detection is the mean Average Precision (mAP), with an IoU threshold of 0.7 for measuring the detection performance of the vehicle category. The evaluation results are divided based on the distance from the object to the sensor, specifically 0–30 m, 30–50 m, and greater than 50 m.

Table 1. Detection results comparisons with recent methods on the KITTI *val* set for car detection. AP_{3D} is calculated with 0.7 IoU threshold on 11 recall positions and 40 recall positions. The best result is displayed in red, the second best in blue, and the third best in green.

Methods	Reference	AP_{3D} (R11)(%)			AP_{3D} (R40)(%)		
		Easy	Mod.	Hard	Easy	Mod.	Hard
PV-RCNN [16]	CVPR2020	-	83.90	-	92.57	84.83	82.69
Voxel R-CNN [9]	AAAI2021	89.41	84.52	78.93	92.38	85.29	82.86
CT3D [5]	ICCV2021	89.54	86.06	78.99	92.85	85.82	83.46
Pyramid-V [10]	ICCV2021	88.44	83.14	78.61	-	-	-
VoTr-TSD [11]	ICCV2021	89.04	84.04	78.68	-	-	-
BtcDet [15]	AAAI2022	-	86.57	-	93.15	86.28	83.86
Focals Conv [20]	CVPR2022	89.52	84.93	79.18	-	-	-
Graph-Vo [7]	ECCV2022	-	-	-	93.33	86.12	83.29
CasA+V [6]	TGRS2022	**89.88**	86.58	**79.38**	93.21	**86.37**	83.93
LoGoNet [19]	CVPR2023	-	-	-	92.04	85.04	**84.31**
OcTr [4]	CVPR2023	89.80	**86.97**	79.28	-	-	-
CasFormer	Ours	89.96	87.04	79.87	**93.24**	86.94	84.55

4.2 Implementation Details

Voxelization. After inputting the raw point cloud into our detector, the first step is to divide the point cloud into regular voxels. Since the KITTI dataset only provides object annotations within the FOV, we set the detection range as follows: X-axis [0, 70.4 m], Y-axis [−40 m, 40 m], and Z-axis [−3 m, 1 m]. Each voxel is sized (0.05 m, 0.05 m, 0.1 m). For the Waymo Open Dataset, the detection distances for the X and Y axes are [−75.2 m, 75.2 m], and the detection distance for the Z-axis is [−2 m, 4 m]. Each voxel is sized (0.1 m, 0.1 m, 0.15 m).

Training and Inference Details. Our network is optimized end-to-end using the Adam optimizer. The initial learning rate is set to 0.001, and we train the network for 100 epochs with updates performed using a cosine annealing strategy. We randomly sampled 512 RoIs as training samples for the detection head. During the inference stage, we retained the top 100 proposals for cascaded refinement and applied NMS with an IoU threshold of 0.1 to filter the bounding boxes and generate the final detection results.

4.3 Detection Results on the KITTI

Table 1 and Table 2 present the performance comparison of our CasFormer with state-of-the-art methods on the vehicle detection task using the KITTI [1] vali-

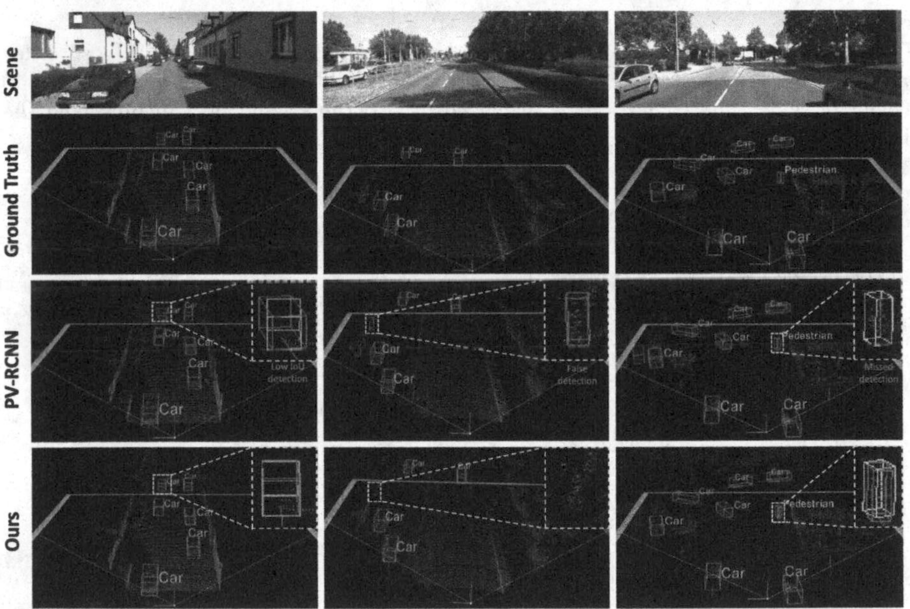

Fig. 2. Illustration of the 3-D detection result on the KITTI *val* set. The ground truth for the vehicle and pedestrian categories are shown in green and blue, respectively, and their prediction bounding boxes are shown in red. (Color figure online)

dation and test sets. At 40 recall thresholds, CasFormer outperforms CT3D [5] by 1.12% and 1.09% on the moderate and hard level of the validation set, respectively. Additionally, at 11 recall thresholds on the hard level of the validation set, CasFormer achieves a performance gain of 0.59% and 0.49% over OcTr [4] and CasA+V [6], respectively. Furthermore, despite the incorporation of the cascaded structure and the relatively slower-converging Transformer [3] module in our CasFormer, its inference speed remains faster than or on par with most methods under the same experimental conditions. Figure 2 showcases the recognition results of our model on the KITTI dataset. Compared to PV-RCNN [16], our model exhibits superior recognition capabilities.

Table 2. Detection results comparisons with recent methods on the KITTI *test* set for car detection. AP_{3D} is calculated with 0.7 IoU threshold on 40 recall positions. The rules for color display of the data are the same as in Table 1.

Methods	Reference	AP_{3D} (R40)(%)			FPS (Hz)
		Easy	Mod.	Hard	
PV-RCNN [16]	CVPR2020	90.25	81.43	76.82	12.5
Voxel R-CNN [9]	AAAI2021	90.90	81.62	77.06	25.2
CT3D [5]	ICCV2021	87.83	81.77	77.16	14.3
Pyramid-V [10]	ICCV2021	87.06	81.28	76.85	9.7
VoTr-TSD [11]	ICCV2021	89.90	82.09	**79.14**	7.2
Focals Conv [20]	CVPR2022	90.20	82.12	77.50	8.9
Graph-Vo [7]	ECCV2022	91.29	82.77	77.20	25.6
CasA+V [6]	TGRS2022	**91.58**	83.06	80.08	11.6
OcTr [4]	CVPR2023	90.88	82.64	77.77	15.6
CasFormer	Ours	91.65	**82.89**	78.43	14.9

4.4 Detection Results on the Waymo

We also conduct experiments on the Waymo Open Dataset to further validate the effectiveness of our CasFormer. Objects in the Waymo Open Dataset are divided into two levels based on the number of points per individual object. LEVEL 1 objects have more than 5 points, while LEVEL 2 objects have 1$\sim$5 points. As shown in Table 3, CasFormer outperforms CT3D by 0.27% 3D mAP on LEVEL 1 30–50 m. And CasFormer outperforms Voxel R-CNN by 1.84% on LEVEL 2 50m-Inf, which indicates that our method can extract richer semantic features with more local context information for sparse objects at a distance.

Table 3. Detection results comparisons with recent methods on the Waymo Open Dataset with 202 validation sequences (∼40k samples) for the vehicle detection. The best result is displayed in red, the second best in blue.

Difficulty	Method	3D mAP(%)			
		Overall	0-30m	30-50m	50m-Inf
LEVEL 1	PointPillars [2]	56.62	81.01	51.75	27.94
	PV-RCNN [16]	70.30	91.92	69.21	42.17
	Voxel-RCNN [9]	75.59	92.49	74.09	53.15
	Pyramid-V [10]	75.83	**92.63**	74.46	53.40
	CT3D [5]	76.30	92.51	**75.07**	**55.36**
	CasFormer (Ours)	**76.13**	92.67	75.34	55.40
LEVEL 2	PV-RCNN [16]	65.36	91.58	65.13	36.46
	Voxel-RCNN [9]	66.59	91.74	67.89	40.80
	CT3D [5]	**69.04**	**91.76**	68.93	**42.60**
	CasFormer (Ours)	69.11	91.77	**68.86**	42.64

4.5 Ablation Studies

In this section, we conducted comprehensive ablation studies on CasFormer to validate the effectiveness of each individual component. All our experiments were conducted on the KITTI *val* set, evaluating the Average Precision (AP) for the car category at 40 recall positions with an IoU threshold of 0.7. In the table below, we show the best results in red.

The Effects of Dynamic Voxel Pyramid. In this section, we validate the effectiveness of the DVP module. As shown in Table 4, method (a) serves as our baseline, while methods (b)–(d) represent different RoI feature selection approaches within method (a). It can be observed that Method (d) with the inclusion of DVP outperforms the other two methods. This is attributed to our

Table 4. Ablation studies for different feature selection methods. "NP", "EP", "DVP" stand for non-expanding pyramid, expanding pyramid, and our dynamic voxel pyramid, respectively.

Methods	NP	EP	DVP	AP_{3D} (R40)(%)		
				Easy	Mod.	Hard
(a)	✗	✗	✗	92.38	85.29	82.86
(b)	✓	✗	✗	92.41	85.42	82.99
(c)	✗	✓	✗	92.49	86.53	83.11
(d)	✗	✗	✓	92.55	85.61	83.23

DVP, which dynamically selects feature layers based on the sparsity of RoIs, thereby capturing rich contextual information around RoIs with low non-empty voxel density.

The Effects of Cascaded Strategy. In this section, our experiments were conducted on three cascaded stages. As shown in Table 5, methods (a)–(c) represent the use of different cascaded strategy. It can be observed that method (c) achieves a performance improvement of 0.81% and 0.84% over methods (a) in the moderate and hard levels, respectively. This improvement is attributed to the capability of our CSA module to better capture the correlations between features across different stages, enabling achieve better object bounding box regression.

Table 5. Ablation studies for different feature aggregation strategy. "DC" and "SA" represent direct concatenation and standard attention, respectively.

Methods	DC	SA	CSA	AP_{3D} (R40)(%)		
				Easy	Mod.	Hard
(a)	✓	✗	✗	92.62	86.13	83.71
(b)	✗	✓	✗	92.97	86.62	84.26
(c)	✗	✗	✓	**93.24**	**86.94**	**84.55**

The Effects of Cascaded Stages. In this section of the experiment, we tested the effect of the number of cascading stages. We employed CSA as the cascading strategy, and the experimental results are shown in Table 6. From the data in the table, it can be observed that the best performance was achieved with three cascading stages at the moderate and hard levels. At the easy level, the best performance was obtained with two cascading stages. Although the performance of these two configurations was very similar, the detector with three cascading stages achieved excellent performance with fewer epochs. Therefore, our detector utilizes three cascading stages.

Table 6. Ablation studies for different cascade stages on the KITTI *val* set.

Stages	Epoch	AP_{3D} (R40)(%)		
		Easy	Mod.	Hard
1	94	93.02	86.35	84.04
2	87	**93.44**	86.71	84.35
3	83	93.24	**86.94**	**84.55**
4	82	93.21	86.89	84.51

5 Conclusion

In this paper, we proposed a two-stage detector named CasFormer, which performs 3-D object detection from point clouds. In our CasFormer, we introduced a cascaded transformer refinement network for the refinement stage. Firstly, we employed a dynamic pyramid module to capture more semantic information from additional feature sources for RoIs with low non-empty voxel density. Then, a Transformer module was applied for the initial refinement. Furthermore, an attention module was used to aggregate the refinement results from different stages, generating high-quality proposals for that stage. Extensive experiments demonstrated the effectiveness of our CasFormer. However, our method still has limitations in real-time detection. In future work, we will focus on improving the computational efficiency of the detection framework.

References

1. Geiger, A., Lenz, P., Urtasun, R.: Are we ready for autonomous driving? The kitti vision benchmark suite. In: 2012 IEEE Conference on Computer Vision and Pattern Recognition, pp. 3354–3361. IEEE (2012)
2. Lang, A.H., Vora, S., Caesar, H., Zhou, L., Yang, J., Beijbom, O.: Pointpillars: fast encoders for object detection from point clouds. In: Proceedings of the IEEE/CVF Conference on Computer Vision and Pattern Recognition, pp. 12697–12705 (2019)
3. Vaswani, A., et al.: Attention is all you need. In: Advances in Neural Information Processing Systems, vol. 30 (2017)
4. Zhou, C., Zhang, Y., Chen, J., Huang, D.: OcTr: octree-based transformer for 3D object detection. In: Proceedings of the IEEE/CVF Conference on Computer Vision and Pattern Recognition, pp. 5166–5175 (2023)
5. Sheng, H., et al.: Improving 3D object detection with channel-wise transformer. In: Proceedings of the IEEE/CVF International Conference on Computer Vision, pp. 2743–2752 (2021)
6. Wu, H., Deng, J., Wen, C., Li, X., Wang, C., Li, J.: Casa: a cascade attention network for 3-D object detection from lidar point clouds. IEEE Trans. Geosci. Remote Sens. **60**, 1–11 (2022)
7. Yang, H., et al.: Graph R-CNN: towards accurate 3D object detection with semantic-decorated local graph. In: Avidan, S., Brostow, G., Cissé, M., Farinella, G.M., Hassner, T. (eds.) ECCV 2022, Part VIII. LNCS, vol. 13668, pp. 662–679. Springer, Cham (2022). https://doi.org/10.1007/978-3-031-20074-8_38
8. Zhao, H., Jiang, L., Jia, J., Torr, P.H., Koltun, V.: Point transformer. In: Proceedings of the IEEE/CVF International Conference on Computer Vision, pp. 16259–16268 (2021)
9. Deng, J., Shi, S., Li, P., Zhou, W., Zhang, Y., Li, H.: Voxel R-CNN: towards high performance voxel-based 3D object detection. In: Proceedings of the AAAI Conference on Artificial Intelligence, vol. 35, no. 2, pp. 1201–1209 (2021)
10. Mao, J., Niu, M., Bai, H., Liang, X., Xu, H., Xu, C.: Pyramid R-CNN: towards better performance and adaptability for 3D object detection. In: Proceedings of the IEEE/CVF International Conference on Computer Vision, pp. 2723–2732 (2021)
11. Mao, J., et al.: Voxel transformer for 3D object detection. In: Proceedings of the IEEE/CVF International Conference on Computer Vision, pp. 3164–3173 (2021)

12. Fan, L., et al.: Embracing single stride 3D object detector with sparse transformer. In: Proceedings of the IEEE/CVF Conference on Computer Vision and Pattern Recognition, pp. 8458–8468 (2022)
13. Sun, P., et al.: Scalability in perception for autonomous driving: waymo open dataset. In: Proceedings of the IEEE/CVF Conference on Computer Vision and Pattern Recognition, pp. 2446–2454 (2020)
14. Sun, P., et al.: Swformer: sparse window transformer for 3D object detection in point clouds. In: Avidan, S., Brostow, G., Cissé, M., Farinella, G.M., Hassner, T. (eds.) ECCV 2022, Part X. LNCS, vol. 13670, pp. 426–442. Springer, Cham (2022). https://doi.org/10.1007/978-3-031-20080-9_25
15. Xu, Q., Zhong, Y., Neumann, U.: Behind the curtain: learning occluded shapes for 3D object detection. In: Proceedings of the AAAI Conference on Artificial Intelligence, vol. 36, no. 3, pp. 2893–2901 (2022)
16. Shi, S., et al.: PV-RCNN: point-voxel feature set abstraction for 3D object detection. In: Proceedings of the IEEE/CVF Conference on Computer Vision and Pattern Recognition, pp. 10529–10538 (2020)
17. Shi, S., Wang, X., Li, H.: Pointrcnn: 3D object proposal generation and detection from point cloud. In: Proceedings of the IEEE/CVF Conference on Computer Vision and Pattern Recognition, pp. 770–779 (2019)
18. Yin, T., Zhou, X., Krahenbuhl, P.: Center-based 3D object detection and tracking. In: Proceedings of the IEEE/CVF Conference on Computer Vision and Pattern Recognition, pp. 11784–11793 (2021)
19. Li, X., et al.: Logonet: towards accurate 3D object detection with local-to-global cross-modal fusion. In: Proceedings of the IEEE/CVF Conference on Computer Vision and Pattern Recognition, pp. 17524–17534 (2023)
20. Chen, Y., Li, Y., Zhang, X., Sun, J., Jia, J.: Focal sparse convolutional networks for 3D object detection. In: Proceedings of the IEEE/CVF Conference on Computer Vision and Pattern Recognition, pp. 5428–5437 (2022)
21. Yan, Y., Mao, Y., Li, B.: Second: sparsely embedded convolutional detection. Sensors **18**(10), 3337 (2018)
22. Cai, Z., Vasconcelos, N.: Cascade R-CNN: delving into high quality object detection. In: Proceedings of the IEEE Conference on Computer Vision and Pattern Recognition, pp. 6154–6162 (2018)
23. Li, Z., et al.: Bevformer: learning bird's-eye-view representation from multi-camera images via spatiotemporal transformers. In: Avidan, S., Brostow, G., Cissé, M., Farinella, G.M., Hassner, T. (eds.) ECCV 2022, Part IX. LNCS, vol. 13669, pp. 1–18. Springer, Cham (2022). https://doi.org/10.1007/978-3-031-20077-9_1

Generalizable and Accurate 6D Object Pose Estimation Network

Shouxu Fu, Xiaoning Li(✉), Xiangdong Yu, Lu Cao, and Xingxing Li

College of Computer Science, Sichuan Nomal University, Chengdu 610101, China
lxn@sicnu.edu.cn

Abstract. 6D object pose estimation is an important task in computer vision, and the task of estimating 6D object pose from a single RGB image is even more challenging. Many methods use deep learning to acquire 2D feature points from images to establish 2D-3D correspondences, and further predict 6D object pose with Perspective-n-Points (PnP) algorithm. However, most of these methods have problems with inaccurate acquisition of feature points, poor generality of the network and difficulty in end-to-end training of the network. In this paper, we design an end-to-end differentiable network for 6D object pose estimation. We propose Random Offset Distraction (ROD) and Full Convolution Asymmetric Feature Extractor (FCAFE) with the Probabilistic Perspective-n-Points (ProPnP) algorithm to improve the accuracy and robustness of 6D object pose estimation. Experiments show that our method achieves a new state-of-the-art result on the LineMOD dataset, with an accuracy of 97.42% in the ADD(-S) metric. Our approach is also very competitive on the Occlusion LineMOD dataset.

Keywords: 6D Object Pose Estimation · Coordinates-based Method · RGB Data

1 Introduction

6D object pose estimation is a hot problem in computer vision and many application areas need to address this key problem. To improve the safety of autonomous driving and to prevent tailgating, one needs to predict the pose of front vehicle. To achieve automatic gripping, the robot arm also needs to estimate the pose of object to be gripped. 6D object pose estimation, which consists of estimating translation and rotation of the target, typically contains predictions in 6 degrees of freedom. The input data is commonly sourced from RGB images, RGB-D images and point cloud data. Estimating 6D object pose from a single RGB image is more promising and in demand due to the ease of data acquisition and inexpensive equipment, however, it is also more challenging.

Estimating 6D object pose from a single RGB image is generally regarded as a geometric problem. Many methods [10,13,14] usually define some 3D feature points to object, then use deep learning to predict the corresponding 2D feature points in the image, and finally construct 2D-3D correspondences and use the PnP algorithm to predict 6D pose.

Q. Liu et al. (Eds.): PRCV 2023, LNCS 14427, pp. 312–324, 2024.
https://doi.org/10.1007/978-981-99-8435-0_25

Some existing methods [4,10,13,14,17] extract 2D key feature points of object bounding boxes, object edges or dense coordinates by using simple convolutional neural networks as encoders, however, it is difficult for simple convolutional neural networks to extract accurate feature points, which affects the 2D-3D correspondence and ultimately the accuracy of the PnP algorithm to estimate object pose. In addition, the selection of feature points also affects the extraction of convolutional neural network. Many works [13,14] use the RANSAC/PnP algorithm to solve for translation and rotation. However, rotation is non-linear and difficult to train, whereas translation is linear and easy to train. The same treatment reduces the speed of converging translation. In addition, using the PnP algorithm to estimate 6D object pose is a two-stage work and network cannot be trained end-to-end. Some tasks [9,13,14,16] operate from the complete image, which increases the workload of network and reduces the generality of network.

In this work, we select dense coordinates to define feature points and design the Full Convolution Asymmetric Feature Extractor (FCAFE) to extract features of object. For FCAFE, we design a U-like fully convolutional architecture with an asymmetric encoder-decoder structure. FCAFE is able to obtain accurate coordinate maps and pixel maps, which in turn improves the accuracy of pose estimation. We decouple pose estimation problem into translation estimation and rotation estimation. The ProPnP algorithm transforms pose estimation problem into a predictive positional probability density problem, as the probability density function is derivable and therefore the network can be trained end-to-end. So we use the ProPnP algorithm instead of the traditional PnP algorithm for end-to-end differentiable training of network. Our network operates from cropped images, which facilitates both the processing of network and the ability of network to adapt to existing detectors. In order to achieve better generality, we propose ROD, which is based on cropped images for processing and is able to simulate different detectors, thus improving the generality of network and enabling network to work with different detectors. Our contributions are summarized as follows:

(1) We propose an end-to-end differentiable 6D pose estimation network and our method achieves a new state-of-the-art results on the LineMOD dataset, with an accuracy of 97.42% in the ADD(-S) metric.
(2) We propose Random Offset Distraction (ROD) and Full Convolution Asymmetric Feature Extractor (FCAFE) to improve the accuracy of 6D object pose estimation.
(3) Our approach is generalizable and can be used with different detectors for different areas.

2 Related Work

2.1 Direct Methods

The direct method typically classifies 6D object pose estimation as a regression or classification task and uses deep learning to estimate the pose directly. PoseCNN

[18] decomposed 6D pose into translation and rotation, predicting translation by direct regression of the object center coordinates and pose by direct regression of quaternions. However, the shortcoming of PoseCNN is that the accuracy is poor. SSD-6D [9] converted 6D pose estimation into a classification problem and introduced a viewpoint library to estimate 6D pose by matching the viewpoint library. GDR-Net [17] established 2D-3D correspondences directly through neural network to simulate the PnP algorithm and achieved end-to-end 6D object pose estimation.

2.2 Geometric Methods

Keypoint-Based Methods. Such methods construct 2D-3D correspondences by predicting the 2D feature points of object in the image and estimate 6D object pose using the PnP algorithm. BB8 [14] recovered 6D object pose by predicting the 2D projections of 8 border points and center of 3D bounding box, paired with the PnP algorithm, however, this method is a two-stage operation and cannot be trained end-to-end. YOLO-6D [16] detected the 2D projection of 3D bounding box and then used the PnP algorithm to recover 6D pose, which was convenient and fast but less accurate. PVNet [13] used a voting mechanism to obtain 2D keypoints with high accuracy and the ability to estimate 6D pose of obscured or truncated objects. DGECN [3] obtained semantic labels and depth maps to establish 2D-3D correspondences through convolutional neural networks. DGECN was able to train network end-to-end by proposing a learnable DG-PnP algorithm as an alternative to RANSAC/PnP.

Dense Coordinate-Based Methods. Such methods recover 6D object pose by regressing object coordinate points and then constructing 2D-3D dense correspondences with the PnP algorithm. DPOD [20] proposed UV maps to establish 2D-3D correspondences in order to recover 6D object pose. CDPN [10] proposed 3D coordinate maps to achieve dense 2D-3D correspondences, and used the PnP algorithm to recover the 6D object pose. EProPnP [4] proposed the ProPnP algorithm, which introduced a probability density function to transform the traditional PnP problem into a probability problem, solving the problem that the traditional PnP algorithm is not differentiable.

Similar to our method is CDPN, which also decoupled translation and rotation and used direct regression method to estimate translation. However, we design a better feature extraction network called FCAFE to generate more accurate 3-channel coordinate maps as well as two-channel weight maps. We use the EProPnP algorithm to replace the traditional PnP algorithm to enable end-to-end differentiable training of network. In addition, CDPN needs to be trained in several stages, while our method only needs one stage, and the training time of our method is much less than CDPN.

3 Method

The target of our method is to predict the 3-DoF translation and the 3-DoF rotation from a single RGB image. The Overview of our method is shown in

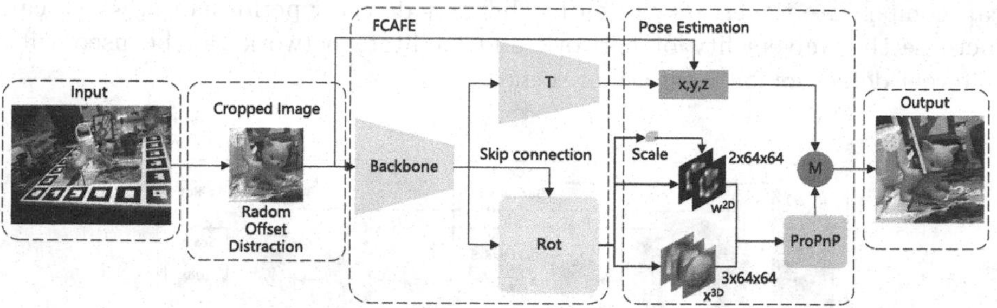

Fig. 1. Overview of our method. Given an input image, the target region is first extracted in RGB image and interfered with by ROD, then the cropped image is fed into FCAFE to extract features. The translation is predicted by direct regression of the coordinate center and the rotation is predicted using the ProPnP algorithm by creating dense 2D-3D correspondences through regression coordinate map and weight map.

Fig. 1. ROD is used for clipped image, and the image in the green box in figure is the new image after adding ROD. The new image is fed into FCAFE to extract features, and translation is predicted by direct regression to the center coordinates of object. For rotation, we first regress 3-channel coordinate maps X^{3D} and 2-channel weight maps W^{2D}, then establish 2D-3D dense correspondences, and finally use the ProPnP algorithm to predict object pose.

3.1 Random Offset Distraction

In most scenes, the target object is far away from the camera and occupies only a small part of the image. It is not easy for network to detect small objects as well as to regress coordinate maps and weight maps. In addition, in order for network to be generalizable, our network needs to be paired with different detectors. However, different detectors have different performance, and it is difficult for network to obtain effective features if the target area extracted by different detectors, which leads to a reduction in the accuracy of estimated pose.

To solve these limitations and improve the accuracy and generality of network, we propose ROD. The image in the second dashed box on the left in Fig. 1 shows the ROD, we randomly modify the center coordinates and size of the detection box, the green box is the modified detection box. Concretely, given the center coordinates of the detection box $C_{x,y}$ and the size S, we use a random number algorithm to obtain a random number that obeys a uniform distribution and set an offset factor to control the size of the random number to obtain a random offset value. The original detection box center coordinates $C_{x,y}$ and size S are added with random offset value to get the offset $C'_{x,y}$ and S', and then the image is recropped in original image.

On the one hand, ROD can improve the robustness of network and increase the accuracy of 6D object pose estimation. On the other hand, this operation can compensate the errors caused by different detector performance, which can increase the universality of network and facilitate network can be used with different detectors for a variety of scenes.

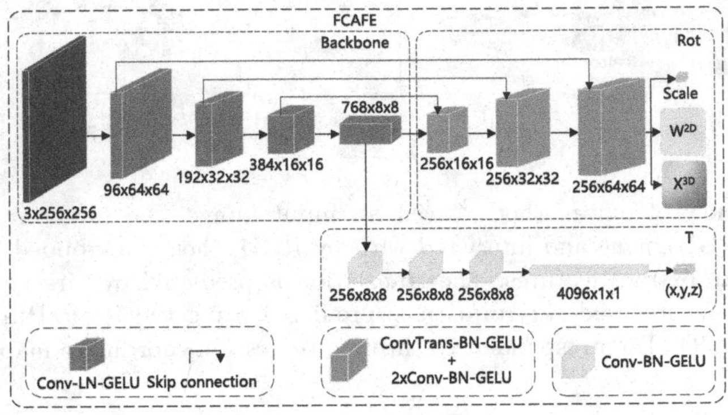

Fig. 2. Full Convolution Asymmetric Feature Extractor.

3.2 Full Convolution Asymmetric Feature Extractor

We propose the Full Convolution Asymmetric Feature Extractor for feature extraction, as shown in the dashed box of FCAFE in Fig. 1, which consists of three parts: Backbone module, T module and Rot module.

Figure 2 shows the detail of FCAFE. We design FCAFE using a U-like fully convolutional architecture with an asymmetric encoder-decoder structure. This asymmetric structure facilitates the generation of feature maps on the one hand and reduces the complexity of network on the other. Backbone module consists of four groups of ConvNextBlock, each outputting feature maps at different scales. ConvNext [11] has better performance compared with ResNet [5], DenseNet [7] and other convolutional neural networks in many upstream tasks, which is suitable for 6D object pose estimation work. Rot module consists of three groups of ConvTrans-BN-GELU, which are used to generate 3-channel coordinate maps, 2-channel weight maps and scaling factors. In order to enhance the feature extraction capability of FCAFE, we output four sizes of feature maps in Backbone module and fuse them into decoder using skip connections. So 3-channel coordinate maps X^{3D} and 2-channel weight maps W^{2D} regressed by Rot module can contain richer low-level information. T module consists of three groups of Conv-BN-GELU for predicting translation of target object.

The 3-channel coordinate maps, 2-channel weight maps, scaling factors and translation of object acquired by FCAFE are used in the Pose Estimation module for further processing.

3.3 Pose Estimation

We decouple 6D pose estimation into translation estimation and rotation estimation. For translation, we predict the object center coordinates based on the cropped image. Concretely, FCAFE obtains detection box center coordinates $C_{x,y}$ and then predicts the relative offset value $(\Delta x, \Delta y)$ between the object center and detection box center. The 2-dimensional translation predicted by FCAFE consists of predicted offset value add center coordinates of detection box. For depth, the z-axis coordinates are predicted by T module.

For rotation, we create 2D-3D dense correspondences and predict it using the ProPnP algorithm by regressing coordinate maps. Concretely, FCAFE in Fig. 1 outputs 3-channel coordinate maps X^{3D}, 2-channel weight maps W^{2D} and global weight scaling factors *Scale*. The 3-channel coordinate maps X^{3D} represents 3-dimensional coordinates of object in the world coordinate system, and 2-channel weight maps W^{2D} represents 2-dimensional pixel coordinates of object with weights. The 2-channel weight maps W^{2D} is subjected to Softmax to enhance the stability of training network. *Scale* represents the global set of predicted pose distributions, ensuring better convergence of the KL loss. In order for network to be differentiable and trained end-to-end, our method establishes 2D-3D dense correspondences and predicts the pose by the ProPnP algorithm.

Finally, we fuse the predicted data in the M module and generate the final pose. We compare the translation predicted directly by the T module with the translation generated additionally by the ProPnP algorithm and score them. We select the highest score as the final 3-DoF translation, and then merge the pose matrix to output 6D object pose. In fact, it is better to decouple the translation estimation if the amount of data is small and the number of training sessions is low. In contrast, if the data volume is large, the number of training sessions is high and the ProPnP algorithm is used, it is better not to decouple the processing.

3.4 Loss Functions

We design three loss functions for translation, 3-channel coordinate maps X^{3D} and the ProPnP algorithm respectively.

The translation is predicted by regression coordinates, so we use the ℓ_2 loss for translation and the L_{trans} loss function is shown as follows:

$$L_{trans} = \ell_2 \left(Pose_{gt}, Pose_{es}\right) \tag{1}$$

where $Pose_{gt}$ represents the ground-truth translation and $Pose_{es}$ represents the predicted translation.

We use $Smooth\ell_1$ loss for the 3-channel coordinate maps X^{3D}. In order to reduce the complexity of network, we calculate the loss only for the effective region of object and ignore insignificant background region. The effective region of object is determined by mask, and the loss function of the 3-channel coordinate maps X^{3D} is shown as follows:

$$L_{rot} = Smooth\ell_1(M_{msk} \cdot M_{coorgt}, M_{msk} \cdot M_{coores}) \tag{2}$$

where M_{msk} represents the mask of object, M_{coorgt} represents the ground-truth 3-channel coordinate maps, M_{coores} represents the predicted 3-channel coordinate maps, and $\cdot$ represents the Hadamard product.

The KL divergence loss is used as the loss of the ProPnP algorithm, which minimizes the training loss by obtaining the probability density of the pose distribution. Loss function is shown as follows:

$$L_{ProPnP} = P + log\frac{1}{K}\sum_{j=1}^{K}\frac{exp - \frac{1}{2}\sum_{i=1}^{N} \| f_i(y_j) \|^2}{q(y_j)} \tag{3}$$

where $P = \frac{1}{2}\sum_{i=1}^{N} ||f_i(y_{gt})||^2$ represents the difference between the ground-truth values and the predicted values of 2D feature points projected into 2D image and N represents the number of feature points. The latter is the Monte Carlo loss, which is used to approximate the probability distribution of the predicted pose, where $q(y_j)$ denotes the probability density function of the integrand $exp - \frac{1}{2}\sum_{i=1}^{N} ||f_i(y_j)||^2$ and K denotes the number of samples.

The total loss function is shown as follows:

$$Loss = \alpha \cdot L_{ProPnP} + \beta \cdot L_{trans} + \gamma \cdot L_{rot} \tag{4}$$

where α, β and γ are hyperparameters, $\alpha = 1$, $\beta = 1$ and $\gamma = 0.02$. L_{ProPnP} is the loss function of the ProPnP algorithm, L_{trans} is translation estimation loss function and L_{rot} is 3-channel coordinate maps X^{3D} loss function.

4 Experiments

4.1 Datasets

We conduct experiments on the LineMOD [6] and Occlusion LineMOD dataset [1]. The LineMOD dataset contains 13 RGB image sequences, and each sequence has about 1.2k images. 15% of the images from each sequence are randomly selected for training and 85% of the images are used for testing. Due to the small amount of real data, we synthesize some images using the training set of the VOC2012 dataset as background. Therefore, we divide the LineMOD dataset into a real dataset and a real+synthetic dataset based on whether synthetic images are added or not. The real dataset has 2.3k images for training and 13k images for testing, and the real + synthetic dataset has 15k images for training and 13k images for testing. The Occlusion LineMOD dataset contains 8 objects, each of which has about 1.2k scenes with cluttered backgrounds and objects that are obscured or truncated. We use the Occlusion LineMOD dataset for testing only.

4.2 Metrics

In this paper, we use three common metrics, which are n°ncm, ADD(-S) and 2D Projection.

n°ncm. nºncm is the evaluation of whether the absolute value of the difference between the predicted pose and the ground-truth pose is less than a threshold (usually 5º5 cm), if it is less than the threshold then the predicted pose is considered correct, otherwise it is considered a failure.

ADD(-S). ADD determines whether the average distance between the predicted pose matrix $[R'|t']$ and the ground-truth pose matrix $[R|t]$ of the corresponding feature points on object surface is less than a threshold of object diameter (commonly 10%), and if it is less than the threshold, predicted pose is considered correct, otherwise it is considered a failure. For symmetrical objects, ADD-S metric is used and the average distance will be calculated based on the nearest point distance.

2D Projection. 2D Projection metric compares the average pixel error of a 3-dimensional model projected onto a 2-dimensional plane. If the difference does not exceed the threshold pixel (usually 5 pixels), the predicted pose is considered correct, otherwise it is considered a failure.

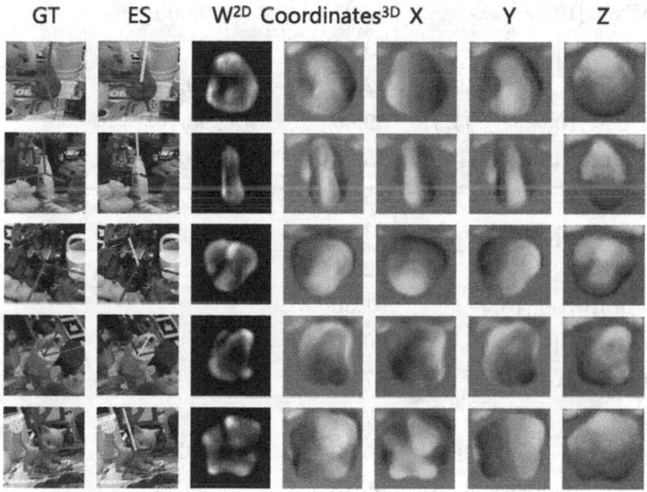

Fig. 3. Visualization results of 6D object pose estimation.

4.3 Results

We conduct experiments on LineMOD real dataset, real+synthetic dataset and Occlusion LineMOD dataset. Table 1 shows that our method compares other RGB-based methods when using different metrics. The first column of Table 1 represents different methods and the other three columns represent the accuracy of each method for estimating 6D object pose at various metrics, with larger values representing higher accuracy and bolded meaning that the method is the best at the current metric. Table 1 shows that our method outperforms other RGB-based methods when using different metrics. Figure 3 shows the visualization of the results of 6D object pose estimation, where the first column represents the ground-truth pose and the second column represents the predicted pose. It can be seen from comparing the two columns that our method accurately predicts the 6D object pose. W^{2D} represents the effective area of object and $Coordinates^{3D}$ represents the 3-dimensional coordinates of object in the world coordinate system. Three columns on the right side of figure represent the X-axis, Y-axis and Z-axis coordinate values of object respectively, and different colors represent different coordinate values. We also conducted experiments on

Table 1. Comparison of other RGB-based methods using commonly used metrics on real + synthetic dataset. "-" means that this method has not been experimented on this metric.

Method	5°5 cm	ADD(-S)(0.1d)	2D Projection(5px)
BB8 [14]	69.00	62.70	89.30
PVNet [13]	-	86.27	99.00
SSD6D [9]	-	79.00	-
PoseCNN [18]	19.40	62.70	70.20
CDPN [10]	94.31	89.86	98.10
Yolo-6D [16]	-	55.95	90.37
DPOD [20]	-	95.15	-
GDRNet [17]	-	93.70	-
Pix2Pose [12]	-	72.40	-
HybridPose [15]	-	91.30	-
RePose [8]	-	96.10	-
Efficientpose [2]	-	97.35	-
RNNPose [19]	-	97.37	-
EproPnP [4]	98.54	95.80	98.78
Ours(160epoch)	98.37	95.30	99.24
Ours(450epoch)	**99.38**	**97.42**	**99.43**

Table 2. Comparison of other methods using commonly used metrics on Occlusion LineMOD dataset.

Occlusion LineMOD	YOLO6D [16]	PoseCNN [18]	PVNet [13]	Ours
ADD(-S)(0.1)	6.42	24.9	**40.77**	37.88
2D Projection(5px)	6.16	17.2	**61.06**	59.01

Occlusion LineMOD dataset using two metrics, and as can be seen in Table 2, our approach is competitive.

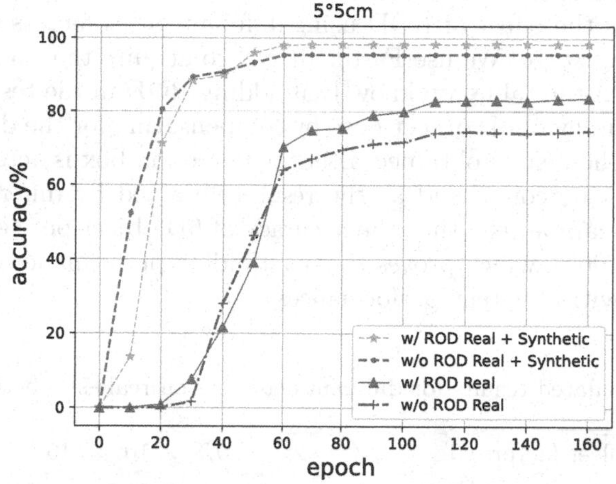

Fig. 4. Comparison with and without ROD. The blue and orange lines represent the results of real + synthetic dataset, and the red and green lines represent the results of real dataset. W/ means ROD is added, and w/o means it is not added. (Color figure online)

4.4 Ablation Study

Ablation Study on FCAFE. Adequate experiments are conducted to evaluate the role of FCAFE proposed in our method. Table 3 shows the results of our experiments using multiple metrics on real dataset and real + synthetic dataset, with the control group using ResNet34 network. It can be seen from Table 3 that using a variety of metrics on both datasets, FCAFE can improve the accuracy of 6D object pose estimation compared to the ResNet34 in all cases.

Table 3. Comparison with and without FCAFE on real dataset and real+synthetic dataset. W/ represents to use FCAFE and w/o represents the control group.

Metrics	0.02d	0.05d	0.1d	2px	5px
Real+w/o	16.17	52.65	82.11	66.77	95.93
Real+w/	**21.87**	**60.18**	**85.11**	**74.49**	**97.39**
Real+Synthetic+w/o	33.07	73.06	92.58	86.75	98.78
Real+Synthetic+w/	**43.52**	**80.69**	**95.30**	**91.99**	**99.24**

Ablation Study on ROD. We conduct adequate experiments to verify the role of ROD. Figure 4 shows the results of the visualisation with and without ROD. We can know that the accuracy of 6D object pose estimation is improved by adding ROD, and the improvement is larger when there is less data.

Table 4 shows the effect of ROD using different offset factors on the results during the testing stage. We use Faster RCNN to acquire the target region and apply ROD. From the table, we know that adding ROD in the testing stage can improve the accuracy of 6D object pose by compensating for the detector errors.

In addition, the results obtained after the detection box is acted by different offset factors can be considered as the results obtained by different detectors. From Table 4, it can be seen that the accuracy of 6D object pose estimation can reach more than 90%, which proves the outstanding performance of our method using detectors with different performances.

Table 4. Simulated results for different detectors on real+synthetic dataset.

offset factors	0	1%	2%	5%	10%	15%
5°5cm	99.38	99.38	99.46	99.40	98.20	91.65

5 Conclusion

In this paper, we develop an end-to-end differentiable network for 6D object pose estimation. We propose Random Offset Distraction and Full Convolution Asymmetric Feature Extractor with the ProPnP algorithm to improve the accuracy and generality of 6D object pose estimation. Our approach exceeds other state-of-the-art RGB-based methods on different metrics and can be adapted with different detectors to suit a wide range of scenarios.

References

1. Brachmann, E., Krull, A., Michel, F., Gumhold, S., Shotton, J., Rother, C.: Learning 6D object pose estimation using 3D object coordinates. In: Fleet, D., Pajdla, T., Schiele, B., Tuytelaars, T. (eds.) ECCV 2014. LNCS, vol. 8690, pp. 536–551. Springer, Cham (2014). https://doi.org/10.1007/978-3-319-10605-2_35
2. Bukschat, Y., Vetter, M.: Efficientpose: an efficient, accurate and scalable end-to-end 6D multi object pose estimation approach. arXiv preprint arXiv:2011.04307 (2020)
3. Cao, T., Luo, F., Fu, Y., Zhang, W., Zheng, S., Xiao, C.: DGECN: a depth-guided edge convolutional network for end-to-end 6d pose estimation. In: Proceedings of the IEEE/CVF Conference on Computer Vision and Pattern Recognition, pp. 3783–3792 (2022)

4. Chen, H., Wang, P., Wang, F., Tian, W., Xiong, L., Li, H.: EPro-PnP: generalized end-to-end probabilistic perspective-n-points for monocular object pose estimation. In: Proceedings of the IEEE/CVF Conference on Computer Vision and Pattern Recognition, pp. 2781–2790 (2022)
5. He, K., Zhang, X., Ren, S., Sun, J.: Deep residual learning for image recognition. In: Proceedings of the IEEE Conference on Computer Vision and Pattern Recognition, pp. 770–778 (2016)
6. Hinterstoisser, S., et al.: Model based training, detection and pose estimation of texture-less 3D objects in heavily cluttered scenes. In: Lee, K.M., Matsushita, Y., Rehg, J.M., Hu, Z. (eds.) ACCV 2012. LNCS, vol. 7724, pp. 548–562. Springer, Heidelberg (2013). https://doi.org/10.1007/978-3-642-37331-2_42
7. Huang, G., Liu, Z., Van Der Maaten, L., Weinberger, K.Q.: Densely connected convolutional networks. In: Proceedings of the IEEE Conference on Computer Vision and Pattern Recognition, pp. 4700–4708 (2017)
8. Iwase, S., Liu, X., Khirodkar, R., Yokota, R., Kitani, K.M.: Repose: fast 6D object pose refinement via deep texture rendering. In: Proceedings of the IEEE/CVF International Conference on Computer Vision, pp. 3303–3312 (2021)
9. Kehl, W., Manhardt, F., Tombari, F., Ilic, S., Navab, N.: SSD-6D: making RGB-based 3D detection and 6D pose estimation great again. In: Proceedings of the IEEE International Conference on Computer Vision, pp. 1521–1529 (2017)
10. Li, Z., Wang, G., Ji, X.: CDPN: coordinates-based disentangled pose network for real-time RGB-based 6-DoF object pose estimation. In: Proceedings of the IEEE/CVF International Conference on Computer Vision, pp. 7678–7687 (2019)
11. Liu, Z., Mao, H., Wu, C.Y., Feichtenhofer, C., Darrell, T., Xie, S.: A convnet for the 2020s. In: Proceedings of the IEEE/CVF Conference on Computer Vision and Pattern Recognition, pp. 11976–11986 (2022)
12. Park, K., Patten, T., Vincze, M.: Pix2pose: pixel-wise coordinate regression of objects for 6D pose estimation. In: Proceedings of the IEEE/CVF International Conference on Computer Vision, pp. 7668–7677 (2019)
13. Peng, S., Liu, Y., Huang, Q., Zhou, X., Bao, H.: PVNet: pixel-wise voting network for 6dof pose estimation. In: Proceedings of the IEEE/CVF Conference on Computer Vision and Pattern Recognition, pp. 4561–4570 (2019)
14. Rad, M., Lepetit, V.: BB8: a scalable, accurate, robust to partial occlusion method for predicting the 3D poses of challenging objects without using depth. In: Proceedings of the IEEE International Conference on Computer Vision, pp. 3828–3836 (2017)
15. Song, C., Song, J., Huang, Q.: Hybridpose: 6D object pose estimation under hybrid representations. In: Proceedings of the IEEE/CVF Conference on Computer Vision and Pattern Recognition, pp. 431–440 (2020)
16. Tekin, B., Sinha, S.N., Fua, P.: Real-time seamless single shot 6D object pose prediction. In: Proceedings of the IEEE Conference on Computer Vision and Pattern Recognition, pp. 292–301 (2018)
17. Wang, G., Manhardt, F., Tombari, F., Ji, X.: GDR-Net: geometry-guided direct regression network for monocular 6D object pose estimation. In: Proceedings of the IEEE/CVF Conference on Computer Vision and Pattern Recognition, pp. 16611–16621 (2021)
18. Xiang, Y., Schmidt, T., Narayanan, V., Fox, D.: PoseCNN: a convolutional neural network for 6D object pose estimation in cluttered scenes. arXiv preprint arXiv:1711.00199 (2017)

19. Xu, Y., Lin, K.Y., Zhang, G., Wang, X., Li, H.: Rnnpose: recurrent 6-DoF object pose refinement with robust correspondence field estimation and pose optimization. In: Proceedings of the IEEE/CVF Conference on Computer Vision and Pattern Recognition, pp. 14880–14890 (2022)
20. Zakharov, S., Shugurov, I., Ilic, S.: DPOD: 6D pose object detector and refiner. In: Proceedings of the IEEE/CVF International Conference on Computer Vision, pp. 1941–1950 (2019)

An Internal-External Constrained Distillation Framework for Continual Semantic Segmentation

Qingsen Yan[1], Shengqiang Liu[1], Xing Zhang[2], Yu Zhu[1], Jinqiu Sun[1], and Yanning Zhang[1(✉)]

[1] Northwestern Polytechnical University, Xi'an, China
ynzhang@nwpu.edu.cn

[2] Xi'an University of Architecture and Technology, Xi'an, China

Abstract. Deep neural networks have a notorious catastrophic forgetting problem when training serialized tasks on image semantic segmentation. It refers to the phenomenon of forgetting previously learned knowledge due to the plasticity-stability dilemma and background shift in the segmentation task. Continual Semantic Segmentation (CSS) comes into being to handle this challenge. Previous distillation-based methods only consider the knowledge of the features at the same level but neglect the relationship between different levels. To alleviate this problem, in this paper, we propose a mixed distillation framework called Internal-external Constrained Distillation (ICD), which includes multi-information-based internal feature distillation and attention-based external feature distillation. Specifically, we utilize the statistical information of features to perform internal distillation between the old model and the new model, which effectively avoids interference at the same scale. Furthermore, for the external distillation of features at different scales, we employ multi-scale convolutional attention to capture the relationships among features of different scales and ensure their consistency across old and new tasks. We evaluate our method on standard semantic segmentation datasets, such as Pascal-VOC2012 and ADE20K, and demonstrate significant performance improvements in various scenarios.

Keywords: Continual Learning · Semantic Segmentation · Feature Distillation

1 Introduction

Semantic segmentation is a fundamental problem in computer vision, where the objective is to assign a label to every pixel in an image. In recent years, this field has witnessed remarkable progress, driven by the advent of deep neural networks and the availability of extensive human-annotated datasets [8,32]. The deep learning-based models [1,14–16,23] typically require prior knowledge of all classes in the datasets and utilize all available data at once during training. However, such a setup would be disconnected from practice, the model should have the ability to continuously learn in realistic scenarios and acquire new

Q. Liu et al. (Eds.): PRCV 2023, LNCS 14427, pp. 325–336, 2024.
https://doi.org/10.1007/978-981-99-8435-0_26

knowledge without having to be retrained from scratch. To address this need, a novel paradigm called Continual Semantic Segmentation (CSS) has emerged, wherein the model can be continuously updated to accommodate the addition of new classes and their corresponding labels. Inevitably, previous methods face the challenge of catastrophic forgetting [9], which refers to the phenomenon of forgetting previously learned knowledge. Moreover, the background represents pixels that have not been assigned to any specific object class in the traditional semantic segmentation task. It shows in CSS that the background may contain both future classes not yet seen by the model, as well as old classes. Therefore, the phenomenon of background shift [2], where future classes or old classes are assigned to background classes, would further exacerbate the problem of catastrophic forgetting.

To address the issue of catastrophic forgetting, many methods [3,18–21,24,29] have been proposed in the field of semantic segmentation. Prototype-based methods [20] in semantic segmentation utilize prototype matching to enforce consistency constraints in the latent space representation, aiming to address the issue of forgetting. Meanwhile, other approaches [24,29] leverage self-training and model adaptation to mitigate this problem. However, there is still room for improvement in the performance of these methods. Replay-based methods [3,18], which require storing image data or object data, offer an effective solution to mitigate catastrophic forgetting. Nevertheless, this approach comes with the additional overhead of storage requirements. Distillation-based methods [19,21] are popular in semantic segmentation, which constrain the model to respond consistently with the old model by constructing some distillation losses, with PLOP [7] being a representative method in this field. Specifically, PLOP proposes a multi-scale pooling distillation scheme that preserves long- and short-range spatial relationships at the feature level. However, it has limitations as it only considers the mean information of features in distillation and also does not take into account the relationships between features of different levels.

Aiming to solve the above problems, we propose a mixed distillation framework called Internal-External Constrained Distillation (ICD), which includes multi-information-based internal feature distillation and attention-based extrinsic feature distillation at different scales to alleviate catastrophic forgetting in CSS. First, we utilize the statistical information of features to perform internal distillation at the same scale between the old model and the new model. Second, for the external distillation of features at different scales, we employ multi-scale convolutional attention to capture the relationships among features of different scales and ensure their consistency across old and new tasks. Finally, we evaluate our method on two benchmark datasets of semantic segmentation datasets, such as Pascal-VOC2012 [8] and ADE20K [32], and demonstrate significant performance improvements in various scenarios. Our contributions can be summarized as follows:

- We propose a novel mixed distillation framework called Internal-external Constrained Distillation (ICD), which efficiently addresses catastrophic forgetting

through multi-scale internal feature distillation and extrinsic feature distillation at different scales.
- We introduce a method called multi-information-based internal feature distillation, which utilizes the statistical information of features to perform distillation between the old model and the new model.
- We propose a novel attention-based extrinsic feature distillation at different scales, employing multi-scale convolutional attention to capture the relationships among features of different scales and perform distillation, thereby mitigating catastrophic forgetting.

2 Related Work

Continual Learning. Deep neural networks suffer form catastrophic forgetting when trained on new data, leading to a substantial performance degradation. To solve this problem, many works have being proposed and can be grouped in three categories according to [6]: regularization-based methods [4,12,13,30], replay-based methods [22,27], and architecture-based methods [17,26]. Regularization-based methods mitigate catastrophic forgetting by introducing additional regularization term to constraint the update of features or parameters, when training the current task. Replay-based methods usually need to store samples in their raw format or generate pseudo-samples using a generative model, which are replayed during the learning process of new tasks. Architecture-based methods can dynamically grow new branches for new tasks while freezing.

Semantic Segmentation. Image semantic segmentation [5,16] achieves remarkable improvements with the advance of deep neural networks. The majority of deep learning-based segmentation methods adopt an encoder-decoder architecture [1,23], which is the most common method for retaining spatial information. Fully Convolutional Networks (FCN) [16] is the first work to use solely convolutional layers for semantic segmentation and can handle input images of any arbitrary size, producing segmentation maps of the same size, but lacking global context information. The recent works have employed various techniques to exploit contextual information, including attention mechanisms [10,25] and fixed-scale aggregation [5,31].

Continual Semantic Segmentation. Despite significant advancements in the two aforementioned areas, continual semantic segmentation remains an urgent problem to solve. Recently, standard datasets [8,32] and general methods [19] for semantic segmentation have been proposed. Among these, Distillation-based methods are popular in continual semantic segmentation. Cermelli *et al.* [2] study the distribution shift of the background class when it includes previous or future classes, indicating that the problem of catastrophic forgetting in segmentation is exacerbated by this background shift. To do so, they propose using an output-level knowledge distillation loss to mitigate forgetting. In [19,21], the authors employ knowledge distillation techniques and freeze the weights of the encoder simultaneously to address the issue of forgetting. PLOP [7] applies a knowledge distillation strategy to intermediate layers. However, it only distills the mean of features, disregarding other information such as extremum, and does not consider the relationship between features in a model.

3 Method

3.1 Problem Definition and Notation

We denote $\mathcal{X}$ as the input space (i.e., the image space) and, without loss of generality, assume that each image $x \in \mathcal{X}$ is composed of a set of pixels $\mathcal{I}$ with a constant cardinality $|\mathcal{I}| = H \times W = N$. The output space is defined as $\mathcal{Y}^N$, and the classes of all tasks are collectively defined as $C^{0:T}$. In the Continual Semantic Segmentation setting, training is realized over multiple *learning steps.* At each learning step t, the ground truth only includes the labels of the current classes C^t, while all other labels (such as old classes $C^{1:t-1}$ or future classes $C^{t+1:T}$) are combined into the background class C^0. Ideally, the model at step t should be able to predict all the classes $C^{1:t}$ seen over time. Typically, the semantic segmentation model at step t can be divided into the composition of a feature extractor $f^t(\cdot)$ and a classifier $g^t(\cdot)$. Features can be extracted at any layer l of the feature extractor $f_l^t(\cdot), l \in \{1, ..., L\}$. Finally, we denote $\widehat{\mathcal{Y}}^t = g^t \circ f^t(\mathcal{I})$ as the output predicted segmentation mask and the Θ^t represents the set of learnable parameters for the current network at step t.

3.2 Internal-External Constrained Distillation Framework

The overview of the proposed mixed distillation framework ICD is shown in Fig. 1. The model is an encoder-decoder-based semantic segmentation network, including a multi-information-based internal feature distillation module and an attention-based extrinsic feature distillation module. The former is a distillation scheme based on intermediate features, matching statistics at the same feature levels between the old and current models. The latter captures the relationships among features of different scales within a single task using multi-scale convolutional attention and performs distillation of attention maps that contain information about external feature relations. Both of the above modules can be shown in Fig. 2. Given an image, the semantic segmentation network will output a prediction map and compute the cross-entropy loss ($\mathcal{L}_{ce}$). The training loss is defined as a combination of the cross-entropy loss and the losses from the proposed internal feature distillation ($\mathcal{L}_{int}$) and extrinsic feature distillation ($\mathcal{L}_{ext}$) modules. Specifically, the training loss is defined as:

$$\mathcal{L}_{tot} = \mathcal{L}_{ce} + \mathcal{L}_{int} + \mathcal{L}_{ext} \tag{1}$$

3.3 Multi-information-based Internal Feature Distillation Module

An existing distillation scheme, called PLOP [7], addresses this issue by introducing pooling along the width or height dimensions and matching global statistics

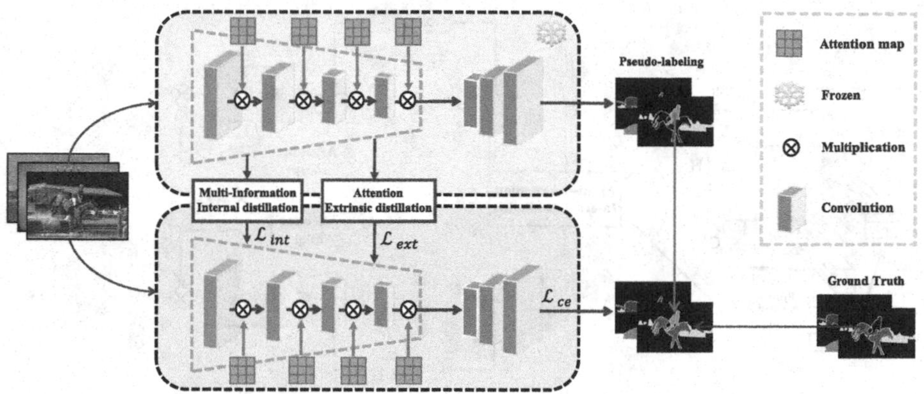

Fig. 1. Overview of the proposed approach, with the old model and the current task model. We perform multi-information-based internal feature distillation and attention-based extrinsic feature distillation between these two models. The old model is frozen and outputs pseudo-labeling, which is combined with ground truth to solve the background shift.

at the same feature levels between the old and current models. Additionally, PLOP computes width and height-pooled slices on multiple regions extracted at different scales to preserve local information. However, PLOP only uses the mean of the feature in distillation and ignores the maximum and minimum values, which may contain important information. Therefore, we propose a multi-information-based internal feature distillation scheme. Let x denote an embedding tensor of size $H \times W \times C$. To extract an embedding $\mathbf{\Phi}$, it is necessary to concatenate the $H \times C$ width-mixed pooling slices and the $W \times C$ height-mixed pooling slices of x:

$$\mathbf{\Phi}(\mathrm{x}) = concat\left(\frac{1}{W}\sum_{w=1}^{W}\mathrm{x}[:,w,:], \frac{1}{H}\sum_{h=1}^{H}\mathrm{x}[h,:,:], \mathbf{\Omega}(\mathrm{x}[:,w,:]), \mathbf{\Omega}(\mathrm{x}[h,:,:])\right) \tag{2}$$

where $concat(\cdot)$ denotes concatenation over the channel axis. $\mathbf{\Omega}(\mathrm{x})$ is the operation of finding the extremum values and combining them based on their corresponding dimensions. Specifically, $\mathbf{\Omega}(\mathrm{x}[:,w,:])$ can be defined as:

$$\mathbf{\Omega}(\mathrm{x}[:,w,:]) = concat(max\,\mathrm{x}[:,w,:], min\,\mathrm{x}[:,w,:]) \tag{3}$$

Its ability to model long-range dependencies across the entire horizontal or vertical axis is beneficial. However, modeling statistics across the entire width or height can blur the local statistics that are important for smaller objects. Hence, to retain both long-range and short-range spatial relationships, we adopt a local feature distillation scheme inspired by PLOP, which involves computing width and height-mixed pooling slices on multiple regions extracted at different scales $\{1/2^s\}_{s=0...S}$. Given an embedding tensor x, the mixed embedding $\Psi^s(\mathrm{x})$

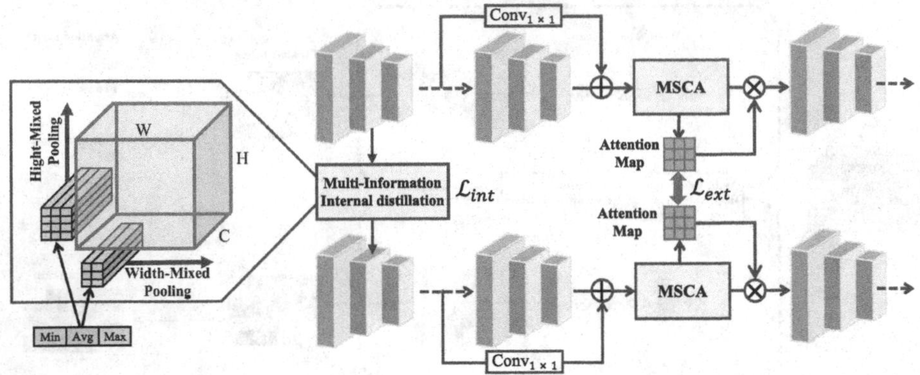

Fig. 2. Illustration of multi-information-based internal feature distillation and attention-based extrinsic feature distillation. The former calculates width-mixed and height-mixed pooling to perform feature distillation at the same level. The latter employs multi-scale convolutional attention (MSCA) to capture the relationships among features of different scales and ensure their consistency across old and new tasks. $\oplus$ is the concat operation and $\otimes$ is the matrix multiplication.

at scale s is computed by concatenating:

$$\Psi^s(\mathrm{x}) = concat(\mathbf{\Phi}(\mathrm{x}^s_{0,0}),\ \ldots\ , \mathbf{\Phi}(\mathrm{x}^s_{s-1,s-1})) \tag{4}$$

where $\mathrm{x}^s_{i,j} = \mathrm{x}[iH/s : (i+1)H/s], jW/s : (j+1)W/s, :]$ is a sub-region of the embedding tensor x of size $H/s \times W/s$ for $\forall i = 0...s-1$, $\forall j = 0...s-1$. Finally, we concatenate the mixed embeddings $\Psi^s(\mathrm{x})$ of each scale s along the channel dimension, obtaining the final embedding:

$$\Psi(\mathrm{x}) = concat(\Psi^1(\mathrm{x}),\ \ldots\ , \Psi^S(\mathrm{x})) \tag{5}$$

We compute mixed embeddings for several layers $l \in \{1, ..., L\}$ of both old and current models and then minimize the $\mathcal{L}_2$ distance between the mixed embeddings computed at several layers. The internal feature distillation loss is defined as:

$$\mathcal{L}_{int} = \frac{1}{L}\sum_{l=1}^{L} ||\Psi(f^t_l(\mathcal{I})) - \Psi(f^{t-1}_l(\mathcal{I}))||^2 \tag{6}$$

3.4 Attention-Based Extrinsic Feature Distillation Module

Although PLOP alleviates forgetting by distilling features between the same layers, cross-layer drift is inevitable, which makes it difficult to distill these features independently. Therefore, to enhance the distillation operation by considering more contextual information, we propose introducing attention mechanisms between layers. Specifically, we begin by concatenating features $\mathbf{\Phi}_c$ from different layers and adjusting their dimensions to match through convolution operations:

$$\mathbf{\Phi}^c_l = Concat\left(\mathbf{\Phi}_l, Conv_{1\times1}\left(\mathbf{\Phi}_{l-1}\right)\right) \tag{7}$$

where l represents the index in the layers, and 1×1 represents a convolution kernel size of 1.

We employ a novel spatial attention method, distinct from the conventional self-attention approach, which achieves spatially-aware feature extraction through multi-scale convolution attention (MSCA) that utilizes element-wise multiplication. We refer the reader to the extended survey in SegNeXt [11] for a detailed explanation of spatial attention. In brief, MSCA can be written as:

$$A_l = Conv_{1\times 1}\left(\sum_{l=0}^{3} Scale_l(DW\text{−}Conv(\Phi_l^c))\right) \tag{8}$$

which converges local information by depth-wise convolution ($DW\text{−}Conv$), and multi-branch depth-wise strip convolutions $Scale_l$ captures multi-scale contextual information. The attention map A_l is obtained by a 1×1 convolution $Conv_{1\times 1}$. Finally, the attention map is used as weights and multiplied with the input to extract spatial relationships:

$$\hat{\Phi}_l^c = A_l \otimes \Phi_l^c. \tag{9}$$

which $\otimes$ denotes matrix multiplication.

Intuitively, this attention map encapsulates rich spatial and contextual information. Therefore, we constrain the $\mathcal{L}_2$ distance $\mathcal{L}_{ext}$ between attention maps from the old and new models as:

$$\mathcal{L}_{ext} = \frac{1}{L}\sum_{l=1}^{L} ||A_l^t - A_l^{t-1}||^2 \tag{10}$$

4 Experiments

4.1 Experimental Setups

Datasets: We evaluate our method on two segmentation datasets: Pascal-VOC 2012 [8] and ADE20k [32]. The PASCAL-VOC 2012 dataset contains 10,582 images in the training set and 1,449 in the validation set, respectively. The pixels of each image can be assigned to 21 different classes (20 plus the background). ADE20K is a large-scale dataset for semantic segmentation that covers daily life scenes. It comprises 20,210 training images and 2,000 validation images, with 150 classes encompassing both stuff and objects.

Baselines: We compare our method against previous CSS methods originally designed for image classification, including Path Integral (PI) [30], Elastic Weight Consolidation (EWC) [12], Riemannian Walks (RW) [4], Learning without Forgetting (LwF) [13], and its multi-class version (LwF-MC) [22]. We also consider previous CSS approaches for segmentation, such as ILT [19], MiB [2], PLOP [7], and CAF [28]. In addition to the state-of-the-art methods, we report results for fine-tuning (FT) and joint training (Joint), which serve as lower and upper bounds.

Table 1. Continual Semantic Segmentation results on Pascal-VOC 2012 in Mean IoU (%). ⋆: results excerpted from [18]

Method	19-1 (2 tasks)						15-5 (2 tasks)						15-1 (6 tasks)					
	Disjoint			Overlapped			Disjoint			Overlapped			Disjoint			Overlapped		
	1–19	20	all	1–19	20	all	1–15	16–20	all	1–15	16–20	all	1–15	16–20	all	1–15	16–20	all
FT*	35.2	13.2	34.2	34.7	14.9	33.8	8.4	33.5	14.4	12.5	36.9	18.3	5.8	4.9	5.6	4.9	3.2	4.5
PI [30]	5.4	14.1	5.9	7.5	14.0	7.8	1.3	34.1	9.5	1.6	33.3	9.5	0.0	1.8	0.4	0.0	1.8	0.5
EWC [12]	23.2	16.0	22.9	26.9	14.0	26.3	26.7	37.7	29.4	24.3	35.5	27.1	0.3	4.3	1.3	0.3	4.3	1.3
RW [4]	19.4	15.7	19.2	23.3	14.2	22.9	17.9	36.9	22.7	16.6	34.9	21.2	0.2	5.4	1.5	0.0	5.2	1.3
LwF [13]	53.0	9.1	50.8	51.2	8.5	49.1	58.4	37.4	53.1	58.9	36.6	53.3	0.8	3.6	1.5	1.0	3.9	1.8
LwF-MC [22]	63.0	13.2	60.5	64.4	13.3	61.9	67.2	41.2	60.7	58.1	35.0	52.3	4.5	7.0	5.2	6.4	8.4	6.9
ILT [19]	69.1	16.4	66.4	67.1	12.3	64.4	63.2	39.5	57.3	66.3	40.6	59.9	3.7	5.7	4.2	4.9	7.8	5.7
MiB [2]	69.6	25.6	67.4	70.2	22.1	67.8	71.8	43.3	64.7	75.5	49.4	69.0	46.2	12.9	37.9	35.1	13.5	29.7
PLOP [7]	75.3	**38.8**	73.6	75.3	**37.3**	73.5	71.0	42.8	64.2	75.7	**51.7**	70.0	57.8	13.6	46.4	65.1	21.1	54.6
CAF [28]	75.5	30.8	73.3	75.5	34.8	73.4	**72.9**	42.1	**65.2**	77.2	49.9	70.4	57.2	15.5	46.7	55.7	14.1	45.3
Ours	**76.9**	34.8	**74.9**	**76.1**	37.1	**74.2**	71.0	**43.4**	64.5	**77.6**	48.9	**70.5**	**61.1**	**18.0**	**50.8**	**65.4**	**26.4**	**56.1**
Joint	77.4	78.0	77.4	77.4	78.0	77.4	79.1	72.6	77.4	79.1	72.6	77.4	79.1	72.6	77.4	79.1	72.6	77.4

Table 2. Continual Semantic Segmentation results on ADE20k in Mean IoU (%).

Method	100-50 (2 tasks)			100-10 (6 tasks)							50-50 (3 tasks)			
	1–100	101–150	all	1–100	100–110	100–120	120–130	130–140	140–150	all	1–50	51–100	101–150	all
FT	0.0	24.9	8.3	0.0	0.0	0.0	0.0	0.0	16.6	1.1	0.0	0.0	22.0	7.3
LwF [13]	21.1	25.6	22.6	0.1	0.0	0.4	2.6	4.6	16.9	1.7	5.7	12.9	22.8	13.9
LwF-MC [22]	34.2	10.5	26.3	18.7	2.5	8.7	4.1	6.5	5.1	14.3	27.8	7.0	10.4	15.1
ILT [19]	22.9	18.9	21.6	0.3	0.0	1.0	2.1	4.6	10.7	1.4	8.4	9.7	14.3	10.8
MiB [2]	37.9	27.9	34.6	31.8	10.4	14.8	12.8	**13.6**	**18.7**	25.9	35.5	22.2	**23.6**	27.0
PLOP [7]	41.9	14.9	32.9	40.6	**15.2**	16.9	18.7	11.9	7.9	31.6	48.6	30.0	13.1	30.4
CAF [28]	37.3	**31.9**	**35.5**	39.0	14.6	22.0	**25.4**	12.1	13.1	31.8	47.5	30.6	23.0	33.7
Ours	**42.5**	19.8	34.9	**42.3**	12.6	**24.2**	20.5	10.2	8.6	**33.3**	**49.4**	**30.9**	21.2	**33.8**
Joint	44.3	28.2	38.9	44.3	26.1	42.8	26.7	28.1	17.3	38.9	51.1	38.3	28.2	38.9

4.2 Experimental Results

We compare our approach with state-of-the-art methods. Table 1 presents quantitative experiments on VOC 19-1, 15-5, and 15-1 on Pascal-VOC 2012. It is evident that our method outperforms most competing approaches in both the overlapped and disjoint settings, often by a substantial margin. On 19-1, our method retains more old information, resulting in a performance improvement of 1.6% points in the disjoint setting and 0.8% points in the overlapped setting, compared to the recent method CAF [28]. On the most challenging 15-1 setting, general continual models (EWC and LwF-MC) and ILT all exhibit very poor performance. However, CAF demonstrates significant improvements. Nevertheless, our method surpasses CAF by a substantial margin on both old and new classes. Specifically, in the overlapped setting, +23.8% on all classes, +17.4% on old classes, and +87.2% on new classes.

The Table 2 presents the experimental results on the ADE20k dataset 100-50, 100-10, and 50-5. Despite ADE20k being a challenging semantic segmentation dataset, our method demonstrates a strong ability to alleviate forgetting and achieves a performance close to the joint model baseline mIoU in old classes. On the 100-10 setting, we initially start with 100 classes, and then gradually

Table 3. Ablation study on disjoint VOC 15-1 in Mean IoU (%).

ce	avg	max	min	att	old	new	all
✓					5.8	4.9	5.6
✓	✓				55.5	16.7	46.3
✓				✓	41.6	10.9	34.3
✓	✓	✓			57.0	15.5	47.1
✓	✓		✓		55.1	17.5	46.2
✓	✓	✓	✓		56.6	19.8	47.8
✓	✓		✓	✓	59.6	19.4	50.0
✓	✓	✓		✓	58.2	17.5	48.5
✓	✓	✓	✓	✓	61.1	18.0	50.8

add the remaining classes in 5 steps, with 10 classes being added in each step. Our approach allows to retain more information from the first 100 classes while effectively learning new knowledge, resulting in an overall improvement of 1.5% points. Similarly, in the 50-50 setting, we add 50 classes in each step over a total of 3 steps. While methods like LwF, LwF-MC, and ILT yield lower results, our method stands out from the competition and outperforms the PLOP and CAF methods.

4.3 Ablation Study

To assess the effect of each component, we conduct an ablation analysis on the Pascal dataset in the challenging 15-1 scenario, and the results are reported in Table 3. As demonstrated previously, the utilization of only cross entropy loss during model training leads to catastrophic forgetting. However, incorporating mean-based internal feature distillation or attention-based extrinsic feature distillation yields significant performance improvements. Building upon the adding of mean-based internal feature distillation, we further integrate max-based and min-based internal feature distillation to assess their effectiveness. As anticipated, the experimental results validate their ability to enhance the model's capacity to alleviate forgetting and acquire new knowledge. Specifically, the inclusion of maximum feature information prevents forgetting (+2.7% on old classes), while the inclusion of minimum feature information enhances learning for new tasks (+4.8% on new classes). Additionally, we conducted experiments to evaluate the compatibility of multi-information-based internal distillation and attention-based external distillation. The results unequivocally indicate that combining these two modules significantly improves model performance. We also conducted experiments to observe the evolution of model performance on the Pascal-VOC 2012 dataset in the 15-1 setting after adding various components. As depicted in Fig. 3, each component effectively mitigates the decline in performance.

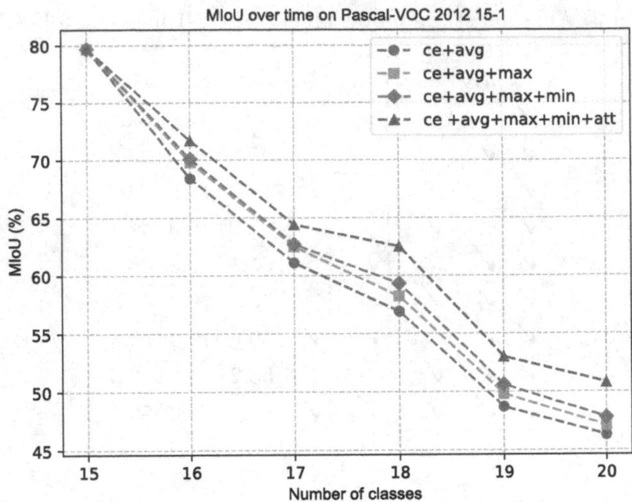

Fig. 3. Evolution in model performance in the Pascal-VOC 2012 dataset on the 15-1 setting after adding various components. Each component effectively mitigates the decline in performance.

5 Conclusion

In this paper, we propose a novel mixed distillation framework called Internal-external Constrained Distillation (ICD) that effectively addresses catastrophic forgetting through multi-information-based internal feature distillation and extrinsic feature distillation at different scales. Our evaluation demonstrates the effectiveness of the proposed techniques, showcasing their strong long-term learning capacity and significant outperformance compared to other methods. Future research endeavors will explore the application of our techniques in different tasks, such as class-incremental open-set domain adaptation. Additionally, the inherent limitations of knowledge distillation impose constraints on the new model's ability to comprehend and effectively express acquired knowledge. We will investigate the combination of our approach with other continual learning methods, as well as explore the integration of output-level techniques to further enhance performance.

Acknowledgement. This work is supported by National Science Foundation of China under Grant No. 62301432, Natural Science Basic Research Program of Shaanxi No. 2023-JC-QN-0685, the Fundamental Research Funds for the Central Universities No. D5000220444 and National Engineering Laboratory for Integrated Aero-Space-Ground-Ocean Big Data Application Technology.

References

1. Badrinarayanan, V., Kendall, A., Cipolla, R.: Segnet: a deep convolutional encoder-decoder architecture for image segmentation. IEEE Trans. Pattern Anal. Mach. Intell. **39**(12), 2481–2495 (2017)
2. Cermelli, F., Mancini, M., Bulo, S.R., Ricci, E., Caputo, B.: Modeling the background for incremental learning in semantic segmentation. In: Proceedings of the IEEE/CVF Conference on Computer Vision and Pattern Recognition, pp. 9233–9242 (2020)
3. Cha, S., Yoo, Y., Moon, T., et al.: SSUL: semantic segmentation with unknown label for exemplar-based class-incremental learning. Adv. Neural. Inf. Process. Syst. **34**, 10919–10930 (2021)
4. Chaudhry, A., Dokania, P.K., Ajanthan, T., Torr, P.H.: Riemannian walk for incremental learning: understanding forgetting and intransigence. In: Proceedings of the European Conference on Computer Vision (ECCV), pp. 532–547 (2018)
5. Chen, L.C., Papandreou, G., Kokkinos, I., Murphy, K., Yuille, A.L.: Deeplab: semantic image segmentation with deep convolutional nets, atrous convolution, and fully connected CRFs. IEEE Trans. Pattern Anal. Mach. Intell. **40**(4), 834–848 (2017)
6. De Lange, M., et al.: A continual learning survey: defying forgetting in classification tasks. IEEE Trans. Pattern Anal. Mach. Intell. **44**(7), 3366–3385 (2021)
7. Douillard, A., Chen, Y., Dapogny, A., Cord, M.: PLOP: learning without forgetting for continual semantic segmentation. In: Proceedings of the IEEE/CVF Conference on Computer Vision and Pattern Recognition, pp. 4040–4050 (2021)
8. Everingham, M., Eslami, S.A., Van Gool, L., Williams, C.K., Winn, J., Zisserman, A.: The pascal visual object classes challenge: a retrospective. Int. J. Comput. Vision **111**, 98–136 (2015)
9. French, R.M.: Catastrophic forgetting in connectionist networks. Trends Cogn. Sci. **3**(4), 128–135 (1999)
10. Fu, J., et al.: Dual attention network for scene segmentation. In: Proceedings of the IEEE/CVF Conference on Computer Vision and Pattern Recognition, pp. 3146–3154 (2019)
11. Guo, M.H., Lu, C.Z., Hou, Q., Liu, Z., Cheng, M.M., Hu, S.M.: Segnext: rethinking convolutional attention design for semantic segmentation. arXiv preprint arXiv:2209.08575 (2022)
12. Kirkpatrick, J., et al.: Overcoming catastrophic forgetting in neural networks. Proc. Natl. Acad. Sci. **114**(13), 3521–3526 (2017)
13. Li, Z., Hoiem, D.: Learning without forgetting. IEEE Trans. Pattern Anal. Mach. Intell. **40**(12), 2935–2947 (2017)
14. Liang, J., Zhou, T., Liu, D., Wang, W.: Clustseg: clustering for universal segmentation. arXiv preprint arXiv:2305.02187 (2023)
15. Liu, D., Cui, Y., Tan, W., Chen, Y.: SG-Net: spatial granularity network for one-stage video instance segmentation. In: Proceedings of the IEEE/CVF Conference on Computer Vision and Pattern Recognition, pp. 9816–9825 (2021)
16. Long, J., Shelhamer, E., Darrell, T.: Fully convolutional networks for semantic segmentation. In: Proceedings of the IEEE Conference on Computer Vision and Pattern Recognition, pp. 3431–3440 (2015)
17. Mallya, A., Lazebnik, S.: Packnet: adding multiple tasks to a single network by iterative pruning. In: Proceedings of the IEEE Conference on Computer Vision and Pattern Recognition, pp. 7765–7773 (2018)

18. Maracani, A., Michieli, U., Toldo, M., Zanuttigh, P.: Recall: replay-based continual learning in semantic segmentation. In: Proceedings of the IEEE/CVF International Conference on Computer Vision, pp. 7026–7035 (2021)
19. Michieli, U., Zanuttigh, P.: Incremental learning techniques for semantic segmentation. In: Proceedings of the IEEE/CVF International Conference on Computer Vision Workshops (2019)
20. Michieli, U., Zanuttigh, P.: Continual semantic segmentation via repulsion-attraction of sparse and disentangled latent representations. In: Proceedings of the IEEE/CVF Conference on Computer Vision and Pattern Recognition, pp. 1114–1124 (2021)
21. Michieli, U., Zanuttigh, P.: Knowledge distillation for incremental learning in semantic segmentation. Comput. Vis. Image Underst. **205**, 103167 (2021)
22. Rebuffi, S.A., Kolesnikov, A., Sperl, G., Lampert, C.H.: ICARL: incremental classifier and representation learning. In: Proceedings of the IEEE Conference on Computer Vision and Pattern Recognition, pp. 2001–2010 (2017)
23. Ronneberger, O., Fischer, P., Brox, T.: U-Net: convolutional networks for biomedical image segmentation. In: Navab, N., Hornegger, J., Wells, W.M., Frangi, A.F. (eds.) MICCAI 2015. LNCS, vol. 9351, pp. 234–241. Springer, Cham (2015). https://doi.org/10.1007/978-3-319-24574-4_28
24. Stan, S., Rostami, M.: Unsupervised model adaptation for continual semantic segmentation. In: Proceedings of the AAAI Conference on Artificial Intelligence, vol. 35, pp. 2593–2601 (2021)
25. Tao, A., Sapra, K., Catanzaro, B.: Hierarchical multi-scale attention for semantic segmentation. arXiv preprint arXiv:2005.10821 (2020)
26. Wang, Y.X., Ramanan, D., Hebert, M.: Growing a brain: fine-tuning by increasing model capacity. In: Proceedings of the IEEE Conference on Computer Vision and Pattern Recognition, pp. 2471–2480 (2017)
27. Yan, Q., Gong, D., Liu, Y., van den Hengel, A., Shi, J.Q.: Learning Bayesian sparse networks with full experience replay for continual learning. In: Proceedings of the IEEE/CVF Conference on Computer Vision and Pattern Recognition, pp. 109–118 (2022)
28. Yang, G., et al.: Continual attentive fusion for incremental learning in semantic segmentation. IEEE Trans. Multimedia (2022)
29. Yu, L., et al.: Semantic drift compensation for class-incremental learning. In: Proceedings of the IEEE/CVF Conference on Computer Vision and Pattern Recognition, pp. 6982–6991 (2020)
30. Zenke, F., Poole, B., Ganguli, S.: Continual learning through synaptic intelligence. In: International Conference on Machine Learning, pp. 3987–3995. PMLR (2017)
31. Zhang, H., et al.: Context encoding for semantic segmentation. In: Proceedings of the IEEE Conference on Computer Vision and Pattern Recognition, pp. 7151–7160 (2018)
32. Zhou, B., Zhao, H., Puig, X., Fidler, S., Barriuso, A., Torralba, A.: Scene parsing through ADE20K dataset. In: Proceedings of the IEEE Conference on Computer Vision and Pattern Recognition, pp. 633–641 (2017)

MTD: Multi-Timestep Detector for Delayed Streaming Perception

Yihui Huang[1] and Ningjiang Chen[1,2,3](✉)

[1] School of Computer and Electronic Information, Guangxi University, Nanning, China
[2] Guangxi Intelligent Digital Services Research Center of Engineering Technology, Nanning, China
[3] Education Department of Guangxi Zhuang Autonomous Region, Key Laboratory of Parallel, Distributed and Intelligent Computing, Guangxi University, Nanning, China
chnj@gxu.edu.cn

Abstract. Autonomous driving systems require real-time environmental perception to ensure user safety and experience. Streaming perception is a task of reporting the current state of the world, which is used to evaluate the delay and accuracy of autonomous driving systems. In real-world applications, factors such as hardware limitations and high temperatures inevitably cause delays in autonomous driving systems, resulting in the offset between the model output and the world state. In order to solve this problem, this paper propose the Multi-Timestep Detector (MTD), an end-to-end detector which uses dynamic routing for multi-branch future prediction, giving model the ability to resist delay fluctuations. A Delay Analysis Module (DAM) is proposed to optimize the existing delay sensing method, continuously monitoring the model inference stack and calculating the delay trend. Moreover, a novel Timestep Branch Module (TBM) is constructed, which includes static flow and adaptive flow to adaptively predict specific timesteps according to the delay trend. The proposed method has been evaluated on the Argoverse-HD dataset, and the experimental results show that it has achieved state-of-the-art performance across various delay settings.

Keywords: Streaming Perception · Object Detection · Dynamic Network

1 Introduction

To ensure user safety and experience, autonomous driving systems need to perceive surrounding environment not only in time, but also in real time. Traditional object detection [1–3] benchmarks focus on offline evaluation, which means that each frame of the video stream is compared with its annotation separately, which requires the system to process the captured frames within the interval time between each frame. Therefore, many researches [4–8] focus on reducing the latency so that the model can finish processing captured frame before next frame is input. However, in real-world applications, hardware limitations [9, 10] and high temperatures [11] inevitably cause processing delay, resulting in changes in the real-time environment after model processes the captured frames. As shown in Fig. 1 (a), the output of the model is always outdated.

Q. Liu et al. (Eds.): PRCV 2023, LNCS 14427, pp. 337–349, 2024.
https://doi.org/10.1007/978-981-99-8435-0_27

To solve the above problem, existing works use methods such as Optical Flow [12, 13], Long Short-Term Memory (LSTM) [14, 15] and ConvLSTM [16–18] to extract features and predict future frames. There are also works that combine video object tracking and detection [19, 20] to extract the spatio-temporal information of the detected object, thereby optimizing the detection accuracy. However, these methods are typically designed for offline environments and do not account for latency in streaming perception scenarios.

Li et al. [21] point out that the difference between standard offline evaluation and real-time applications: the surrounding world may have changed by the time an algorithm finishes processing a frame. To address the problem, 'Streaming Average Precision (sAP)' is introduced, which integrates latency and accuracy to evaluate real-time online perception. This metric evaluates the entire perception stack in real-time, forcing model to predict current world state, as shown in Fig. 1(b).

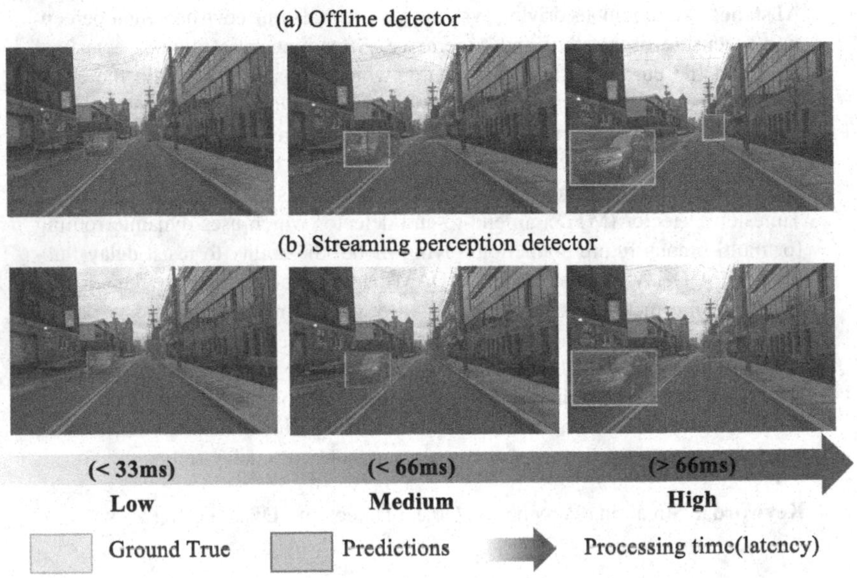

Fig. 1. Visualization of theoretical output of offline detectors vs. streaming perception detectors under various latency environments. Offline detectors (a) and streaming perception detectors (b) produce very different predictions when encountering processing delays.

StreamYOLO [22] suggests that the performance gap in streaming perception tasks is due to the current processing frame not matching the next frame, so they simplified streaming perception to the task of predicting the next frame. The work introduces Dual-Flow Perception (DFP) Module to merge the features of the previous and current frames, and Trend-Aware Loss (TAL) to adjust adaptive weights for each object based on its motion trend, achieving state-of-the-art sAP performance. However, it can only predict the next frame, and its performance is optimal when the processing time is less than inter-frame time.

Based on the work of StreamYOLO, DaDe [23] suggested that automatic driving system should consider additional delays. It proposes Feature Queue Module to store historical features, and Feature Selection Module supervises processing delay. Based on the processing delay, the Feature Select Module adaptively selects different historical features for fusion to create object motion trends, predicting future with different timesteps. The study improves the baseline to handle unexpected delays, achieving the highest sAP performance with one additional timestep, but performance degradation increases when more timesteps are considered.

Although the above methods improve performance to some extent, the predicted timestep is still inadequate to keep up with the real-time environment when processing delay is substantial.

Dynamic networks [24] present a novel option for solving this problem. Unlike static models with fixed calculation graphs and parameters, dynamic networks can dynamically adjust structure during inference, providing properties that static models lack. Common dynamic networks include dynamic depth networks that distinguish sample difficulty by early exit [25, 26] or skipping layers [27, 28], and dynamic width networks [29, 30] that selectively activate neurons and branches according to input during the inference process. Both methods adjust depth or width of the architecture by activating computing units to reduce model costs. In contrast, dynamic routing networks [31, 32] perform dynamic routing in a supernetwork with multiple possible paths, by assigning features to different paths, allowing the computational graph to adapt to each sample during inference.

In this paper, a Multi-Timestep Detector (MTD) is proposed. The main idea is to transform the detection problem of delayed streaming perception into a future prediction problem with multiple timesteps. A Timestep Branch Module (TBM) is constructed to investigate the generality of backbone features on similar tasks. By redesigning the training pipeline, multiple detection head branches are trained to predict different timesteps in the future. The Delay Analysis Module (DAM) continuously monitors preprocessing and inference delays, calculates the delay trend, and analyzes the target timestep, enabling the model to choose the best branch for detection. Experiments are conducted on the Argoverse-HD [21] dataset, and MTD demonstrates significant performance improvements over various delay settings compared to previous methods. The contributions of the work are as follows:

- By analyzing the processing time fluctuations of the model, the inference delay trend is introduced for more reliable calculation of the delay trend. Based on the idea of dynamic routing, Timestep Branch Module is proposed, where the delay trend is used as the routing basis. Under various delay settings, these methods can effectively improve the accuracy of streaming perception.
- To the best of our knowledge, MTD is the first end-to-end model that uses dynamic routing for delayed streaming perception. Under various delay settings, our method improves the sAP [21] performance by 1.1–2.9 on the Argoverse-HD dataset compared with baselines.

2 The Methods

This section describes how MTD is designed to predict future at different timesteps. The Delay Analysis Module is constructed to calculate delay trend. The Timestep Branch Module dynamically routes within the network based on delay trend and produces outputs that align with the real environment. The training pipeline is designed to avoid adding extra computations during inference and maintain real-time performance. Figure 2 illustrates the model's inference process.

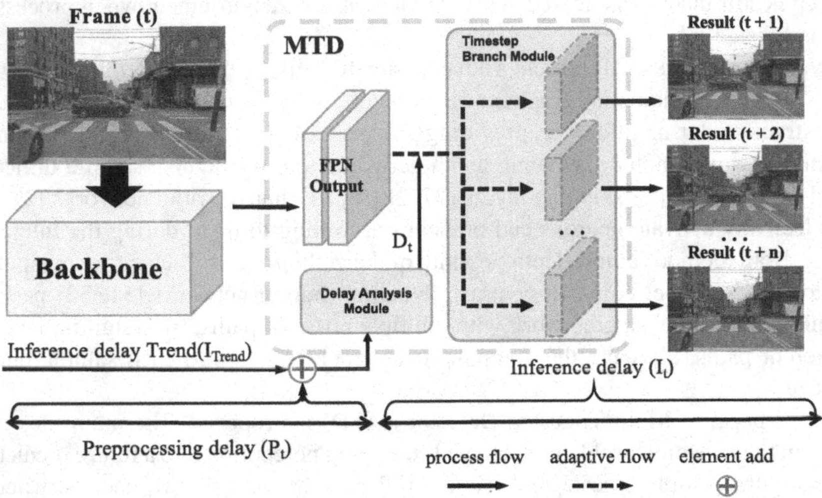

Fig. 2. MTD extracts features using a shared backbone. Delay Analysis Module (DAM) calculates delay trend D_t according to current preprocessing delay P_t and inference delay trend I_{Trend}. Timestep Branch Module (TBM) selects detection head adaptively to predict future at different timestep based on input D_t. The inference path of the detector is divided into process flow and adaptive flow, in which the process flow is a fixed inference path, and the adaptive flow is an optional path, which is dynamically routed by the TBM according to the delay trend.

2.1 Delay Analysis Module (DAM)

The Delay Analysis Module calculates delay trend for subsequent modules inference. In the work of DaDe[23], the delay trend D_t of the current frame t is simply expressed as the sum of the preprocessing delay P_t and the last frame inference delay I_{t-1}. This approach is simple and effective, but performs best only when the model's processing time trend is flat.

Figure 3 visualizes frame processing time under none delay setting. It is observed that the model's processing time is mostly smooth, with occasional small peaks. As shown in Fig. 3 (a)–(c), model delay fluctuations often occur within a single frame, leading to the delay analysis method in [23] incorrectly assessing the delay after each anomaly frame. Consequently, subsequent modules struggle to hit the correct timestep.

To solve this issue, Inference delay trend I_{Trend} is introduced to reduce the sensitivity of inference delay perception. The formula for calculating I_{Trend} is as follows:

$$I_{Trend} = \begin{cases} I_{t-1}, \frac{I_{t-1}}{I_{t-2}} < \tau \\ I_{t-2}, \frac{I_{t-1}}{I_{t-2}} \geq \tau \end{cases} \quad (1)$$

where the inference delay trend I_{Trend} is the ratio of the inference delay from the previous two frames, I_{t-1} and I_{t-2}, which evaluates the fluctuation of the processing delay. Hyperparameter τ a is obtained by grid search.

By calculating inference delay trend and preprocessing delay of the current frame, the delay trend D_t is defined as follows:

$$D_t = P_t + I_{Trend} \quad (2)$$

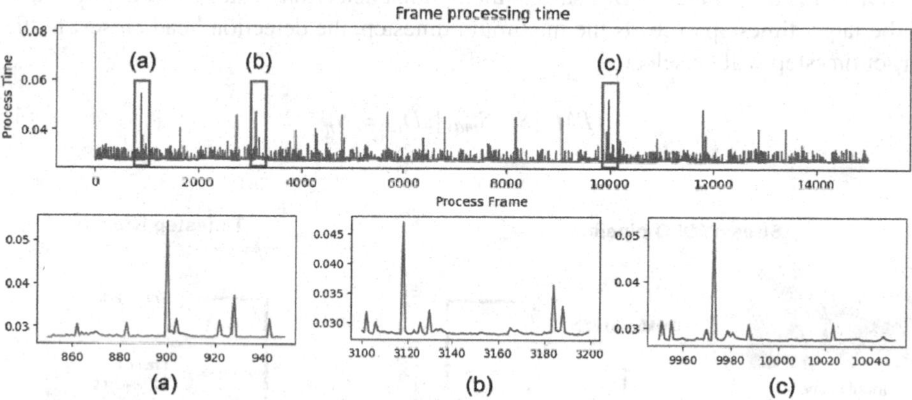

Fig. 3. Visualization of model frame processing time. The processing time of each frame is recorded in an environment without additional delay.

2.2 Timestep Branch Module (TBM)

The feature extracted by the backbone has generalization capabilities for similar tasks. Specifically, the feature contains moving trend and basic semantic information that can be used not only to predict the next frame but also to predict future frames at farther timesteps.

To this end, Timestep Branch Module (TBM) is designed for multi-branch decoding of features. To demonstrate the effectiveness of this method, StreamYOLO's detection head [22] was repurposed as a template for multiple branches, rather than designing a more complex one.

In StreamYOLO, the training input is a triplet consisting of previous frame F_{t-1}, current frame F_t, and the ground truth F_{t+1}. So the training dataset is reconstructed as $\{(F_{t-1}, F_t, F_{t+1})\}_{t=1}^{N}$, where N is the total number of samples. To adapt this method for multi-branch structure model training, future frames with various timesteps is used as the predicted ground truth for different detection heads. Dataset is reconstructed as $\{\{(F_{t-1}, F_t, F_{t+s})\}_{t=1}^{N}\}_{i=2}^{S}$, where S is the total number of detection heads and i is the index of detection heads. The pre-trained head from StreamYOLO is skipped and started the index of i from 2. During training process, the model trains each newly added detection head iteratively based on dataset. To prevent other parts of the model from being affected by backpropagation, the backbone and other detection heads are frozen during training. Figure 4 visualizes the training process.

During the inference process, TBM receives delay trend D_t from DAM as input, and calculates the target timestep n according to delay trend D_t and inter-frame time T. The formula for target timestep n is as follows:

$$n = \left\lfloor \frac{D_t}{T} \right\rfloor \tag{3}$$

Once n is determined, TBM selects the optimal detection head S_n using Formula 4. If the target timestep exceeds the maximum timestep, the detection head closest to the target timestep will be selected.

$$TBM\left([S_1, S_{max}], D_t\right) = S_n \tag{4}$$

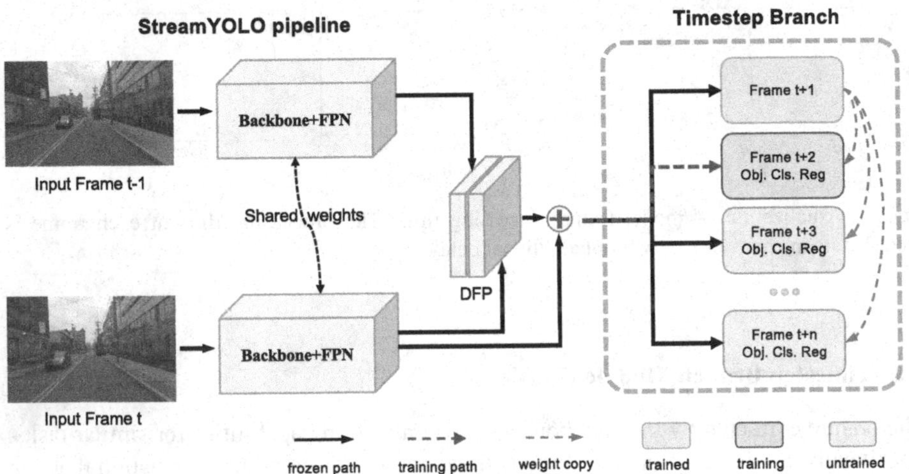

Fig. 4. Training process of MTD. To initialize all detection heads of the TBM, the detection head from StreamYOLO is used as a template and set the same weight. The training process begins with the t + 2 branch and gradually traverses all subsequent heads.

3 Experiments and Evaluation

3.1 Experimental Settings

Dataset

Argoverse-HD [21] dataset provides high-resolution and high-frame-rate driving video image data, and provides sufficient streaming perception annotation data to evaluate the real-time performance of object detection algorithm. The train/val split in [21–23] is followed for evaluation and comparison with other relevant studies.

Evaluation Metrics

Streaming Average Precision (sAP). sAP [21] is a metric for streaming perception. It takes into account both latency and accuracy, and evaluates the model output under the framework of current world state. Similar to MSCOCO [33], The sAP toolkit [34] evaluates average mAP over Intersection-over-Union (IoU) thresholds from 0.5 to 0.95, and sAP_S, sAP_M, sAP_L for different object size.

Missed Timestep. To assess the performance of DAM, a new metric called Missed Timestep is introduced. The delay trend can be obtained from Formula 2, while the actual delay trend is defined as Formula 5, which is the sum of the preprocessing time and inference time of the current frame.

$$AD_t = P_t + I_t \tag{5}$$

By applying these formulas to Formula 3, the timestep n of the delay trend D_t and timestep m of the actual delay trend AD_t can be obtained. When n and m are equal, the model selected the best detection head. Conversely, when n and m are not equal, the model missed the timestep and selected a sub-optimal detection head. The lower the total number of Missed Timesteps, the better the performance of DAM.

Implementation Details

MTD is fine-tuned based on StreamYOLO [22] pre-training weight. Two additional detection heads are trained to cover common delays. The model is trained on four 1080Ti GPUs, and each detection head is trained for 5 epochs. Batchsize is set to 32. The model performance is evaluated on a 3090Ti GPU.

Delay Settings

Delay settings (none, low, medium, and high) are simulated by manually injecting processing delays. The medium delay simulated the fluctuation of delay across different timesteps to assess the model's performance in cases of severe delay fluctuations. Figure 5 illustrates the processing time distribution under each setting. Inter-frame time is 33 ms at 30 FPS, so processing time should be less than 33 ms to maintain real-time performance. As shown in Table 1, under none delay settings, all three models can maintain real-time performance. As the delay increases, evaluation requires the use of t + n frames that correspond to the real environment as the ground truth for comparison, and the model's prediction difficulty gradually increases.

Table 1. Mean delay and standard deviation in various delay settings. All delays are in milliseconds.

Environment	None		Low		Medium		High	
StreamYOLO	27.8	1.63	59.2	2.82	69.6	4.31	90.4	3.47
DaDe	27.9	1.89	59.3	2.47	69.5	4.52	90.3	3.67
Ours	28.1	2.1	59.5	3.24	68.7	4.46	89.8	3.23

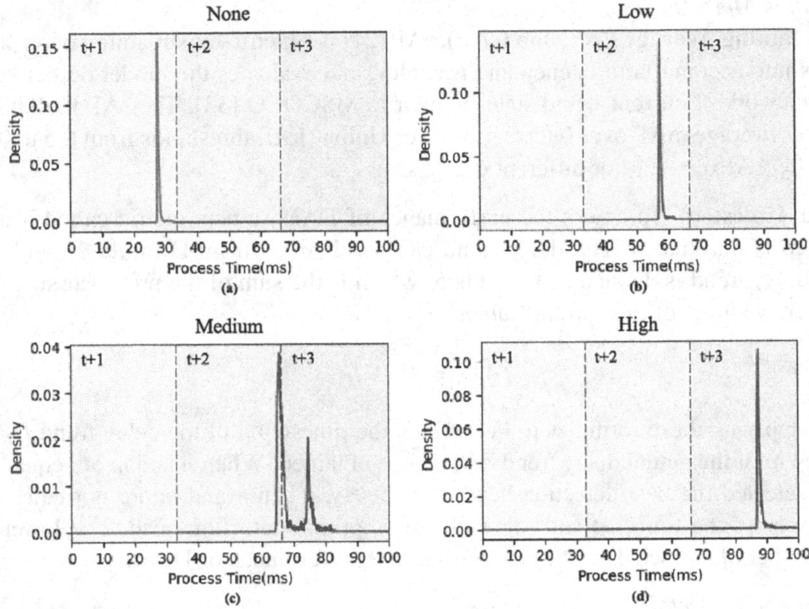

Fig. 5. Visualization of delay distribution for various delay environments. Vertical dotted line separates the inter-frame space.

3.2 Experimental Results

Evaluation of Delay Analysis Methods

The Delay Analysis Module calculates delay trends for subsequent inference modules according to inference time and preprocessing time. Incorrect delay trends can cause the Timestep Branch Module to hit the wrong timestep, resulting in poor inference performance. The performance of the Delay Analysis Module is assessed by counting the number of missed timesteps. According to the definition of Formula 1, the value of τ is used as a threshold to control the model's sensitivity to delay fluctuations. A grid search is performed on the value of τ, and the results are shown in Table 2. The Delay Analysis Module has the least number of missed timesteps when the value of τ is 1.00. Table 3 show the comparisons of our delay analysis method and DaDe's method in terms of missed timesteps under various delay settings. Our method reduces the number of missed timesteps by 14.4% to 30% compared to DaDe [23].

Table 2. Grid search for τ in Formula 1

τ	None	Low	Medium	High
0.70	72	12	1434	2
0.80	57	10	1369	2
0.90	46	10	1278	2
		...		
0.99	45	10	1166	2
1.00	**45**	**10**	**1162**	**2**
1.01	45	10	1164	2
		...		
1.10	47	10	1246	2
1.20	48	9	1306	2
1.30	54	10	1357	2

Table 3. Performance comparisons of delay analysis method (MTD vs DaDe).

Environment	Processed Frame	Missed Timestep	Decrease Percent
	None (27 ms ± 1)		
DaDe	15035/15062	58	22.4%
MTD ($\tau = 1$)		45	
	Low (59 ms ± 1)		
DaDe	8285/15062	13	23.0%
MTD ($\tau = 1$)		10	
	Medium (69 ms ± 1)		
DaDe	7164/15062	1359	14.4%
MTD ($\tau = 1$)		1162	
	High (90 ms ± 1)		
DaDe	5580/15062	3	30.0%
MTD ($\tau = 1$)		2	

Comparison of Computations

The additional detection head increases the total parameters of the MTD, but the increased parameters are distributed on different paths, so the processing time and Gflops during inference are unchanged (see Table 4), which makes the processing time gap between the MTD and the baselines almost negligible. (27.84 ms vs 28.1 ms).

Table 4. Comparison of model parameters, Gflops and processing time.

Environment	Params	Gflops	Processing Time (ms)
StreamYOLO	54.84M	383.47	27.84
DaDe	54.84M	383.47	27.92
MTD	69.95M (**+15.11**)	383.47	28.1 (**+0.26**)

Comparisons with SOTA Methods

A performance comparison is performed between MTD and two baselines (DaDe [23] and StreamYOLO [22]) under various delay settings, with the outcomes detailed in Table 5. Without extra delay, the performance of all detectors is similar. Under low delay, StreamYOLO's performance starts to lag behind the others as its frame processing time exceeds the frame interval time (33 ms). As the processing delay increases, StreamYOLO cannot predict the future of farther timesteps, resulting in a widening performance gap. DaDe constructs long-term motion trends by fusing historical frame features with longer time intervals, resulting in a slower decline in performance under low-latency settings. However, as the delay trend moves towards medium and high delay, its performance declines more and more faster, even lower than StreamYOLO under high delay. This issue arises from the fused features being temporally distant, causing feature-level fusion ineffective in creating a motion trend, hence reducing the model's inference performance. Under different delay settings, our method improves sAP performance by 1.1–2.9, achieving the best performance across all metrics, further validating the effectiveness of our approach.

Table 5. Performance comparison with baselines in Argoverse-HD dataset.

Environment	sAP	sAP_{50}	sAP_{75}	sAP_S	sAP_M	sAP_L
			None (27 ms ± 1)			
StreamYOLO	36.9	58.1	37.6	14.7	37.4	64.2
DaDe	36.9	58.0	37.6	14.6	37.4	64.4
MTD	36.9	58.1	37.7	15.0	37.0	64.5
			Low (59 ms ± 1)			
StreamYOLO	26.2	48.0	24.4	8.8	24.5	41.5
DaDe	27.8	49.0	26.8	9.8	27.2	42.3
MTD	**29.1**	**51.5**	**28.0**	**10.2**	**28.3**	**46.6**

(*continued*)

Table 5. *(continued)*

Environment	sAP	sAP_{50}	sAP_{75}	sAP_S	sAP_M	sAP_L
Medium (69 ms ± 1)						
StreamYOLO	25.3	46.8	23.1	8.2	23.4	39.9
DaDe	25.3	46.9	23.5	8.2	23.6	39.7
MTD	**26.4**	**49.0**	**24.2**	**8.7**	**24.7**	**43.1**
High (90 ms ± 1)						
StreamYOLO	22.7	43.3	20.5	6.9	20.7	35.8
DaDe	22.4	42.2	20.1	7.0	20.5	34.5
MTD	**24.1**	**46.1**	**21.4**	**7.9**	**21.6**	**39.1**

4 Conclusion

This paper presents a Multi-timestep Detector[1] (MTD) that includes a Delay Analysis Module (DAM) to monitor the model's delay trend in real-time, and a Timestep Branch Module (TBM) that employs dynamic routing in multiple branches for multi-timestep prediction. The experimental results shows that MTD is particularly suitable for delayed streaming perception scenarios and achieves state-of-the-art performance under various delay settings. Moreover, the TBM demonstrates that the features extracted by the backbone are universally applicable to predicting future video frames with various timesteps. In the future, it is plan to investigate whether this method has generalization significance in other related tasks, such as video sequence generation and human pose prediction.

Acknowledgment. This work is funded by the Natural Science Foundation of China (No. 62162003).

References

1. Girshick, R., Donahue, J., Darrell, T., Malik, J.: Rich feature hierarchies for accurate object detection and semantic segmentation. In: IEEE Conference on Computer Vision and Pattern Recognition, pp. 580–587 (2014)
2. Girshick, R.: Fast R-CNN. In: IEEE International Conference on Computer Vision, pp. 1440–1448 (2015)
3. Ren, S., He, K., Girshick, R., Sun, J.: Faster R-CNN: towards real-time object detection with region proposal networks. In: Advances in Neural Information Processing Systems, vol. 28(2015)
4. Redmon, J., Divvala, S., Girshick, R., Farhadi, A.: You only look once: unified, real-time object detection. In: IEEE Conference on Computer Vision and Pattern Recognition, pp. 779–788(2016)

[1] The code is available at: https://github.com/Yulin1004/MTD.

5. Redmon, J., Farhadi, A.: YOLO9000: better, faster, stronger. In: IEEE Conference on Computer Vision and Pattern Recognition, pp. 7263–7271 (2017)
6. Farhadi, A., Redmon, J.: Yolov3: an incremental improvement. In: IEEE Conference on Computer Vision and Pattern Recognition, pp. 1–6 (2018)
7. Ge, Z., Liu, S., Wang, F., Li, Z., Sun, J.: YOLOX: exceeding YOLO series in 2021. arXiv preprint arXiv:2107.08430 (2021)
8. Liu, W., et al.: SSD: single shot multibox detector. In: Leibe, B., Matas, J., Sebe, N., Welling, M. (eds.) ECCV 2016. LNCS, vol. 9905, pp. 21–37. Springer, Cham (2016). https://doi.org/10.1007/978-3-319-46448-0_2
9. Kato, S., Brandt, S., Ishikawa, Y., Rajkumar, R.: Operating systems challenges for GPU resource management. In:International Workshop on Operating Systems Platforms for Embedded Real-Time Applications, pp. 23–32 (2011)
10. Bateni, S., Wang, Z., Zhu, Y., Hu, Y., Liu, C.: Co-optimizing performance and memory footprint via integrated CPU/GPU memory management, an implementation on autonomous driving platform. In: 2020 IEEE Real-Time and Embedded Technology and Applications Symposium, pp. 310–323 (2020)
11. Benoit-Cattin, T., Velasco-Montero, D., Fernández-Berni, J.: Impact of thermal throttling on long-term visual inference in a CPU-based edge device. Electronics **9**(12), 2106 (2020)
12. Zhu, X., Xiong, Y., Dai, J., Yuan, L., Wei, Y.: Deep feature flow for video recognition. In: IEEE Conference on Computer Vision and Pattern Recognition, pp. 2349–2358 (2017)
13. Zhu, X., Wang, Y., Dai, J., Yuan, L., Wei, Y.: Flow-guided feature aggregation for video object detection. In: IEEE International Conference on Computer Vision, pp. 408–417 (2017)
14. Hochreiter, S., Schmidhuber, J.: Long short-term memory. Neural Comput.Comput. **9**(8), 1735–1780 (1997)
15. Xiao, F., Lee, Y.J.: Video object detection with an aligned spatial-temporal memory. In: Ferrari, V., Hebert, M., Sminchisescu, C., Weiss, Y. (eds.) ECCV 2018. LNCS, vol. 11212, pp. 494–510. Springer, Cham (2018). https://doi.org/10.1007/978-3-030-01237-3_30
16. Shi, X., Chen, Z., Wang, H., Yeung, D.-Y., Wong, W., Woo, W.: Convolutional LSTM network: a machine learning approach for precipitation nowcasting. In: Advances in Neural Information Processing Systems, vol. 28(2015)
17. Chang, Z., et al.: MAU: a motion-aware unit for video prediction and beyond. In: Advances in Neural Information Processing Systems, vol. 34, pp. 26950–26962 (2021)
18. Hu, J.-F., Sun, J., Lin, Z., Lai, J.-H., Zeng, W., Zheng, W.-S.: APANet: auto-path aggregation for future instance segmentation prediction. IEEE Trans. Pattern Anal. Mach. Intell.Intell. **44**(7), 3386–3403 (2022)
19. Mao, H., Kong, T.: CaTDet: cascaded tracked detector for efficient object detection from video. Proc. Mach. Learn. Syst. **1**, 201–211 (2019)
20. Feichtenhofer, C., Pinz, A., Zisserman, A.: Detect to track and track to detect. In: IEEE International Conference on Computer Vision, pp. 3038–3046 (2017)
21. Li, M., Wang, Y.-X., Ramanan, D.: Towards streaming perception. In: Vedaldi, A., Bischof, H., Brox, T., Frahm, J.-M. (eds.) ECCV 2020. LNCS, vol. 12347, pp. 473–488. Springer, Cham (2020). https://doi.org/10.1007/978-3-030-58536-5_28
22. Yang, J., Liu, S., Li, Z., Li, X., Sun, J.: Real-time object detection for streaming perception. In: IEEE/CVF Conference on Computer Vision and Pattern Recognition, pp. 5385–5395 (2022)
23. Jo, W., Lee, K., Baik, J., Lee, S., Choi, D., Park, H.: DaDe: delay-adaptive detector for streaming perception.arXiv preprint arXiv:2212.11558 (2022)
24. Han, Y., Huang, G., Song, S., Yang, L., Wang, H., Wang, Y.: Dynamic neural networks: a survey. IEEE Trans. Pattern Anal. Mach. Intell.Intell. **44**(11), 7436–7456 (2022)
25. Teerapittayanon, S., McDanel, B., Kung, H.T.: BranchyNet: fast inference via early exiting from deep neural networks. In: International Conference on Pattern Recognition, pp. 2464–2469 (2016)

26. Liu, X., Mou, L., Cui, H., Lu, Z., Song, S.: Finding decision jumps in text classification. Neurocomputing **371**, 177–187 (2020)
27. Wang, X., Yu, F., Dou, Z.-Y., Darrell, T., Gonzalez, J.E.: SkipNet: learning dynamic routing in convolutional networks. In: Ferrari, V., Hebert, M., Sminchisescu, C., Weiss, Y. (eds.) ECCV 2018. LNCS, vol. 11217, pp. 420–436. Springer, Cham (2018). https://doi.org/10.1007/978-3-030-01261-8_25
28. Shen, J., Wang, Y., Xu, P., Fu, Y., Wang, Z., Lin, Y.: Fractional skipping: towards finer-grained dynamic CNN inference. In: AAAI Conference on Artificial Intelligence, pp. 5700–5708 (2020)
29. Mullapudi, R.T., Mark, W.R., Shazeer, N., Fatahalian, K.: HydraNets: specialized dynamic architectures for efficient inference. In: IEEE Conference on Computer Vision and Pattern Recognition, pp. 8080–8089 (2018)
30. Ehteshami Bejnordi, A., Krestel, R.: Dynamic channel and layer gating in convolutional neural networks. In: Schmid, U., Klügl, F., Wolter, D. (eds.) KI 2020. LNCS (LNAI), vol. 12325, pp. 33–45. Springer, Cham (2020). https://doi.org/10.1007/978-3-030-58285-2_3
31. Li, Y., et al.: Learning dynamic routing for semantic segmentation. In: IEEE/CVFConference on Computer Vision and Pattern Recognition, pp. 8553–8562 (2020)
32. Liu, L., Deng, J.: Dynamic deep neural networks: optimizing accuracy-efficiency trade-offs by selective execution. In: AAAI Conference on Artificial Intelligence, vol. 32, no. 1 (2018)
33. Lin, T.-Y., Maire, M., Belongie, S., Hays, J., Perona, P., Ramanan, D., Dollár, P., Zitnick, C.L.: Microsoft COCO: common objects in context. In: Fleet, D., Pajdla, T., Schiele, B., Tuytelaars, T. (eds.) ECCV 2014. LNCS, vol. 8693, pp. 740–755. Springer, Cham (2014). https://doi.org/10.1007/978-3-319-10602-1_48
34. sAP: Code for Towards Streaming Perception. https://github.com/mtli/sAP. Accessed 20 June 2023

Semi-Direct SLAM with Manhattan for Indoor Low-Texture Environment

Zhiwen Zheng[1], Qi Zhang[2(✉)], He Wang[1], and Ru Li[1(✉)]

[1] School of Computer and Information Technology, Shanxi University, Shanxi 030006, China
liru@sxu.edu.cn

[2] University of Bath, Bath BA2 7AY, UK
qz727@bath.ac.uk

Abstract. Simultaneous Localization and Mapping (SLAM) with the incorporation of the Manhattan World (MW) assumption has been significantly discovered in recent years. While previous methods relied on the MW assumption to estimate camera rotation accurately, they faced limitations due to the requirement of a suitable planar environment. These constraints restricted the applicability of such systems. To overcome these limitations, we propose a novel approach that addresses the strict requirements of MW-based systems and significantly enhances tracking robustness in low-texture scenes. Our system leverages planar information in the environment to identify the presence of an MW scene. By decoupling the process, we achieve drift-free rotation estimation when the system detects an MW scene. Simultaneously, utilizing a semi-direct approach that combines point and line features to estimate translations in MW scenes, while performing full-camera pose estimation in non-MW scenes. Furthermore, we introduce a more precise loop closure detection strategy by exploiting the relative relationship between the Manhattan axes (MA) and line features in the scene. This strategy enhances the accuracy of identifying loop closures, which are crucial for SLAM systems. To evaluate the performance of our approach, we conducted experiments using public benchmarks. The results demonstrate improved pose estimation and loop closure performance compared to state-of-the-art methods. Overall, our proposed method alleviates the strict requirements of previous MW-based systems, enhances tracking robustness in low-texture scenes, and achieves improved performance in terms of pose estimation and loop closure detection.

Keywords: RGB-D SLAM · Indoor environments · Manhattan World · Semi-Direct approach

This work was supported in part by research grants from the Science and Technology Cooperation and Exchange Special Project of Shanxi Province (No.202204041101016), the 1331 Engineering Project of Shanxi Province, and the Key Research and Development Project of Shanxi Province (NO.202102020101008).

Q. Liu et al. (Eds.): PRCV 2023, LNCS 14427, pp. 350–362, 2024.
https://doi.org/10.1007/978-981-99-8435-0_28

1 Introduction

In a wide range of applications such as autonomous robots, self-driving cars, and augmented and virtual reality, Simultaneous Localization and Mapping (SLAM) algorithms are crucial for providing accurate pose estimation and maps. Camera pose estimation enables the localization of vehicles, robots, and mobile devices, while the map of the scene provides an environmental representation for the interaction between humans or robots and the environment.

In SLAM systems, the accumulation of small errors in pose estimation over time ultimately leads to the drift of camera pose trajectory. It is proved that inaccurate rotation estimations are the main cause of trajectory errors [1], and a possible method to reduce trajectory errors is to employ the Manhattan World (MW) assumption for rotation estimation [2,3]. This method models the environment as a MW [4] by using parallel and perpendicular relationships between structural features in the scene. MW is widely present in artificial environments [5], such as indoor scenes and outdoor parking lots. However, the textural features of most artificial structures, such as walls and ceilings, are not obvious in most of indoor scenes. While using MW assumption can result in drift-free rotation in such scenes, estimating translation remains the biggest restraining factor.

In this paper, we propose a method that uses plane information in the scene to determine whether MW exists and decouples MW when detected to obtain drift-free rotation, and using the semi-direct approach to estimate camera translation. In scenes without MW, the semi-direct approach is used entirely for camera pose estimation. Therefore, our method not only mitigates the strict requirements of MW assumption but also has stronger adaptability and can efficiently perform tracking in low-texture scenes, making it more robust than existing MW-based methods. In addition, since the Manhattan Axes (MA) is fixed in the same scene, we use the relative position relationship between MA and line segments in the scene for loop detection, which can better utilize loop closure information when appearing during runtime due to long-time operation. The main contributions of this paper are as follows:

1. Improving existing MW-based tracking methods by decoupling the MW for drift-free rotation estimation under certain conditions, while using a semi-direct approach for tracking when MW is not present.
2. A new loop detection method that screens correct loop closure keyframes by determining the relative position relationship between structural features in the scene and MA.
3. A general SLAM system for indoor tracking using a real-time RGB-D camera that is used for high-precision positioning in structured scenes or low-texture environments.

We have conducted experiments on both synthetic and real-world datasets with the combination of our proposed method and a baseline RGBD-SLAM [14]. The real-time ATE of our proposed method has decreased by 12.23%, 42.65% on ICL-NUIM and TUM RGB-D datasets respectively when comparing with the baseline method. Our method also showed better performance compared to state-of-the-art methods on both synthetic and real-world datasets.

2 Related Work

Visual odometry can be divided into two main branches. The first is a feature-based algorithm, and point-only feature methods [6–9] suffer from poor performance in low-textured regions. Structural features such as lines and planes can complement point features in low-texture environments. The already widely used ORB-SLAM point feature framework [6,8,9] can be easily extended to line and plane [10,11]. Another branch is called direct method, in [12] the authors propose a dense direct method, called DTAM, to estimate camera pose by directly aligning the full photometric image between each keyframe. However, this method needs to use GPU parallelization for computation since it processes the entire image. A new monocular odometry is proposed in [18], where the author proposed DSO based on sparse direct method to estimate camera motion, thus achieving real-time performance on CPU. In [19], the author proposed a semi-direct approach called SVO, which combines feature-based methods with sparse direct methods. Subsequently, Gomez-Ojeda *et al.* [21] introduced line features into the semi-direct approach SVO, which improved the adaptability of the semi-direct approach to the environment.

Existing research [1,22] shows that the main reason of long-term drift in SLAM systems is the rotation estimation error. Artificial environments such as corridors and ceilings in indoor settings often contain surfaces with orthogonal structures, [4] creation of the term MW to describe such an environment. SP-SLAM [11] is inspired by MW assumption and adds constraints between planes in a scene to enhance its robustness. In [22], the author uses the sphere mean-shift algorithm to track the dominant direction of MW and obtain drift-free rotation, while using density alignments for translation estimation.

These methods require at least two mutually orthogonal planes in each frame. RGBD-SLAM [14] alleviates the strict requirement of the system for scenes by incorporating line segments as planar normal vectors into calculations and uses line and planar features for translation estimation. Yunus [15] proposes switching to using feature-based methods based on points, lines, and planes in scenarios where MW assumption is not valid. Joo [16] extends the sphere mean-shift algorithm to a more general Atlanta World scenario. [17] combining the estimation of MW with the reprojection error of point and line features based on RGB-D cameras and optimizing camera posture through local map optimization. Peng [20] uses vanishing points to re-identify Manhattan frame.

3 System Overview

In this section, we describe the detailed implementation of our SLAM framework. Our system is built on top of the tracking and loop closing components of RGBD-SLAM. Therefore, it comprises two threads running in parallel. The overview of our system is illustrated in Fig. 1.

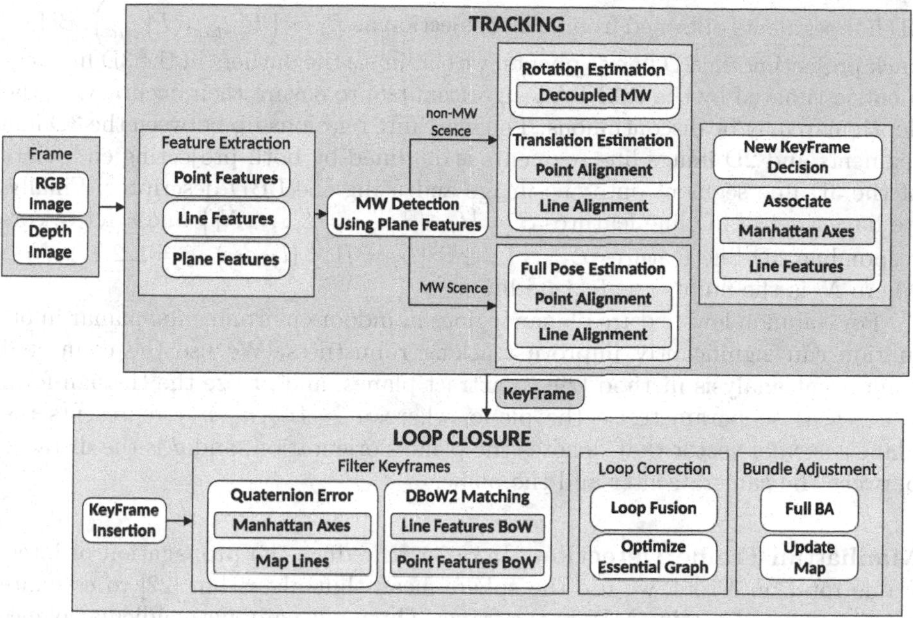

Fig. 1. Overview of the proposed framework. The tracking thread preprocesses the RGB-D input, and after extracting features, divides the scene into MW scenes and non-MW scenes for pose estimation respectively. The loop closure thread detects loop closures using the relative positional relationship between MA and the map line.

3.1 Tracking

The tracking thread is responsible for estimating the 6-DoF camera pose for each input frame. It is obtained by calculating the error between point features, line features and MW in adjacent keyframes.

Feature Extraction. Our proposed system both utilizes structured information in the environment including points, lines and planes for tracking to ensure the system can track low texture scenes.

To extract feature points, we use ORB features [25]. The 2D feature points extracted from frame F_i are represented as $x_j = (u_j, v_j)$, and we can obtain a feature point set $\mathcal{U} = \left\{x_j \in \mathbb{R}^2 \mid j = 1, \ldots, N_p\right\}$, where N_p is the number of point features. Then, we define Π as the projection function which is determined by the camera parameters. By using the inverse projection function Π^{-1} and depth d_{x_j}, we can obtain the corresponding 3D points, i.e. $P_j = \Pi^{-1}\left(x_j, d_{x_j}\right), P_j \in \mathbb{R}^3$. Matching is performed by projecting 3D points onto images and comparing the Hamming distance of their respective descriptors.

For line data, we extract line segments using the LSD line segment detection algorithm [26] and LBD descriptor [27]. We represent the 2D line segments extracted from image frame F_i as $l_j = \left(x^l_{j,start}, x^l_{j,end}\right)$ and their corresponding

3D line segments obtained from back projection as $L_j = \left(P^l_{j,start}, P^l_{j,end}\right)$. Before back projecting the 2D line segments, we eliminate the outliers in the 3D line segments estimated by the RANSAC algorithm [28] to ensure their accuracy, as the depth map may be discontinuous. The matching relationship between the 3D line segments and 2D image line segments is obtained by both projecting endpoints of the 3D line segment onto the image and using the LBD descriptor. Finally, we have a set of 2D line features $\mathcal{S} = \{l_j \in \mathbb{R}^2 \mid j = 1, \ldots, N_l\}$ and a set of corresponding 3D line features $\mathcal{L} = \{L_j \in \mathbb{R}^3 | L_j = \Pi^{-1}(l_j, d_{l_j}), j = 1, 2, \ldots, N_l\}$, where N_l is the number of feature lines.

For common low-texture planar regions in indoor environments, planar information can significantly improve tracking robustness. We use the connected component analysis method [29] to extract planes, and utilize the Hessian form $\pi = (n, d)$ to parameterize the plane, where $n = (n_x, n_y, n_z)$ represents the plane's normal vector that denotes the plane's orientation, and d is the distance between the camera center and the plane.

Manhattan Frame Detection. In order to reduce the propagation of inter-frame rotation errors, we use the sphere mean-shift algorithm [22] to estimate rotation when detecting MW in the frame. Three mutually perpendicular planes can form the MW assumption, which can be represented by the normal vector of the plane as (n_1, n_2, n_3). We use the plane information extracted from the scene to determine whether MW exists, as we calculate the angles between the plane normal vectors n in the scene, where $n \in (n_0 \ldots n_r)$ is the normal vector of the plane extracted from the frame and r is the total number of planes extracted from the frame. If any three plane normal vectors are found to be mutually perpendicular, MW is detected. While if only two mutually perpendicular planes are found, and their normal vectors are n_1 and n_2, the normal vector of the third plane n_3 can be calculated by the cross product of n_1 and n_2. In summary, only two mutually perpendicular planes are needed to obtain the MW.

Pose Estimation. The camera pose $\mathbf{T} \subset SE(3)$ consists of a rotation matrix $\mathbf{R} \in SO(3)$ and a translation vector $\mathbf{t} \in \mathbb{R}^3$. If the MW exists, we decouple the MW to obtain a drift-free rotation, which serves as the initial value for subsequent pose estimation. If MW cannot be detected, we estimate the 6D camera pose fully through semi-direct approach.

In the non-MW scenes, we use the semi-direct approach of points and lines to estimate the 6D camera pose. We represent the image intensity in the k-th frame as $I_k : \Omega \subset \mathbb{R}^2 \mapsto \mathbb{R}$, where Ω represents the image area. The photometric error of aligning sparse feature images between two consecutive frames is defined as δI and δI_l:

$$\begin{cases} \delta I(\mathbf{T}, x) = I_k\left(\Pi \cdot \left(\mathbf{T} \cdot \Pi^{-1}(x, d_x)\right)\right) - I_{k-1}(x) \quad \forall x \in \mathcal{U} \\ \delta I_l(\mathbf{T}, l) = \frac{1}{N} \sum_{n=0}^{N} \left| I_k\left(\Pi \cdot \left(\mathbf{T} \cdot \Pi^{-1}(w_n, d_{w_n})\right)\right) - I_{k-1}(w_n) \right| \quad \forall w_n \in \mathcal{L} \end{cases} \tag{1}$$

Since it is impossible to directly align the entire region of line feature correspondence between two frames, we use segmented sampling to minimize the residual between N image blocks uniformly distributed along the line segment. Where w_m, with $m = 2, \ldots, N-1$ referring to the intermediate points defined homogeneously along the line segments. When $m = 0$ or $m = N$, w_m refers to the two endpoints of the line segment. Then, we minimize the photometric error between all image blocks corresponding to the point and line features to obtain the relative transformation between two consecutive frames:

$$T_{k,k-1} = \underset{T_{k,k-1}}{agrmin} \left\{ \sum_{i \in \mathcal{U}} \left\| \delta I \left(\mathbf{T}, x_i \right) \right\| + \sum_{j \in \mathcal{L}} \left\| \delta I_l \left(\mathbf{T}, l_j \right) \right\| \right\} \tag{2}$$

In MW scenes, we can decouple pose estimation into two parts: estimating non-drifting rotation by MW, and estimating the translation using the semi-direct approach. This method differs from directly tracking the camera frame by frame, as it allows us to estimate the rotation R_{cm} from the Manhattan coordinate system to the camera coordinate system by modeling the indoor environment as MW. By doing so, we can reduce the drift caused by frame-by-frame camera tracking. We use:

$$R_{k,w} = R_{k,m} \cdot R_{m,w} \tag{3}$$

where $R_{k,w}$ represents the camera rotation of the k-th frame. And $R_{m,w}$ represents the coordinates of the first frame, i.e., $R_{1,m} = R_{m,w}^T$. The MW rotations, denoted as $R_{k,m}$, are obtained using the sphere mean-shift algorithm. Now that we have the rotation matrix R_{cw}, we want to find the complete camera pose. To do this, we use this rotation as an initial value to input into the pose matrix $\mathbf{T}$, then substitute it into in Eq. 2, and calculate the final camera pose under the premise of having a drift-free rotation.

3.2 Loop Closing

Loop closure detection thread consists of two steps including detect loop closure keyframes and correct the loop to optimize the pose graph. Our system mainly improves the detection and validation steps of the loop closure to increase the probability of detecting the correct loop closure successfully. In the first step of loop detection, as computing the Bag-of-Word (BoW) [30] of the current frame with all previous keyframes consumes too much computing resources, we use the relative positioning relationship between the MA and line features to filter the keyframes.

Two frames in the same scene share the same MA. The relative position of the line features in the scene and the MA also remains unchanged. Therefore, we can determine if a loop closure occurred by comparing the normal vector of MA and relative positions of the line features and MA between two frames. We define three mutually orthogonal unit vectors $\left\{n_a^i\right\}_{a=1}^3$ as the MA of frame F_i, where $MA_a^i = \{n_1^i, n_2^i, n_3^i\}$. Firstly, we compare whether the two frames share the same MA:

$$\left|MA_a^i \cdot MA_a^j\right| > \varphi, a \in \{1,2,3\} \tag{4}$$

Then, we calculate the relative positional deviation between the line features and the MA:

$$E_a = \frac{1}{m} \sum_{(L_t, MA_a) \in \mathcal{X}} \left\| log\left(\mathbf{R}(L_t)^{-1} \cdot \mathbf{R}(MA_a)\right)\right\|, a \in \{1,2,3\} \tag{5}$$

where $t = 1, 2, \ldots, m$ represents the indices of the matching pairs that meet the requirements. Here m represents the number of matched pairs in $\mathcal{X}$. And $\mathcal{X}$ represents the set of matching pairs of 3D feature line segments with MA, where the angle with MA is greater than the threshold θ.

$$\mathcal{X} = \{(L_t, MA_a) \mid \exists a = \{1,2,3\}\, s.t. \forall L_t \in \mathcal{L}, |L_t \cdot MA_a| > \theta\} \tag{6}$$

$R(\cdot)$ represents the use of the Rodrigues formula to convert 3D line features and MA into rotation matrices. And $log(\cdot)$ represents the logarithmic quaternion error in $SO(3)$.

We compare the positional deviation between the current frame and keyframe to find keyframes with similar positional deviation:

$$\left|E_a^i - E_a^j\right| < \delta, a \in \{1,2,3\} \tag{7}$$

Finally, we use point-line feature visual DBoW2 to search for loop frames on these keyframes, and calculate the similar transformation between the current frame and the loop frames.

4 Experiments

In this section, we evaluate our SLAM system on two publicly available datasets, ICL-NUIM [23] and TUM RGBD [24]. ICL-NUIM dataset is a synthetic indoor dataset with rich textures that comprises two main scenes: the living room *lr* and office *of*. It contains a multitude of structured areas with numerous MW, and we can evaluate the tracking performance of our system in the MW scenes. The TUM RGBD dataset provides real indoor sequences depicting scenes with diverse texture and structural information. This enables us to evaluate the system with different scenes, including the existence of MW and the texture richness of the sequence.

Additionally, we compare our approach with other methods such as feature-based methods ORB-SLAM2 [8] and SP-SLAM [11], MW-based methods RGBD-SLAM [14], MH-SLAM [15] and S-SLAM [13]. And the MSC-VO [17] uses MA to constrain structural features. The effectiveness of our model is evaluated using the root mean square error (RMSE) of the absolute trajectory error (APE) measured in meters [24]. We utilize the EVO library to compute errors and generate trajectory graphs. Our experiments were conducted using an Intel Core i9-13900HX@2.20 GHz/16 GB RAM.

4.1 Pose Estimation

Table 1 shows the comparative results of our system, alongside other advanced SLAM systems. We achieved the best performance in ten out of the sixteen sequences.

ICL-NUIM Dataset. The ICL-NUIM dataset contains a large number of MW assumption, so our method and other MW-based methods such as RGBD-SLAM and MH-SLAM have achieved good results. The presence of rich textures in both environments also enables smooth tracking using feature-based methods, such as ORB-SLAM and SP-SLAM. The performance of our method on the sequences *of-kt0* and *of-kt1* is slightly inferior to MH-SLAM. The absolute trajectory error increases from 0.025 to 0.027 and 0.013 to 0.018, respectively, because MH-SLAM estimates translation using feature-based methods, which allows it to perform better in scenes with rich texture. The feature extraction of these two sequences is shown in Fig. 2. The sequence *lr-kt3* exhibits substantial wall areas, as depicted in Fig. 2a. The only shortage of our method is that it obtains insufficient photometric error gradients, leading to inferior performance compared to systems that leverage feature-based tracking methods. Nevertheless, our method outperforms previous state-of-the-art methods in most sequences.

TUM RGB-D Dataset. In *fr1* and *fr2* sequences with cluttered scenes and few or no MW involved, MW-based systems like RGBD-SLAM and S-SLAM cannot achieve ideal results. On the contrary, our approach can robustly estimate the

Table 1. Comparison of RMSE (m) for ICL-NUIM and TUM RGB-D sequences. × indicates that the method fails in the tracking process.

		Methods						
Dataset	Sequence	Ours	RGBD-SLAM	S-SLAM	MH-SLAM	MSC-VO	ORB-SLAM2	SP-SLAM
ICL NUIM	lr-kt0	**0.005**	0.006	×	0.007	0.016	0.008	0.019
	lr-kt1	**0.008**	0.015	0.016	0.011	0.011	0.011	0.015
	lr-kt2	**0.011**	0.020	0.045	0.015	0.015	0.016	0.017
	lr-kt3	0.034	0.012	0.046	**0.011**	0.037	0.014	0.022
	of-kt0	0.027	0.041	×	**0.025**	0.031	**0.025**	0.031
	of-kt1	0.018	0.020	×	**0.013**	0.023	0.029	0.018
	of-kt2	**0.008**	0.011	0.031	0.015	0.021	0.012	0.027
	of-kt3	**0.011**	0.014	0.065	0.013	0.018	0.012	0.012
TUM RGB-D	fr1/xyz	**0.010**	0.022	×	**0.010**	0.011	**0.010**	**0.010**
	fr2/xyz	0.007	0.015	×	0.008	**0.005**	0.009	0.009
	fr3/s-nt-far	**0.021**	0.022	0.281	0.040	0.066	×	0.031
	fr3/s-nt-near	**0.017**	0.025	0.065	0.023	×	×	0.024
	fr3/s-t-near	0.016	0.055	0.014	0.012	0.011	0.011	**0.010**
	fr3/s-t-far	0.020	0.034	0.014	0.022	0.012	**0.011**	0.016
	fr3/cabinet	**0.015**	0.035	×	0.023	×	×	×
	fr3/l-cabinet	**0.054**	0.071	×	0.083	0.120	×	0.074

(a) lr-kt3 (b) of-kt0 (c) of-kt1

Fig. 2. Partial sequences in the ICL dataset. (a) The scene in the *lr-kt3* sequence where the camera is close to the wall. (b) and (c) represent the distribution of features in the *of* sequence.

pose in non-MW scenes, and has performance comparable to feature-based ORB-SLAM and SP-SLAM, as well as MSC-VO that uses MA to constrain structured features. For the *fr3* sequence, our method outperforms other methods in four of the six sequences with limited texture or no texture. Low-texture sequence images as shown in Fig. 3. On the other hand, MW-based RGBD-SLAM and MH-SLAM are capable of achieving stable tracking in scenes with low texture by utilizing structural information, our system exhibits superior tracking accuracy. Our system still performs well in the *s-t-near* and *s-t-far* sequences with highly textures, as demonstrated in Fig. 3a, but it is apparent that feature-based methods have a stronger advantage.

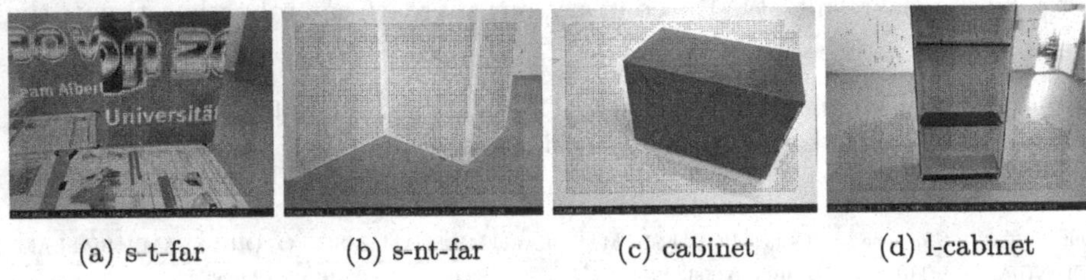

(a) s-t-far (b) s-nt-far (c) cabinet (d) l-cabinet

Fig. 3. Partial sequences in the TUM dataset. (a) The distribution of features in the sequences *s-t-near* and *s-t-far*, which are located in the same scene but only differ in the distance between the camera and the wall. (b), (c) and (d) are low-texture sequence scenes in *fr3*.

4.2 Loop Closure

Moreover, we evaluated the actual effect of our loop detection in the *fr3-l-office* sequence of the TUM RGB-D dataset. As datasets containing closed-loops are relatively rare, we used the *fr3-l-office* sequence to evaluate the effect of our loop detection algorithm. This sequence contains explicit loop closure conditions, as well as planar information. We use the EVO library to draw the comparison graph between the trajectories obtained by our system and RGBD-SLAM, with and without loop closure detection, and the ground truth trajectory. RGBD-SLAM is the baseline for our system, and it uses the same loop closure detection module as the most popular ORB-SLAM2 system, which is also used by S-SLAM, MH-SLAM and SP-SLAM. Moreover, MSC-VO does not have a loop closure detection module. Therefore, in this section, we compare our results with RGBD-SLAM. The loop closure detection module in RGBD-SLAM is the original module without any modification, and we only control its usage in the experiments.

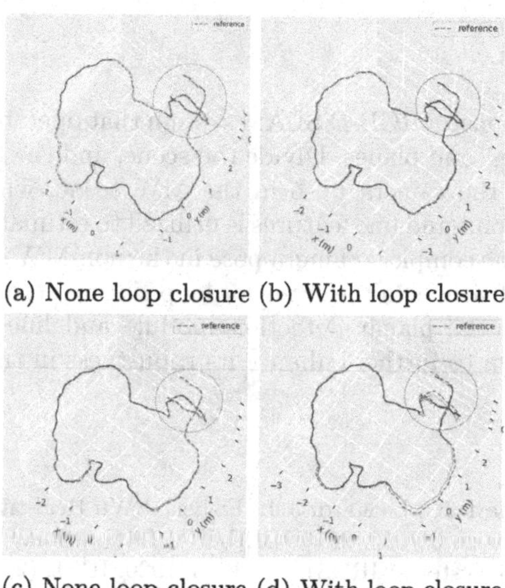

(a) None loop closure (b) With loop closure

(c) None loop closure (d) With loop closure

Fig. 4. Visualization of the camera trajectory. (a) and (b) illustrate the trajectory maps before and after adding the loop closure module to our system, while (c) and (d) show that of RGB-D SLAM. The dashes shows the ground truth of camera trajectories, and the curves indicate the camera trajectories. Meanwhile, The trajectory colors transition from red to yellow to green to blue, representing a decreasing deviation of the trajectories in sequence. (Color figure online)

Figure 4a and Fig. 4b represent the results of our system. It can be seen that after adding our loop closure detection module, the color of the trajectory curve of the detected loop area has changed from red to blue. Meanwhile, RGBD-SLAM exhibits inferior performance upon the integration of the loop closure detection module, as observed in Figs. 4c and 4d. This is caused by inaccurate loop closure keyframes detection.

Furthermore, we calculate the Absolute Trajectory Errors (ATE) of the motion trajectory, as shown in Table 2. It can be illustrated that the accuracy of our system is further improved after adding a loop optimization thread.

Table 2. Comparison of ATE for *fr3-l-office* sequence.

		Ours		RGBD-SLAM	
Sequence	ATE	without loop	with loop	without loop	with loop
fr3/l-office	Max	0.103	**0.099**	0.121	0.238
	Mean	0.030	**0.023**	0.037	0.066
	Median	0.027	**0.022**	0.033	0.052
	Min	**0.001**	**0.001**	0.002	0.011
	Rmse	0.033	**0.026**	0.042	0.079

5 Conclusions

In this paper, we propose a RGB-D SLAM system that operates in indoor scenes, based on points, lines, and planes. Divide the scene, and the MW-based method is used to estimate the camera pose in the MW scene, while the semi-direct approach based on point and line features is utilized to estimate the translation in the MW scene and the complete camera pose in the non-MW scene. Additionally, we use MA and line features in the scene for loop closure detection. In the future, we plan to improve the planar detection module and line feature extraction method of the system to further enhance its robustness in tracking.

References

1. Straub, J., Bhandari, N., Leonard, J.J., Fisher, J.W.: Real-time Manhattan world rotation estimation in 3D. In: 2015 IEEE/RSJ International Conference on Intelligent Robots and Systems (IROS), pp. 1913–1920. IEEE (2015)
2. Kim, P., Li, H., Joo, K.: Quasi-globally optimal and real-time visual compass in manhattan structured environments. IEEE Rob. Autom. Lett. **7**(2), 2613–2620 (2022)
3. Ge, W., Song, Y., Zhang, B., Dong, Z.: Globally optimal and efficient manhattan frame estimation by delimiting rotation search space. In: Proceedings of the IEEE/CVF International Conference on Computer Vision, pp. 15213–15221 (2021)
4. Coughlan, J.M., Yuille, A.L.: Manhattan world: compass direction from a single image by bayesian inference. In: Proceedings of the Seventh IEEE International Conference on Computer Vision, pp. 941–947. IEEE (1999)
5. Li, H., Zhao, J., Bazin, J.C., Liu, Y.H.: Quasi-globally optimal and near/true realtime vanishing point estimation in manhattan world. IEEE Trans. Pattern Anal. Mach. Intell. **44**(3), 1503–1518 (2020)
6. Campos, C., Elvira, R., Rodríguez, J.J.G., Montiel, J.M., Tardós, J.D.: Orb-slam3: an accurate open-source library for visual, visual-inertial, and multimap slam. IEEE Trans. Rob. **37**(6), 1874–1890 (2021)
7. Klein, G., Murray, D.: Parallel tracking and mapping for small AR workspaces. In: 2007 6th IEEE and ACM International Symposium on Mixed and Augmented Reality, pp. 225–234. IEEE (2007)
8. Mur-Artal, R., Tardós, J.D.: Orb-slam2: an open-source slam system for monocular, stereo, and rgb-d cameras. IEEE Trans. Rob. **33**(5), 1255–1262 (2017)

9. Mur-Artal, R., Montiel, J.M.M., Tardos, J.D.: ORB-SLAM: a versatile and accurate monocular SLAM system. IEEE Trans. Rob. **31**(5), 1147–1163 (2015)
10. Gomez-Ojeda, R., Moreno, F.A., Zuniga-Noël, D., Scaramuzza, D., Gonzalez-Jimenez, J.: PL-SLAM: a stereo SLAM system through the combination of points and line segments. IEEE Trans. Rob. **35**(3), 734–746 (2019)
11. Zhang, X., Wang, W., Qi, X., Liao, Z., Wei, R.: Point-plane slam using supposed planes for indoor environments. Sensors **19**(17), 3795 (2019)
12. Newcombe, R.A., Lovegrove, S.J., Davison, A.J.: DTAM: dense tracking and mapping in real-time. In: 2011 International Conference on Computer Vision, pp. 2320–2327. IEEE (2011)
13. Li, Y., Brasch, N., Wang, Y., Navab, N., Tombari, F.: Structure-slam: low-drift monocular slam in indoor environments. IEEE Rob. Autom. Lett. **5**(4), 6583–6590 (2020)
14. Li, Y., Yunus, R., Brasch, N., Navab, N., Tombari, F.: RGB-D SLAM with structural regularities. In: 2021 IEEE International Conference on Robotics and Automation (ICRA), pp. 11581–11587. IEEE (2021)
15. Yunus, R., Li, Y., Tombari, F.: Manhattanslam: robust planar tracking and mapping leveraging mixture of manhattan frames. In: 2021 IEEE International Conference on Robotics and Automation (ICRA), pp. 6687–6693. IEEE (2021)
16. Joo, K., Oh, T.H., Rameau, F., Bazin, J.C., Kweon, I.S.: Linear rgb-d slam for atlanta world. In: 2020 IEEE International Conference on Robotics and Automation (ICRA), pp. 1077–1083. IEEE (2020)
17. Company-Corcoles, J.P., Garcia-Fidalgo, E., Ortiz, A.: MSC-VO: exploiting manhattan and structural constraints for visual odometry. IEEE Rob. Autom. Lett. **7**(2), 2803–2810 (2022)
18. Yang, S., Scherer, S.: Direct monocular odometry using points and lines. In: 2017 IEEE International Conference on Robotics and Automation (ICRA), pp. 3871–3877. IEEE (2017)
19. Forster, C., Pizzoli, M., Scaramuzza, D.: SVO: fast semi-direct monocular visual odometry. In: 2014 IEEE International Conference on Robotics and Automation (ICRA), pp. 15–22. IEEE (2014)
20. Peng, X., Liu, Z., Wang, Q., Kim, Y.T., Lee, H.S.: Accurate visual-inertial slam by manhattan frame re-identification. In: 2021 IEEE/RSJ International Conference on Intelligent Robots and Systems (IROS), pp. 5418–5424. IEEE (2021)
21. Gomez-Ojeda, R., Briales, J., Gonzalez-Jimenez, J.: PL-SVO: semi-direct monocular visual odometry by combining points and line segments. In: 2016 IEEE/RSJ International Conference on Intelligent Robots and Systems (IROS), pp. 4211–4216. IEEE (2016)
22. Zhou, Y., Kneip, L., Rodriguez, C., Li, H.: Divide and conquer: efficient density-based tracking of 3D sensors in manhattan worlds. In: Lai, S.-H., Lepetit, V., Nishino, K., Sato, Y. (eds.) ACCV 2016. LNCS, vol. 10115, pp. 3–19. Springer, Cham (2017). https://doi.org/10.1007/978-3-319-54193-8_1
23. Handa, A., Whelan, T., McDonald, J., Davison, A.J.: A benchmark for RGB-D visual odometry, 3D reconstruction and SLAM. In: 2014 IEEE International Conference on Robotics and Automation (ICRA), pp. 1524–1531. IEEE (2014)
24. Sturm, J., Engelhard, N., Endres, F., Burgard, W., Cremers, D.: A benchmark for the evaluation of RGB-D SLAM systems. In: 2012 IEEE/RSJ International Conference on Intelligent Robots and Systems, pp. 573–580. IEEE (2012)
25. Rublee, E., Rabaud, V., Konolige, K., Bradski, G.: ORB: an efficient alternative to SIFT or SURF. In: 2011 International Conference on Computer Vision, pp. 2564–2571. IEEE (2011)

26. Von Gioi, R.G., Jakubowicz, J., Morel, J.M., Randall, G.: LSD: a line segment detector. Image Process. Line **2**, 35–55 (2012)
27. Zhang, L., Koch, R.: An efficient and robust line segment matching approach based on LBD descriptor and pairwise geometric consistency. J. Vis. Commun. Image Represent. **24**(7), 794–805 (2013)
28. Fischler, M.A., Bolles, R.C.: Random sample consensus: a paradigm for model fitting with applications to image analysis and automated cartography. Commun. ACM **24**(6), 381–395 (1981)
29. Trevor, A.J., Gedikli, S., Rusu, R.B., Christensen, H.I.: Efficient organized point cloud segmentation with connected components. In: Semantic Perception Mapping and Exploration (SPME), vol. 1 (2013)
30. Gálvez-López, D., Tardos, J.D.: Bags of binary words for fast place recognition in image sequences. IEEE Trans. Rob. **28**(5), 1188–1197 (2012)

L2T-BEV: Local Lane Topology Prediction from Onboard Surround-View Cameras in Bird's Eye View Perspective

Shanding Ye, Tao Li, Ruihang Li, and Zhijie Pan(✉)

College of Computer Science and Technology, Zhejiang University, Hangzhou, China
{ysd,litaocs,12221089,zhijie_pan}@zju.eud.cn

Abstract. High definition maps (HDMaps) serve as the foundation for autonomous vehicles, encompassing various driving scenario elements, among which lane topology is critically important for vehicle perception and planning. Existing work on lane topology extraction predominantly relies on manual processing, while automated methods are limited to road topology extraction. Recently, road representation learning based on surround-view with bird's-eye view (BEV) has emerged, which directly predicts localized vectorized maps around the vehicle. However, these maps cannot represent the topological relationships between lanes. As a solution, we propose a novel method, L2T-BEV, which learns local lane topology maps in BEV. This method utilizes the EfficientNet to extract features from surround-view images, followed by transforming these features into the BEV space through the Inverse Perspective Mapping (IPM). Nonetheless, the IPM transformation often suffers from distortion issues. To alleviate this, we add a learnable residual mapping function to the features after the IPM transformation. Finally, we employ a transformer network with learnable positional embedding to process the fused images, generating higher-precision lane topology. We validated our method on the NuScenes dataset, and the experimental results demonstrate the feasibility and excellent performance.

Keywords: Autonomous driving · HDMap · Lane topology map · BEV

1 Introduction

Since their inception, HDMaps have received widespread attention and experienced rapid development in recent years, facilitating vehicles in achieving autonomous driving in specific scenarios [1,2]. As a common basic technology for intelligent connected vehicle transportation, HDMaps cater not only to human drivers but also to autonomous vehicles. For self-driving cars at Level 3 and above, HDMaps are a crucial component [3].

This work was supported by the Key Research and Development Program of Zhejiang Province in China (No. 2023C01237), and the Natural Science Foundation of China(No.U22A202101).

Q. Liu et al. (Eds.): PRCV 2023, LNCS 14427, pp. 363–375, 2024.
https://doi.org/10.1007/978-981-99-8435-0_29

HDMaps typically contain a wealth of driving scene elements. Analyzing element attributes reveals that HDMap elements can be categorized into dynamic and static elements [4]. Dynamic information [5] generally includes traffic light signals, traffic flow, and traffic events, which aid vehicles in avoiding road congestion, accidents, and improving driving efficiency and safety. Static information [6] primarily encompasses positioning data [7], semantic information [8], and lane topology [9]. Lane topology, serving as the interface between perception and planning, effectively assists vehicles in performing planning tasks such as predicting other vehicles' trajectories, self-tracing driving, lane change planning, and turning, among others. This paper mainly discusses the generation of lane topology.

Traditional lane topology is often manually annotated on basemaps constructed offline using SLAM methods [10], which is labor-intensive and time-consuming. As a result, leveraging machine learning to directly extract road topology from data sources fulfills this demand [11]. Although some existing works can automatically extract road topology from satellite aerial images, such as BoundaryNet [12] and csBoundary [13], ground building or tree cover may impact the final road topology generation. Additionally, some methods generate road topology based on vehicle GPS trajectory information [14]. However, the road topology extracted by these two methods is not lane-level and lacks the precision required for autonomous driving tasks. Currently, only a limited number of studies focus on generating lane-level road topology, also known as lane topology [9,15]. However, these works only consider the forward-looking bird's-eye view (BEV), limiting the surround-view perception required by autonomous vehicles.

In recent years, road representation learning based on surround-view BEV has emerged as a research hotspot, with numerous studies predicting local vectorized maps surrounding a vehicle using BEV instance segmentation [16–18]. Although these maps encompass road boundary lines, lane demarcation lines, and pedestrian crosswalk lines, they fail to represent the topological relationships between lanes. Furthermore, at intersections, the absence of road boundaries and lane demarcation lines prevents these methods from characterizing local driving scenarios, necessitating alternative approaches for determining drivable areas. In summary, these methods generate maps from the perspective of human-driven vehicles, while more efficient options could exist for autonomous vehicles. We argue that lane topology is the most suitable representation of road information in HDMaps, as it includes lane centerlines and their connecting relationships [9].

Building on this foundation, we propose a local lane topology map learning method from a BEV perspective, called L2T-BEV. This method directly utilizes onboard surround-view cameras to predict lane centerline graphs and their interconnecting relationships. Lane centerlines are represented by Bezier curves, with their start and end points indicating the direction of traffic flow. The connection relationships between lane centerlines are modeled using an assignment matrix [15]. Specifically, we first employ an efficient network to extract features from surround-view images, then combine intrinsic and extrinsic camera parameters to transform surround-view image features into BEV space using the IPM technique [19]. Due to distortion issues in IPM-transformed images, we employ a learnable Positional Encoding (PE) [20] to encode features in the BEV space, mitigating the impact of distortion on predictions. Finally, a transformer network is utilized to learn from these encoded features, and the Hungarian matching algorithm is applied to the output results, as described in reference [15]. The predicted lane centerlines and their connection relationships define the lane topology within a local HDMap. Our contributions can be summarized as follows:

- We introduce L2T-BEV, a learning method for predicting local lane topology directly from onboard surround-view cameras in the BEV perspective.
- We evaluate the performance of L2T-BEV on the large-scale public NuScenes dataset [21] to demonstrate the efficacy of our approach.

2 Related Works

2.1 Road Network Extraction

In recent years, road network generation techniques based on machine learning have been developed and widely applied to improve the accuracy of road topology and reduce manual costs. Máttyus et al. proposed DeepRoadMapper [22] to directly estimate road topology from aerial remote sensing images and extract road networks; however, some inappropriate connections persist. To address this issue, Batra et al. introduced the orientation learning and connectivity refinement approach [23] to resolve discontinuity problems. From an iterative growth perspective, Bastani proposed RoadTracer [24], which employs a search algorithm guided by CNN-based decision functions to automatically extract road networks and address occlusion issues in aerial remote sensing images. However, iterative growth methods require significant time to create road networks vertex by vertex. To further reduce incompleteness and curvature loss, BoundaryNet [12] utilized four learning networks to extract and complete road boundaries from laser scanning point clouds and satellite images [25]. In summary, the road networks extracted from aerial remote sensing or satellite images are road-level, suitable for road path planning. However, considering the precision and local dynamic driving planning tasks required for autonomous driving, such as lane changing, lane-level road topology is more suitable for self-driving vehicles.

2.2 BEV Representation

In recent years, research on road representation learning based on BEV has gained prominence [9,15,16], with a focus on both single-view and surround-view BEV. Can et al. [15] employed Bezier curves to represent lane centerlines and predicted lane topology using a DETR-like network in a single front-view BEV. However, their approach is limited by the single-view perception and cannot handle map elements in other views surrounding the vehicle. As for surround-view BEV, some recent studies have predicted rasterized maps by performing BEV semantic segmentation [26,27]. However, these rasterized maps lack the vectorized lane structure information required for downstream autonomous driving tasks, such as motion prediction. To address this issue, HDMapNet [16] encodes surround-view images and LiDAR point clouds, predicting local vector map elements in BEV perspective, including road boundary lines, lane dividers, and crosswalks. Nonetheless, the aforementioned surround-view BEV studies do not represent the topology between lanes, which is crucial for many downstream tasks in autonomous driving, such as motion prediction and vehicle navigation. Therefore, in this paper, we focus on directly predicting lane topology from vehicle surround-view images.

3 Methodology

In this section, we present L2T-BEV, a learning method specifically designed for predicting local lane topology from surround-view cameras under the BEV perspective. We begin by providing an overview of the L2T-BEV architecture. Subsequently, we elaborate on the lane topology representation model used in our approach. Following this, we discuss the process of extracting features from six vehicle-mounted cameras and transforming them into BEV. We then explain the rationale behind employing a fusion encoding-based transformer module for lane topology prediction. Lastly, we describe the composition of the loss function used in our method.

3.1 Overview

When planning driving tasks, self-driving vehicles must comprehend local scene information in addition to global road data. A crucial aspect of local scene perception is generating local lane topology from perception data. This paper focuses on employing surround-view images to create lane topology maps from a BEV perspective for local autonomous driving tasks. The surround-view images are captured by six on-board cameras, addressing the issue of insufficient viewing angles from a single camera and providing a more cost-effective solution for large-scale deployment on autonomous vehicles than expensive LiDAR sensors.

In our approach, we employ the EfficientNet [28] capable of fusing different levels of feature information as the image feature extraction network. The two-dimensional image space exhibits a “disappearance phenomenon,” making it unsuitable for vehicular task planning. To better serve autonomous vehicles, we utilize the camera's intrinsic and extrinsic parameters and apply the IPM [19] method to transform the image from the image coordinate system to the vehicle coordinate system from a BEV perspective. However, IPM transformations are distorted and lack depth information, preventing the direct generation of lane topology. To address this challenge, we use a fused-encoded Transformer network to process the transformed images, producing higher-precision lane topologies. Finally, a post-processing method combines the predicted lane topologies into a local lane topology network. Figure 1 provides a detailed overview of our method's architecture.

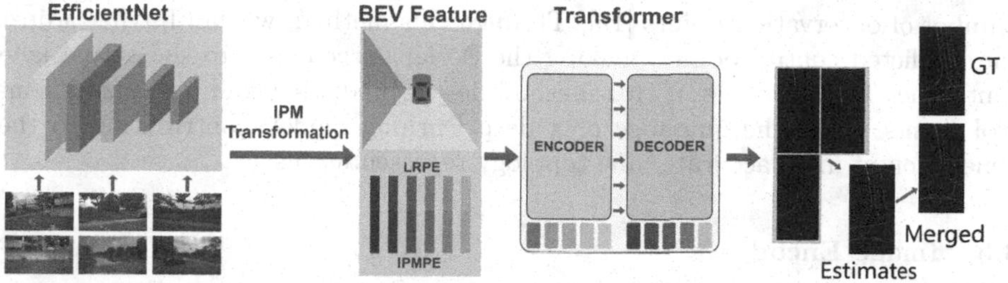

Fig. 1. The EfficinetNet network extracts image features from surround-view images, which are then transformed into the BEV space using the IPM. These features are subsequently combined with LRPE and IPMPE and input into a transformer network for lane topology prediction. After post-processing, a local lane topology map can be obtained.

3.2 Lane Topologies Representation

Lane topology is a vital foundational element in HDMap, facilitating autonomous vehicles in achieving better local driving planning, such as lane changes and behavior prediction. L2T-BEV aims to predict the local lane network map surrounding autonomous vehicles, which is a directed acyclic graph $G = (V, E)$, where vertices V are annotated with spatial coordinates, and edges E represent lane centerline segments, signifying the connection relationship between two vertices. This connection relationship can be represented by an incidence matrix [15] A. Specifically, $A[X, Y] = 1$ indicates that the endpoint of lane centerline segment X is the same as the starting point of lane centerline segment Y. Note that due to the limited representation range of onboard surround-view images, generally, when $A[X, Y] = 1$, $A[X, Y]$ will not equal 1.

Current lane centerline representation models mainly include polynomial fitting, spline curves, and Bézier curves. In this paper, we focus on using Bézier curves to represent the lane centerline, as Bézier curves possess several significant advantages, such as ease of computation, strong local control, and effortless multi-segment stitching. The Bézier curve model considers the lane centerline as a weighted sum of a set of control points. That is, given a set of control points $\boldsymbol{P} = \{P_0, P_1, \cdots, P_n\}$, the lane centerline can be represented as:

$$C(t) = \sum_{j=0}^{n} \binom{n}{j} (1-t)^{n-j} t^j P_j \tag{1}$$

where $t \in [0, 1]$ is the control factor of the Bézier curve. Thus, by adjusting the control point positions, a smooth curve that adapts to different lane curvatures can be generated, making Bézier curves an ideal choice for representing lane centerlines.

Next, to find the optimal control points, we first employ a deep learning model to predict the control points of the Bézier curve, and then transform it into solving a least squares problem, that is, minimizing the difference between the predicted control points and the actual observation points given a certain number of observation points [15]. Through this method, we can further adjust the predicted control points, making the Bézier curve closer to the actual lane centerline. The rationale of this method lies in predicting and optimizing control points under the guidance of a deep learning model, contributing to the generation of more accurate lane topology representations.

3.3 Image Encoder

The image encoder consists of two components: feature extraction and Inverse Perspective Mapping (IPM) transformation. Feature extraction is designed to obtain high-dimensional information from images, while IPM transformation converts the feature map from the image coordinate system to the ego-car coordinate system within BEV perspective, which is more suitable for vehicle driving task planning.

In recent years, a variety of image feature extraction networks have been introduced, such as ResNet, Inception series, and EfficientNet [28]. This study employs EfficientNet to extract lane centerline feature information due to its ability to maintain high accuracy with low computational complexity and fewer parameters. The original N images with dimensions $N \times K \times H \times W$ are first processed through the EfficientNet and the neural view transformer [16], producing camera coordinate features $F^{cam} \subseteq \mathbb{R}^{N \times K_{cam} \times H_{cam} \times W_{cam}}$.

As the internal and external parameters of onboard cameras are usually known, this paper applies IPM transformation to convert image features from the camera coordinate system to the ego-car coordinate system to obtain the BEV feature $F^{bev} \subseteq \mathbb{R}^{N \times K_{bev} \times H_{bev} \times W_{bev}}$. However, the accuracy of IPM transformation can be affected by factors such as vehicle vibrations and changes in ambient temperature. To improve post-transformation accuracy, we adopt a learnable residual mapping function [29] for the features following IPM transformation, as presented in Eq. 2:

$$\boldsymbol{F}^{fuse} = F^{bev} + \phi F^{cam} \tag{2}$$

where ϕ models the relation between image features in the camera coordinate system and the BEV space. Finally, the fused BEV features $\boldsymbol{F}^{fuse}$ are fed into a transformer model to predict the lane graph.

3.4 Lane Graph Detection

In this study, we employ a transformer-based network to predict lane topologies. The transformer model first receives the fused feature $\boldsymbol{F}^{fuse}$ processed by the image encoder, and subsequently outputs a series of candidate proposal vectors. These vectors are then fed into the lane topology detector for estimating the lane graph.

Typically, transformer models require positional embedding (PE) to enhance spatial attention. However, given the potential distortion and susceptibility to external environmental interference of IPM transformation, directly encoding features after IPM transformation may not facilitate the transformer model's perception of lane topology. Consequently, we adopt a fusion embedding method based on PE. This approach incorporates both IPM positional embedding (IPMPE) and learnable positional embedding (LRPE). Similar to [15], IPMPE directly encodes IPM features by employing sinusoidal functions to process normalized cumulative locations. As a dynamically tuned positional embedding, LRPE can be automatically optimized based on input data during training. This adaptability enables learnable embedding to capture more fine-grained position information relevant to the task, enhancing the transformer model's flexibility.

Referring to [15], the lane topology detector comprises four components: detection head, control head, association head, and association classifier. To obtain lane topology, the transformer's output vectors are first fed into the detection head and control head. The detection head outputs the probability of a lane centerline's presence corresponding to each query vector, while the control head generates Bézier curve control points representing the lane centerline geometry. After these two steps, we acquire a set of candidate lane centerlines $G_{candi} = (V_{candi}, E_{candi})$. Subsequently, the association head processes these candidate centerlines into low-dimensional associated feature vectors F_{assoc}. The association matrix M_{assos} composed of these association feature vectors, is then fed into the association classifier, which calculates the association probabilities between centerline pairs. Based on these probabilities, we can determine the relationships between lane centerlines and establish connections among them, ultimately constructing a comprehensive network of lane centerlines.

3.5 Loss Function

Based on the lane topology representation model, we divide the network training loss $Loss_{train}$ into five parts: lane centerline classification loss (L_{label}), lane centerline endpoint loss (L_{end}), Bézier curve control point loss (L_{bezer}), lane centerline association loss (L_{assoc}), and lane centerline endpoint association loss (L_{enda}), as shown in Eq. 3:

$$L_{total} = \lambda_1 L_{label} + \lambda_2 L_{end} + \lambda_3 L_{bezer} + \lambda_4 L_{assoc} + \lambda_5 L_{enda} \tag{3}$$

where L_{label} and L_{assoc} are calculated using cross-entropy, while the remaining loss terms employ 1-norm loss computation.

4 Experiments

4.1 Experimental Settings

Dataset. We use the NuScenes dataset [21] to evaluate and verify our proposed L2T-BEV. The NuScenes dataset is a large-scale, diverse, and high-quality dataset designed for autonomous driving research. It contains 1,000 scenes, each 20 s long, with a total of 1.4 million images captured from six cameras mounted on the data collection vehicle. The rich annotations provided in the NuScenes dataset include 3D bounding boxes, semantic maps, and high-definition lane-level maps.

Metrics. Following [15], we take the **precision-recall, detection ratio** and **connectivity** as the evaluating metrics to measure the L2T-BEV performance. Specifically, the precision-recall scores are calculated based on matched centerlines at different distance thresholds, using a matching process that considers the direction of control points and uses minimum L1 loss. And this metric measures how well the estimates fit the matched ground truth centerlines and how accurately the captured subgraph is represented. The detection ratio is the ratio of the number of unique GT centerlines that are matched to at least one estimated centerline to the total number of GT centerlines. It was proposed to address the limitations of the precision-recall metric in dealing with missed centerlines. These two metrics provide a comprehensive evaluation of the performance of centerline detection methods. The connectivity metric is used to evaluate the accuracy of the association between centerlines in a road network. It uses the incidence matrices of the estimated and GT centerlines to compute precision and recall.

Implementation Details. In this study, we select 698 scenes from the NuScenes dataset as the training set and 148 scenes as the evaluation set. The BEV area is set as [–15 m, 15 m] along the x-direction and [–30 m, 30 m] along the z-direction, with a precision of 15 cm. We train our model on 4 GTX 1080Ti GPUs with a batch size of 12. The Efficient-b4 [28] model was chosen as the backbone network, pre-trained on the ImageNet. To detect lane centerlines, we use 100 query vectors in the transformer model. We adopt the Adam optimizer for model training, with a learning rate of 1e–5. The weights for the various loss terms in Eq. 3 are set to $\lambda_1 = 3$, $\lambda_2 = 1$, $\lambda_3 = 3$, $\lambda_4 = 2$, and $\lambda_5 = 1$.

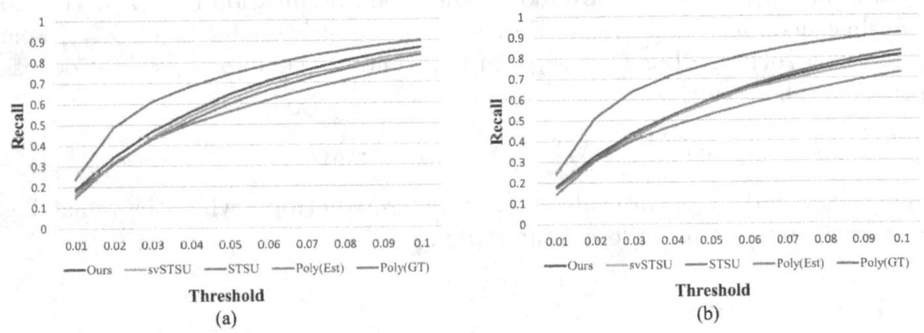

Fig. 2. Precision/Recall vs thresholds. Thresholds are uniformly sampled in [0.01, 0.1] (normalized coordinates) with 0.01 increments (0.01 corresponds to 50cm).

4.2 Comparison with Other Methods

In this study, we compare our proposed L2T-BEV method with four other techniques, namely Poly(Est) [15], Poly(GT) [15], STSU [15], and svSTSU. The first three methods utilize monocular camera images as input, while the latter, svSTSU, employs surround-view camera images. Poly(Est) and Poly(GT) correspond to Poly-RNN [30] using estimated initial points and ground truth (GT) initial points, respectively, for generating centerline curves. For a fair comparison, we only apply the STSU method for lane graph prediction without conducting object detection in this work. In order to adapt STSU for surround-view images within the svSTSU method, we initially process each image separately using a backbone, concatenate the encoding results in the feature channel dimension, and finally feed the resultant data to the transformer.

Table 1. Lane graph prediction results. M-Prec and M-Recall indicate mean of the sampled points of precision-threshold and recall-threshold curves, see Fig. 2. C-Prec and C-Rec refer to connectivity precision and recall, while C-IOU is connectivity TP/(TP + FP + FN).

Method	M-Pre	M-Rec	Detect	C-Pre	C-Rec	C-IOU
Poly(GT)	70.0	72.3	76.4	55.2	50.9	36.0
Poly(Est)	54.8	51.2	40.5	60.3	15.9	14.3
STSU	58.3	52.5	60.0	58.2	53.5	38.7
svSTSU	59.2	56.0	55.2	61.7	57.2	42.2
Ours	61.7	57.6	59.1	63.9	69.3	49.9

As demonstrated in Table 1 and Fig. 2, apart from the detection rate, the performance metrics of svSTSU are higher than those of STSU, indicating that the STSU method can be applied to process surround-view camera images. In contrast, our proposed method outperforms svSTSU in all the metrics, with a detection rate very close to that of STSU, which demonstrates the effectiveness of the IPM-based BEV feature fusion and PE fusion approach introduced in this study. The visualization results of lane topology in the BEV space are shown in Fig. 3. Compared to svSTSU, our method produces a more faithful lane graph.

4.3 Ablation

Due to the distortion problem inherent in IPM transformations and their susceptibility to external interference, we propose a fusion of PE and BEV features to mitigate this impact. In this section, we tried using three sets of experiments to validate the effectiveness of LRPE and fused BEV features. The baseline method refers to the BEV features input into the transformer model without considering LRPE and learnable residual mapping function. The experimental results are shown in Table 2. We observed that the LRPE and BEV features have a substantial influence on detection and connectivity scores, with relatively minimal effect on performance of precision-recall.

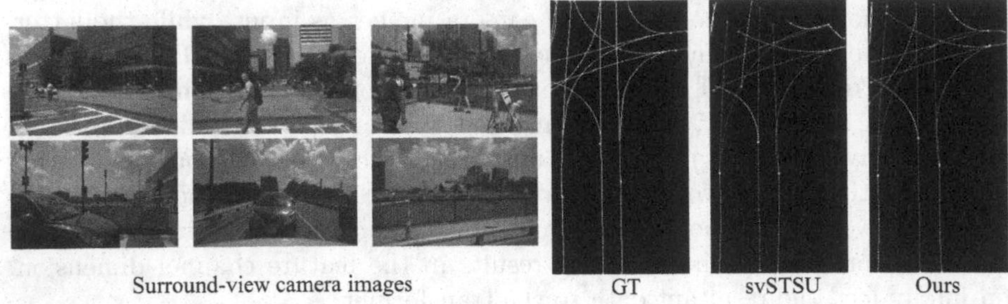

Fig. 3. Sample centerline estimates using surround-view camera images. Our method produces more faithtul lane graph representation.

Table 2. Ablations are carried out on three models that test the performance contribution of the positional embeddings and fused BEV features.

Method	M-Pre	M-Rec	Detect	C-Pre	C-Rec	C-IOU
Baseline	60.5	56.7	55.5	62.5	58.8	43.5
Baseline + LRPE	61.4	57.0	57.3	62.4	63.4	45.9
Baseline + LRPE + FusBEV	61.7	57.6	59.1	63.9	69.3	49.9

4.4 Limitations and Future Work

Our method focuses on predicting endpoints and Bezier curve control points of lane centerlines before establishing lane topology relationships. However, this approach has a limitation: it does not provide adequate topological association indicators during lane centerline association. As a result, minimal circles [9] are proposed to enhance accuracy in lane topology associations, but performance gains are limited. In future studies, we plan to generate lane centerline topology maps using a segmentation-based approach. Specifically, we will first segment the lane centerlines and select key points within these centerlines. These key points can facilitate fitting the lane centerlines and enable the determination of connection relationships between centerlines by analyzing relationships between the key points.

5 Conclusion

We introduced a novel method, L2T-BEV, which predicts local lane topology directly from onboard surround-view cameras in the BEV perspective. For generating a higher-precision lane topology, we proposed to leverage the learnable residual mapping function and the learnable positional embedding to alleviate the distortion issues in IPM transformations. Our validation of this method on the NuScenes dataset yielded promising results, attesting to its feasibility and excellent performance. Ablation studies also demonstrate the effects of various design choices.

References

1. . Wang, H., Xue, C., Zhou, Y., Wen, F., Zhang, H.: Visual semantic localization based on HD map for autonomous vehicles in urban scenarios. In: 2021 IEEE International Conference on Robotics and Automation (ICRA), Xi'an, China, pp. 11255–11261 (2021). https://doi.org/10.1109/ICRA48506.2021.9561459
2. Chiang, K.-W., Zeng, J.-C., Tsai, M.-L., Darweesh, H., Chen, P.-X., Wang, C.-K.: Bending the curve of HD maps production for autonomous vehicle applications in Taiwan. IEEE J. Sel. Topics Appl. Earth Obs. Remote Sens. **15**, 8346–8359 (2022). https://doi.org/10.1109/JSTARS.2022.3204306
3. Chiang, K.W., Wang, C.K., Hong, J.H., et al.: Verification and validation procedure for high-definition maps in Taiwan. Urban Inf. **1**, 18 (2022). https://doi.org/10.1007/s44212-022-00014-0
4. Liu, J.N., Zhan, J., Guo, C., Li, Y., Wu, H.B., Huang, H.: Data logic structure and key technologies on intelligent high-precision map. Acta Geodaetica et Cartographica Sinica **48**(8), 939–953 (2019). https://doi.org/10.11947/j.AGCS.2019.20190125
5. Maiouak, M., Taleb, T.: Dynamic maps for automated driving and UAV geofencing. IEEE Wirel. Commun. **26**(4), 54–59 (2019). https://doi.org/10.1109/MWC.2019.1800544
6. HERE. https://www.here.com/. Accessed 8 Apr 2023
7. Kim, C., Cho, S., Sunwoo, M., Resende, P., Bradaï, B., Jo, K.: Updating point cloud layer of high definition (HD) map based on crowd-sourcing of multiple vehicles installed LiDAR. IEEE Access **9**, 8028–8046 (2021). https://doi.org/10.1109/ACCESS.2021.3049482
8. Jang, W., An, J., Lee, S., Cho, M., Sun, M., Kim, E.: Road lane semantic segmentation for high definition map. In: IEEE Intelligent Vehicles Symposium (IV). Changshu, China 2018, pp. 1001–1006 (2018). https://doi.org/10.1109/IVS.2018.8500661
9. Can, Y.B., Liniger, A., Paudel, D.P., Van Gool, L.: Topology preserving local road network estimation from single onboard camera image. In: 2022 IEEE/CVF Conference on Computer Vision and Pattern Recognition (CVPR), New Orleans, LA, USA, pp. 17242–17251 (2022). https://doi.org/10.1109/CVPR52688.2022.01675
10. Kiran, B.R., et al.: Real-time dynamic object detection for autonomous driving using prior 3D-maps. In: Proceedings of the European Conference on Computer Vision (ECCV) Workshops (2018). https://doi.org/10.1007/978-3-030-11021-5_35
11. Bao, Z., Hossain, S., Lang, H., Lin, X.: High-definition map generation technologies for autonomous driving: a review (2022). arXiv preprint arXiv:2206.05400
12. Ma, L., Li, Y., Li, J., Junior, J.M., Gonçalves, W.N., Chapman, M.A.: BoundaryNet: extraction and completion of road boundaries with deep learning using mobile laser scanning point clouds and satellite imagery. IEEE Trans. Intell. Transp. Syst. **23**(6), 5638–5654 (2022). https://doi.org/10.1109/TITS.2021.3055366

13. Xu, Z., et al.: csBoundary: city-scale road-boundary detection in aerial images for high-definition Maps. IEEE Rob. Autom. Lett. **7**(2), 5063–5070 (2022). https://doi.org/10.1109/LRA.2022.3154052
14. Gao, S., Li, M., Rao, J., Mai, G., Prestby, T., Marks, J., Hu, Y.: Automatic urban road network extraction from massive GPS trajectories of taxis. In: Werner, M., Chiang, Y.-Y. (eds.) Handbook of Big Geospatial Data, pp. 261–283. Springer, Cham (2021). https://doi.org/10.1007/978-3-030-55462-0_11
15. Can, Y.B., Liniger, A., Paudel, D.P., Van Gool, L.: Structured bird's-eye-view traffic scene understanding from onboard images. In: 2021 IEEE/CVF International Conference on Computer Vision (ICCV), Montreal, QC, Canada, pp. 15641–15650 (2021). https://doi.org/10.1109/ICCV48922.2021.01537
16. Li, Q., Wang, Y., Wang, Y., Zhao, H.: HDMapNet: an online HD map construction and evaluation framework. In: International Conference on Robotics and Automation (ICRA), Philadelphia, PA, USA, pp. 4628–4634 (2022). https://doi.org/10.1109/ICRA46639.2022.9812383
17. Liu, Y.C., Wang, Y., Wang, Y.L., Zhao, H.: Vectormapnet: end-to-end vectorized hd map learning. arXiv preprint arXiv:2206.08920 (2022)
18. Liao, B.C., et al.: MapTR: structured modeling and learning for online vectorized HD map construction. arXiv preprint arXiv:2208.14437 (2022)
19. Deng, L., Yang, M., Li, H., Li, T., Hu, B., Wang, C.: Restricted deformable convolution-based road scene semantic segmentation using surround view cameras. IEEE Trans. Intell. Transp. Syst. **21**(10), 4350–4362 (2020). https://doi.org/10.1109/TITS.2019.2939832
20. Raisi, Z., Naiel, M.A., Younes, G., Wardell, S., Zelek, J.: 2LSPE: 2D learnable sinusoidal positional encoding using transformer for scene text recognition. In: 2021 18th Conference on Robots and Vision (CRV), Burnaby, BC, Canada, pp. 119–126 (2021). https://doi.org/10.1109/CRV52889.2021.00024
21. Caesar, H., et al.: nuScenes: a multimodal dataset for autonomous driving. In: 2020 IEEE/CVF Conference on Computer Vision and Pattern Recognition (CVPR), Seattle, WA, USA, pp. 11618–11628 (2020). https://doi.org/10.1109/CVPR42600.2020.01164
22. Máttyus, G., Luo, W., Urtasun, R.: DeepRoadMapper: extracting road topology from aerial images. In: 2017 IEEE International Conference on Computer Vision (ICCV), Venice, Italy, pp. 3458–3466 (2017). https://doi.org/10.1109/ICCV.2017.372
23. Batra, A., Singh, S., Pang, G., Basu, S., Jawahar, C.V., Paluri, M.: Improved road connectivity by joint learning of orientation and segmentation. In: 2019 IEEE/CVF Conference on Computer Vision and Pattern Recognition (CVPR), Long Beach, CA, USA, pp. 10377–10385 (2019). https://doi.org/10.1109/CVPR.2019.01063
24. Bastani, F., et al.: RoadTracer: automatic extraction of road networks from aerial images. In: 2018 IEEE/CVF Conference on Computer Vision and Pattern Recognition, Salt Lake City, UT, USA, pp. 4720–4728 (2018). https://doi.org/10.1109/CVPR.2018.00496
25. Zhang, J., Hu, X., Wei, Y., Zhang, L.: Road topology extraction from satellite imagery by joint learning of nodes and their connectivity. IEEE Trans. Geosci. Remote Sens. **61**, 1–13 (2023). https://doi.org/10.1109/TGRS.2023.3241679
26. Zhou, B., Krähenbühl, P.: Cross-view transformers for real-time map-view semantic segmentation. In: 2022 IEEE/CVF Conference on Computer Vision and Pattern Recognition (CVPR), New Orleans, LA, USA, pp. 13750–13759 (2022). https://doi.org/10.1109/CVPR52688.2022.01339

27. Hu, A., et al.: FIERY: future instance prediction in bird's-eye view from surround monocular cameras. In: 2021 IEEE/CVF International Conference on Computer Vision (ICCV), Montreal, QC, Canada, pp. 15253–15262 (2021). https://doi.org/10.1109/ICCV48922.2021.01499
28. Tan, M., Le, Q.V.: EfficientNet: rethinking model scaling for convolutional neural networks (2019). ArXiv preprint arXiv:1905.11946
29. Xu, Z.H., Liu, Y.X., Sun, Y.X., Liu, M., Wang, L.J.: CenterLineDet: CenterLine Graph detection for road lanes with vehicle-mounted sensors by transformer for HD map generation (2023). ArXiv preprint arXiv:2209.07734
30. Acuna, D., Ling, H., Kar, A., Fidler, S.: Efficient interactive annotation of segmentation datasets with Polygon-RNN++. In: 2018 IEEE/CVF Conference on Computer Vision and Pattern Recognition, Salt Lake City, UT, USA, pp. 859–868 (2018). https://doi.org/10.1109/CVPR.2018.00096

CCLane: Concise Curve Anchor-Based Lane Detection Model with MLP-Mixer

Fan Yang[1], Yanan Zhao[1(✉)], Li Gao[1], Huachun Tan[1(✉)], Weijin Liu[2], Xue-mei Chen[1], and Shijuan Yang[1]

[1] Beijing Institute of Technology, Beijing 100081, China
{zyn,tanhc}@bit.edu.cn

[2] China Changfeng Science Technology Industry Group Corp, Beijing 100081, China
https://www.bit.edu.cn/, http://www.ascf.com.cn/

Abstract. Lane detection needs to meet the real-time requirements and efficiently utilize both local and global information on the feature map. In this paper, we propose a new lane detection model called CCLane, which uses the pre-set curve anchor method to better utilize the prior information of the lane. Based on the Cross Layer Refinement method for extracting local information at different levels, we propose a way to combine MLP-Mixer and spatial convolution to obtain global information and achieve information transmission between lanes, which flexibly and efficiently integrates local and global information. We also extend the DIoU loss function to lane detection and design the LDIoU loss function. The method is evaluated on two widely used lane detection datasets, and the results show that our method performs well.

Keywords: Lane detection · Line anchor · MLP

1 Introduction

Lane detection is an important task in the field of computer vision, and it is also one of the tasks needed to achieve autonomous driving. Lane detection can provide information such as the road condition and the drivable area, and can also help with self-localization. In the early stages, lane detection tasks were achieved by digital image processing methods. Thanks to the development of deep learning, more and more lane detection methods use Convolutional Neural Networks and other methods to extract features. According to the way lane lines are modeled, the methods based on deep learning can be divided into several categories: segmentation-based methods, polynomial curve-based methods, row-wise-based methods, and anchor-based methods.

Anchor-based methods usually makes good use of the prior information about the shape of the lane lines and performs better in the absence of visual cues. Most

This project is supported by the National Key R&D Program of China under Grant 2018YFB0105205-02 and 2020YFC0833406, Key R&D Program of Shandong Province under Grant 2020CXGC010118, Jinan Space-air Information Industry Special Project under Grant QYYZF1112023105.

Q. Liu et al. (Eds.): PRCV 2023, LNCS 14427, pp. 376–387, 2024.
https://doi.org/10.1007/978-981-99-8435-0_30

anchor-based methods use straight lines at different positions as pre-set anchors. However, lane lines are not always straight, and in scenes such as curves and ramps, lane lines have a certain degree of curvature. Straight line anchors can affect the convergence of training and the final detection performance.

In this paper, we extended the work of CLRNet [28] and proposed curve anchors generated by spline curves based on Cross Layer Refinement method of the network. The curve anchors with different radii and shapes are closer to real lane lines, and can better utilize the prior information of lane lines, enabling easier convergence during training and better detection results. Then, we abandoned the attention mechanism adopted by many methods and used MLP-Mixer [19] to obtain global information, which can simplify the parameter of the model, and the model size can be flexibly adjusted by adjusting the depth of MLP-Mixer [19]. In addition, we also adopted the method of spatial convolution to achieve information transmission between different lane lines. The entire model only uses Convolutional Layer and fully connected layers, and is deployment-friendly.

We validated the effectiveness of our method on two commonly used lane detection datasets, TuSimple [20] and CULane [14]. The experiments showed that our method achieved state-of-the-art F1 score on the TuSimple [20] dataset, and competitive results on the CULane [14] dataset, with an F1 score only 0.3% lower than the state-of-the-art method. The smallest model of this method can reach 45 frames per second, and the largest model can also reach 26 frames per second, meeting the real-time requirements. The main contributions can be summarized as follows:

- We proposed a method for pre-set curve anchors that better utilize the prior information of lane lines and improve the model's ability to detect lane lines.
- We adopted a combination of MLP-Mixer [19] and spatial convolution to obtain global information and achieve information transfer between lane lines, making lane line detection more robust.
- We propose a loss function LDIoU for lane lines, which considers the whole of lane lines at different scales.
- The model achieved the state-of-the-art performance on the TuSimple [20] dataset with the highest F1 score and achieved competitive results on the CULane [14] dataset.

2 Related Work

2.1 Segmentation-Based Methods

Many early lane detection methods used a similar approach as semantic segmentation, by determining whether each pixel is foreground or background and then decoding each lane line instance in post-processing. SCNN [14] extended traditional deep layer-by-layer convolution to feature maps to achieve message passing between rows and columns of pixels in a single layer, achieving state-of-the-art performance at the time. Compared with SCNN [14], ENet SAD [7] reduces parameters by 20 times while maintaining performance and increases running speed by 10 times. CurveLane NAS [23] achieved outstanding performance in challenging curved environments by adopting a feature fusion search module. PINet [8] adopted a keypoint-based approach, treating the clustering problem of predicted keypoints as an instance segmentation problem, and had a smaller network than other methods.

2.2 Polynomial Curve-Based Methods

In contrast to segmentation-based methods, curve-based methods usually have high inference speed because they represent lane lines as curves and obtain each lane line by regressing the curve equation. The pioneering work of this method [21] used a differentiable least square fitting module to represent lane lines and finally obtained the best fitting curve parameters. PolyLaneNet [18] used a similar method, outputting the confidence scores of each lane and the areas of these polynomials. LSTR [13] used a network based on Transformer trained to predict parameters, considering non-local interaction of information. BézierLaneNet [3] proposed using Bézier curves to represent lane lines and BSNet [2] utilized B-spline curves to fit lane lines.

2.3 Row-Wise-Based Methods

The lane detection task can be transformed into a row-wise classification task using methods like E2E-LMD [24] and UFLD [15]. These methods significantly reduce the computational cost by using row-wise classification patterns. UFLD [15] uses a large receptive field on global features and can handle challenging scenes. UFLDv2 [16] further develops this method by using sparse coordinate representation for lanes on a series of mixed (row and column) anchors. CondLaneNet [12] introduces a conditional lane detection strategy, which first detects lane instances and then dynamically predicts the geometry of each instance. Additionally, a recursive instance module is designed to overcome the problem of detecting lane lines with complex topological structures.

2.4 Line-Anchor-Based Methods

Line-CNN [9] is a pioneer in line anchor representation, using pixel features at the starting point of a line anchor to predict offsets. LaneATT [17] introduces attention mechanisms to facilitate information exchange between different lane lines and enhance global features, achieving good results with fast speed. CLRNet [28] proposes a Cross Layer Refinement method that first detects lanes with high-level semantic features and then refines them based on low-level features. This approach can leverage more contextual information to detect lanes and also improve the localization accuracy of lane lines using local features. CLRNet [28] greatly reduces the number of line anchors and the computational burden of the network, achieving state-of-the-art detection results. In view of the outstanding performance of methods based on line anchors, this paper develops this method based on CLRNet [28].

There are other models that are different from the typical methods mentioned above. The common challenge faced by these methods is how to effectively utilize the features of lane lines, including prior information, global information, and local information, to maintain robust detection performance in various complex environments, while also meeting real-time requirements and maintaining satisfactory performance.

3 Method

Our network architecture is shown in Fig. 1. Firstly, a series of prior lane lines are defined on the input image. These lane lines are distributed at different positions on the image, with a variety of tilt angles and radians. After the backbone, FPN [10] is connected to further fusion the feature information of different levels, and then output 4 layers of feature maps containing different level feature information. The highest-level feature map connects the MLP-Mixer [19] to extract global information for recalling and roughly locating lane lines. The lowest-level feature map connects the spatial convolution module to achieve information transmission between different lane lines, and to obtain further global information. After traversing the 4 feature maps, get ROI at the corresponding position based on the lane anchor in each feature map. The three different types of information are fused together and fed into several fully connected layers to obtain the final output.

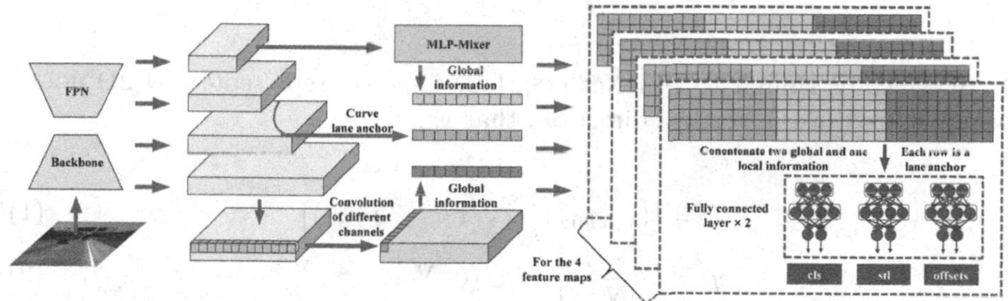

Fig. 1. The structure of CCLane. The features are extracted from the backbone network and fused with the FPN [10]. Use the pre-set curve lane anchor as local information. The MLP-Mixer [19] was adopted to extract global information and carry out convolution operation in different directions of the feature map to transfer information between lane lines.

3.1 The Lane Representation

The pre-set lane anchor expresses the possible existence of lane lines in the image. The closer the prior lane lines are to the actual lanes in terms of position and shape, the easier the network converges, and the better the final detection performance. In previous methods, lane anchor was usually straight lines distributed at different positions in the image. However, lane lines are not always straight, they also have different radians in curved and ramp scenes. Therefore, as shown in Fig. 2, we use some curves with different radians as lane anchor and mix them with straight lane anchor in a certain proportion to jointly serve as the prior lane lines.

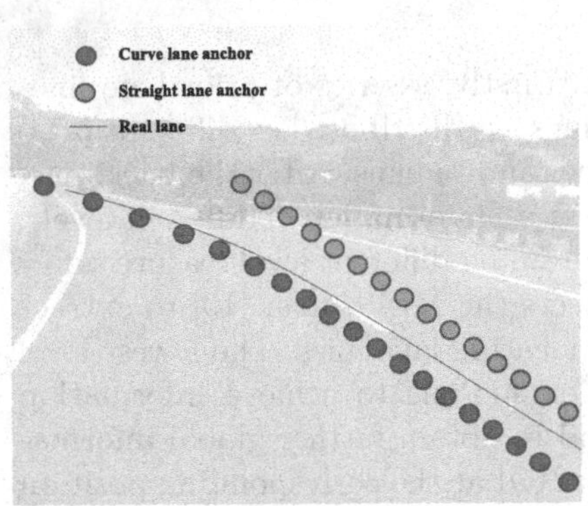

Fig. 2. Lane anchor. The lane anchor for curves can be closer to real lane lines.

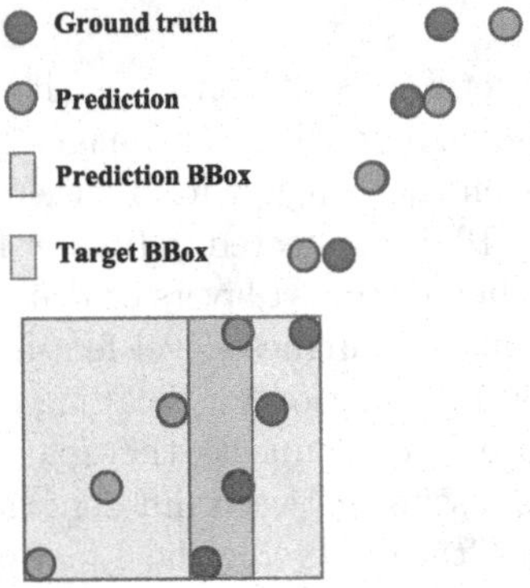

Fig. 3. LDIoU. The lane lines can be considered at different scales.

Specifically, similar to CLRNet [28], lane lines are represented by 2D points evenly spaced in the vertical direction, that is:

$$P = \{(x_0, y_0), \cdots, (x_{N-1}, y_{N-1})\} \tag{1}$$

$$y_i = i \cdot \frac{H}{N-1}, i \in (0, N-1) \tag{2}$$

Where H is the height of the image, and N is the number of points used to represent a lane line.

For each point, the y-coordinate is fixed and invariant. For each point of the prior lane anchor, the x-coordinate will be sampled on the image. The predicted lane line will be overlaid with an offset on the x-coordinate of the prior lane line. Each prior lane anchor contains foreground and background probabilities, lane line length, starting point of the lane line, the angle between the lane line and the horizontal direction, and the horizontal offset of each 2D point in the image. For curved lane anchor, the start points, the end point, and the four dividing points near the top of the straight lane anchor will be taken, and different amounts of offsets will be applied to the middle two points. Finally, the minimum least squares polynomial is used to fit the curve.

3.2 Global Information Extraction Module

A lane line often spans a large part of the image, and there is an association between lane lines. When lane lines are obscured or worn, the missing parts of the lane lines in the image can be detected by using the overall information of the lane lines or the information of nearby lane lines.

Many methods use attention mechanisms to achieve information transmission between different lane lines [17, 28], or use complex special convolution methods to achieve message transmission between rows and columns of pixels in a layer [14]. These methods bring huge computational demands.

To address the above problem, we use two methods to obtain global information. Inspired by MLP-Mixer [19], we divide the highest level feature map into many small regions, which can extract information within each region and between regions through mixing. This structure is based on a Multilayer Perceptron, and can be easily adjusted in depth to control the network size. MLP-Mixer [19] consists of multiple stacked units, each containing two parts. The first part is the token-mixing MLP, which acts on each small block and extracts information between blocks, i.e., $\mathbb{R}^S \to \mathbb{R}^S$. The second part is the channel-mixing MLP, which acts on each channel and extracts information between channels, i.e., $\mathbb{R}^C \to \mathbb{R}^C$.

In addition, unlike regular convolutional operations that are performed along the channel direction, we apply two convolutional operations on the lowest level feature map output by the FPN [10]. One is along the direction of the lane line (vertical direction of the image), and the other is along the direction between the lane lines (horizontal direction of the image). This enables the exchange of information between different lane lines.

3.3 Local Information Extraction Module

The local features of the lane markings are crucial for precise lane positioning. Following the idea of the line-anchor-based method, we use models like Faster R-CNN [5] to obtain ROIs from the feature map and extract local features corresponding to each lane anchor. Similar to the implementation method of CLRNet [28], we first uniformly sample the lane anchor on the feature map and then perform bilinear interpolation, i.e. $\chi_p \in \mathbb{R}^{C \times N_p}$. The difference is that our lane anchor mixes straight and curved lines. In addition, we use Cross Layer Refinement method, starting from the high-level features to the last low-level feature, and each time superimpose the offset of the predicted lane x-coordinates to make the predicted lane more accurate. The formula is expressed as:

$$\lambda = \lambda_0 + \lambda_{l_1} + \lambda_{l_2} + \lambda_{l_3} + \lambda_{l_4} \tag{3}$$

λ includes the starting coordinate of the lane marking, the horizontal angle θ, and the values of N sampling points. λ_0 represents the initial value in lane anchor, and λ_{l_i} represents the offset predicted in the i layer.

3.4 Loss Function

As shown in Fig. 3, we use a series of discrete points to represent the lane marking, and the difference between the lane anchor and the ground truth value can be regressed using commonly used distance losses such as $smooth - L_1$. As stated in [26], this overly simplified assumption may lead to inaccurate positioning. We designed a loss function LDIoU that considers the overall lane marking in different scales to address this issue.

We extend the concept of DIoU [29] to lane lines. A lane line is represented as a set of N points, M discrete points forming a subset of N, and their enclosing rectangular box as the BBox. In this way, a lane line can be subdivided into many small segments at different scales. When we perform inference on feature maps of different scales, we use different values of M. A larger M is used on higher-level feature maps to recall and roughly locate the lane lines as much as possible. A smaller M is used on lower-level feature maps to more finely regress the lane lines.

3.5 Training and Infercence Details

Similar to [28], we adopt the idea of [4] to assign positive samples. During training, each ground truth lane line is dynamically assigned one or more predicted lanes as positive samples.

The loss function includes three parts: the classification loss for predicting whether the lane line is foreground or background, the regression loss for the starting coordinates, length, and angle to the horizontal direction of the positive sample lane lines, and the regression loss for the horizontal offset of the sampled points on the lane lines. For the classification loss, we use *Focal Loss* [11]. For the first regression loss, we use $smooth - L_1$ [5]. For the second regression loss, LDIoU is introduced and combined with DIoU [28]. The overall loss function can be represented as:

$$\mathcal{L} = w_{cls}\mathcal{L}_{cls} + w_{stl}\mathcal{L}_{stl} + w_{LDIoU}\mathcal{L}_{LDIoU} + w_{LIoU}\mathcal{L}_{LIoU} \quad (4)$$

In addition, in the middle of the network, a segmentation loss is introduced for each layer of feature maps output by FPN [10] to guide the network to learn lane-related features. Like other line-anchor-based methods, NMS was used in inference to eliminate lane lines with very similar predicted results.

We used ResNet [6] and DLA [25] as our pre-trained backbone networks, with FPN [10] as the neck and outputting four feature maps. We represented each lane line with 72 points, and used $N_P = 36$ sampling points for extracting local information. The data augmentation method is similar to [28], and in addition, we also use methods such as adding Gaussian noise and random cropping. During training, we used the AdamW optimizer with an initial learning rate of 6×10^{-4}. The loss function weights were set to 2.0, 0.2, 0.5, 1.0. We trained the model for 200 epochs on TuSimple [20] and 25 epochs on CULane [14]. Our experiments were conducted using PyTorch on four NVIDIA TITAN Xp devices. These are the basic experimental parameters, which may vary for different groups.

4 Experiment

4.1 Datasets and Evaluation Metric

We conducted experiments on the widely used benchmark datasets CULane [14] and TuSimple [20]. CULane [14] is a challenging dataset that contains various scenarios such as normal, crowded, dazzling light, shadows, occlusions, arrows, curves, and night scenes. It consists of 88,880 training images and 34,680 testing images. The images have a resolution of 590×1640. TuSimple [20] is a dataset of highway scenes that includes 3,268 images for training, 358 images for validation, and 2,782 images for testing. The images have a resolution of 720×1280.

For CULane [14], we use the F1 score as the evaluation criterion. If the IoU between the predicted and true values is greater than 0.5, these predicted values are considered correct predictions. The formula for calculating F1 score is as follows:

$$F1 = \frac{2 \times \text{ Precision} \times \text{ Recall}}{\text{Precision} + \text{ Recall}} \tag{5}$$

Where $Precision = \frac{TP}{FP+FN}$, $Recall = \frac{TP}{TP+FN}$. TP refers to the number of correct predictions, FP refers to the number of false positive predictions, and FN refers to the number of false negative predictions.

For TuSimple [20], the official evaluation criterion is accuracy, which is calculated as follows:

$$\text{accuracy} = \frac{\sum_{clip} C_{clip}}{\sum_{clip} S_{clip}} \tag{6}$$

Where C_{clip} refers to the number of points that were correctly predicted, and S_{clip} refers to the total number of ground truth points. We also calculate the F1 score on TuSimple [20].

4.2 Results

Table 1 displays the performance of our method and other state-of-the-art methods. The performance of different methods on the TuSimple [20] dataset is very similar, with the best performance approaching saturation. Once again, we have set a new F1 score record on this dataset and achieved the state-of-the-art performance. Our model did not adopt attention mechanisms that require high computational burden and high computational cost to obtain useful information, as many other models did. Instead, we used relatively simple MLP structures to achieve high precision and low false positive and false negative rates with a lower computational cost.

The performance on the CULane [14] dataset is shown in Table 2. Our method achieved the best F1 score on the smaller backbone network ResNet-18, indicating that our method can achieve excellent performance with a simple network structure and small computational cost. Moreover, in the "Cross" category, the false positive rate of our method is significantly lower than that of other

Table 1. Performance of different methods on TuSimple [20]

Method	F1%	Acc%	FP%	FN%	F1%	Acc%	FP%	FN%	F1%	Acc%	FP%	FN%
	ResNet-18				ResNet-34				ResNet-101			
LaneATT [17]	96.71	95.57	3.56	3.01	96.77	95.63	3.53	2.92				
CLRNet [28]	97.89	96.84	2.28	1.92	97.82	96.87	2.27	2.08	97.62	96.83	2.37	2.38
UFLDv2 [16]	96.16	95.65	3.06	4.61	96.22	95.56	3.18	4.37				
CondLaneNet [12]	97.01	95.84	2.18	3.8	96.98	95.37	2.2	3.82	97.24	96.54	2.01	3.5
GANet [22]	97.71	95.95	1.97	2.62	97.68	95.87	1.99	2.64	97.45	96.44	2.63	2.47
PolyLaneNet [18]	90.62	93.36	9.42	9.33								
RESA [27]					96.93	96.82	3.63	2.84				
BSNet [2]	97.79	96.63	2.03	2.39	97.79	96.78	2.12	2.32	97.84	96.73	1.98	2.35
CCLane	97.88	96.51	1.65	2.6	97.72	96.54	2.31	2.23	97.91	96.81	2.09	2.06

Table 2. Performance of different methods on CULane [14]

Method	Total	Normal	Crowd	Dazzle	Shadow	No line	Arrow	Curve	Cross	Night
ResNet-18										
LaneATT [17]	75.13	91.17	72.71	65.82	68.03	49.13	87.82	63.75	1020	68.58
CLRNet [28]	79.58	93.3	78.33	73.71	79.66	53.14	90.25	71.56	1321	75.11
UFLDv2 [16]	74.7	91.7	73	64.6	74.7	47.2	87.6	68.7	1998	70.2
CondLaneNet [12]	78.14	92.87	75.79	70.72	80.01	52.39	89.37	72.4	1364	73.23
GANet [22]	78.79	93.24	77.16	71.24	77.88	53.59	89.62	75.92	1240	72.75
BézierLaneNet [3]	73.67	90.22	71.55	62.49	70.91	45.3	84.09	58.98	996	68.7
BSNet [2]	79.64	93.46	77.93	74.25	81.95	54.24	90.05	73.62	1400	75.11
CCLane	79.82	93.41	78.64	73.55	81.77	52.7	89.57	71.21	907	74.69
ResNet-34										
LaneATT [2]	76.68	92.14	75.03	66.47	78.15	49.39	88.38	67.72	1330	70.72
CLRNet [28]	79.73	93.49	78.06	74.57	79.92	54.01	90.59	72.77	1216	75.02
RESA [27]	74.5	91.9	72.4	66.5	72	46.3	88.1	68.6	1896	69.8
UFLDv2 [16]	75.9	92.5	74.9	65.7	75.3	49	88.5	70.2	1864	70.6
CondLaneNet [12]	78.74	93.38	77.14	71.17	79.93	51.85	89.89	73.88	1387	73.92
GANet [22]	79.39	93.73	77.92	71.64	79.49	52.63	90.37	76.32	1368	73.67
BézierLaneNet [3]	75.57	91.59	73.2	69.2	76.74	48.05	87.16	62.45	888	69.9
BSNet [2]	79.89	93.75	78.01	76.65	79.55	54.69	90.72	73.99	1445	75.28
CCLane	79.6	93.33	77.85	75.1	81.74	54.23	90.05	70.91	1235	75.34
ResNet-101										
CLRNet [28]	80.13	93.85	78.78	72.49	82.33	54.5	89.79	75.57	1262	75.51
CondLaneNet [12]	79.48	93.47	77.44	70.93	80.91	54.13	90.16	75.21	1201	74.8
GANet [22]	79.63	93.67	78.66	71.82	78.32	53.38	89.86	77.37	1352	73.85
BSNet [2]	80	93.75	78.44	74.07	81.51	54.83	90.48	74.01	1255	75.12
CCLane	79.9	93.71	78.26	72.98	80.24	54.43	90.69	74.1	1088	74.95
DLA-34										
CLRNet [28]	80.47	93.73	79.59	75.3	82.51	54.58	90.62	74.13	1155	75.37
LaneAF [1]	77.41	91.8	75.61	71.78	79.12	51.38	86.88	72.7	1360	73.03
BSNet [2]	80.28	93.87	78.92	75.02	82.52	54.84	90.73	74.71	1485	75.59
CCLane	80.22	93.72	78.99	73.87	81.32	53.4	89.99	70.18	970	75.41

non-polynomial curve methods, and even lower than that of polynomial curve-based methods on the smaller ResNet-18 backbone network. This indicates that our method has strong recognition capabilities for real lane lines and good robustness against other confusions.

5 Ablation Study

We conducted extensive experiments on the CULane [14] dataset to validate our method. All the ablation studies were based on the ResNet-18 backbone network version with the training of 1 epoch. The experiments were conducted in three aspects: lane representation, global information extraction, and loss function.

Table 3. Ablation experiment on CULane [14]

Method	Total	Normal	Crowd	Dazzle	Shadow	No line	Arrow	Curve	Cross	Night
L	56.78	73.44	51.79	49.24	49.84	35.43	62.99	49.62	2402	53.17
L&Re	70.45	84.14	69.17	60.33	74.15	47.05	79.46	60.31	2100	66.29
L&Re&G	73.56	88.26	72.26	62.41	74.31	48.99	82.49	64.53	1761	67.81
C	57.48	71.84	53.92	42.08	54.25	35.46	61.85	45.89	2147	55.79
C&Re	72.33	87.6	70.65	60.48	72.04	46.9	81.7	57.29	1744	66.76
C&Re&G	74.12	88.90	71.72	65.89	73.43	49.24	82.49	62.32	1364	69.26
C&Re&G&D	72.79	87.57	70.92	64.34	73.65	48.92	81.19	61.69	2052	68.41

In Table 3, L represents lane anchor for straight lines, C represents lane anchor for curves, Re represents using the Cross Layer Refinement method, G represents including global information extraction, and D represents using the LDIoU loss function. From the experimental results, using the Cross Layer Refinement method has the greatest impact on the model's performance, with the Total F1 score increasing by more than 10%, demonstrating the effectiveness of this method for lane detection. Using the lane anchor for curves as prior information for lane detection can increase the model's Total F1 score by about 1%, indicating that pre-set anchors closer to the real lane lines can make the model converge more easily and quickly. Adding global information based on local information can increase the model's Total F1 score by about 2%, improving the model's performance in various scenarios. The effectiveness of LDIoU is limited, and further research is needed for its use at different granularity levels.

6 Conclusion

In this work, we have proposed a lane detection model based on curve lane anchors. By using the lane anchor for curves, which are closer to real lane lines, as prior information for lane detection. We extracted global information on the feature map using the simple and efficient MLP and Convolutional Layer that transmit spatial information. Additionally, we proposed an optional LDIoU loss function that extends the DIoU [29] loss function to lane detection. Our method had the advantages of high efficiency and good performance. On the TuSimple [20] dataset, our method achieved the highest F1 score compared to other advanced models. On the CULane [14] dataset, our method also showed competitive performance.

References

1. Abualsaud, H., Liu, S., Lu, D.B., Situ, K., Rangesh, A., Trivedi, M.M.: Laneaf: robust multi-lane detection with affinity fields. IEEE Rob. Autom. Lett. **6**(4), 7477–7484 (2021)
2. Chen, H., Wang, M., Liu, Y.: BSNET: lane detection via draw b-spline curves nearby. arXiv preprint arXiv:2301.06910 (2023)
3. Feng, Z., Guo, S., Tan, X., Xu, K., Wang, M., Ma, L.: Rethinking efficient lane detection via curve modeling. In: Proceedings of the IEEE/CVF Conference on Computer Vision and Pattern Recognition, pp. 17062–17070 (2022)
4. Ge, Z., Liu, S., Wang, F., Li, Z., Sun, J.: Yolox: exceeding yolo series in 2021. arXiv preprint arXiv:2107.08430 (2021)
5. Girshick, R.: Fast r-cnn. In: Proceedings of the IEEE International Conference on Computer Vision, pp. 1440–1448 (2015)
6. He, K., Zhang, X., Ren, S., Sun, J.: Deep residual learning for image recognition. In: Proceedings of the IEEE Conference on Computer Vision and Pattern Recognition, pp. 770–778 (2016)
7. Hou, Y., Ma, Z., Liu, C., Loy, C.C.: Learning lightweight lane detection cnns by self attention distillation. In: Proceedings of the IEEE/CVF International Conference on Computer Vision, pp. 1013–1021 (2019)
8. Ko, Y., Lee, Y., Azam, S., Munir, F., Jeon, M., Pedrycz, W.: Key points estimation and point instance segmentation approach for lane detection. IEEE Trans. Intell. Transp. Syst. **23**(7), 8949–8958 (2021)
9. Li, X., Li, J., Hu, X., Yang, J.: Line-CNN: end-to-end traffic line detection with line proposal unit. IEEE Trans. Intell. Transp. Syst. **21**(1), 248–258 (2019)
10. Lin, T.Y., Dollár, P., Girshick, R., He, K., Hariharan, B., Belongie, S.: Feature pyramid networks for object detection. In: Proceedings of the IEEE Conference on Computer Vision and Pattern Recognition, pp. 2117–2125 (2017)
11. Lin, T.Y., Goyal, P., Girshick, R., He, K., Dollár, P.: Focal loss for dense object detection. In: Proceedings of the IEEE International Conference on Computer Vision, pp. 2980–2988 (2017)
12. Liu, L., Chen, X., Zhu, S., Tan, P.: Condlanenet: a top-to-down lane detection framework based on conditional convolution. In: Proceedings of the IEEE/CVF International Conference on Computer Vision, pp. 3773–3782 (2021)
13. Liu, R., Yuan, Z., Liu, T., Xiong, Z.: End-to-end lane shape prediction with transformers. In: Proceedings of the IEEE/CVF Winter Conference on Applications of Computer Vision, pp. 3694–3702 (2021)
14. Pan, X., Shi, J., Luo, P., Wang, X., Tang, X.: Spatial as deep: spatial CNN for traffic scene understanding. In: Proceedings of the AAAI Conference on Artificial Intelligence, vol. 32 (2018)
15. Qin, Z., Wang, H., Li, X.: Ultra fast structure-aware deep lane detection. In: Vedaldi, A., Bischof, H., Brox, T., Frahm, J.-M. (eds.) ECCV 2020. LNCS, vol. 12369, pp. 276–291. Springer, Cham (2020). https://doi.org/10.1007/978-3-030-58586-0_17
16. Qin, Z., Zhang, P., Li, X.: Ultra fast deep lane detection with hybrid anchor driven ordinal classification. IEEE Trans. Pattern Anal. Mach. Intell. (2022)
17. Tabelini, L., Berriel, R., Paixao, T.M., Badue, C., De Souza, A.F., Oliveira-Santos, T.: Keep your eyes on the lane: Real-time attention-guided lane detection. In: Proceedings of the IEEE/CVF Conference on Computer Vision and Pattern Recognition, pp. 294–302 (2021)

18. Tabelini, L., Berriel, R., Paixao, T.M., Badue, C., De Souza, A.F., Oliveira-Santos, T.: Polylanenet: lane estimation via deep polynomial regression. In: 2020 25th International Conference on Pattern Recognition (ICPR), pp. 6150–6156. IEEE (2021)
19. Tolstikhin, I.O., et al.: MLP-mixer: an all-MLP architecture for vision. Adv. Neural. Inf. Process. Syst. **34**, 24261–24272 (2021)
20. TuSimple: Tusimple benchmark. https://github.com/TuSimple/tusimple-benchmark/. Accessed 1 May 2023
21. Van Gansbeke, W., De Brabandere, B., Neven, D., Proesmans, M., Van Gool, L.: End-to-end lane detection through differentiable least-squares fitting. In: Proceedings of the IEEE/CVF International Conference on Computer Vision Workshops (2019)
22. Wang, J., et al.: A keypoint-based global association network for lane detection. In: Proceedings of the IEEE/CVF Conference on Computer Vision and Pattern Recognition, pp. 1392–1401 (2022)
23. Xu, H., Wang, S., Cai, X., Zhang, W., Liang, X., Li, Z.: CurveLane-NAS: unifying lane-sensitive architecture search and adaptive point blending. In: Vedaldi, A., Bischof, H., Brox, T., Frahm, J.-M. (eds.) ECCV 2020. LNCS, vol. 12360, pp. 689–704. Springer, Cham (2020). https://doi.org/10.1007/978-3-030-58555-6_41
24. Yoo, S., et al.: End-to-end lane marker detection via row-wise classification. In: Proceedings of the IEEE/CVF Conference on Computer Vision and Pattern Recognition Workshops, pp. 1006–1007 (2020)
25. Yu, F., Wang, D., Shelhamer, E., Darrell, T.: Deep layer aggregation. In: Proceedings of the IEEE Conference on Computer Vision and Pattern Recognition, pp. 2403–2412 (2018)
26. Yu, J., Jiang, Y., Wang, Z., Cao, Z., Huang, T.: Unitbox: an advanced object detection network. In: Proceedings of the 24th ACM International Conference on Multimedia, pp. 516–520 (2016)
27. Zheng, T., et al.: RESA: recurrent feature-shift aggregator for lane detection. In: Proceedings of the AAAI Conference on Artificial Intelligence, vol. 35, pp. 3547–3554 (2021)
28. Zheng, T., et al.: CLRNET: cross layer refinement network for lane detection. In: Proceedings of the IEEE/CVF Conference on Computer Vision and Pattern Recognition, pp. 898–907 (2022)
29. Zheng, Z., Wang, P., Liu, W., Li, J., Ye, R., Ren, D.: Distance-iou loss: faster and better learning for bounding box regression. In: Proceedings of the AAAI Conference on Artificial Intelligence, vol. 34, pp. 12993–13000 (2020)

Uncertainty-Aware Boundary Attention Network for Real-Time Semantic Segmentation

Yuanbing Zhu[1,2], Bingke Zhu[1,4](✉), Yingying Chen[1,4], and Jinqiao Wang[1,2,3,4]

[1] Foundation Model Research Center, Institute of Automation, Chinese Academy of Sciences, Beijing, China
zhuyuanbing2021@ia.ac.cn, {yingying.chen,jqwang}@nlpr.ia.ac.cn

[2] School of Artificial Intelligence, University of Chinese Academy of Sciences, Beijing, China

[3] The Peng Cheng Laboratory, Shenzhen, China

[4] Wuhan AI Research, Wuhan, China
bingke.zhu@nlpr.ia.ac.cn

Abstract. Remarkable progress has been made in real-time semantic segmentation by leveraging lightweight backbone networks and auxiliary low-level training tasks. Despite several techniques have been proposed to mitigate accuracy degradation resulting from model reduction, challenging regions often exhibit substantial uncertainty values in segmentation results. To tackle this issue, we propose an effective structure named Uncertainty-aware Boundary Attention Network(UBANet). Specifically, we model the segmentation uncertainty via prediction variance during training and involve it as a regularization item into optimization objective to improve segmentation performance. Moreover, we employ uncertainty maps to investigate the role of low-level supervision in segmentation task. And we reveal that directly fusing high- and low-level features leads to the overshadowing of large-scale low-level features by the encompassing local contexts, thus hindering the synergy between the segmentation task and low-level tasks. To address this issue, we design a Low-level Guided Feature Fusion Module that avoids the direct fusion of high- and low-level features and instead employs low-level features as guidance for the fusion of multi-scale contexts. Extensive experiments demonstrate the efficiency and effectiveness of our proposed method by achieving the state-of-the-art latency-accuracy trade-off on Cityscapes and CamVid benchmark.

Keywords: Uncertainty Estimation · Real-time Semantic Segmentation

1 Introduction

With the development of deep neural networks, semantic segmentation has achieved significant improvement over traditional methods [3,5,21].

This work was supported by National Key R&D Program of China under Grant No.2021ZD0114600, National Natural Science Foundation of China (62006230, 62206283, 62076235).

Q. Liu et al. (Eds.): PRCV 2023, LNCS 14427, pp. 388–400, 2024.
https://doi.org/10.1007/978-981-99-8435-0_31

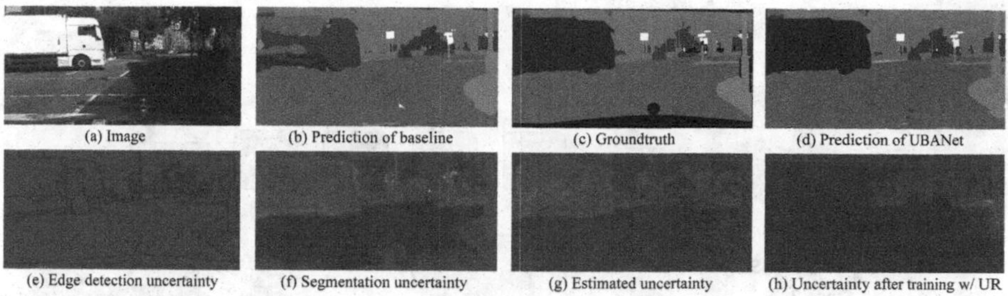

Fig. 1. Visualization of uncertainty maps. Uncertainty values are higher in brighter locations. (e) and (f) are produced by Monte-Carlo dropout [10]. (g) is the uncertainty estimation result, and (h) is produced by the network trained with Uncertainty Regularization (UR).

Nevertheless, the computationally intensive operations restrict their practicality for scenarios with real-time constraints. Previous real-time methods [4,8,14,19,22, 26,27,29] mainly focus on reducing computation cost of deep networks by designing lightweight backbones tailored for segmentation tasks [14,29] and maintaining localization ability of down-sampled image features [19,22,27,29] by employing dual-branch architecture, aiming to find better trade-off between accuracy and latency. And significant improvement has been achieved by these great works so far.

More recently, auxiliary low-level supervision [8,15,23] has been developed to enhance the encoding capability of spatial details of segmentation models. Comparing with the dual-branch method, the main distinction of these low-level supervisions is to introduce structure information without incurring extra computational cost during inference. These inference-free detailed signals are particularly suitable for real-time task. Thus we consider the collaboration of low-level features and context features should be more investigated in real-time semantic segmentation task.

Uncertainty estimation is commonly used as a reliability assessment of a model prediction [10,13], and is typically employed to rectify pseudo-label within self-learning frameworks [31] under semi-supervised setting. In this paper, we thoroughly analyze the incorporation of low-level task and segmentation task from the perspective of uncertainty estimation, and yields two key observations. First, the segmentation results still exhibit high uncertainty on the boundaries of the objects (as in Fig. 1(a)(f)), which is not reduced by introducing low-level tasks. Second, the challenging area remains high uncertainty values, even within object contents (as in Fig. 1(b)(f)).

The first observation indicates a defect in directly fusing high-resolution details with low-frequency contexts, as it can lead to the detailed features being easily overshadowed by the encompassing local contexts. To address this issue, we introduce Low-level guided Attention Feature Fusion Module(LAFFM) as shown in Fig. 3, which aims to avoid the direct attendance of low-level supervised feature but employ it as prior guidance for the aggregation of contexts.

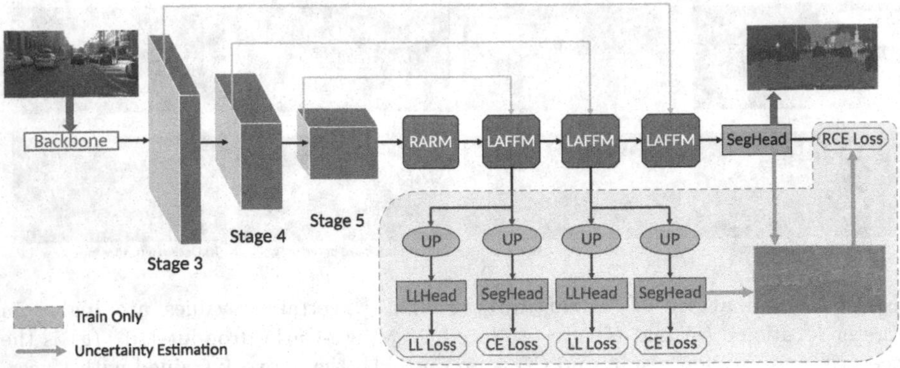

Fig. 2. Overall architecture of the proposed UBANet. The branches in the dashed pink box are only hold during the training, and the LLHead and SegHead denote the low-level head and segmentation head built with FCN [21]. The orange line denotes the uncertainty estimation flow.

The second observation indicates that segmentation performance can be improved by reducing predictive uncertainty. In light of this, we propose to model the predictive uncertainty during training via the variance of semantic prediction maps as in Fig. 1(g), and involve it as a regularization term into the optimization objective. Furthermore, we incorporate uncertainty regularization and LAFFM into an encoder-decoder architecture and propose Uncertainty-aware Boundary Attention Network(UBANet). By taking extensive experiments, we find that UBANet can take advantage of low-level features more effectively and achieve better latency-accuracy trade-off than previous works. To summarize, our contributions are as follows:

- A novel, low-level guided, uncertainty-based regularization method is proposed to assist the semantic segmentation task.
- An effective attention-based feature fusion module is proposed to investigate a novel view of feature incorporation.
- With the components above, we compile Uncertainty-aware Boundary Attention Network (UBANet), a network architecture for real-time semantic segmentation.
- Extensive experiments demonstrate that UBANet achieves the state-of-the-arts latency-accuracy trade-off on real-time semantic segmentation benchmarks, including Cityscapes [7] and CamVid [1].

2 Related Work

Generic Semantic Segmentation. Modern mainstream semantic segmentation networks are predominantly based on the encoder-decoder architecture [3,9,21,28,30]. Since FCN [21] replace the linear layer by a convolutional layer in classification networks, semantic segmentation network can be trained in an end-to-end way. PSPNet [30] proposed the pyramid pooling module which captures and aggregates context clues by global pooling at varying scales. DeepLab series [2,3] employ dilate convolutions with different rates to increase receptive field. Most recently, Transformer-based networks (such as [5,12]) have achieved superior performance on various semantic segmentation benchmark.

Real-time Semantic Segmentation. To fulfill the real-time demand, many high-efficiency networks were proposed [4,8,14,26,27,29]. ICNet [29] proposes a cascade feature fusion unit for secondary refinement of coarse prediction. DFANet [14] proposes a network to aggregate spatial and semantic information on sub-net and sub-stage. BiseNet [26,27] and DDRNet [19] employs a two-pathway network to encode semantic and spatial information separately. STDC-Seg [8] incorporate edge detection tasks with a specially designed backbone for segmentation. SFNet [16] proposed a feature alignment module to predict the offset between multi-scale feature maps. Also, [4,17] adopt neural architecture search algorithm to design efficient networks for real-time segmentation. Most recently, PIDNet [25] apply the concept of a PID controller to the segmentation task and introduces a three-branch network.

Uncertainty Estimation in Semantic Segmentation. In semantic segmentation, the description of epistemic uncertainty usually reliy on sampling-based processes such as Monte Carlo Dropout [10,13]. However, models typically require multiple forward propagations to obtain uncertainty information. To address this issue, several methods employ feature similarity [18], feature variance [31] and category-level entropy [24] to model uncertainty for pseudo label refinement in self-training framework under semi-supervised setting. Most

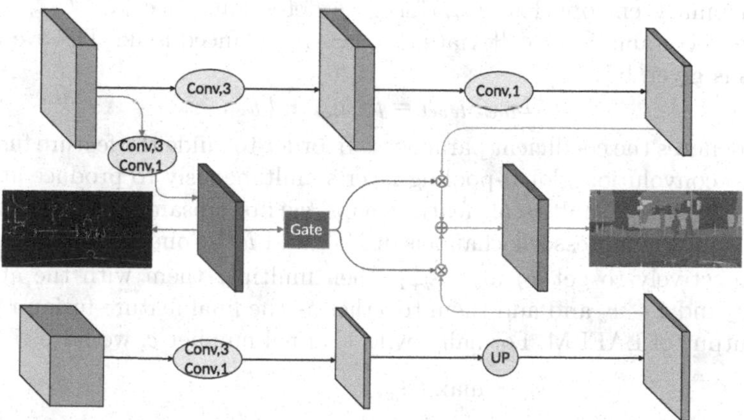

Fig. 3. Illustration of Low-level guided Attention Feature Fusion Module. "Conv,k" denotes the convolution layer with kernel size k.

recently, [11] combine pretrained deep feature and moment propagation to estimate uncertainty result.

3 Method

We employ Monte-Carlo dropout [10] to visualize uncertainty maps for both segmentation task and edge detection task in [8], as shown in Fig. 1. The visualization yields two observations: (1)Segmentation uncertainty is primarily located along object boundaries, which is expected to be reduced by edge supervision, as the edge uncertainty is complementary to segmentation uncertainty in Fig. 1(e). (2) The challenging regions exhibit higher uncertainty value, such as the large white track in Fig. 1(b)(f). Based on observation (1), we assume that the inconsistent uncertainty behavior results from the ineffective feature fusion module. Accordingly, we introduce the Low-level guided Attention Feature Fusion Module in Sect. 3.1. In response to observation (2), we propose to model the uncertainty during training by utilizing the variance of auxiliary feature maps, as depicted in Sect. 3.2.

3.1 Low-Level Guided Attention Feature Fusion Module

Low level supervision is regarded as an effective oracle for preserving spatial information [8,15]. However, as we observed in Sect.3, directly fusing high-resolution details with low-frequency contexts may lead to the detailed features being overshadowed by the encompassing local contexts.

Therefore, we propose Low-level guided Attention Feature Fusion Module (LAFFM), which alleviates the direct attendance of low-level supervised features, but utilizes them as the guidance for feature fusion, as shown in Fig. 3. Formally, the feature maps output obtained from last few stages of encoder are denote as $\{F_i\}_{i=1}^{N}$. Specifically, we use two convolutional layers to capture detail feature maps D_i from F_i, then feed it to the subsequent low-level task, where we utilize weighted binary entropy loss L_{bce} for edge detection. Dice loss L_{dice} are also employed in conjunction to alleviate the class-imbalanced issue. The overall low-level loss is given by

$$L_{low-level} = \mu L_{dice} + L_{bce}, \tag{1}$$

where μ denotes the coefficient parameter. In order to guide the feature fusion, D_i is fed into convolution-global-pooling layer simultaneously to produce attention weight w. To handle multi-scale feature maps, we first upsample F_{i+1} to the same size as F_i, then compress the channels of F_{i+1} and F_i by one single convolutional layer respectively to get $\widetilde{F}_i$ and $\widetilde{F}_{i+1}$, then multiply them with the attention weight w_i and $1 - w_i$ and add them together as the final feature fusion result R_i as the output of LAFFM, Formally, with channel number c, we have

$$\begin{aligned} w_i &= \max_{c} D_{i,c}, \\ R_i &= (1 - w_i) \cdot \widetilde{F}_i + w_i \cdot \widetilde{F}_{i+1}. \end{aligned}$$

Fig. 4. Feature visualization of the LAFFM module. It is clear that the network can encode more spatial information with detail guidance as shown in (b)(c)(basic structure of lane line, object boundaries). And with proposed LAFFM(d), the network can produce more fine-granted features.

3.2 Uncertainty Modeling

Inspired by [10,13,31], we aim to estimate uncertainty during the training procedure by calculating the variance of segmentation predictions. Intuitively, in a Bayesian convolutional neural network with parameter θ, the predictive uncertainty for pixel Y can be approximated through T inferences [13]:

$$\mathrm{Var}(Y) \approx E\hat{Y}^2 - (E\hat{Y})^2 + E\hat{\sigma}^2, \tag{2}$$

$$= \frac{1}{T}\sum_{t=1}^{T}\hat{y}_t^2 - (\frac{1}{T}\sum_{t=1}^{T}\hat{y}_t)^2 + \frac{1}{T}\sum_{t=1}^{T}\hat{\sigma}_t^2, \tag{3}$$

where $\hat{y}$ represents the segmentation prediction, and σ is predicted by a additional branch with learning objective [13]:

$$L_{BNN}(\theta) = \frac{1}{D}\sum_{i} e^{-s_i}||y_i - \hat{y}_i||^2 + s_i. \tag{4}$$

Here, $s_i = \log\hat{\sigma}_i^2$ is the predictive variable. However, Monte Carlo dropout [10,13] necessitates multiple forward propagations, leading to an inefficient training process. To address this issue, we employ the variance of the predictions from different N_{head} auxiliary heads to compute $\mathrm{Var}(Y)$ in Eq. (2).

In practice, we approximate predictive variance by computing the Kullback-Leible divergence (as in [31]) of segmentation prediction $\hat{Y}_{pred}$ and auxiliary head output $\hat{Y}_{aux}$(as depicted in Fig. 1(g)):

$$\mathrm{Var}(Y) \approx D_{KL}(\hat{\mathbf{Y}}_{pred}||\hat{\mathbf{Y}}_{aux}), \tag{5}$$

$$= \sum |\hat{y}_{pred}|\log|\hat{y}_{pred}| - |\hat{y}_{pred}|\log|\hat{y}_{aux}|. \tag{6}$$

Similar to Eq. 4, we fix the predictive uncertainty as a regularization item $d_i := \text{normalize}(D_{KL}(\hat{\mathbf{Y}}_{pred}||\hat{\mathbf{Y}}_{aux}))$ and propose regularized cross entropy loss:

$$L_{rce} = \frac{1}{N}\sum_{i} e^{\frac{d_i}{\tau}} \sum_{c=1}^{C} y_{ic}\text{log}\hat{y}_{ic}, \tag{7}$$

where τ is the temperature. Note that in proposed regularized cross entropy loss, when uncertainty value increases($\frac{d_i}{\tau}$ increases), we refer to our observation (2) in Sect. 3, indicating that current training sample is challenging. Thus we aim for the network to focus more on current training sample, resulting in a larger value of L_{rce}. Note that we abandon the penalty term s_i in Eq. 4, because d_i is computed and fixed for one single forward propagation and does not require any constraint.

3.3 Uncertainty-Aware Boundary Attention Network

Pyramid Pooling Module (PPM) is proposed by [30], and [19] proposes a powerful variant, namely, Deep Aggregation PPM (DAPPM). We slightly modify the DAPPM in [19] by reducing input channel, reducing pooling scale number and adding attention gate, then propose a more efficient context refinement module, namely, Residual Attention feature Refinement Module (RARM), as shown in Fig. 5.

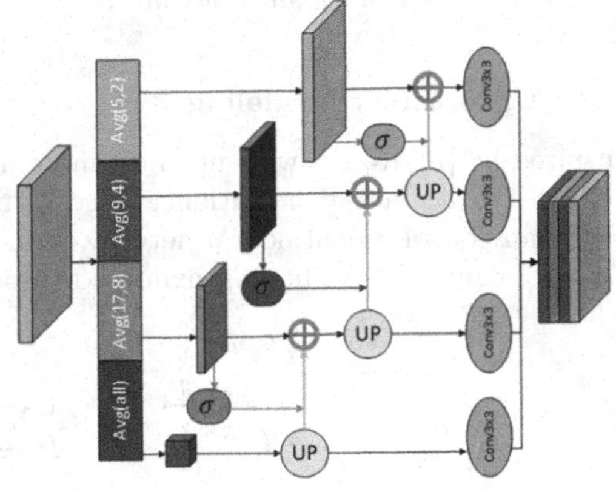

Fig. 5. Illustration of Residual Attention feature Refinement Module. "Avg(k,s)" denotes average pooling with kernel size k and stride s.

We integrate LAFFM, RARM and uncertainty regularization into an encoder-decoder architecture, proposing Uncertainty-aware Boundary Attention Network (UBANet) as in Fig. 2. To avoid extra computation cost, we adopt STDCNet [8] as the encoder, rather than a two-pathway network. We remove the time-consuming original Attention Refinement Module (ARM), and adopt only the final feature map to construct semantics.

We set down a semantic head for the immediate outputs of each LAFFM and compute the cross entropy loss L_{ce} following [30]. During training, the semantic prediction of auxiliary head is passed to the decode head to generate uncertainty map(as shown in Fig. 2) and then compute the regularized cross entropy loss L_{rce}. Additionally, we feed the detail feature map D_i to an low-level head for edge detection task and compute $L_{low-level}$ introduced in Sect. 3.1, where we apply the laplacian kernel on the semantic label map following [8] to generate groundtruth of the low-level task. The loss function of whole network is given as

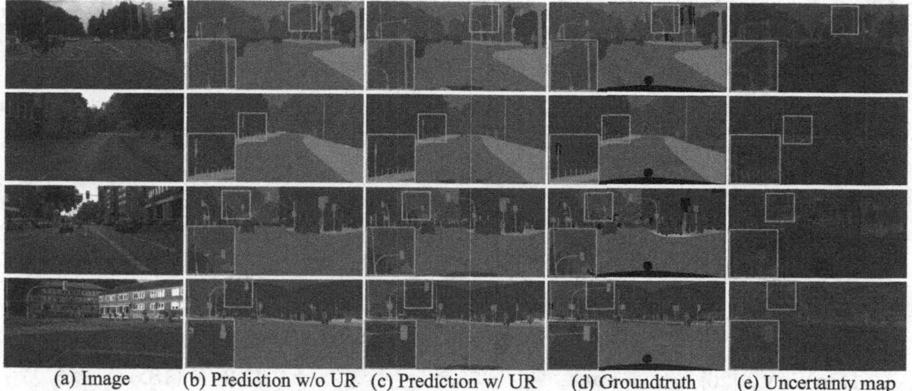

Fig. 6. Qualitative results and visual explanations of Uncertainty Regularization(UR). The network struggles to handle object boundaries and challenging areas effectively (b). In contrast, with the implementation of estimated uncertainty map (e), the network can manage object boundaries more competently and become increasingly aware of the shapes (c).

follows:

$$L = \lambda L_{rce} + N \times (L_{ce} + L_{low-level}), \tag{8}$$

where M and N denote the number of semantic head and detail head respectively, generally detail outputs of last two LAFFMs are used for low-level task($N = 2$). Note that in the minuscule version of UBANet, we only use detail head on the output of the last LAFFM, *i.e.*, $N = 1$.

4 Experiments

4.1 Dataset

Cityscapes [7] is a dataset for semantic understanding that contains 30 categories and more than 50 urban scenes in different weather conditions. The images used for the semantic segmentation task consist of 19 categories and 5000 high resolution (resolution 2048×1024) images, which are officially divided into 2975/500/1525 images for training/validation/test respectively.
CamVid [1] is a video collection that captured from perspective of driving automobile, which include pixel-wise annotated semantic map with resolution 960× 720. The 701 frames of the video are well-organized and empirically split into 367/101/233 for training/validation/test.

4.2 Implementation Details

We use PyTorch as our experiment platform. Our runtime environment are Pytorch 1.8.1, CUDA 11.4, cuDNN 8.0. All networks are trained with the same

Table 1. Comparison with state-of-the-art real-time methods on Cityscapes *val*. The number marked with "*" means we measure the #GFLOPS by experiment framework [6], and "*none*" indicates these methods have no backbone.

Methods	Backbone	mIoU%	#Params	FPS	#GFLOPS
ICNet [29]	PSPNet50	67.7	-	30.3	28.3
DF1-Seg [17]	DF1	74.1	-	106.4	-
DF2-Seg1 [17]	DF2	75.9	-	67.2	-
DF2-Seg2 [17]	DF2	76.9	-	56.3	-
BiseNet [27]	Xception39	69.0	5.8M	105.8	55.3
BiseNet [27]	ResNet18	74.8	49.0M	65.5	55.3
BiseNetV2 [26]	*none*	75.8	-	47.3	118.5
BiseNetV2 [26]	*none*	75.8	-	47.3	118.5
FasterSeg [4]	*none*	73.1	4.4M	136.2	-
STDC-Seg75 [8]	STDC1	74.5	14.2M	74.6	38.1*
STDC-Seg75 [8]	STDC2	77.0	22.2M	73.5	53.0*
SFNet-DF1 [16]	DF1	75.0	-	121	-
DDRNet23-Slim [19]	*none*	76.1	5.7M	101.0	38.3
DDRNet23-Base [19]	*none*	78.9	20.1M	38.3	143.1
RTFormer-slim [22]	*none*	76.3	4.8M	110.0	-
RTFormer-Base [22]	*none*	**79.3**	16.8M	39.1	-
UBANet-Slim-75	STDC1	76.7	10.8M	**137.5**	34.1
UBANet-Base-75	STDC2	**77.3**	15.3M	101.7	54.1
UBANet-Slim	STDC1	78.1	10.8M	70.7	60.7
UBANet-Base	STDC2	**79.3**	15.3M	54.2	96.1

setting, specifically, 860 epochs, batch size of 16, SGD optimizer with learning rate $1e-2$, momentum of 0.9, weight decay of $5e^{-4}$ for Cityscapes [7], 200 epochs, batch size of 12, SGD optimizer with learning rate $5e^{-3}$, momentum of 0.9, weight decay of $5e^{-4}$ for CamVid [1]. We use poly decay policy to adapt learning rate after each iteration, Online Hard Example Mining(OHEM) is also used. All input images firstly are random resize, rotate, flip to $[0.5, 2.0]\times$ of its original resolution, then are randomly cropped into 1024×512 / 960×720 for Cityscapes and CamVid datasets respectively. For fair comparison, TensorRT acceleration is not used.

4.3 Comparisons

Cityscapes . We follow [8] to measure the inference speed of UBANet with input resolution 2048×1024 and 1536×769 and denoted by UBANet and UBANet-75 respectively. All the inference experiments are conducted by a single RTX 3090 GPU. In Table 1, we compare UBANet with previous state-of-the-art model. Note that the minuscule UBANet-Base-75 achieves very competitive

Table 2. Comparison with state-of-the-art real-time methods on CamVid ***test***. "*none*" indicates these methods have no backbone. Note that our result is reported *without* Cityscapes pretraining [19].

Methods	Backbone	mIoU %	FPS
ICNet [29]	PSPNet50	67.1	27.8
DFANet [14]	*none*	64.7	120
BiseNet [27]	Xception39	68.7	116.3
BiseNetV2 [26]	*none*	72.4	124.5
SwiftNetRN [8]	ResNet18	72.6	-
SwiftNetRN pyr [8]	ResNet18	73.9	-
STDC-Seg [8]	STDC1	73.0	125.6
STDC-Seg [8]	STDC2	73.9	100.5
SFNet [16]	ResNet18	73.8	63.9
SFNet [16]	DF2	70.4	134
Faster-Seg [4]	*none*	71.1	-
UBANet-Slim	STDC1	74.0	**170.1**
UBANet-Base	STDC2	**74.4**	105.3

result(76.7 mIoU %) with extremely high speed(138 FPS). And it is noteworthy that UBANet-Base achieve state-of-the-art result(79.3 mIoU %) with faster inference speed and less computation cost.

CamVid. We evaluate the inference speed with input resolution of 720 × 960. As shown in Table 2, notably, UBANet-Slim achieves higher mIoU % with the state-of-the-art method such as [8], while being 70% faster. Moreover, UBANet-Base surpasses other state-of-the-art methods by 0.5 % mIoU, attaining 74.4 mIoU % without need for additional extra training data(*e.g.*, Cityscapes [19,26]).

Supplement. The supplement provides qualitative analysis and additional experiments on ADE20k and Pascal Context.

4.4 Ablation Study

Effectiveness of Uncertainty Regularization. Firstly, we visualize estimated uncertainty maps in Fig. 6(e). Incorporating with proposed uncertainty regularization(discussed in Sect. 3), the network becomes more sensitive to details and can better handle challenging areas, resulting in more accurate predictions(as shown in Fig. 6(c)(d)). We also conduct the quantitative experiment on uncertainty regularization, as presented in Table 3. Furthermore, we conduct parameter study of τ and λ_{rce} in Table 4. Importantly, since Uncertainty Regularization is applied only during the training phrase and does not affect inference speed, we investigate its broader applications in the supplement.

Effectiveness of LAFFM. We employ Grad-CAM [20] to visualize the attention areas in Stage 3 of the network. As illustrated in Fig. 4 (b) and (c), when

Table 3. Ablation experiments on each components in UBANet. "UR" denotes uncertainty regularization. "LAFFM", and "RARM" are described in Sect. 3.

Method	UR	LAFFM	RARM	mIoU %
Baseline				71.99
	✓			72.65
		✓		75.6
			✓	74.48
	✓	✓	✓	76.72

Table 4. Experimental results of hyperparameter sensitivity in UBANet. Hyperparameters include Uncertainty regularization temperature τ, coefficient of regularized loss λ_{rce} and coefficient of dice loss μ_{dice}. The default parameters are marked with "*". The evaluation setup is equivalent to the baseline in Table 3.

τ_{UR}	mIoU%	λ_{rce}	mIoU%	μ_{dice}	mIoU%
1	70.2	1*	72.7	1.0*	75.6
0.5	71.4	0.5	72.5	0.5	75.6
0.05*	72.7	0.1	72.1	0.1	75.1
0.01	72.7	0.05	71.4	0.05	75.3
0.001	71.5	0.001	71.1	0.01	75.0

trained with detail guidance [8], the network is able to generate more detailed information. Distinctly, with proposed LAFFM, the network can produce fine-grained features, *i.e.*, the attention areas are more accurately located and consist of smaller pieces, as shown in Fig. 4 (d), demonstrating that LAFFM can take advantage of low-level features effectively. Furthermore, we conduct parameter study of μ_{dice} to investigate the functionality of dice loss in Table 4.

Collaboration of the Network Components. We conduct ablation experiments on the effectiveness of each components of proposed UBANet as in Table 3. For convenience, we adopt STDC1 as the backbone, and directly upsample and add the feature maps as the feature fusion baseline.The result shows that LAFFM brings about a 3.6% mIoU improvement over baseline, indicating its better ability to employ low-level features. Furthermore, RARM is effective in accumulating global context, with over 2.4% mIoU performance gain. Also, uncertainty regularization brings more than 0.6 % mIoU improvement, which partly confirms our assumption in Sect.3.

5 Conclusion

In this paper, we propose a UBANet for real-time semantic segmentation, which incorporates low-level supervision and uncertainty regularization within encoder-decoder architecture efficiently. Experiments on standard real-time scene parsing benchmarks demonstrate that UBANet achieves state-of-the-art trade-off between latency and efficiency. However, our minuscule UBANet still need more than 30 GFLOPS computation cost, which is unacceptable for mobile devices, and we leave the problem to the future.

References

1. Brostow, G.J., Shotton, J., Fauqueur, J., Cipolla, R.: Segmentation and recognition using structure from motion point clouds. In: ECCV (1), pp. 44–57 (2008)
2. Chen, L.C., Papandreou, G., Kokkinos, I., Murphy, K., Yuille, A.L.: Deeplab: semantic image segmentation with deep convolutional nets, atrous convolution, and fully connected crfs. arXiv:1606.00915 (2016)
3. Chen, L.C., Zhu, Y., Papandreou, G., Schroff, F., Adam, H.: Encoder-decoder with atrous separable convolution for semantic image segmentation. In: ECCV (2018)
4. Chen, W., Gong, X., Liu, X., Zhang, Q., Li, Y., Wang, Z.: Fasterseg: searching for faster real-time semantic segmentation. In: ICLR (2020)
5. Cheng, B., Schwing, A.G., Kirillov, A.: Per-pixel classification is not all you need for semantic segmentation (2021)
6. Contributors, M.: Mmsegmentation: openmmlab semantic segmentation toolbox and benchmark (2020). https://github.com/open-mmlab/mmsegmentation
7. Cordts, M., et al.: The cityscapes dataset for semantic urban scene understanding. In: CVPR (2016)
8. Fan, M., et al.: Rethinking bisenet for real-time semantic segmentation. In: CVPR, pp. 9716–9725 (June 2021)
9. Fu, J., et al.: Dual attention network for scene segmentation. In: CVPR (2019)
10. Gal, Y., Ghahramani, Z.: Dropout as a bayesian approximation: Representing model uncertainty in deep learning (2016)
11. Goan, E., Fookes, C.: Uncertainty in real-time semantic segmentation on embedded systems. In: Proceedings of the IEEE/CVF Conference on Computer Vision and Pattern Recognition (CVPR) Workshops, pp. 4490–4500 (June 2023)
12. Jin, Y., Han, D., Ko, H.: Trseg: trransformer for semantic segmentation. Pattern Recogn. Lett. **148**, 29–35 (2021)
13. Kendall, A., Gal, Y.: What uncertainties do we need in bayesian deep learning for computer vision? (2017)
14. Li, H., Xiong, P., Fan, H., Sun, J.: Dfanet: deep feature aggregation for real-time semantic segmentation (2019). 10.48550/ARXIV.1904.02216, https://arxiv.org/abs/1904.02216
15. Li, X.,et al.: Improving semantic segmentation via decoupled body and edge supervision. In: ECCV (2020)
16. Li, X., et al.: Semantic flow for fast and accurate scene parsing. In: ECCV (2020)
17. Li, X., Zhou, Y., Pan, Z., Feng, J.: Partial order pruning: for best speed/accuracy trade-off in neural architecture search. CoRR abs/1903.03777 (2019). https://arxiv.org/abs/1903.03777
18. Mukhoti, J., van Amersfoort, J., Torr, P.H., Gal, Y.: Deep deterministic uncertainty for semantic segmentation. arXiv preprint arXiv:2111.00079 (2021)
19. Pan, H., Hong, Y., Sun, W., Jia, Y.: Deep dual-resolution networks for real-time and accurate semantic segmentation of traffic scenes. IEEE Trans. Intell. Transp. Syst. **24**(3), 3448–3460 (2022)

20. Selvaraju, R.R., Cogswell, M., Das, A., Vedantam, R., Parikh, D., Batra, D.: Grad-CAM: visual explanations from deep networks via gradient-based localization. Int. J. Comput. Vision **128**(2), 336–359 (oct 2019). https://doi.org/10.1007/s11263-019-01228-7, https://doi.org/10.10072Fs11263-019-01228-7
21. Shelhamer, E., Long, J., Darrell, T.: Fully convolutional networks for semantic segmentation. IEEE TPAMI **39**(4), 640–651 (2017)
22. Wang, J., et al.: Rtformer: efficient design for real-time semantic segmentation with transformer. In: NeurIPS (2022)
23. Wang, L., Li, D., Zhu, Y., Tian, L., Shan, Y.: Dual super-resolution learning for semantic segmentation. In: Proceedings of the IEEE/CVF Conference on Computer Vision and Pattern Recognition (CVPR) (June 2020)
24. Wang, Y., Peng, J., Zhang, Z.: Uncertainty-aware pseudo label refinery for domain adaptive semantic segmentation. In: Proceedings of the IEEE/CVF International Conference on Computer Vision, pp. 9092–9101 (2021)
25. Xu, J., Xiong, Z., Bhattacharyya, S.P.: Pidnet: a real-time semantic segmentation network inspired from pid controller (2022)
26. Yu, C., Gao, C., Wang, J., Yu, G., Shen, C., Sang, N.: Bisenet v2: bilateral network with guided aggregation for real-time semantic segmentation. In: IJCV, pp. 1–18 (2021)
27. Yu, C., Wang, J., Peng, C., Gao, C., Yu, G., Sang, N.: Bisenet: bilateral segmentation network for real-time semantic segmentation. In: ECCV, pp. 325–341 (2018)
28. Yuan, Y., Wang, J.: Ocnet: object context network for scene parsing. CoRR abs/1809.00916 (2018)
29. Zhao, H., Qi, X., Shen, X., Shi, J., Jia, J.: Icnet for real-time semantic segmentation on high-resolution images. In: ECCV, pp. 405–420 (2018)
30. Zhao, H., Shi, J., Qi, X., Wang, X., Jia, J.: Pyramid scene parsing network. In: CVPR (2017)
31. Zheng, Z., Yang, Y.: Rectifying pseudo label learning via uncertainty estimation for domain adaptive semantic segmentation. Int. J. Comput. Vision **129**(4), 1106–1120 (2021)

URFormer: Unified Representation LiDAR-Camera 3D Object Detection with Transformer

Guoxin Zhang[1], Jun Xie[2], Lin Liu[3], Zhepeng Wang[2(✉)], Kuihe Yang[1], and Ziying Song[3]

[1] School of Information Science and Engineering, Hebei University of Science and Technology, Shijiazhuang 050018, China
[2] Lenovo Research, Beijing 100094, China
wangzpb@lenovo.com
[3] School of Computer and Information Technology, Beijing Jiaotong University, Beijing 100044, China

Abstract. Current LiDAR-camera 3D detectors adopt a 3D-2D design pattern. However, this paradigm ignores the dimensional gap between heterogeneous modalities (*e.g.*, coordinate system, data distribution), leading to difficulties in marrying the geometric and semantic information of two modalities. Moreover, conventional 3D convolution neural networks (3D CNNs) backbone leads to limited receptive fields, which discourages the interaction between multi-modal features, especially in capturing long-range object context information. To this end, we propose a **U**nified **R**epresentation **Transformer**-based multi-modal 3D detector (**URFormer**) with better representation scheme and cross-modality interaction, which consists of three crucial components. First, we propose Depth-Aware Lift Module (**DALM**), which exploits depth information in 2D modality and lifts 2D representation into 3D at the pixel level, and naturally unifies inconsistent multi-modal representation. Second, we design a Sparse Transformer (**SPTR**) to enlarge effective receptive fields and capture long-range object semantic features for better interaction in multi-modal features. Finally, we design Unified Representation Fusion (**URFusion**) to integrate cross-modality features in a fine-grain manner. Extensive experiments are conducted to demonstrate the effectiveness of our method on KITTI benchmark and show remarkable performance compared to the state-of-the-art methods.

Keywords: Multi-modal fusion · 3D object detection · Unified representation · Transformer

1 Introduction

LiDAR and camera are essential sensors for autonomous driving perception, owing to that they can conveniently become the eyes of intelligent vehicles.

Q. Liu et al. (Eds.): PRCV 2023, LNCS 14427, pp. 401–413, 2024.
https://doi.org/10.1007/978-981-99-8435-0_32

LiDAR points are too sparse and have poor semantic information even if collected by a high-quality LiDAR, which results in sub-optimal performance on long-range objects [17,20]. On the other hand, the image loses the important depth information for 3D perception while providing dense and rich semantic information, which makes it easier to capture long-range and small objects in high resolution [24]. Based on the natural complementarity of the two sensors, it motivate researchers to design multi-sensor fusion approaches.

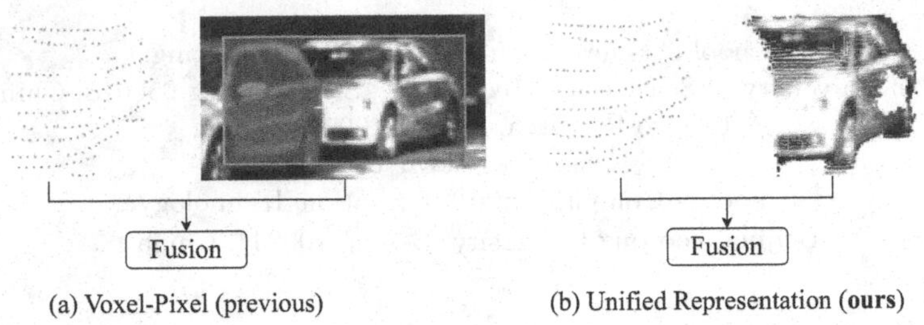

(a) Voxel-Pixel (previous)   (b) Unified Representation (**ours**)

Fig. 1. An illustration of the dimensional gap in representation. (a) Most previous state-of-the-art methods adopt voxel-pixel fusion, which ignores the heterogeneity between the two 3D and 2D features. Redundant noise in the image (indicated in red) is detrimental to performance. (b) We adopt a unified representation using consistent 3D features to fuse the two sensors.

The previous LiDAR-camera methods [21] can be broadly categorized into two paradigms: sequence fusion [11,19] and feature-level fusion [22,26]. The former processes the pipeline of the two modalities separately, and the results are finally grouped together for tuning. This approach decouples the two modalities, fusing them in result level generally. Nevertheless, its performance is extremely dependent on the performance of each stage. The latter [1,22,26] feed both modalities into the model simultaneously for feature extraction and integrate both features in the intermediate stage.

Inconsistent Representation. Most methods usually alternately transform the data representation, *e.g.*, multi-view [1,26] and voxel-pixel [2,16] methods. Although these approaches improve performance certainly, information fusion under chaotic representations inevitably brings about viewpoint inconsistencies and dimensional gaps (shown in Fig. 1(a)). These two drawbacks might limit the maximum performance of multi-modal architecture.

Insufficient Perception. Most feature-level fusion approaches [22,25] utilize region-based local fusion techniques. In the refinement stage, the region of interest (RoI) feature fails to perceive the large contextual information correctly. This is because most approaches rely on convolution operations for feature extraction, which results in a limited receptive field that hinders to capture of large context features effectively combined with another modality. However, RoIs are highly dependent on context information to correct their locations. This drawback leads to sub-optimal performance on long-range objects.

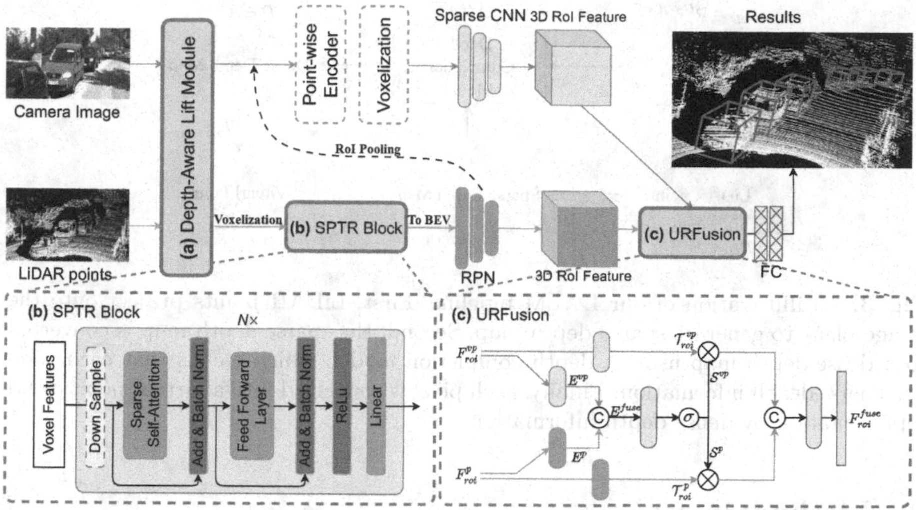

Fig. 2. An overview of our URFormer. Blue and green paths represent camera flow and LiDAR flow, respectively. (a) DALM converts the image into virtual points using point clouds, images, and calibration, where a depth completion model [6] is required to generate the dense depth map (detail in Sec. 2.1). (b) SPTR block leverages a sparse self-attention to encode voxel features with rich context information. N SPTR blocks are used (we set it to 9) and downsampling is used only three times (8 times downsampling) in order to extract multi-scale features. (c) Thanks to the unified representation framework, URFusion adaptively fuses cross-modal RoI features in a uniform representation.

To address the aforementioned issues, we propose a unified representation transformer-based framework, termed URFormer, which not only unifies chaos LiDAR-camera representation but also enlarges the limited receptive field in the point cloud branch. **First**, we propose **DALM** (Depth-Aware Lift Module) that exploits mere depth information in images, adopting a virtual point-generating strategy to make inconsistent representation into uniform representation (as shown in Fig. 1(b)). **Second**, to overcome the insufficient perception problem by limited perceptive field, we employ self-attention mechanism to extract voxel features instead of traditional sparse convolution. We design a **SPTR** (Sparse Transformer) that enables the network to acquire a larger effective receptive field. **Third**, to effectively fuse the contextual information of the two branches, we propose a cross-attention-based fusion module **URFusion** (Unified Representation Fusion), which combines uniform RoI features from LiDAR and camera branches and adaptively selects informative features. **Finally**, we conduct extensive experiments to demonstrate the effectiveness of our method. Note that we surpassed the state-of-the-art methods and achieved 83.40% AP on the KITTI car benchmark.

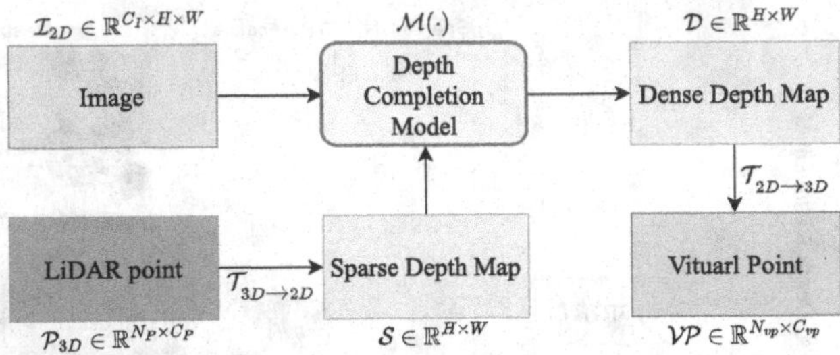

Fig. 3. An illustration of our DALM pipeline. First, LiDAR points project onto the image plane to generate sparse depth map. Second, the sparse depth map is converted to a dense depth map using a depth completion model, which means that each pixel has coarse depth information. Finally, each pixel is converted into a virtual point cloud representation by dense depth information.

2 Methodology

URFormer is a multimodal detector that converts heterogeneous representations into a uniform form. As shown in Fig. 2, it consists of three key parts: (a) a DALM for converting the image into a consistent representation of the point cloud; (b) multiple SPTR blocks for extracting voxel feature with sparse self-attention; (c) an URFusion for adaptively fusing 3D RoI features from two branches. We will detail them in the following section.

2.1 Depth-Aware Lift Module

LiDAR and the camera are two completely different sensors that contain multiple discrepancies and heterogeneities. In terms of coordinate system, LiDAR points are mostly characterized as 3D Cartesian or spherical coordinate system, while the images collected by the camera are usually represented by 2D Cartesian coordinate system. This phenomenon leads to the fact that encoding these two representations requires different architectures (*e.g.,* 3D CNN and 2D CNN), which is non-trivial for sensor fusion. Furthermore, in data distribution, LiDAR points are sparse, whereas pixels are dense. The point clouds need to be encoded with sparse vectors and the image can be processed by a matrix easily. This natural structure inconsistency results in a data structure gap. In summary, the key challenge for LiDAR-camera fusion is alleviating the discrepancy between the point cloud and image representations. To this end, we design the Depth-Aware Lift Module (DALM), which utilizes the feature that LiDAR points can provide depth information for the camera to lift dimension images from 2D pixels to 3D points (denoted as virtual points in the following).

Next, we explain the detail of DALM (shown in Fig. 3). First, the camera and LiDAR scan the current scene to generate a frame of raw data, which are the

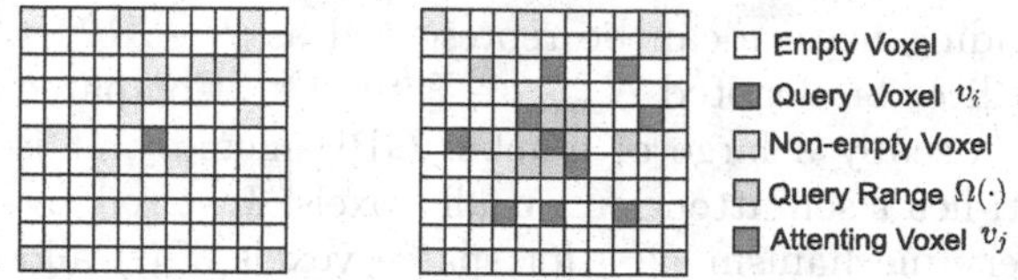

Fig. 4. A 2D example of attention range in our sparse self-attention.

images $\mathcal{I}_{2D} \in \mathbb{R}^{C_I \times H \times W}$ and point cloud $\mathcal{P}_{3D} \in \mathbb{R}^{N_P \times C_P}$, respectively. Moreover, in general, calibration information exists between vehicle sensors, and here we use a translation matrix $\mathcal{T}_{3D \longrightarrow 2D} \in \mathbb{R}^{3 \times 4}$ to represent the correspondence between LiDAR and camera. Therefore, the relationship between each pixel and point can be expressed as the following:

$$(u, v) = \mathcal{T}_{3D \longrightarrow 2D}(P_{3D}), (u, v) \in \mathcal{I}_{2D} \tag{1}$$

where (u, v) are pixel coordinate only corresponding to a subset of $\mathcal{I}_{2D}$. Next, we project the depth information of $\mathcal{P}_{3D}$ onto the image plane by Eq. 1 and then a subset of pixels of the image gets depth information to obtain a sparse depth map $\mathcal{S} \in \mathbb{R}^{H \times W}$. Second, we leverage a depth completion model [6] $\mathcal{M}$ to complement the depth of the remaining pixels. Sequentially, a dense depth map $\mathcal{D} \in \mathbb{R}^{H \times W}$ is generated by the depth completion model [6]. Finally, we leverage the inverse project matrix $\mathcal{T}_{2D \longrightarrow 3D} \in \mathbb{R}^{3 \times 4}$ to inverse-project $\mathcal{D}$ back into 3D space and we obtain a frame of virtual point cloud $\mathcal{VP} \in \mathbb{R}^{N_{vp} \times C_{vp}}$. Mathematically, the processing procedure can be represented as follows:

$$\mathcal{D} = \mathcal{M}(\mathcal{I}_{2D}, \mathcal{T}_{3D \longrightarrow 2D}(\mathcal{P}_{3D})) \tag{2}$$

$$\mathcal{VP} = \mathcal{T}_{2D \longrightarrow 3D}(\mathcal{I}_{2D}, \mathcal{D}), \mathcal{VP} \in \mathbb{R}^{N_{vp} \times C_{vp}} \tag{3}$$

where C_{vp} consists of 8 channels (x, y, z, r, g, b, u, v). (x, y, z) represent virtual point 3D Cartesian coordinate; (r, g, b) are color information; (u, v) are corresponding pixel coordinate.

2.2 Sparse Transformer

To extract 3D geometric features, voxel-based detectors use 3D sparse convolution to encode point cloud features generally. Although the convolution operator is very efficient, it only computes in a local range at each step, which leads to a limited perceptive field. However, we require a greater range of context-aware capabilities in order for multimodal features to interact more effectively. Inspired by [9], we present SPTR by applying sparse voxels to the transformer. Benefiting from the sparse self-attention mechanism, our SPTR can capture large context information and enlarge the effective receptive field.

Sparse Self-attention. First, following [3], we voxelize a frame of raw point with voxel size (0.05m, 0.05m, 0.1m). The 3D scene is split in N_{voxel} non-empty voxels, corresponding feature can be represented as $\mathcal{F} \in \mathbb{R}^{N_{voxel} \times C_P}$, and the corresponding indices be denoted $N_{index} \in \mathbb{R}^{N_{voxel} \times 3}$. Compared to 2D images, 3D voxel tensors occupy a large of number GPU memory. It is too expensive to perform multi-head self-attention on all voxels. Inspired by [9], we design an attention query mechanism $\Omega(\cdot)$ for sparse voxels. $\Omega(\cdot)$ allows not only to expand the perceptive field but also to cover neighboring voxels to ensure that fine-grained features are preserved. Specifically, we note that $\Omega(i)$ is the query range of i-th voxel v_i. $v_j \subseteq \Omega(i)$ are attending voxel that are queried by v_i. As shown in Fig. 4, $\Omega(i)$ query voxels in both local and dilated ranges as attending voxels v_j. We perform multi-head self-attention for v_i and v_j to obtain attention feature F_{att}. Mathematically, the self-attention can be expressed as:

$$F_{att} = \sum_{j \subseteq \Omega(i)} \sigma\left(Q_i K_j^T / \sqrt{d}\right) \cdot V_j \tag{4}$$

where $\sigma(\cdot)$ is the softmax function for normalizing, and the Q_i, K_j, V_j are the corresponding embedding for self-attention which can be represented as follows:

$$Q_i = W_Q \cdot v_i, K_j = W_K \cdot v_j, V_j = W_V \cdot v_j \tag{5}$$

where W_Q, W_K, W_V are learnable linear project matrices for query, key, and value respectively. After this, we use the simple residual structure and feed-forward layer [18]. However, the transformer architecture usually has the same input and output dimensions, which prevents further broadening of the perceptual field. To this end, we add downsampling modules to part of the SPTR blocks to expand the context-aware information. Finally, we obtain multi-scale downsample features $\left(F_{att}^{1\times}, F_{att}^{2\times}, F_{att}^{4\times}, F_{att}^{8\times}\right)$ for the following detection pipeline.

2.3 Unified Representation Fusion

Benefiting from our uniform representation framework, we can easily extract heterogeneous modal features grid-by-grid, which are of the unified form. However, the raw point cloud and virtual points are sparse and dense, respectively, which can lead to subtle deviations in the point-to-point correspondence. Furthermore, most of the previous approaches leverage element-by-element addition or concatenation to integrate multi-modal features. If it is applied here, it may cause the features to be misaligned. To this end, we design URFusion (Unified Representation Fusion) to adaptively align heterogeneous features for unified representation fusion (detail shown in Fig. 2(c)). Specifically, this module utilizes cross-attention to effectively query and fuse the heterogeneous features, thereby facilitating information exchange and integration among different modalities.

We set the fusion stage at the refinement stage, where RoI pooling is utilized to aggregate the features of two heterogeneous modalities, and we obtain a pair of RoI features, namely F_{roi}^{p} and F_{roi}^{vp}. To fuse them, we adopt learnable alignment to query complementary features. First, we employ two linear project layers to achieve the corresponding embedding features (E_p, E_{vp}) and concatenate them to generate cross-attention weight. Formally, it can be represented as follows:

$$E^p = MLP(F^p_{roi}), E^{vp} = MLP(F^{vp}_{roi}) \tag{6}$$

$$E^{fuse} = CONCAT(E^p, E^{vp}) \tag{7}$$

where $MLP(\cdot)$ denotes the linear project layer. E^{fuse} is the combined embedding feature for predicting attention weight. Sequentially, we utilize linear project to regulate $(F^p_{roi}, F^{vp}_{roi})$ to a suitable dimension $(\mathcal{T}^p_{roi}, \mathcal{T}^{vp}_{roi})$ for employing the attention mechanism.

$$\mathcal{T}^p_{roi} = MLP(F^p_{roi}), \mathcal{T}^{vp}_{roi} = MLP(F^{vp}_{roi}) \tag{8}$$

We obtain attention score $\mathcal{S}$ via $MLP(\cdot)$, which can be split to $(\mathcal{S}^p, \mathcal{S}^{vp})$. Mathematically, the workflow is expressed as

$$(\mathcal{S}^p, \mathcal{S}^{vp}) = \sigma(MLP(E^{fuse})) \tag{9}$$

$$F^{fuse}_{roi} = CONCAT(\mathcal{S}^p \cdot MLP(\mathcal{T}^p_{roi}), \mathcal{S}^{vp} \cdot MLP(\mathcal{T}^{vp}_{roi})) \tag{10}$$

where $\sigma(\cdot)$ represent sigmoid function for normlizing attention weight and F^{fuse}_{roi} denote the final unified fusion feature. In fact, our URFusion adaptively combines the features of both modalities. Important features are given enhanced focus by the attention mechanism, while unimportant ones are given reduced concern. Therefore, it can effectively fuse to compensate for heterogeneity.

3 Experiments

3.1 Dataset and Evaluation Metrics

We evaluate our URFormer on the KITTI dataset [4], one of the most popular self-driving perception datasets, which contains 7481 frames of labeled training set and 7518 frames of unlabeled *test* set. Following the previous work, we divided the training set into 3712 samples and 3769 samples for the *train* set and *val* set, respectively.

We employ 3D AP and BEV AP as our main evaluation metrics. Generally, there are 11 recall positions or 40 recall positions for AP measurements. For a fair comparison, we adopt the popular 40 recall positions. KITTI benchmark [4] consists of three categories car, pedestrian, and cyclist, while we only evaluate popular car following our baseline [3]. Based on the truncation level of the objects and the height of the bounding box, KITTI divides the objects into 3 difficulty levels of easy, moderate, and hard.

Table 1. Performance comparison with the state-of-the-art methods on KITTI *test* set for car 3D detection on KITTI server. 'L' and 'C' represent LiDAR and Camera, respectively. Best in bold.

Method	Modality	$AP_{3D}(\%)$			$AP_{BEV}(\%)$		
		Easy	Mod	Hard	Easy	Mod	Hard
MV3D [1]	L&C	74.97	63.63	54.00	86.62	78.93	69.80
SECOND [23]	L	84.65	75.96	68.71	88.07	79.37	77.50
Part-A^2 [15]	L	85.94	77.86	72.00	89.52	84.76	81.47
MMF [8]	L&C	86.81	76.75	68.41	89.49	87.47	79.10
EPNet [7]	L&C	89.81	79.28	74.59	94.22	88.47	83.69
VoxSet [5]	L	88.53	82.06	77.46	92.70†	89.07†	86.29†
PointPainting [19]	L&C	82.11	71.70	67.08	92.45†	88.11†	83.36†
PV-RCNN [13]	L	90.25	81.43	76.82	**94.98**	90.65	86.14
Fast-CLOCs [10]	L&C	89.11	80.34	76.98	93.02	89.49	**86.40**
FocalsConv [2]	L&C	90.55	82.28	77.59	92.67†	89.00†	86.33†
VoxelRCNN [3]	L	**90.90**	81.62	77.06	94.85†	88.83†	86.13†
Ours(URFormer)	L&C	89.64	**83.40**	**78.62**	94.40	**91.22**	86.35

† released by KITTI official.

3.2 Implementation Detail

In our experiments, we adopt offline processing for DALM in order to conduct the experiments more efficiently, and we adopt [6] as the depth completion model of DALM. For our SPTR blocks, we leverage the linear layer to adjust the voxel channel to 16 to get $F_{att}^{1\times}$ before entering the SPTR block. Sequentially, the SPTR block is downsampled to obtain the multi-scale features $(F_{att}^{2\times}, F_{att}^{4\times}, F_{att}^{8\times})$, whose number of channels are (32,64,64) respectively, and the linear layers are applied for each channel adjustment. For data augmentation, we follow [22,23], simultaneously augmenting data (gt-sampling, global noise, global rotation, and global flip).

3.3 Main Results

Results on KITTI Test Set. We compare our URFormer with other state-of-the-art methods on KITTI 3D object benchmark. As shown in Table 1, we report our approach on *test* set, which is achieved by submitting our results to the KITTI official server. To further take full advantage of our SPTR, we train our framework using all training data. Notably, our URFormer outperforms the baseline [3] by 1.78%, 1.56% on the AP_{3D} of car moderate and hard, respectively. Furthermore, we also found that our method improves the AP_{BEV} by 1.02% and 0.22% on moderate and hard respectively. Note that the easy detection

Table 2. Performance comparison with state-of-the-art methods on KITTI *val* set for car class. The results are reported by the AP with 0.7 IoU threshold. 'L' and 'C' represent LiDAR and Camera, respectively. Best in bold.

Method	Moiality	AP_{3D}(%)			AP_{BEV}(%)		
		Easy	Mod	Hard	Easy	Mod	Hard
PointRCNN [14]	L	88.88	78.63	77.38	–	–	–
SECOND [23]	L	87.43	76.48	69.10	–	–	–
Part-A^2 [15]	L	89.47	79.47	78.54	90.42	88.61	87.31
MV3D [1]	L&C	71.29	62.68	56.56	86.55	78.10	76.67
EPNet [7]	L&C	92.28	82.59	80.14	95.51	91.47	91.16
PV-RCNN [13]	L	92.57	84.83	82.69	95.76	91.11	88.93
CT3D [12]	L	92.85	85.82	83.46	96.14	91.88	89.63
FocalsConv [2]*	L&C	92.44	85.73	83.49	95.17	91.60	89.44
VoxelRCNN [3]	L	92.38	85.29	82.86	95.52	91.25	88.99
Ours(URFormer)	L&C	**95.53**	**88.27**	**85.61**	**96.40**	**92.18**	**91.55**

* reproduced by official code.

performance lightly drops. We believe this is due to noise generated by inaccurate depth complement.

Results on KITTI Validation Set. To further demonstrate the effectiveness of URFormer, we also report the performance of KITTI validation, as shown in Table 2. Note that we train our model on *train* set that contains only 3712 samples. However, URFormer demonstrates excellent performance, surpassing most state-of-the-art methods. Notably, our method improves significantly the single-modal baseline by 3.15%, 2.98%, and 2.75% on AP_{3D}. Thanks to our unified multimodal representation, multi-modal complementarity is fully exploited. Likewise, URFormer outperforms the baseline on AP_{BEV} marginally.

3.4 Abaltion Studies

To further explore the performance of our model, we conduct extensive experiments to analyze the effectiveness of our approach.

Distance Analysis. To figure out how URFormer enhances the detection performance, we conduct experiments to verify the performance at different distance segments. As shown in Table 3, although our method only improves near performance marginally, it has a heavy improvement in long-distance. One aspect, this proves our hypothesis that a large perceptive field contributes to long-range context capture. On another hand, virtual points are able to compensate for the lack of the raw point cloud at long distances.

Impact of Components. We next explore the effects of each component of the URFormer. We remove the different URFormer components one by one and

Table 3. Performance comparison at different distances on KITTI *val* set for car class. The results are reported by the AP with 0.7 IoU threshold. 'L' and 'C' represent LiDAR and Camera, respectively. Best in bold.

Method	$AP_{3D}(\%)$				$AP_{BEV}(\%)$			
	Mean	0–20 m	20–40 m	40-inf	Mean	0–20 m	20–40 m	40-inf
Baseline [3]	71.61	95.94	79.45	39.45	79.81	96.09	89.34	54.00
Ours	**74.00**	**96.00**	**80.34**	**45.66**	**82.23**	**96.34**	**90.12**	**60.23**
Improvement	*+2.38*	*+0.06*	*+0.89*	*+6.21*	*+2.42*	*+0.25*	*+0.78*	*+6.23*

Table 4. The Ablation study to evaluate the effect of our method in KITTI *val* set on car category.

Method	$AP_{3D}(\%)$			$AP_{BEV}(\%)$		
	Easy	Mod.	Hard	Easy	Mod.	Hard
URFormer	**95.53**	**88.27**	**85.61**	**96.40**	**92.18**	**91.55**
w/o DALM	93.55	85.79	82.86	94.62	90.19	89.23
w/o SPTR	94.90	87.70	84.23	95.43	91.50	88.20
w/o URFusion	94.50	87.21	84.71	95.57	91.06	90.12

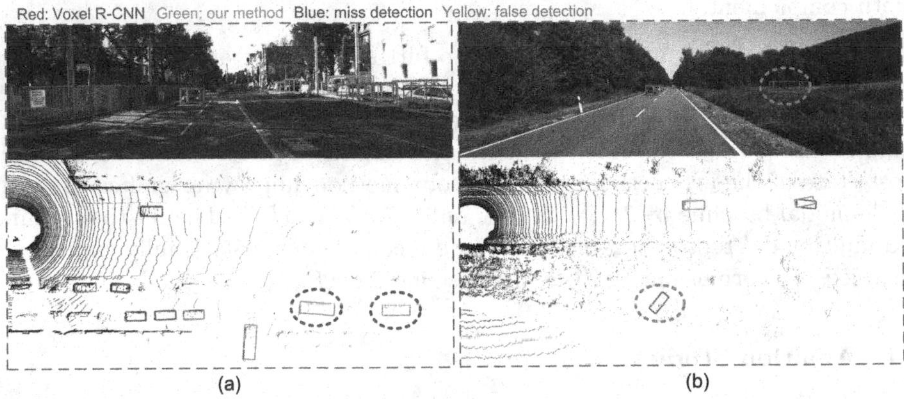

Fig. 5. Qualitative results for comparison between baseline and our method on KITTI *test* set. The first row is the camera perspective view and the second row is the BEV map generated by LiDAR. The blue and yellow circles represent missed detection and false detection, respectively. (Color figure online)

report their impact, as shown in Table 4. Among them, we find that DALM contributes the most to the overall pipeline by providing unified representation. We note that SPTR improves AP in hard, mostly composed of high-occluded objects, which means it elevates the contextual information capture. Also, URFusion emphasizes multi-modal fusion effectiveness, the moderate AP is promoted by it significantly.

Qualitative Results. We also provide qualitative results and some analysis. In Fig. 5(a), in a street scene, baseline (red box) misses some long-range objects, and our method (green box) accurately predicts these vehicles. In Fig. 5(b), the baseline incorrectly predicts a mid-range bush due to the lack of long-range contextual information. Whereas our method, with a larger range of contextual information, circumvents this situation.

4 Conclusion

In this paper, we propose a unified representation multi-modal algorithm for 3D object detection, termed URFormer, which converts pixels to virtual points via DALM, unifying the representation of heterogeneous sensors. Besides, URFormer leverages the SPTR blocks to enlarge the effective receptive field, which allows URFormer to acquire more contextual information to interact with multimodal features. Furthermore, we design a URFusion for the fusion of heterogeneous features in a more fine-grain manner. Extensive experiments on KITTI demonstrate that our approach effectively improves detection performance.

Acknowledgements. This work was supported by the Fundamental Research Funds for the Central Universities (2023YJS019) and the STI 2030-Major Projects under Grant 2021ZD0201404 (Funded by Jun Xie and Zhepeng Wang from Lenovo Research).

References

1. Chen, X., Ma, H., Wan, J., Li, B., Xia, T.: Multi-view 3D object detection network for autonomous driving. In: Proceedings of the IEEE Conference on Computer Vision and Pattern Recognition, pp. 1907–1915 (2017)
2. Chen, Y., Li, Y., Zhang, X., Sun, J., Jia, J.: Focal sparse convolutional networks for 3D object detection. In: Proceedings of the IEEE/CVF Conference on Computer Vision and Pattern Recognition, pp. 5428–5437 (2022)
3. Deng, J., Shi, S., Li, P., Zhou, W., Zhang, Y., Li, H.: Voxel R-CNN: towards high performance voxel-based 3D object detection. In: Proceedings of the AAAI Conference on Artificial Intelligence, vol. 35, pp. 1201–1209 (2021)
4. Geiger, A., Lenz, P., Urtasun, R.: Are we ready for autonomous driving? The KITTI vision benchmark suite. In: 2012 IEEE Conference on Computer Vision and Pattern Recognition, pp. 3354–3361. IEEE (2012)
5. He, C., Li, R., Li, S., Zhang, L.: Voxel set transformer: a set-to-set approach to 3D object detection from point clouds. In: Proceedings of the IEEE/CVF Conference on Computer Vision and Pattern Recognition, pp. 8417–8427 (2022)
6. Hu, M., Wang, S., Li, B., Ning, S., Fan, L., Gong, X.: Penet: towards precise and efficient image guided depth completion. In: 2021 IEEE International Conference on Robotics and Automation (ICRA), pp. 13656–13662. IEEE (2021)

7. Huang, T., Liu, Z., Chen, X., Bai, X.: EPNet: enhancing point features with image semantics for 3D object detection. In: Vedaldi, A., Bischof, H., Brox, T., Frahm, J.-M. (eds.) ECCV 2020. LNCS, vol. 12360, pp. 35–52. Springer, Cham (2020). https://doi.org/10.1007/978-3-030-58555-6_3
8. Liang, M., Yang, B., Chen, Y., Hu, R., Urtasun, R.: Multi-task multi-sensor fusion for 3D object detection. In: Proceedings of the IEEE/CVF Conference on Computer Vision and Pattern Recognition, pp. 7345–7353 (2019)
9. Mao, J., et al.: Voxel transformer for 3D object detection. In: Proceedings of the IEEE/CVF International Conference on Computer Vision, pp. 3164–3173 (2021)
10. Pang, S., Morris, D., Radha, H.: Fast-clocs: fast camera-lidar object candidates fusion for 3D object detection. In: Proceedings of the IEEE/CVF Winter Conference on Applications of Computer Vision, pp. 187–196 (2022)
11. Qi, C.R., Liu, W., Wu, C., Su, H., Guibas, L.J.: Frustum pointnets for 3D object detection from RGB-D data. In: Proceedings of the IEEE Conference on Computer Vision and Pattern Recognition, pp. 918–927 (2018)
12. Sheng, H., et al.: Improving 3D object detection with channel-wise transformer. In: Proceedings of the IEEE/CVF International Conference on Computer Vision, pp. 2743–2752 (2021)
13. Shi, S., et al.: PV-RCNN: point-voxel feature set abstraction for 3D object detection. In: Proceedings of the IEEE/CVF Conference on Computer Vision and Pattern Recognition, pp. 10529–10538 (2020)
14. Shi, S., Wang, X., Li, H.: PointRCNN: 3D object proposal generation and detection from point cloud. In: Proceedings of the IEEE/CVF Conference on Computer Vision and Pattern Recognition, pp. 770–779 (2019)
15. Shi, S., Wang, Z., Shi, J., Wang, X., Li, H.: From points to parts: 3D object detection from point cloud with part-aware and part-aggregation network. IEEE Trans. Pattern Anal. Mach. Intell. **43**(8), 2647–2664 (2020)
16. Song, Z., Jia, C., Yang, L., Wei, H., Liu, L.: Graphalign++: an accurate feature alignment by graph matching for multi-modal 3D object detection. IEEE Trans. Circuits Syst. Video Technol. (2023)
17. Song, Z., Wei, H., Jia, C., Xia, Y., Li, X., Zhang, C.: Vp-net: voxels as points for 3-D object detection. IEEE Trans. Geosci. Remote Sens. **61**, 1–12 (2023). https://doi.org/10.1109/TGRS.2023.3271020
18. Vaswani, A., et al.: Attention is all you need. In: Advances in Neural Information Processing Systems, vol. 30 (2017)
19. Vora, S., Lang, A.H., Helou, B., Beijbom, O.: Pointpainting: sequential fusion for 3D object detection. In: Proceedings of the IEEE/CVF Conference on Computer Vision and Pattern Recognition, pp. 4604–4612 (2020)
20. Wang, L., et al.: SAT-GCN: self-attention graph convolutional network-based 3D object detection for autonomous driving. Knowl.-Based Syst. **259**, 110080 (2023)
21. Wang, L., et al.: Multi-modal 3D object detection in autonomous driving: a survey and taxonomy. IEEE Trans. Intell. Veh. **8**(7), 3781–3798 . https://doi.org/10.1109/TIV.2023.3264658
22. Wu, X., et al.: Sparse fuse dense: towards high quality 3D detection with depth completion. In: Proceedings of the IEEE/CVF Conference on Computer Vision and Pattern Recognition, pp. 5418–5427 (2022)
23. Yan, Y., Mao, Y., Li, B.: Second: sparsely embedded convolutional detection. Sensors **18**(10), 3337 (2018)

24. Yang, L., et al.: Bevheight: a robust framework for vision-based roadside 3D object detection. In: Proceedings of the IEEE/CVF Conference on Computer Vision and Pattern Recognition (CVPR), pp. 21611–21620, June 2023
25. Yin, T., Zhou, X., Krähenbühl, P.: Multimodal virtual point 3D detection. In: Advances in Neural Information Processing Systems, vol. 34, pp. 16494–16507 (2021)
26. Zhang, X., et al.: Ri-fusion: 3D object detection using enhanced point features with range-image fusion for autonomous driving. IEEE Trans. Instrum. Meas. **72**, 1–13 (2022)

A Single-Stage 3D Object Detection Method Based on Sparse Attention Mechanism

Songche Jia[1(✉)] and Zhenyu Zhang[1,2]

[1] School of Information Science and Engineering, Xinjiang University, Urumqi 830017, China
107552103624@stu.xju.edu.cn, zhangzhenyu@xju.edu.cn

[2] Xinjiang Key Laboratory of Multilingual Information Technology, Urumqi 830017, China

Abstract. The Bird's Eye View (BEV) feature extraction module is an important part of 3D object detection based on point cloud data. However, the existing methods ignore the correlation between objects, resulting in a large amount of irrelevant information participating in feature extraction, which makes the detection accuracy low. To solve this problem, this paper proposes a BEV feature extraction method named Dynamic Extraction Of Effective Features (DEF) and designs a single-stage 3D object detection model. This feature extraction method first uses convolution operations to extract local features. Then the weight of elements in the BEV feature map is redistributed by spatial attention, highlighting the position of critical elements in the feature map. Then, a sparse two-level routing attention mechanism is used globally to screen out top-k routing regions with the strongest correlation with the target region to avoid interference from irrelevant information. Finally, a token-to-token attention operation is applied to the joint top-k routing regions to extract effective features. The results on the benchmark KITTI dataset show that our method can effectively improve the detection accuracy of 3D objects.

Keywords: 3D object detection · Feature extraction · Sparse attention mechanism

1 Introduction

Point cloud data has richer spatial information than image data, so it can better reflect the position of objects in space. Point cloud data obtained from lidar scanning has been widely used in areas such as autonomous driving, robot control, and augmented virtual reality. As an essential perception task in autonomous driving, 3D object detection has also begun widely using point cloud data to locate and classify objects.

At present, existing single-stage methods [8,22,24,26] usually use stacked convolutional layers or a global self-attention mechanism to extract Bird's Eye

Q. Liu et al. (Eds.): PRCV 2023, LNCS 14427, pp. 414–425, 2024.
https://doi.org/10.1007/978-981-99-8435-0_33

View (BEV) features. However, stacking convolutional layers for downsampling results in poor quality of low-level spatial features, and the 2D convolution kernel cannot take into account global information well, thus affecting detection accuracy. The global self-attention mechanism emphasizes the interaction of global information between tokens. It does not distinguish relevant information from irrelevant information according to object categories, so not all extracted features are effective.

As shown in Fig. 1, the car marked with the red bounding box is the detection object. The information contained in other classes of objects, such as roads, trees, and buildings that exist in the graph and the corresponding point cloud data is irrelevant information for cars. Therefore, during feature extraction, the network model should pay more attention to objects belonging to the car class and as little attention to objects of other classes as possible.

Fig. 1. Example of 3D object detection based on point cloud data.

In order to enable the single-stage model to extract more effective features, this paper proposes a method for extracting BEV features based on Bi-Level Routing Attention (BRA) [27]: Dynamic Extraction Of Effective Features (DEF). DEF first uses a small convolutional neural networks to extract local features quickly, and sends it to the spatial attention module to redistribute the element weights of the feature map. Then the BRA module dynamically filters out key-value pairs that are irrelevant to the target region globally to reduce the number of regions involved in information interaction. Finally, a fine-grained attention operation is applied to the top-k routing regions. At the same time, the top-k highly correlated regions will not be disturbed by the filtered information during information interaction.

2 Related Work

Currently, there are mainly two types of 3D object detection models based on point cloud data: 1) Two-stage detection models that first generate region proposals in the first stage and then refine regions of interest in the second stage; 2) Single-stage detection models that directly classify objects and predict bounding box locations.

Among the two-stage detection models, PointRCNN [19] uses PointNet++ [14] as the backbone for feature extraction, generates region proposals after segmenting foreground objects. STD [23] uses the spherical anchor box mechanism and designs the IoU branch, which reduces the number of positive anchor boxes and effectively reduces the amount of calculation. PV-RCNN [17] uses key points to gather the context information of multi-scale voxels, effectively improving detection accuracy. PV-RCNN++ [18] proposes a proposal-centric keypoint sampling strategy and a vectorpool feature aggregation method to solve the problem of the slow computing speed of PV-RCNN. Voxel-RCNN [3] proposes a method to gather multi-scale voxel features without requiring original point cloud features, which improves the detection speed of the model. VoTr [12] uses Voxel Transformer to encode point clouds and improve region proposal network. CT3D [15] uses a channel-wise Transformer to refine proposals.

Among the single-stage detection models, VoxelNet [26] uses a voxel feature encoding layer to encode the sparse point cloud data into 3D voxel features, and performs feature extraction. SECOND [22] is the first network model that uses sparse convolution [10] and submanifold sparse convolution [6] to extract voxel features, which has dramatically improved the computing speed. PointPillars [8] divides point clouds into pillars instead of voxels and then localizes objects using a typical 2D convolutional backbone network. Point-GNN [20] proposes a graph neural network to learn point cloud features better. CIA-SSD [24] proposes an IoU-aware confidence rectification to improve the misalignment between localization accuracy and classification confidence. Furthermore, a lightweight spatial-semantic feature aggregation module is proposed to extract features better. SE-SSD [25] initializes teacher SSD and student SSD with the pre-trained CIA-SSD model and proposes a single-stage self-ensemble detection framework. RDIoU [16] proposes a new rotation-decoupled IoU method. UP3DETR [21] proposes a pretext task named random block detection to unsupervisedly pre-train 3D detection models, which effectively improves the detection accuracy.

In general, two-stage models will perform better in detection accuracy. However, intractable problems include complex model structures, too many parameters, and slow inference speed. In contrast, the model structure of single-stage models is lighter. Moreover, in recent research results, the detection accuracy of the single-stage models is already higher than some advanced two-stage models. Although single-stage models have developed rapidly and achieved remarkable results, they still fail to distinguish relevant information from irrelevant information based on object categories when extracting 2D features.

3 Method

3.1 The Overall Design of the Framework

Figure 2 shows the pipeline of the model in this paper, which mainly consists of three parts: 1) SPConvNet performs voxelization encoding on the original point clouds and extracts 3D features using sparse convolution and submanifold sparse convolution. 2) DEF uses a Conv group and a SA-SSA group to extract

BEV features. 3) Performing category classification and location regression in the Multi-task detection head.

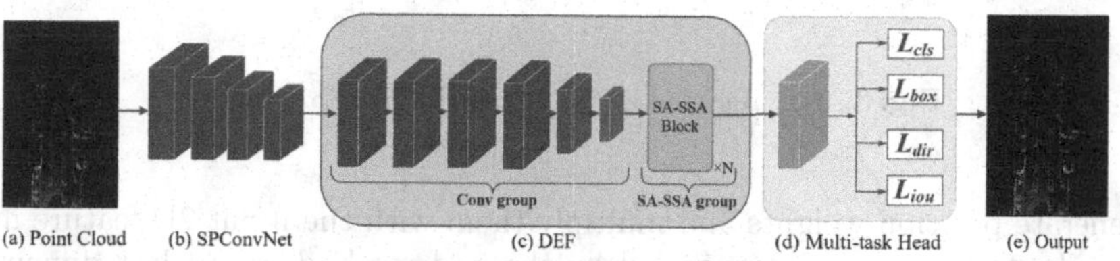

Fig. 2. The pipeline of the model in this paper.

3.2 SPConvNet

To encode the raw point cloud data, this paper first voxelizes it and computes the average coordinates and intensities of all points in each voxel as initial features for each sparse voxel. Then use sparse convolution (SC) [10] and submanifold sparse convolution (SSC) [6] to form a sparse convolutional network SPConvNet, as shown in Fig. 2(b). Specifically, SPConvNet consists of 4 3D convolutional blocks. The first convolutional block has 2 SSC layers and no SC layer, the feature dimension is 16. The other 3 convolution blocks have 2 SSC layers and 1 SC layer. The SC layer is executed before 2 SSC layers for 2x downsampling on the 3D feature map. The feature dimensions are (32, 64, 64). Finally, features in the height dimension (z-axis) are concatenated to generate the BEV feature map and fed into the DEF (Fig. 2(c)).

3.3 DEF

Dynamic Extraction Of Effective Features (DEF) proposed in this paper is shown in Fig. 2(c). DEF consists of a convolutional group and a SA-SSA group. The convolution group rapidly extract the features in the local range of the feature map. First, to avoid the loss of spatial information, this paper keeps the output features of the first 4 blue convolutional layers in the same dimensionality (the number of channels and resolution of the feature map) as the input features. Then, the 2D feature map is subjected to a 2x downsampling operation by a green convolutional layer with a stride of 2, reducing the size of the feature map by half. Finally, to reduce the calculation amount of the subsequent SA-SSA block and obtain higher-level semantic information, the new 2D feature map is subjected to a 2x downsampling operation through the red convolution layer.

The SA-SSA group consists of N consecutive Spatial Attention-Sparse Self-Attention (SA-SSA) blocks, and the structure of each SA-SSA block is shown in Fig. 3. First, for the network model to focus on the position containing the most critical information in the feature map, this paper uses spatial attention

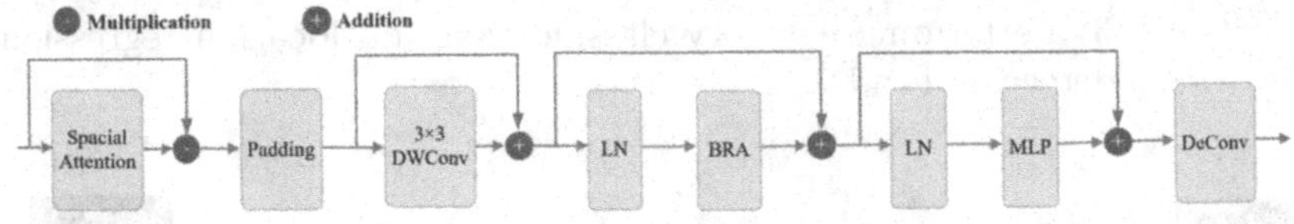

Fig. 3. The internal structure of the SA-SSA block.

to generate position weights and multiply them with the input 2D feature map. Then, the feature map is expanded by the pad method to the last dimension (W dimension), so that the last two dimensions (H and W) of the 2D feature map are consistent. The relative position information is then implicitly encoded using a 3×3 depthwise separable convolution, and the two feature maps are added using a residual connection. Then, let the 2D feature maps pass through a BRA module to dynamically extract effective features for object categories with a sparse attention mechanism. A 2-layer MLP module with expansion ratio e is then applied for position-wise embedding, and the residual connection is performed after the BRA module and the MLP module, respectively. Finally, this paper employs a 2D DeConv layer to upsample features by 4x.

BRA. Bi-Level Routing Attention (BRA) is a dynamic, query-aware sparse attention mechanism. The main idea is to filter out the most irrelevant key-value pairs in a coarse region level so that only a portion of routed regions remain. Then, fine-grained token-to-token attention is applied to the remaining regions [27].

First, this paper divides the BEV feature map $\mathbf{X}$ with resolution H×W into S×S non-overlapping regions (S is a hyperparameter), so that the height and width of each region are $\frac{H}{S}$ and $\frac{W}{S}$, respectively. And each region contains $\frac{HW}{S^2}$ feature vectors. At this point, $\mathbf{X}$ will be reshaped as $\mathbf{X}^r \in \mathbb{R}^{S^2 \times \frac{HW}{S^2} \times C}$. Then use weights $\mathbf{W}^q, \mathbf{W}^k, \mathbf{W}^v$ to perform linear projection to get $\mathbf{Q}, \mathbf{K}, \mathbf{V} \in \mathbb{R}^{S^2 \times \frac{HW}{S^2} \times C}$.

Then, region averaging is performed on Q and K to derive region-level query and key $\mathbf{Q}^r, \mathbf{K}^r \in \mathbb{R}^{S^2 \times C}$. Then get the adjacency matrix $\mathbf{A}^r \in \mathbb{R}^{S^2 \times S^2}$ by matrix multiplication:

$$\mathbf{A}^r = \mathbf{Q}^r \left(\mathbf{K}^r\right)^{\mathrm{T}}. \tag{1}$$

The position of each element in the adjacency matrix is used to measure the degree of semantic correlation between two regions. The critical step of the BRA module is to retain the indices of the top-k regions most relevant to each region by pruning the adjacency matrix (k is a hyperparameter). So the routing index matrix $\mathbf{I}^r \in \mathbb{N}^{S^2 \times k}$ is derived using row-by-row top-k operations:

$$\mathbf{I}^r = topkIndex\left(\mathbf{A}^r\right). \tag{2}$$

Finally, for each query token in region i, it will attend to all key-value pairs contained in k routing regions according to the index value $\mathbf{I}^r_{(i,1)}, \mathbf{I}^r_{(i,2)}, ..., \mathbf{I}^r_{(i,k)}$ of the row. So first gather the tensors of keys and values separately to get $\mathbf{K}^g, \mathbf{V}^g \in \mathbb{R}^{S^2 \times \frac{kHW}{S^2} \times C}$:

$$\mathbf{K}^g = gather\left(\mathbf{K}, \mathbf{I}^r\right), \mathbf{V}^g = gather\left(\mathbf{V}, \mathbf{I}^r\right). \tag{3}$$

Then apply the fine-grained attention operation:

$$\mathbf{O} = Attention\left(\mathbf{Q}, \mathbf{K}^g, \mathbf{V}^g\right) = softmax\left(\frac{\mathbf{Q}\left(\mathbf{K}^g\right)^{\mathrm{T}}}{\sqrt{C}}\right)\mathbf{V}^g. \tag{4}$$

3.4 Loss Function

This paper follows the settings of the bounding box encoding and loss functions in [22,24]. Specifically, this paper uses the Focal loss, Smooth-L1 loss, and cross-entropy loss to supervise 3D bounding box classification (L_{cls}), 3D bounding box regression (L_{box}) and IoU prediction regression (L_{iou}), and direction classification (L_{dir}), respectively. The overall loss for training is defined as:

$$L = \gamma_1 L_{cls} + \gamma_2 L_{box} + \gamma_3 L_{dir} + \gamma_4 L_{iou}. \tag{5}$$

Setting hyperparameters $\gamma_1 = 1.0, \gamma_2 = 2.0, \gamma_3 = 0.2, \gamma_4 = 1.0$.

4 Experiments

4.1 Datasets

The KITTI dataset [5] contains 7481 training samples and 7518 testing samples. This paper follows the same way as previous models [8,22] to further divide the original training samples into a training set (3712 samples) and a validation set (3769 samples). In this paper, the detection results of the car class are evaluated by the average precision (AP) with an IoU threshold of 0.7 on three difficulty levels (Easy, Moderate, and Hard).

4.2 Implementation Details

This paper crops the original point clouds and sets the voxel size according to the settings in [24]. In the convolution group of DEF, six convolution layers all have a convolution kernel size of 3×3 and the number of channels is 256. The strade of six convolution layers are (1,1,1,1,2,2). In the SA-SSA block, the spatial attention part applies average pooling and maximum pooling operations to the BEV feature maps, uses a 3×3 convolutional layer to change the channel and normalizes the spatial weights with the sigmoid function. For the BRA module, the region partition factor S=5 and top-k=20. The expansion ratio e=3 of the MLP module. The last 2D DeConv layer uses 4×4 convolution kernels and 256 output channels with a stride of 4. Finally, the consecutive factor N=2.

The development environment used in the experiment is as follows: 1) Ubuntu-18.04 operating system. 2) The CPU is Intel(R) Xeon(R) Platinum 8358P. 3) The GPU is a single NVIDIA GeForce RTX 3090.

4.3 Results on Datasets

The detection results of our model on the test server and validation set are shown in Tables 1 and 2, respectively. "L" means Lidar, and "I" means color image. The top-1 of the two-stage and single-stage models are in bold, respectively.

Table 1. Comparison with other models on the car class of the KITTI test server by evaluating 40 sampling recall points.

Tpye	Method	Modality	Car 3D AP(R40)			
			Easy	Mod	Hard	mAP
2-stage	PointRCNN [19]	L	86.69	75.64	70.70	77.68
	PV-RCNN [17]	L	90.25	81.43	76.82	82.83
	Voxel-RCNN [3]	L	**90.90**	81.62	77.06	**83.19**
	CLOCs [13]	L+I	88.94	80.67	77.15	82.25
	CT3D [15]	L	87.83	81.77	77.16	82.25
	PV-RCNN++ [18]	L	90.14	81.88	77.15	83.06
	Pyramid-PV [11]	L	88.39	**82.08**	**77.49**	82.65
1-stage	SECOND [22]	L	83.34	72.55	65.82	73.90
	PointPillars [8]	L	82.58	74.31	68.99	75.29
	CIA-SSD [24]	L	89.59	80.28	72.87	80.91
	SE-SSD [25]	L	**91.49**	**82.54**	77.15	**83.73**
	SVGA-Net [7]	L	87.33	80.47	75.91	81.24
	SASA [1]	L	88.76	82.16	77.16	82.69
	Our model	L	88.49	82.19	**77.40**	82.69

In Table 1, the model in this paper achieves average precision (AP) of 88.49%, 82.19%, and 77.40%, respectively. The model is 1.91% higher than the baseline CIA-SSD on the most important Moderate level. In addition, among all the single-stage models in Table 1, our model ranks first and second in the AP of the Hard and Moderate levels, respectively, and is comparable to the two-stage models in the table.

Although our model is slightly lower than SE-SSD by 0.35% on the Moderate level, SE-SSD is trained on two CIA-SSDs, and a complex self-ensemble strategy is used. The model in this paper does not adopt the self-ensemble strategy, so it is more reasonable to compare with CIA-SSD. Because the proposed model shows better performance than CIA-SSD, it can be inferred that if the proposed model is used to replace CIA-SSD in SE-SSD for self-ensemble training, it should show better performance than the existing SE-SSD. The proposed model has shown comparable performance with SE-SSD, so it can be seen that the proposed model in this paper is effective enough.

When evaluating 11 recall positions, compared with single-stage detection models in Table 2, our model ranks first on the most crucial Moderate level,

Table 2. Comparison with other models on the car class of the KITTI validation set by evaluating 11 sampling recall points and 40 sampling recall points, respectively.

Tpye	Method	Modality	Car 3D AP(R11)				Car 3D AP(R40)			
			Easy	Mod	Hard	mAP	Easy	Mod	Hard	mAP
2-stage	PV-RCNN [17]	L	89.35	83.69	78.70	83.91	92.57	84.83	82.69	86.70
	Voxel-RCNN [3]	L	89.41	84.52	78.93	84.29	92.38	85.29	82.86	86.84
	CLOCs [13]	L+I	–	–	–	–	89.49	79.31	77.36	82.05
	CT3D [15]	L	**89.54**	**86.06**	78.99	**84.86**	**92.85**	**85.82**	**83.46**	**87.38**
	Pyramid-PV [11]	L	89.37	84.38	78.84	84.20	–	–	–	–
	VFF-PV [9]	L+I	89.45	84.21	**79.13**	84.26	92.31	85.51	82.92	86.91
1-stage	SECOND [22]	L	87.43	76.48	69.10	77.67	–	–	–	–
	PointPillars [8]	L	–	77.98	–	77.98	86.46	77.28	74.65	79.46
	CIA-SSD [24]	L	90.04	79.81	78.80	82.88	–	–	–	–
	FocalsConv [2]	L+I	89.82	85.22	**85.19**	**86.74**	92.26	85.32	**82.95**	86.84
	SVGA-Net [7]	L	**90.59**	80.23	79.15	83.32	–	–	–	–
	Our model	L	89.65	**85.27**	79.46	84.79	**92.75**	**86.06**	82.70	**87.17**

exceeding the baseline CIA-SSD by 5.46%. Moreover, it surpasses CIA-SSD by 1.91% on mean Average Precision (mAP). Meanwhile, compared with two-stage detection models, our model is only 0.07% lower than the best-performing model CT3D in Table 2 on mAP. When evaluating 40 recall positions, our model performs best on Easy and Moderate levels compared to single-stage models in Table 2. Combined with the results on the test server, which verifies the effectiveness of the method in this paper on the task of 3D object detection.

4.4 Ablation Study

In this part, ablation experiments are carried out on the whole DEF, key submodules in DEF, and the hyperparameters top-k and S in the BRA module, respectively. And report the 3D average precision of 40 sampling recall points on the car class of the KITTI validation set. The results are shown in Tables 3, 4, and 5, respectively.

Table 3. Ablation study on the whole DEF.

Method	Car 3D AP(R40)		
	Easy	Mod	Hard
RPN	93.08	83.56	79.66
SSFA	**93.38**	84.08	81.19
ViT	91.92	93.10	79.28
DEF	92.75	**86.06**	**82.70**

Effect of DEF Module. In order to verify the effectiveness of the proposed DEF feature extraction method, this paper uses RPN [22], SSFA [24], and ViT [4] to replace DEF in the pipeline as a whole, and the other parts of the model remain the same. In Table 3, DEF outperforms the best-performing baseline SSFA by 1.98% and 1.51% on Moderate and Hard levels. It is verified that DEF is effective in extracting BEV features.

Table 4. Ablation study on components in DEF: Conv group, spatial attention and BRA module.

Conv group	Spatial Attention	BRA	Car 3D AP(R40)		
			Easy	Mod	Hard
–	–	–	92.06	83.64	79.91
✓	–	–	92.33	84.57	80.94
✓	✓	–	92.56	85.06	81.41
✓	✓	✓	**92.75**	**86.06**	**82.70**

Effect of Key Sub-modules in DEF. The Conv group and Spatial Attention are removed, and the BRA module is replaced with multi-head attention as the baseline of this ablation study. The results in Table 4 show that the AP of the model is significantly improved after adding the Conv group. The three levels of Easy, Moderate, and Hard are 0.27%, 0.93%, and 1.03% higher than the baseline, respectively. In order to verify the effectiveness of the Spatial Attention module, this paper adds Conv group and spatial attention to the baseline. The experimental results show that the AP has increased by 0.23%, 0.49%, and 0.47%, respectively. Finally, the multi-head attention is replaced by BRA, which achieves the best AP of the model in this paper and verifies the effectiveness of BRA in feature extraction.

Table 5. Ablation study on different top-k and region partition factor S in the BRA module.

S	k	Car 3D AP(R40)
		Mod
10	35	84.48
10	50	85.10
10	65	85.59
10	80	85.80
10	95	85.72
5	15	85.71
5	20	**86.06**

Effect of Different Top-K and Region Partition Factor S. As shown in Table 5, When S=10, the BEV feature map (50×50) is divided into 100 regions, each with a resolution of 5×5. When the value of k gradually increases, it can be found that the AP gradually improves and reaches 85.80% when k=80 (four-fifths of the total number of regions). When reducing the value of S and setting S=5, the BEV feature map is divided into 25 regions, and the resolution of each region is increased to 10×10. When k=20 (four-fifths of the total number of regions), the AP reaches the maximum value of 86.06%.

4.5 Visualization

Figure 4 shows the visualization of some predicted bounding boxes generated by the model in this paper, and the predicted bounding boxes are shown in green. (a)(b)(e)(f) are the projections on the color images, and (c)(d)(g)(h) are the projections on the lidar point clouds.

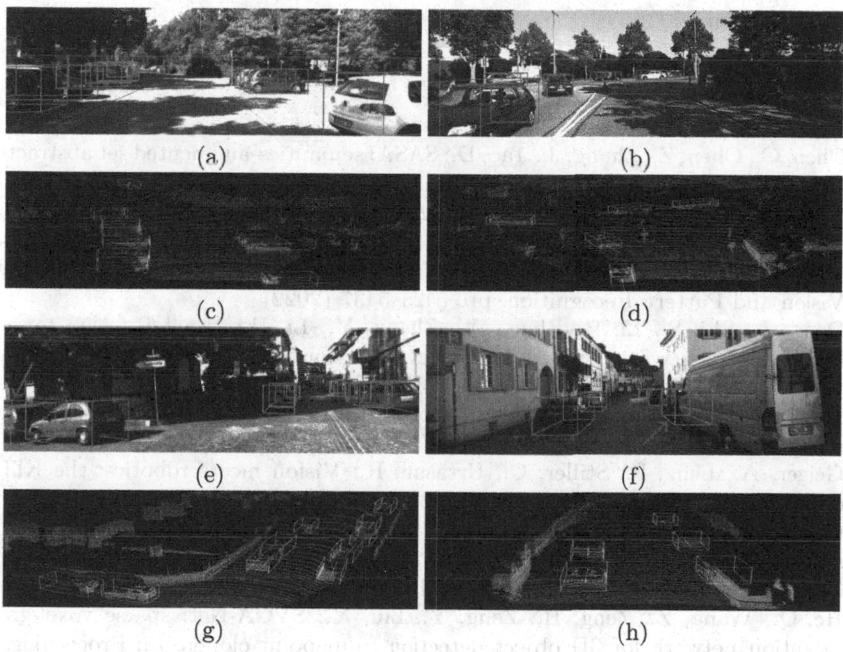

Fig. 4. Snapshots of 3D detection results on the KITTI validation set.

As shown in Fig. 4, the model in this paper can effectively detect the remote small object pointed by the red arrow in Fig. 4(a), the truncated and severely occluded objects pointed by the red arrow in Fig. 4(e) and 4(f), and the interference of the white truck in Fig. 4(f) is excluded. Thanks to the redistribution of element weights by spatial attention, the screening of strongly relevant information by the BRA module, and the expansion of the receptive field range when the

BEV feature map is segmented into regions with larger resolutions, improving detection accuracy on both Moderate and Hard levels.

5 Conclusion

This paper proposes a 3D object detection method based on a sparse attention mechanism, focusing on the extraction of effective features of objects. The method adopts the BEV feature extraction idea from local to global, quickly extracts local features through convolution operation and spatial attention, and provides critical location information for sparse attention. The routing index matrix is built globally through the BRA module, and the top-k routing areas with the strongest correlation with the target region are screened out, which solves the problem that a large amount of irrelevant information affects feature extraction. The detection accuracy of the model is improved by using strongly correlated information for fine-grained attention operations. Experimental results on the KITTI dataset show that our method is effective in 3D object detection.

References

1. Chen, C., Chen, Z., Zhang, J., Tao, D.: SASA: semantics-augmented set abstraction for point-based 3D object detection. In: Proceedings of the AAAI Conference on Artificial Intelligence, vol. 36, pp. 221–229 (2022)
2. Chen, Y., Li, Y., Zhang, X., Sun, J., Jia, J.: Focal sparse convolutional networks for 3D object detection. In: Proceedings of the IEEE/CVF Conference on Computer Vision and Pattern Recognition, pp. 5428–5437 (2022)
3. Deng, J., Shi, S., Li, P., Zhou, W., Zhang, Y., Li, H.: Voxel R-CNN: towards high performance voxel-based 3D object detection. In: Proceedings of the AAAI Conference on Artificial Intelligence, vol. 35, pp. 1201–1209 (2021)
4. Dosovitskiy, A., et al.: An image is worth 16×16 words: transformers for image recognition at scale. arXiv preprint arXiv:2010.11929 (2020)
5. Geiger, A., Lenz, P., Stiller, C., Urtasun, R.: Vision meets robotics: the KITTI dataset. Int. J. Robot. Res. **32**(11), 1231–1237 (2013)
6. Graham, B., Engelcke, M., Van Der Maaten, L.: 3D semantic segmentation with submanifold sparse convolutional networks. In: Proceedings of the IEEE Conference on Computer Vision and Pattern Recognition, pp. 9224–9232 (2018)
7. He, Q., Wang, Z., Zeng, H., Zeng, Y., Liu, Y.: SVGA-Net: sparse voxel-graph attention network for 3D object detection from point clouds. In: Proceedings of the AAAI Conference on Artificial Intelligence, vol. 36, pp. 870–878 (2022)
8. Lang, A.H., Vora, S., Caesar, H., Zhou, L., Yang, J., Beijbom, O.: PointPillars: fast encoders for object detection from point clouds. In: Proceedings of the IEEE/CVF Conference on Computer Vision and Pattern Recognition, pp. 12697–12705 (2019)
9. Li, Y., et al.: Voxel field fusion for 3D object detection. In: Proceedings of the IEEE/CVF Conference on Computer Vision and Pattern Recognition, pp. 1120–1129 (2022)
10. Liu, B., Wang, M., Foroosh, H., Tappen, M., Pensky, M.: Sparse convolutional neural networks. In: Proceedings of the IEEE Conference on Computer Vision and Pattern Recognition, pp. 806–814 (2015)

11. Mao, J., Niu, M., Bai, H., Liang, X., Xu, H., Xu, C.: Pyramid R-CNN: towards better performance and adaptability for 3D object detection. In: Proceedings of the IEEE/CVF International Conference on Computer Vision, pp. 2723–2732 (2021)
12. Mao, J., et al.: Voxel transformer for 3D object detection. In: Proceedings of the IEEE/CVF International Conference on Computer Vision, pp. 3164–3173 (2021)
13. Pang, S., Morris, D., Radha, H.: CLOCs: camera-lidar object candidates fusion for 3D object detection. In: 2020 IEEE/RSJ International Conference on Intelligent Robots and Systems (IROS), pp. 10386–10393. IEEE (2020)
14. Qi, C.R., Yi, L., Su, H., Guibas, L.J.: PointNet++: deep hierarchical feature learning on point sets in a metric space. In: Advances in neural information processing systems, vol. 30 (2017)
15. Sheng, H., et al.: Improving 3D object detection with channel-wise transformer. In: Proceedings of the IEEE/CVF International Conference on Computer Vision, pp. 2743–2752 (2021)
16. Sheng, H., et al.: Rethinking IoU-based optimization for single-stage 3D object detection. In: Avidan, S., Brostow, G., Cissé, M., Farinella, G.M., Hassner, T. (eds.) Computer Vision – ECCV 2022. ECCV 2022. Lecture Notes in Computer Science, vol. 13669, pp. 544–561. Springer, Cham (2022). https://doi.org/10.1007/978-3-031-20077-9_32
17. Shi, S., et al.: PV-RCNN: point-voxel feature set abstraction for 3D object detection. In: Proceedings of the IEEE/CVF Conference on Computer Vision and Pattern Recognition, pp. 10529–10538 (2020)
18. Shi, S., et al.: PV-RCNN++: point-voxel feature set abstraction with local vector representation for 3D object detection. Int. J. Comput. Vision **131**(2), 531–551 (2023)
19. Shi, S., Wang, X., Li, H.: PointRCNN: 3D object proposal generation and detection from point cloud. In: Proceedings of the IEEE/CVF Conference on Computer Vision and Pattern Recognition, pp. 770–779 (2019)
20. Shi, W., Rajkumar, R.: Point-GNN: graph neural network for 3D object detection in a point cloud. In: Proceedings of the IEEE/CVF Conference on Computer Vision and Pattern Recognition, pp. 1711–1719 (2020)
21. Sun, M., Huang, X., Sun, Z., Wang, Q., Yao, Y.: Unsupervised pre-training for 3D object detection with transformer. In: Yu, S., et al. Pattern Recognition and Computer Vision. PRCV 2022. Lecture Notes in Computer Science, vol. 13536, pp. 82–95. Springer (2022). https://doi.org/10.1007/978-3-031-18913-5_7
22. Yan, Y., Mao, Y., Li, B.: Second: sparsely embedded convolutional detection. Sensors **18**(10), 3337 (2018)
23. Yang, Z., Sun, Y., Liu, S., Shen, X., Jia, J.: STD: sparse-to-dense 3D object detector for point cloud. In: Proceedings of the IEEE/CVF International Conference on Computer Vision, pp. 1951–1960 (2019)
24. Zheng, W., Tang, W., Chen, S., Jiang, L., Fu, C.W.: CIA-SSD: confident IoU-aware single-stage object detector from point cloud. In: Proceedings of the AAAI conference on artificial intelligence, vol. 35, pp. 3555–3562 (2021)
25. Zheng, W., Tang, W., Jiang, L., Fu, C.W.: SE-SSD: self-ensembling single-stage object detector from point cloud. In: Proceedings of the IEEE/CVF Conference on Computer Vision and Pattern Recognition, pp. 14494–14503 (2021)
26. Zhou, Y., Tuzel, O.: VoxelNet: end-to-end learning for point cloud based 3D object detection. In: Proceedings of the IEEE Conference on Computer Vision and Pattern Recognition, pp. 4490–4499 (2018)
27. Zhu, L., Wang, X., Ke, Z., Zhang, W., Lau, R.: BiFormer: vision transformer with bi-level routing attention. arXiv preprint arXiv:2303.08810 (2023)

WaRoNav: Warehouse Robot Navigation Based on Multi-view Visual-Inertial Fusion

Yinlong Zhang[1,2,3,4], Bo Li[1], Yuanhao Liu[2,3,4], and Wei Liang[2,3,4](✉)

[1] School of Information Science and Engineering, Shenyang University of Technology, Shenyang 110020, China
zhangyinlong@sia.cn

[2] Key Laboratory of Networked Control Systems, Shenyang Institute of Automation, Chinese Academy of Sciences, Shenyang 110016, China
{liuyuanhao,weiliang}@sia.cn

[3] State Key Laboratory of Robotics, Shenyang Institute of Automation, Chinese Academy of Sciences, Shenyang 110016, China

[4] Institutes for Robotics and Intelligent Manufacturing, Chinese Academy of Sciences, Shenyang 110169, China

Abstract. An accurate and globally consistent navigation system is crucial for the efficient functioning of warehouse robots. Among various robot navigation techniques, the tightly-coupled visual-inertial fusion stands out as one of the most promising approaches, owing to its complementary sensing and impressive performance in terms of response time and accuracy. However, the current state-of-the-art visual-inertial fusion methods suffer from limitations such as long-term drifts and loss of absolute reference. To address these issues, this paper proposes a novel globally consistent multi-view visual-inertial fusion framework, called WaRoNav, for warehouse robot navigation. Specifically, the proposed method jointly exploits a downward-looking QR-vision sensor and a forward-looking visual-inertial sensor to estimate the robot poses and velocities in real-time. The downward camera provides absolute robot poses with reference to the global workshop frame. Furthermore, the long-term visual-inertial drifts, inertial biases, and velocities are periodically compensated at spatial intervals of QR codes by minimizing visual-inertial residuals with rigid constraints of absolute poses estimated from downward visual measurements. The effectiveness of the proposed method is evaluated on a developed warehouse robot navigation platform. The experimental results show competitive accuracy against state-of-the-art approaches with the maximal position error of $0.05m$ and maximal attitude error of $2\,^{\circ}$, irrespective of the trajectory lengths.

Keywords: Warehouse Robot · Navigation · Multiple View · Visual-Inertial Fusion

1 Introduction

Warehouse robots have become a popular choice for transporting bulky and heavy cargo, such as packages and shelves, along predefined trajectories to specific locations

Q. Liu et al. (Eds.): PRCV 2023, LNCS 14427, pp. 426–438, 2024.
https://doi.org/10.1007/978-981-99-8435-0_34

[1]. With the rise of smart manufacturing, warehouse robots are playing an increasingly crucial role in enabling the flexibility of manufacturing and logistics [2].

An accurate and consistent navigation system is important in guiding the warehouse robot and ensuring the production safety. Basically, the warehouse robot navigation could be categorized into vision based, laser based, magnetism based, wire based and color tape based methods [3]. Among them, the vision-based guidance system has been widely explored in both academic and industrial fields because of its advantages in flexible configurations, abundant visual information, low costs, etc. However, the vision system has the inherent drawbacks of long-term drifts & computational complexity, tracking failures in presence of textureless environments and body fast rotations, scale uncertainties [4]. These issues have hindered its wide application in the general warehouse robot navigation scenarios. Comparably, the inertial sensor could offer the complementary properties by providing the body linear accelerations and angular rates at a quick sampling rate, which could compensate for the inherent visual navigation drawbacks. Henceforth, the combination of visual and inertial measurements, to some extent, could offer a promising warehouse robot navigation solution [5].

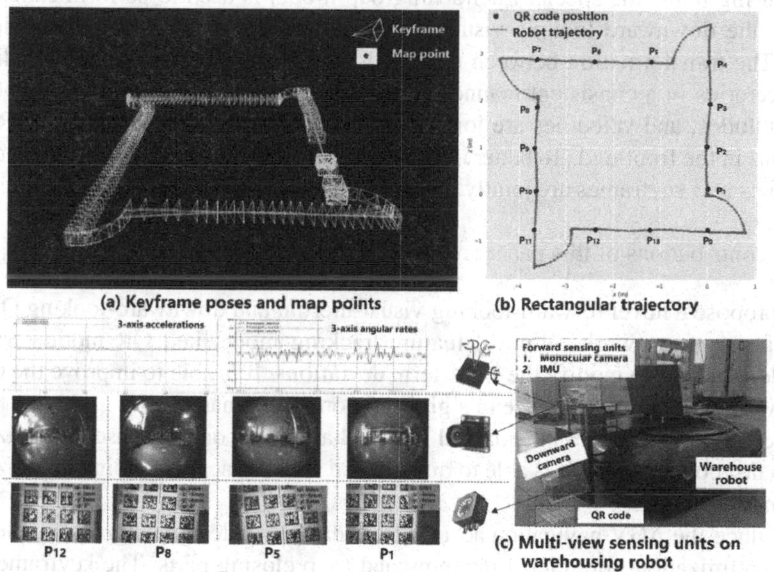

Fig. 1. Illustration on WaRoNav. (a) the screenshot of the estimated robot keyframes and the map points; (b) the recovered rectangular trajectory; (c) the forward visual-inertial sensing units and the downward vision sensor that scans the QR codes on the ground.

The forward-looking visual-inertial sensing for warehouse robot navigation still have some instinctive issues such as the loss of absolute reference frame, long-term drifts, inertial random walks [6]. To overcome these problems, this paper seeks to make full use of the downward-looking visual measurements to compensate for the forward visual-inertial drifts, as illustrated in Fig. 1. The downward camera captures the QR codes pasted on the ground and estimates the absolute positions and attitudes with respect to the global reference frame in workshops [7]. Since the downward-looking visual navigation has the relatively higher accuracy, it could considerably reduce the accumulated forward-looking visual-inertial drifts, and correct the positions of the estimated local map points over the adjacent QR code distances.

Despite the merits of downward-looking and forward-looking visual-inertial sensing, the fusion still has several issues that need to be solved for the accurate robot navigation. Firstly, the forward-looking and downward-looking cameras are supposed to be rigidly fixed and precisely calibrated before fusion. Yet the traditional calibrations fail to find the appropriate covisible viewpoints on checkerboard [8]. Besides, the solution on minimizing the downward and forward-looking visual-inertial residuals may get stuck in the local minimum or even cause system dumping due to the inappropriate fusion strategy.

This paper proposes a novel framework for warehouse robot navigation that addresses the issues discussed above. The WaRoNav framework, as shown in Fig. 2, fuses information from a globally consistent forward-looking visual-inertial sensor and a downward-looking QR vision sensor. To achieve online calibration of the multi-view sensing units, the special Euclidean group model is used to perform calibrations between the downward-looking visual sensor and the forward-looking visual-inertial sensor. The transformation between the multi-view references is obtained by aligning the trajectories in a consistent manner. To reduce accumulated errors, the robot positions, attitudes, and velocities are jointly estimated using the visual-inertial-QR measurements in the front-end. To better improve the navigation performance, the recovered map points and keyframes are jointly optimized in a tightly-coupled graph optimization framework.

The contributions of this paper are summarized as follows:

(i) We propose a novel forward-looking visual-inertial and downward-looking QR sensor fusion framework. A novel feature tracking mode, i.e., QR motion tracking mode, is added to modify the short-term deviations. It is able to improve the overall AGV navigation performance in a globally consistent manner.
(ii) We exploit the SE(3) mathematical model that allows online sensor initializations with fast convergence. It is able to quickly and reliably achieve online initializations on multiple sensors.
(iii) To reduce the AGV navigation accumulated drifts, we design a tightly-coupled unified optimization in the local mapping and loop closing parts. The keyframes with QR codes are rigidly inserted to periodically correct the long-term visual-inertial drifts. The loop closing could be ensured without the false matches by using the geometrical constraints.

2 System Overview

The scheme of the proposed approach is illustrated in Fig. 2. WaRoNav firstly performs the initialization to initiate the parameters: inertial biases $\{\mathbf{b}^{\alpha}, \mathbf{b}^{\omega}\}$, absolute scale s, body initial velocity $\mathbf{v_b}$, gravity direction $\mathbf{R}_{wg}$, the frame transformation matrix $\mathbf{T}_{BC_F}$

between the body frame (inertial frame) and forward-looking camera frame (short as C_F frame), and $\mathbf{T}_{C_F C_D}$ between the forward-looking camera frame and downward-looking camera frame (short as C_D frame), and the initial 3D points $\{p_1^w, p_2^w, \cdots, p_M^w\}$ by virtue of the Maximum-A-Posteriori visual-inertial alignment strategy.

After initialization, the robot navigation will be split into the front-end and back-end. The front-end yields the initial robot poses and the captured 3D points. The back-end updates these estimates in the graph optimization manner. It should be noted that the downward camera captures the QR info and estimates the robot positions more accurately, but the QR codes are merely available intermittently (i.e., there is a certain distance between the adjacent QR codes). Comparably, the forward-looking visual-inertial sensors could constantly collect the forward sequence of images and the inertial measurements while the robot is running, although the presence of long-term drifts and unexpected random walks.

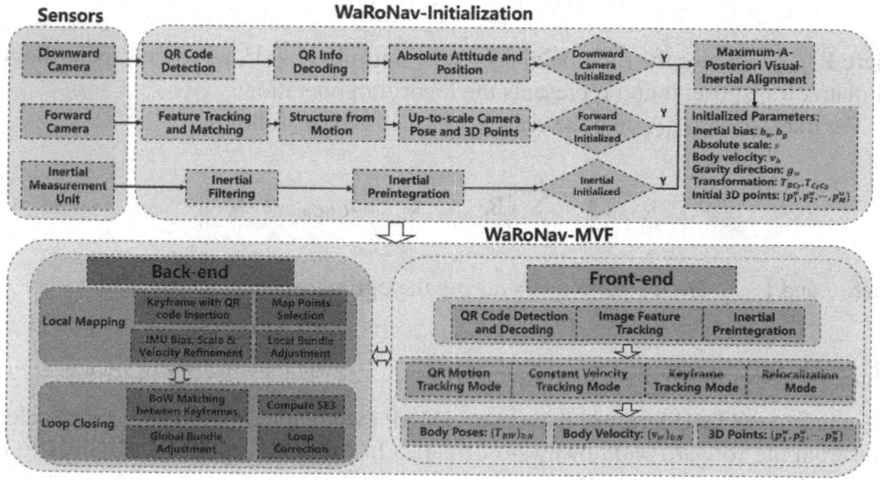

Fig. 2. The workflow of the proposed WaRoNav. The scheme consists of system initialization, tracking and local mapping.

3 WaRoNav Initialization

The robot state estimation depends heavily on the initialization among the multi-view visual-inertial sensors. In this section, we will first initialize the transformation $\mathbf{T}_{C_D C_F}$ (which includes the rotation $\mathbf{R}_{C_D C_F}$ and translation $\mathbf{t}_{C_D C_F}$) between the downward camera frame $\{C_D\}$ and the forward camera frame $\{C_F\}$, as well as the scale s of the forward camera. Afterwards, the inertial biases $\{\mathbf{b}^\alpha, \mathbf{b}^\omega\}$, body velocities $\{\mathbf{v}_{0:k}\}$, and gravity direction $\mathbf{R}_{wg}$ will be optimized in the maximum-a-posteriori formulation.

3.1 Initialization on $\{C_F\}$ and $\{C_D\}$

The downward camera frame and forward camera frame are initialized in a loosely-coupled manner. The forward camera poses are initialized in the pure ORB-initialization procedure [5]. The features between the successive frames are detected and matched. Afterwards, the fundamental matrix or the homography matrix with RANSAC outlier removal [9] is yielded to select the correct matches, which will be triangulated to obtain the corresponding 3D positions in the forward camera frame. Following this, the perspective-n-point (PNP) will be performed to estimate the forward camera poses, which are represented by $\{\mathbf{R}_{t_i}^{C_F}, t_{t_i}^{C_F}\}_{0:k}$. Similarly, the downward camera poses $\{\mathbf{R}_{t_i}^{C_D}, t_{t_i}^{C_D}\}_{0:k}$ will also be computed.

The rotation matrix $\mathbf{R}_{C_DC_F}$ is first calculated as follows:

$$\mathbf{R}_{C_DC_F} = \min_{\tilde{\mathbf{R}}_{C_DC_F}} \sum_{i=0}^{k} \log\left(\tilde{\mathbf{R}}_{C_DC_F}\mathbf{R}_{t_i}^{C_D}\left(\mathbf{R}_{t_i}^{C_F}\right)^T\right) \tag{1}$$

where $\tilde{\mathbf{R}}_{C_DC_F}$ denotes the rotation matrix to be optimized. $\{\cdot\}^T$ represents the orthogonal matrix transpose; $\log(\cdot)$ represents the logorithm operation.

The translation $\mathbf{t}_{\mathbf{C_DC_F}}$ and the scale s are calculated as follows:

$$\{s, \mathbf{t}_{C_DC_F}\} = \sum_{i=0}^{k} \left\| \mathbf{R}_{C_DC_F} \cdot \mathbf{t}_{t_i}^{C_F} + \tilde{\mathbf{t}}_{C_DC_F} - \tilde{s} \cdot \mathbf{t}_{t_i}^{C_D} \right\| \tag{2}$$

where $\tilde{s}$ and $\tilde{\mathbf{t}}_{C_DC_F}$ are the parameters during the optimization.

3.2 IMU Initialization

We initialize the IMU terms in the sense of MAP estimation, from the trajectory calculated from the forward and downward cameras. In this work, the IMU variables to be initialized are $\chi_k = \{\mathbf{b}^\alpha, \mathbf{b}^\omega, \mathbf{R}_{wg}, \mathbf{v}_{0:k}\}$.

$\mathbf{R}_{wg} \in SO(3)$ denotes the rotation from the gravity direction to the body frame, such that $\mathbf{g} = R_{wg}(0,0,G)^T$, where G is the magnitude of gravity.

The posterior distribution of χ_k is $p(\chi_k | I_{0:k})$. The inertial terms can be obtained by optimizing the following function.

$$\begin{aligned} \chi_k{}^* &= argmax_{\chi_k} p(\chi_k | I_{0:k}) \\ &= argmin_{\chi_k}(-log(p(\chi_k))) - \sum_{i=1}^{k} log\left(p\left(I_{i-1,i} | \chi_k\right)\right) \end{aligned} \tag{3}$$

$p(\chi_k)$ denotes the prior; $p(I_{i-1,i} | \chi_k)$ denotes the corresponding likelihood. $log(\cdot)$ stands for the logarithm operation.

Equation 3 is also equivalent to

$$\chi_k{}^* = \arg min_{\chi_k}\left(\|r_p\|_{\Sigma_p}^2 + \sum_{i=1}^{k} \|r_{I_{i-1,i}}\|_{\Sigma_{I_{i-1.i}}}^2\right) \tag{4}$$

where r_p is the image feature reprojection error, which is used as the regularization parameter to impose on inertial residuals being closed to 0; $r_{I_{i-1,i}}$ is the residual of IMU preintegrated measurements between the adjacent keyframes $i-1$ and i; Σ_p and $\Sigma_{I_{i-1,i}}$ are the corresponding covariance matrix of r_p and $r_{I_{i-1,i}}$ respectively.

3.3 3D Point Initialization

Apart from computing the inertial terms and frame transformations, the initial positions of features $\mathbf{X}_{3d}^W = \{X_{3d}^W(1), X_{3d}^W(2), \cdots, X_{3d}^W(n)\}$ with respect to the world frame can also be computed. The corresponding features on the first frame $\{C_{F1}\}$ and the second frame $\{C_{F2}\}$ of the forward camera are denoted by $\left\{p_{2d}^{C_{F1}}(1), p_{2d}^{C_{F1}}(2), \cdots, p_{2d}^{C_{F1}}(n)\right\}$ and $\left\{p_{2d}^{C_{F2}}(1), p_{2d}^{C_{F2}}(2), \cdots, p_{2d}^{C_{F2}}(n)\right\}$ respectively, which satisfy

$$\begin{aligned} \hat{p}_{2d}^{C_{F1}}(i) = \mathbf{K}[\mathbf{R}_{WC_{F1}}, \mathbf{t}_{WC_{F1}}]\hat{X}_{3d}^W(i) \\ \hat{p}_{2d}^{C_{F2}}(i) = \mathbf{K}[\mathbf{R}_{WC_{F2}}, \mathbf{t}_{WC_{F2}}]\hat{X}_{3d}^W(i) \end{aligned} \tag{5}$$

The hat on $\hat{X}$ and $\hat{p}$ represent the positions in their homogeneous form. $\mathbf{K}$ represents the camera intrinsic matrix. $\mathbf{R}_{WC_{F1}}$ and $\mathbf{R}_{WC_{F2}}$ represent the rotation from the world frame $\{W\}$ to the camera frame $\{C_{F1}, C_{F2}\}$; $\mathbf{t}_{WC_{F1}}$ and $\mathbf{t}_{WC_{F2}}$ represent the translation from the world frame $\{W\}$ to the forward camera frame $\{C_{F1}, C_{F2}\}$ respectively, which could be computed by using the poses from the downward camera $\mathbf{T}_{WC_D}$ and the transformation matrix $\mathbf{T}_{C_D C_F}$.

4 WaRoNav Multi-view Fusion

4.1 WaRoNav-MVF Frontend

In this work, the front end computes the 3D points, body poses, velocities and biases between the current and last frames. Note that there are four tracking modes, i.e., QR motion tracking, constant velocity tracking, keyframe tracking and relocalization.

In the presence of a QR code, the system will decode the corresponding label qr and compute its code position $\{x_{qr}, y_{qr}, z_{qr}\}$ as well as its planar translation & angular displacements $\{\triangle x, \triangle y, \triangle\phi\}$. Afterwards, the AGV current pose is updated. Then the body velocities and inertial biases are updated with the rigid constraints on the camera pose between the adjacent frames estimated from the QR code.

Between the QR spatial intervals, the robot states will be updated by the traditional pose tracking method [10]. When the reprojection error of features between the adjacent frames is within the predefined threshold, constant velocity tracking will be performed. Otherwise, keyframe tracking will be performed. The features between the current frame and the latest keyframe will be searched and matched. If there are adequate numbers of correct feature matches, the corresponding states will be updated. Or else, the relocalization will be implemented between the current frame and the historical keyframes, in the presence of feature tracking failures. The map points will be built using the current frame and the matched keyframe.

4.2 WaRoNav-MVF Backend

In the WaRoNav backend, the local mapping and loop closing will be performed. They are designed to correct the robot poses and the corresponding map point positions. The descriptions of these procedures are described in the following sections.

Local Mapping. In the local mapping, a certain number of keyframes (denoted by $\{F_1, F_2, \cdots, F_M\}$) and the corresponding map points (denoted by $\{p_1, p_2, \cdots, p_L\}$) seen by these keyframes are optimized using the local bundle adjustment model, as depicted in Fig. 3. Unlike [5], the poses of some specific keyframes among $\{F_1, F_2, \cdots, F_M\}$ that are also observed by the downward camera stay unchanged. In other words, these poses are fixed in the BA process. The reason is that the poses deduced from QR codes on the ground have much higher accuracy than the poses estimated from the forward visual-inertial measurements.

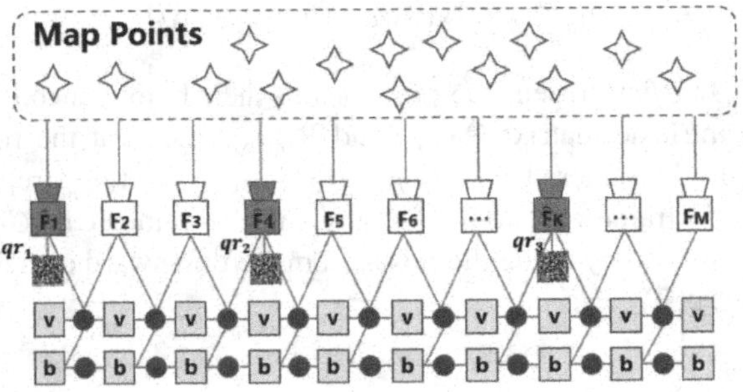

Fig. 3. Illustration on the WaRoNav local bundle adjustment. The map point positions and the camera poses will be optimized in the BA manner. Several poses denoted by orange color $\{\hat{F}_1, \hat{F}_4, \hat{F}_K\}$ corresponding to $\{qr_1, qr_2, qr_3\}$ are the fixed poses in the optimization process.

In the local bundle adjustment, we use the graph optimization model to update the keyframe poses and map point positions. In the model, the poses that are deduced from the QR codes have much higher confidence. Thus, we set these keyframe poses fixed. The other poses of keyframes without QR codes and the map points will be optimized in the model. Additionally, note that the inertial terms, i.e., velocity and biases are also optimized, together with the reprojection errors.

It should also be noted that for each QR code, we rigidly choose a certain number of keyframes (often 2 or 3 keyframes) in the cost function. These constraints could ensure that the optimal values are not stuck in the local minimum.

Loop Closing. In the digital workshop, the robot is programmed to move in regular trajectories. Loop closures could be periodically performed to correct the accumulated drifts. In this work, we perform loop closing upon the keyframes. The bag-of-words (BoWs) of the current keyframe will be compared with BoWs of the historical keyframes. To reduce the occurrence of false matches on keyframes, we add rigid constraints on kyeframe positions.

Assume the current keyframe and the putative match are F_c and F_p. The corresponding positions of these keyframes are $P(F_c)$ and $P(F_p)$ respectively. If they satisfy Eq. 6, it can be inferred that the keyframe match $\{P(F_c), P(F_p)\}$ is genuine.

$$\left\|P(F_c) - P(F_p)\right\| \leq \delta_{lc} \tag{6}$$

Afterwards, the 6 DOF rigid transformation (Sim3 model) will be computed. The poses of the current frame and the nighbouring keyframes that observe the common map points of $P(F_c)$ will be updated on the covisibility graph.

5 Experiments

5.1 Platform

In the experiments, the proposed WaroNav is evaluated on the developed platform. There are 80 QR codes horizontally and vertically pasted on the floor. The distance between the adjacent QR codes is 1200 mm.

The warehouse robot is equipped with the froward sensing units, i.e., monocular camera and IMU, which are integrated in the Intel Realsense T265 module. Also, the robot is equipped with the downward camera, which scans the QR codes on the ground. The robot is programmed to run the trajectories in accordance with the commands wireless sent from the server. The resolution of the image from the forward camera is 848 × 800. The IMU is Bosch *BMI055*, which collects three-axis accelerations and three-axis angular velocities. The sampling rates of forward camera and IMU are 20 Hz and 100 Hz respectively. The downward camera collects the ground QR codes with the image resolution 640 × 480 and sampling rate 20 Hz. The distance between the downward camera and the ground is approximately 120 mm.

5.2 Results and Analysis

In the tests, the AGV runs three types of generic trajectories, namely, L-shape, U-shape and rectangular trajectories in the digital workshop. To evaluate the effectiveness of WaRoNav, we use the evaluation tool (EVO) [11]. SE(3) is exploited to align the estimated trajectory with the ground truth. The root mean square error (RMSE) of translations and rotations are computed and compared with the state-of-the-art, i.e., Vins-Mono [12], ORB-SLAM3 (Monocular-Inertial) [5] and LBVIO [13]. The warehouse robot trajectory ground truth are obtained from the Leica TS60 tracker, which has millimeter accuracy [14].

The forward camera, IMU and downward camera will be initialized within 2 s after the downward camera detects the first QR code. The global pose of the AGV will be estimated and updated. For each type of trajectory, approximately 100 tests are performed, which are described and analyzed as follows.

L-Shape Trajectory Tests. In the L-shape trajectory tests, the robot is programmed to follow the linear path and then turn left at the QR codes. Figure 4 shows the estimated trajectories from the proposed WaRoNav, Vins-Mono [12], ORB-SLAM3 (Monocular-Inertial) [5], and LBVIO [13]. It can be seen that there is a turning arc (approximately 90°C) at the corner of the trajectory, because the forward and downward sensing units are fixed at the front of the AGV. It can also be seen that WaRoNav considerably corrects the drifts from the forward visual-inertial measurements at the QR code spots. Thus, the deviations could be periodically compensated. In contrast, ORB-SLAM3 [5], Vins-Mono [12] and LBVIO [13] yield obvious discrepancies against the ground truth over time. These discrepancies are caused by scale drifts, unexpected robot jitters, inadequate landmarks, etc.

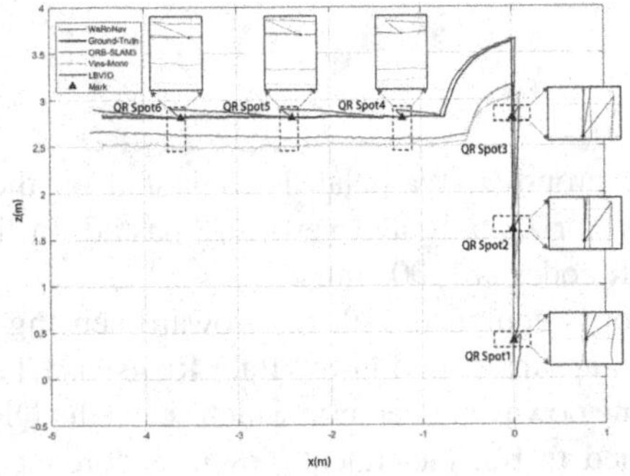

Fig. 4. Comparisons on the estimated AGV L-shape trajectories between ours (WaRoNav) and the Vins-Mono [12] and ORB-SLAM3 (Monocular-Inertial) [5], LBVIO [13].

Table 1 presents the orientation & translation errors for the robot L-shape trajectories. The tests for the L-shape trajectory have been repeated 100 times. From Table 1, it can be seen that WaRoNav yields the smallest RMSE on both orientation and translation errors, which are less than 45 mm and 2°. Comparably, there exist different levels of accumulated errors for Vins-Mono [12] and ORB-SLAM3 [5], especially on the robot turning intervals, which can be attributed to the long-term visual-inertial drifts and the

Table 1. Comparisons on the RMSE of translations and rotations of L-shape trajectories between WaRoNav (Ours) and the Vins-Mono [12], ORB-SLAM3 (Monocular-Inertial) [5] and LBVIO [13].

Trajectory	Length	Metric	Ours	ORB-SLAM3	Vins-Mono	LBVIO
L-shape	8400 mm	Trans. err	**38.55** mm	434.11 mm	472.55 mm	172.72 mm
		Rot. err	**1.78°**	3.97°	3.54°	2.53°
	16800 mm	Trans. err	**39.62** mm	583.21 mm	654.14 mm	180.97 mm
		Rot. err	**1.81°**	5.23°	6.85°	3.76°
	25200 mm	Trans. err	**41.97** mm	782.16 mm	862.58 mm	186.43 mm
		Rot. err	**1.82°**	8.95°	8.76°	4.22°
	33600 mm	Trans. err	**43.28** mm	873.58 mm	956.82 mm	238.19 mm
		Rot. err	**1.85°**	12.14°	13.97°	4.92°

loss of the absolute reference corrections. Although LBVIO [13] achieves relatively higher accuracy on the QR code spots, the severe drifts are still obvious in the whole navigation process.

U-Shape Trajectory Tests. In the U-shape trajectory tests, Vins-Mono [12] and ORB-SLAM3 (Monocular-Inertial) [5] yield obviously larger translation and rotation errors, especially at the turning time, which deviate considerably from the ground truth, as illustrated in Fig. 5. LBVIO [13] is able to correct the drifts periodically, but the zig-zag patterns during the correction are still obvious because of its drawbacks in the inherent local sliding-filter fusion mode. Comparably, WaRoNav is able to maintain a relatively smaller tracking error, due to the periodical corrections from the QR codes and the tightly-coupled global optimization framework.

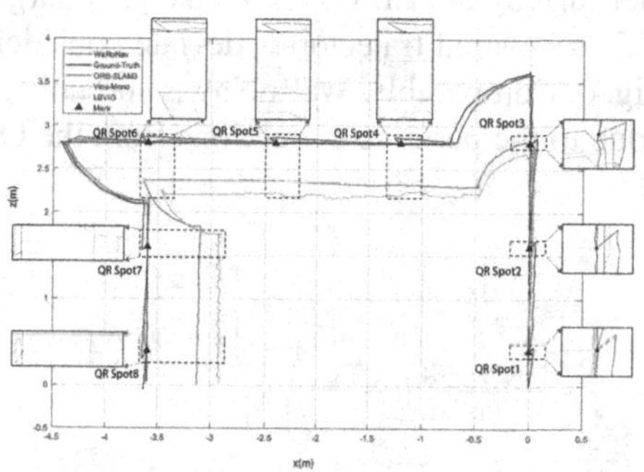

Fig. 5. Comparisons on the estimated AGV L-shape trajectories between ours (WaRoNav) and the Vins-Mono [12], ORB-SLAM3 (Monocular-Inertial) [5] and LBVIO [13].

Table 2 shows the corresponding orientation & translation errors for the AGV U-shape trajectories. The tests for the U-shape trajectory have been repeated 100 times. From Table 2, we can see that the RMSEs of WaRoNav on both orientations and translation errors grow relatively smaller, which are less than 49 mm and 2°. In contrast, the accumulated errors of Vins-Mono [12] and ORB-SLAM3 [5] are relatively higher, reaching up to 990 mm and 14° on the recovered translation and rotation. It should be noted that severe exposure exists when the AGV is running toward the workshop window directions. In this scenario, the number of features being tracked in the forward-looking image sequences is relatively small, which causes overall AGV tracking degradation. LBVIO [13] exhibits relatively smaller errors on both translations and rotations, which are less than 194 mm and 4° respectively. But it is still inferior to the proposed WaRoNav.

Table 2. Comparisons on the RMSE of translations and rotations of U-shape trajectories between ours (WaRoNav) and the Vins-Mono [12], ORB-SLAM3 (Monocular-Inertial) [5] and LBVIO [13].

Trajectory	Length	Metric	Ours	ORB-SLAM3	Vins-Mono	LBVIO
U-shape	12000 mm	Trans. err	**41.87** mm	483.82 mm	540.12 mm	171.25 mm
		Rot. err	**1.72°**	7.13°	8.34°	2.97°
	24000 mm	Trans. err	**43.91** mm	589.75 mm	684.46 mm	187.38 mm
		Rot. err	**1.89°**	8.87°	9.96°	3.11°
	36000 mm	Trans. err	**47.62** mm	825.89 mm	912.74 mm	189.62 mm
		Rot. err	**1.91°**	9.92°	11.46°	3.35°
	48000 mm	Trans. err	**48.94** mm	890.74 mm	985.16 mm	193.74 mm
		Rot. err	**1.95°**	11.33°	13.87°	3.86°

Rectangular Trajectory Tests. In rectangular trajectory tests, Vins-Mono [12], ORB-SLAM3 (Monocular-Inertial) [5] and LBVIO [13] also yield larger translation and rotation errors over time. The recovered trajectories deviate considerably from the ground truth, as illustrated in Fig. 6. Comparably, WaRoNav is still able to maintain a relatively smaller tracking error, due to the periodic corrections from the QR codes.

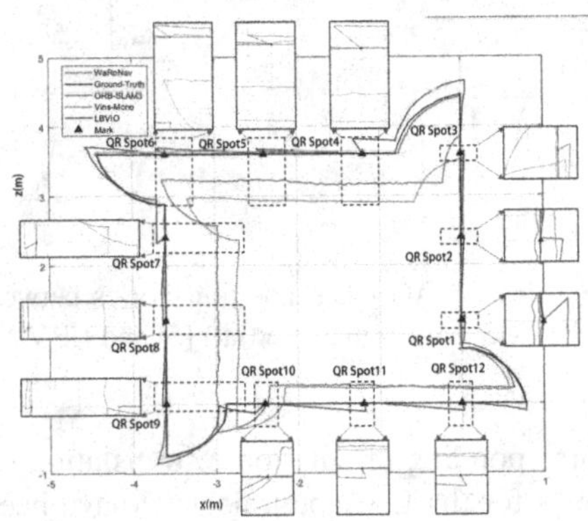

Fig. 6. Comparisons of the estimated AGV rectangular trajectories between ours (WaRoNav) and Vins-Mono [12], ORB-SLAM3 (Monocular-Inertial) [5], LBVIO [13].

Table 3 shows the orientation & translation errors for the AGV rectangular trajectories. The tests for the rectangular trajectory have been repeated 100 times. It can be seen that the RMSEs of WaRoNav on both orientations and translation errors grow relatively smaller, which are less than 39 mm and 2°. Comparably, there exist severe accumulated errors for Vins-Mono and ORB-SLAM3, especially on the robot turning intervals, which can be attributed to the long-term visual-inertial drifts and the loss of the absolute reference corrections. Note that LBVIO [13] could reduce the errors on the QR spots to some extent, but the errors on the non-QR sports still cannot be overlooked.

Table 3. Comparisons on the RMSE of translations and rotations of rectangular trajectories between ours (WaRoNav) and the Vins-Mono [12], ORB-SLAM3 (Monocular-Inertial) [5] and LBVIO [13].

Trajectory	Length	Metric	Ours	ORB-SLAM3	Vins-Mono	LBVIO
Rectangle	16800 mm	Trans. err	**37.35** mm	564.36 mm	595.73 mm	151.77 mm
		Rot. err	**1.91°**	6.93°	7.76°	3.23°
	33600 mm	Trans. err	**38.41** mm	575.22 mm	612.94 mm	155.49 mm
		Rot. err	**1.93°**	7.15°	8.67°	4.27°
	50400 mm	Trans. err	**38.62** mm	599.89 mm	613.42 mm	167.62 mm
		Rot. err	**1.97°**	8.92°	9.46°	3.76°
	67200 mm	Trans. err	**38.77** mm	619.94 mm	654.53 mm	179.76 mm
		Rot. err	**1.98°**	9.72°	10.68°	4.98°

6 Conclusion

In this paper, we have presented an innovative multi-view visual-inertial fusion framework for warehouse robot navigation. WaRoNav is able to correct the long-term visual-inertial estimation drifts periodically, offer the robot poses & velocities in the absolute reference frame and preserve the global consistency by jointly using the measurements from the forward camera and IMU, as well as the downward camera that captures the QR codes on the floor. WaRoNav has been extensively evaluated on the developed platform. The experimental results have shown competitive accuracy against the state-of-the-art with position and attitude errors less than 0.05 m and 2 ° respectively, irrespective of the trajectory lengths.

Acknowledgment. This work was supported by the National Natural Science Foundation of China (62273332) and the Youth Innovation Promotion Association of the Chinese Academy of Sciences (2022201).

References

1. Hu, H., Jia, X., Liu, K., Sun, B.: Self-adaptive traffic control model with behavior trees and reinforcement learning for AGV in industry 4.0. IEEE Trans. Ind. Inf. **17**(12), 7968–7979 (2021)
2. Li, X., Wan, J., Dai, H.-N., Imran, M., Xia, M., Celesti, A.: A hybrid computing solution and resource scheduling strategy for edge computing in smart manufacturing. IEEE Trans. Ind. Inf. **15**(7), 4225–4234 (2019)
3. De Ryck, M., Versteyhe, M., Debrouwere, F.: Automated guided vehicle systems, state-of-the-art control algorithms and techniques. J. Manuf. Syst. **54**, 152–173 (2020)
4. Labbé, M., Michaud, F.: RTAB-map as an open-source lidar and visual simultaneous localization and mapping library for large-scale and long-term online operation. J. Field Rob. **36**(2), 416–446 (2019)
5. Campos, C., Elvira, R., Rodríguez, J.J.G., Montiel, J.M., Tardós, J.D.: Orb-slam3: an accurate open-source library for visual, visual-inertial, and multimap slam. IEEE Trans. Rob. **37**(6), 1874–1890 (2021)
6. Cao, S., Lu, X., Shen, S.: GVINS: tightly coupled GNSS-visual-inertial fusion for smooth and consistent state estimation. IEEE Trans. Rob. **38**(4), 2004–2021 (2022)

7. Lv, W., Kang, Y., Zhao, Y.-B., Wu, Y., Zheng, W.X.: A novel inertial-visual heading determination system for wheeled mobile robots. IEEE Trans. Control Syst. Technol. **29**(4), 1758–1765 (2020)
8. Eckenhoff, K., Geneva, P., Bloecker, J., Huang, G.: Multi-camera visual-inertial navigation with online intrinsic and extrinsic calibration. In; 2019 International Conference on Robotics and Automation (ICRA), pp. 3158–3164. IEEE (2019)
9. Xia, Y., Ma, J.: Locality-guided global-preserving optimization for robust feature matching. IEEE Trans. Image Process. **31**, 5093–5108 (2022)
10. Mur-Artal, R., Tardós, J.D.: Visual-inertial monocular slam with map reuse. IEEE Rob. Autom. Lett. **2**(2), 796–803 (2017)
11. Chou, C.C., Chou, C.F.: Efficient and accurate tightly-coupled visual-lidar slam. IEEE Trans. Intell. Transp. Syst. **23**(9), 14509–14523 (2021)
12. Qin, T., Li, P., Shen, S.: Vins-mono: a robust and versatile monocular visual-inertial state estimator. IEEE Trans. Rob. **34**(4), 1004–1020 (2018)
13. Lee, J., Chen, C., Shen, C., Lai, Y.: Landmark-based scale estimation and correction of visual inertial odometry for VTOL UAVs in a GPS-denied environment. Sensors **22**(24), 9654 (2022)
14. Karimi, M., Oelsch, M., Stengel, O., Babaians, E., Steinbach, E.: Lola-slam: low-latency lidar slam using continuous scan slicing. IEEE Rob. Autom. Lett. **6**(2), 2248–2255 (2021)

Enhancing Lidar and Radar Fusion for Vehicle Detection in Adverse Weather via Cross-Modality Semantic Consistency

Yu Du[1], Ting Yang[1], Qiong Chang[2], Wei Zhong[1], and Weimin Wang[1](✉)

[1] Dalian University of Technology, Dalian, China
{d01u01,yangting}@mail.dlut.edu.cn, {zhongwei,wangweimin}@dlut.edu.cn
[2] Tokyo Institute of Technology, Tokyo, Japan
q.chang@c.titech.ac.jp

Abstract. The fusion of multiple sensors such as Lidar and Radar provides richer information and thus improves the perception ability for autonomous driving even in adverse weather. However, Lidar paints 3D geometric point cloud of the scene while Radar provides 2D scan images of Radar cross-section (RCS) intensity, which brings challenges for efficient multi-modality fusion. To this end, we propose an enhanced Lidar-Radar fusion framework, Cross-Modality Semantic Consistency Networks (CMSCNet), to mitigate the modal gap and improve detection performance. Specifically, we implement the semantic consistency loss by leveraging the concept of Knowledge Distillation for cross-modality feature learning. Moreover, we investigate effective point cloud processing strategies of adaptive ground removal and Laser-ID slicing when transforming Lidar point cloud to the BEV representation that closely resembles the Radar image. We evaluate the proposed method on Oxford Radar RobotCar dataset of simulated fog point cloud and RADIATE dataset of real adverse weather data. Extensive experiments validate the advantages of CMSCNet for Lidar and Radar fusion, especially on real adverse data.

Keywords: Lidar-Radar Fusion · Knowledge Distillation · Object Detection · Adverse Weather

1 Introduction

Object detection is a crucial perception task for autonomous driving. Recent deep neural network models driven by massive amounts of training data have achieved remarkable detection performance with 3D point cloud data by Lidar sensors [4,6,13,14,21]. To further improve the automatic driving level, researchers strive to expand the capability of robust detection under all-weather conditions. However, the robustness and performance of most models trained on Lidar point cloud data drastically degrade as the quality of data worsens due to the poor visibility in adverse weather (snowy, foggy, and rainy).

Q. Liu et al. (Eds.): PRCV 2023, LNCS 14427, pp. 439–451, 2024.
https://doi.org/10.1007/978-981-99-8435-0_35

Different from optical sensors, millimeter wave Radar uses frequency modulated continuous waves (FMCW) to perceive the surroundings. Owing to its longer wavelength, Radar waves can penetrate or diffract small particles in adverse weather, enabling the Radar sensor to perceive the environment stably regardless of the weather. However, relying solely on Radar for detection poses significant challenges due to the sparsity, noise, and lack of vertical resolution in Radar scan data. Thus, Lidar with high-quality 3D information can be the complementary modality with Radar even under certain adverse weather. The representative Lidar-Radar fusion network MVDNet [10] shows high accuracy even in foggy conditions, demonstrating the remarkable resilience of the fusion model towards adverse weather. They propose to fuse Lidar and Radar features adaptively using an attention module. However, they do not explicitly consider the modality gap between the two types of data during feature learning.

Radar operates in the range-azimuth plane with a 360 ° horizontal view but lacks vertical height information. On the other hand, Lidar enables the creation of 3D representations along horizontal and vertical directions. By transforming the Lidar data into a Bird's Eye View (BEV) format, two modal features can be aligned in a unified representation space. In MVDNet work [10], self- and cross-attention mechanisms are used to establish connections between concatenated Lidar and Radar features. However, notable differences exist between independently extracted features from different modalities, which can potentially hinder the subsequent fusion process. To address this, we propose Cross-Modality Semantic Consistency loss to enhance Lidar and Radar feature learning for better cross-modal consistency. The proposed loss is implemented by leveraging the concept of Knowledge Distillation to explicitly constrain the feature learning of two modality data for object detection. Furthermore, as Lidar and Radar data are fused in the BEV perspective, we observe that the ground portion of Lidar point cloud data may interfere with the fusion process involving Radar data. Thus, we employ adaptive ground removal techniques to aid in the detection process. Additionally, we explore the 2D projection of the 3D point cloud and discover that Laser-ID based slicing provides a more natural and efficient data format for BEV fusion.

The contributions of this work are as follows:

- We propose CMSCNet, a semantic consistency constrained cross-modality fusion network to take complementary advantages of Lidar and Radar in all-weather conditions. Knowledge Distillation is innovatively introduced to multi-modal fusion to mitigate the modality gap between extracted features.
- We introduce two strategies for Lidar BEV generation to enhance the fusion of Lidar point clouds and Radar images. Different from the traditional height-based BEV dividing method, we slice point clouds according to Laser IDs to retain their complete information. In addition, ground points within a certain range are filtered out to decrease noisy data.
- We conduct extensive experiments and ablation studies on fogged Oxford Radar RobotCar (ORR) and RAdar Dataset In Adverse weaThEr (RADIATE) datasets to validate the superiority of the proposed fusion method for

object detection in adverse weather, as well as the effectiveness of the ground-removal and ID-based slicing strategy.

2 Related Work

Radar is usually equipped as an integral part of an Advanced Driver Assistance System (ADAS). Although Radar has been widely equipped in vehicles, there are only a few works on Radar-based perception compared to other sensors. In this section, we revisit works of Radar-only and Radar-fused perception.

Radar Only. Major et al. [8] propose the first Radar-based deep neural object detection network in the self-developed dataset. Wang et al. [16] propose a deep Radar object detection network (RODNet) with range-azimuth frequency heatmaps (RAMaps) Radar data as input. Due to the low accuracy, there are few detection networks that rely on Radar alone. More commonly, Radar is used as a complement to other sensors to enhance the performance of the perception.

Radar-Camera CenterFusion [9] is a middle-fusion approach for 3D object detection. Radar feature maps are generated using Radar predictions and the initial detection results obtained from images, and then Radar features and image features jointly predict bounding boxes. Radar predictions and their corresponding objects are associated through a novel frustum-based Radar association method. The authors of [5] present the Gated Region of Interest Fusion (GRIF) Network to fuse Radar sparse point clouds and monocular camera images. GRIF-Net adaptively selects the appropriate sensor data to improve robustness using an explicit gating mechanism. Low-level Radar and image fusion achieve comparable results to Lidar on the nuScenes dataset [3], showing the potential of Radar in autonomous driving.

Radar-Lidar RadarNet [18] fuse Lidar and Radar data with a two-sensor fusion mechanism: a voxel-based early fusion and an attention-based late fusion. It utilizes both geometric and dynamic information from Radar clusters. MVDNet [10] is the first work that performs object detection with the fusion of panoramic Lidar and Radar in simulated foggy weather conditions. Besides feature fusion, it investigates temporal fusion for two data sequences. Based on MVDNet, ST-MVDNet [7] proposes a novel training framework that addresses the issue of sensor absence in multi-modal vehicle detection by incorporating clear and missing patterns. The Mean Teacher framework facilitates parallel learning of teacher and student models, resulting in a substantial increase in parameter count.

3 Proposed Method

The architecture of the proposed CMSCNet is illustrated in Fig. 1. CMSCNet takes Millimetre-Wave Radar scan images and preprocessed Lidar multi-dimensional BEV images as inputs. The VGG16 [15] backbone is utilized for the initial feature extraction. CMSC Loss constrains Lidar and Radar feature extractors to learn similar features on focal channels and pixels. The obtained feature maps with aligned spatial size are concatenated together along the channel dimension and later coupled by self- and cross- attention [10]. The final detection results are predicted by the fused feature through the detection head.

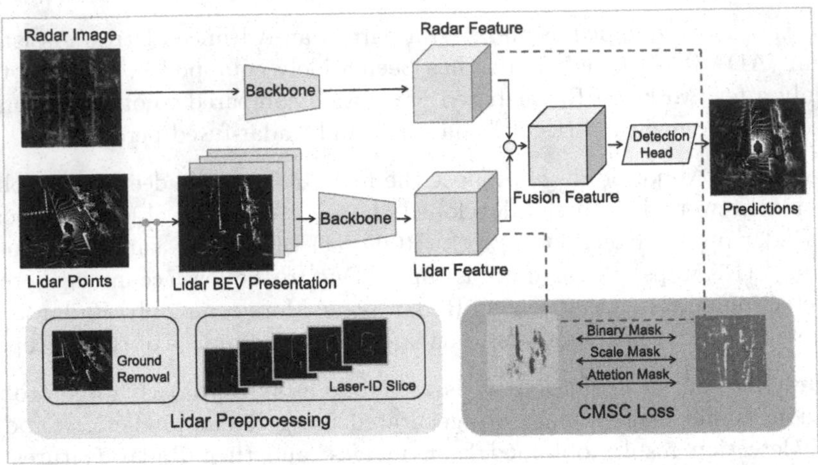

Fig. 1. The architecture of CMSCNet.

3.1 Losses

Cross-Modality Semantic Consistency Loss. To fuse Lidar and Radar at feature-level, MVDNet [10] introduced self- and cross- attention blocks. However, it only adaptively sets weights for fusion and does not constrain the semantic consistency of features learned for each modality. An external constraint is expected to make the extractors focus on the qualified information. As an efficient approach for cross-modality learning, we introduce the concept of Knowledge Distillation to enhance the correlation of feature maps extracted from the network. In this work, CMSC loss L_{CMSC} is implemented using focal knowledge distillation loss [20], which is specifically designed for object detection tasks. L_{CMSC} is defined as :

$$L_{CMSC} = L_{fea} + L_{att} \tag{1}$$

where L_{fea} is the feature loss and L_{att} is the attention loss. Note that the loss is only for the constraints of semantic consistency and no student model is actually learned.

In contrast to conventional global feature distillation methods, L_{fea} is divided into two components: foreground loss L_{fg} and background loss L_{bg}. By segmenting the image into foreground and background, our approach guides both parts to focus on crucial pixels and channels. The hyper-parameters α and β are used to maintain the balance between foreground and background loss. We formulate it as:

$$L_{fea} = \alpha L_{fg} + \beta L_{bg} \tag{2}$$

Following the approach proposed in [20], the foreground and background feature losses L_{fg} and L_{bg} are composed of three components, Binary Mask M, Scale Mask S, Attention Mask A^S and A^C. Through Foreground Binary Mask M^F and Background Binary Mask M^B, the feature map is separated into foreground area and background area, as illustrated in Fig. 2. To mitigate the influence of varying object pixel areas on the loss, we take Scale Mask S for scale normalization by dividing the number of pixels that belong to the target box or background. Besides, by performing the sum of absolute values on the feature map, we can get a spatial attention map and a channel attention map separately. With the consequent application of Softmax function with temperature parameters, Spatial and Channel Attention Mask A^S and A^C are obtained, which can effectively guide the network to focus on focal pixels and channels. Lidar features F^{Li} and Radar feature F^{Ra} are obtained from respective extractors. The feature loss can be formulated as:

$$L_{fg} = \sum_{k=0}^{C}\sum_{i=0}^{H}\sum_{j=0}^{W} M_{i,j}^{F} S_{i,j} A_{i,j}^{S} A_{k}^{C} (F_{i,j,k}^{Li} - F_{i,j,k}^{Ra})^2 \tag{3}$$

$$L_{bg} = \sum_{k=0}^{C}\sum_{i=0}^{H}\sum_{j=0}^{W} M_{i,j}^{B} S_{i,j} A_{i,j}^{S} A_{k}^{C} (F_{i,j,k}^{Li} - F_{i,j,k}^{Ra})^2 \tag{4}$$

where H, W, and C denote height, width, and channel of the feature. i, j, k denote the horizontal coordinate, vertical coordinate and channel in the feature map.

Besides, the attention loss L_{att} is calculated with the Spatial and Channel Attention Mask A^S and A^C as the same with [20]:

$$L_{att} = \gamma \cdot (l(A_{Li}^{S}, A_{Ra}^{S}) + l(A_{Li}^{C}, A_{Ra}^{C})) \tag{5}$$

where l denotes L1 loss and γ is a hyper-parameter.

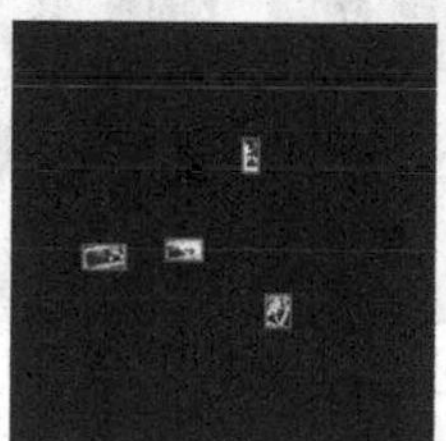

(a) Foreground Mask

(b) Background Mask

Fig. 2. Illustration of the Binary Mask. The pixels inside the outer rectangle (red) of the rotated bounding boxes (yellow) are identified as foreground.

Detection Loss. In training region proposal network (RPN), we use Binary Cross-Entropy (BCE) loss for the classification L_{RPN}^{cls} and smooth L1 loss for the bounding box regression L_{RPN}^{reg} as [11]:

$$L_{RPN} = L_{RPN}^{cls} + L_{RPN}^{reg} \tag{6}$$

In addition to classification loss L_{RCNN}^{cls} and regression loss L_{RCNN}^{reg} , orientation loss L_{RCNN}^{dir} is incorporated within the framework. The loss for classification and bounding-box regression is the same as in Eq. 6. The rotating bounding box orientation is represented by a binary positive and negative direction and a specific angle of 180°C. We also use a cross-entropy loss L_{dir} for direction classification.

$$L_{RCNN} = L_{RCNN}^{cls} + L_{RCNN}^{reg} + L_{RCNN}^{dir} \tag{7}$$

Total Loss. We apply multi-task loss for classification and regression of detection bounding boxes. Besides, we introduce Focal and Global Distillation Loss [20] to enhance the cross-modal correlation between Lidar and Radar. The total loss is defined as:

$$L_{total} = L_{CMCS} + L_{RPN} + L_{RCNN} \tag{8}$$

3.2 Ground Removal

The points belonging to the ground are the majority part of point clouds which are less useful in most tasks. Moreover, these points may be noisy for car feature extraction, especially in BEV form.

Therefore, we propose to utilize Random Sample Consensus (RANSAC) to filter them out to avoid the interference of background point clouds.

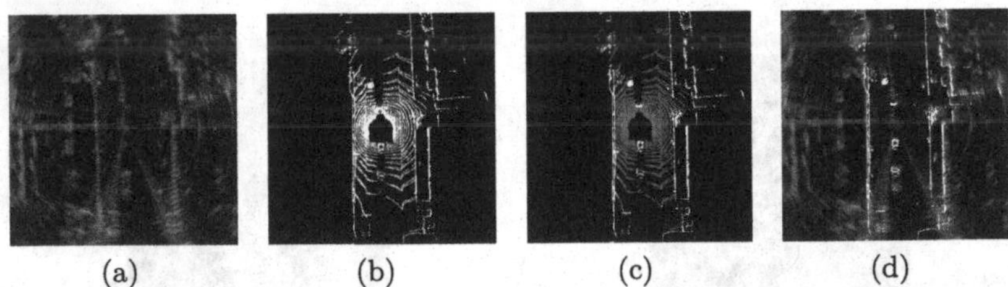

Fig. 3. Visualization of Radar and BEV of Lidar. From left to right are: Radar scan(a), BEV of Lidar point cloud(b), BEV of Lidar point cloud with points of ground marked as red(c) and ground-removed Lidar data overlayed on the Radar scan(d).

As shown in Fig. 3, we can find that the BEV of Lidar data with ground removal looks more similar to the original Radar data. We also find that most ground points distribute at a certain distance due to the Lidar scanning rays' character. If we remove the distant points, it would lead to the detection degradation caused by the point reduction of the distant objects. To avoid this, we set a distance threshold to adaptively filter the ground point.

3.3 Laser-ID Based Slicing

In point cloud related tasks, one common operation is to project the point cloud into a BEV view to be applicable for CNN to efficiently extract features. For example, points are projected into multiple slices along the height direction in PIXOR [19]. Although it is simple and intuitive to divide by height, we consider that division by height does not dependably conform to the operation of scanning Lidar ray. As shown in Fig. 4, a continuous scan (*ID_j* and *ID_k*) on a surface may be truncated into multiple discontinued projections by the height-based method, while it keeps as a continuous scanline when projected according to the Laser-ID. We divide all Lidar points into 32 slices based on the 32-beam Laser IDs and concatenated a intensity channel.

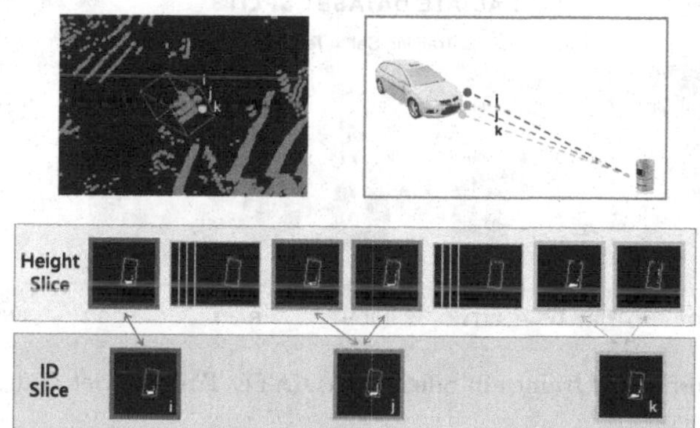

Fig. 4. The projection results of 3 continuous Lidar rings (*ID_i* ,*ID_j* ,*ID_k*) by height-based and our proposed Laser-ID based slice method. From the comparison, we can find ID-based BEV slice is a denser and more efficient height-based one.

4 Experiments

4.1 Datasets Preparation

Oxford Radar RobotCar Dataset. The Oxford Radar Robotcar Dataset [1] (ORR) is collected by a scanning system equipped with one panoramic Navtech CTS350-X Millimetre-Wave FMCW scanning Radar and two 32-beam HDL-32E Lidar sensors. We take the same data generation strategy of MVDNet [10]. Radar images and Lidar BEV representation covers the range of 64m in each direction with a resolution of 320 × 320. The number of Lidar BEV channels, i.e. the slices in the vertical direction, depends on the projection method as described in Sec. 3.3. The time synchronization and alignment strategy is applied for the correspondence of Lidar and Radar data. For ease of comparison, we build the same train and test set with fog simulation.

RADIATE Dataset. The RADIATE [12] dataset is collected in snow, fog, rain and clear (including night) weather conditions and various driving scenarios with the same Radar sensor as ORR used. Train and Test set splits are shown in Fig. 5. We combine 4 vehicle classes (car, van, bus and truck) into one class as the vehicle detection targets. For the difference in sensor scanning frequency, we select the closest Lidar frame as the corresponding one of Radar image from the timestamp list. The Radar coordinate system is also transformed to the Lidar coordinate system with the calibration data. The original Radar image is downsampled to 576 × 576 by max pooling, and the Lidar BEV image is processed to the same resolution.

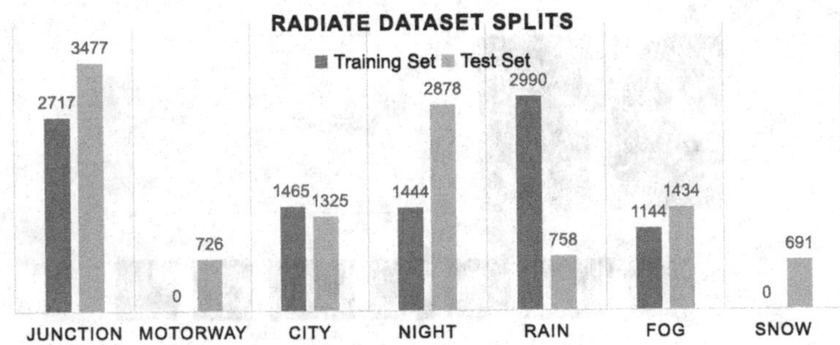

Fig. 5. Number of frames in official RADIATE Training Set and Test Set.

4.2 Implementation Details

Following experiment settings in MVDNet [10], we train the model. Our approach is implemented based on Detectron2 [17], an open-source codebase widely utilized for object detection tasks. In CMSC loss, we empirically set $\alpha = 5 \times 10^{-5}$ and $\beta = 2.5 \times 10^{-5}$ for the feature focal loss, and $\gamma = 5 \times 10^{-5}$ for the attention focal loss. For Laser-ID based slicing, we remove the 7m and 5m ground points from the ORR dataset and RADIATE dataset respectively.

Table 1. Results on ORR dataset. The network is trained on the clear and foggy set. AP (average precision) is evaluated with IOU of 0.5 and 0.65, following the common COCO protocol. MVDNet and MVDNet-Fast [10] are retrained in single-frame mode by the public codes. We present two results, CMSCNet and a simple version CMSCNet-Fast corresponding to MVDNet-Fast. The best results are in **bold**.

Train	Clear+Foggy					
Test	Clear		Foggy		Average	
IoU	0.5	0.65	0.5	0.65	0.5	0.65
PIXOR [19]	72.7	68.3	62.6	58.9	67.7	63.6
PointPillar [6]	85.7	83.0	72.8	70.3	78.0	67.7
DEF [2]	86.6	78.2	81.4	72.5	84.0	75.3
MVDNet [10]	88.9	86.8	82.7	78.9	85.8	82.9
CMSCNet (Ours)	**90.6**	**87.8**	**84.5**	**81.0**	**87.6**	**84.4**
MVDNet-Fast [10]	88.7	85.8	80.4	77.3	84.55	81.55
CMSCNet-Fast (Ours)	**89.7**	**86.9**	**83.9**	**80.2**	**86.8**	**83.55**

4.3 Quantitative Results

The quantitative results on ORR and RADIATE are reported in Table 1 and Table 2 respectively. We compare our methods with Lidar-only methods: PIXOR [19], PointPillar [6] and Lidar-Radar fusion methods: DEF [2], MVDNet [10]. MVDNet and MVDNet-Fast [10] are obtained by our retraining with the provided open-source code and hyper-parameters with the temporal fusion branch off. Similarly, we also experiment with CMSCNet-Fast, which is the same as [10] in that the sensor fusion module has fewer dimensions. Besides, we compare our fusion network with the Radar-only and Lidar-only detection method and the state-of-the-art Lidar-Radar model [10] on RADIATE dataset. The Radar-only and Lidar-only method are simplified versions of our network, by only taking single sensor data as input and discarding the fusion part.

From Table 1, we can find our method generally overperforms other methods on ORR dataset. For the inference time consumption, our model takes 60ms for direction on each pair on an NVIDIA GeForce RTX 3090 GPU while MVDNet takes 43 ms. Although the performance of our proposed CMSCNet is a little lower than MVDNet on RADIATE dataset in some scenes, as shown in Table 2, we can find CMSCNet achieves better performance in the challenging snow and night scenes. Compared to the single-modal approach, fusion exhibits performance degradation on Motorway or City scenes. The reason behind this mainly relates to the range of sensors and the existence of objects in RADIATE dataset. Different from ORR dataset, LiDAR data in some scenes contain more significant corruption and limited scanning range while the annotation is labeled based on Radar images. Thus, the fusion with less informative data(LiDAR) works worse than the one w/o fusing (Radar-only) to detect objects in far places.

Table 2. Detection results on RADIATE dataset. All methods are trained with the official set RADIATE Training Set in Good and Bad Weather and tested with RADIATE Test Set. The best results are in **bold**.

	Junction	Motorway	City	Night	Rain	Fog	Snow	overall
Lidar only	2.9	0.3	4.6	4.6	5	14.6	0.1	0.7
Radar only	57.5	**41.5**	**28.8**	31.3	31.3	34.8	3.1	3.1
MVDNet [10]	**67.5**	30.3	25.3	49.4	32.3	**42.3**	7.0	35.1
CMSCNet (Ours)	67.3	38.3	27.2	**50.6**	**32.6**	32.3	**7.6**	**36.0**

4.4 Qualitative Results

Except for the quantitative evaluation, we visualize some detection results in Fig. 6. From these figures, we can find that MVDNet [10] tends to generate false positives under the distraction by large particles like snow compared with other scenes. We consider that the design of MVDNet is more suitable for foggy

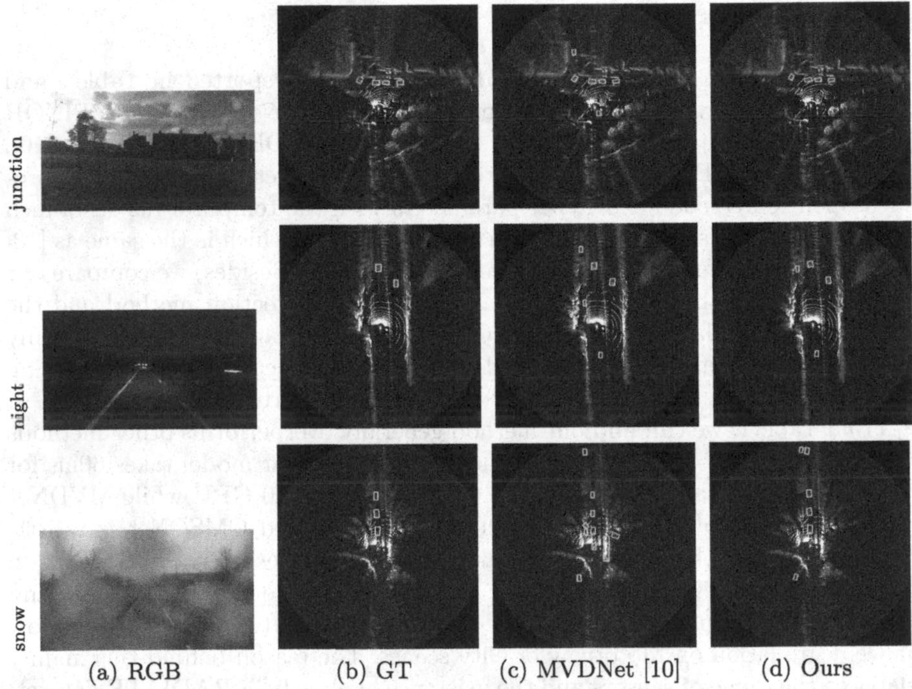

Fig. 6. Qualitative results on RADIATE dataset. (a) shows the weather conditions at that time. (b) (c) (d) are the overlapped images of Radar and Lidar BEV projection map with GT (yellow) or predicted (blue) car boxes. Due to the adverse weather, MVDNet is prone to false detection compared with our CMSCNet. Our network shows better performance in a wide range of situations.

weather and may meet challenges in severe weather conditions even the model is fine-trained on the new real dataset, while the proposed CMSCNet performs more generally.

Table 3. The ablation results on Oxford Radar Robotcat Dataset. It shows that all modules contribute to the detection performance.

Train	Clear+Foggy			
Test	Clear		Foggy	
IoU	0.5	0.65	0.5	0.65
w/o ID slice	89.8 (–0.8)	87.6 (–0.2)	84.3 (–0.2)	80.4 (–0.6)
w/o De-Ground	90.1 (–0.5)	87.2 (–0.6)	83.5 (–1.0)	79.7 (–1.3)
w/o CMSC loss	90.1 (–0.5)	87.2 (–0.6)	84.1 (–0.4)	80.3 (–0.7)
CMSCNet (Default)	**90.6**	**87.8**	**84.5**	**81.0**

4.5 Ablation Study

In Table 3, we show ablation studies results on ORR dataset to investigate how each module of CMSCNet contribute to the detection performance.

In the first experiment, we apply the traditional height-based slicing method, resulting in a decrease on AP compared to our ID-based method. The approach of height-based BEV generation splits point clouds along the vertical direction, which disrupts the continuity of the original Lidar laser points. In the next experiment, we only remove the De-Grounding module and retain the point cloud belonging to the ground. According to the results, removing CMSC loss also causes degradation for the lack of semantic consistency between independently learned features. Overall, the detection results show the effectiveness of the Laser-ID based slicing, Ground Removal and CMSC loss. in slight adverse weather and less relied on in snow weather.

5 Conclusion and Future Work

We propose a cross-modality fusion model named CMSCNet, which takes panoramic Lidar point clouds and Radar Range-Azimuth images as inputs, to robustly detect vehicles in various weather conditions. To reduce the modal differences between sensors, we introduce Knowledge Distillation loss to correlate the feature learning and solve multi-modal discrepancies. Furthermore, to represent point cloud data more naturally and effectively, Lidar is processed to remove the ground by RANSAC and then divided into 32-dimensional BEV images according to Laser IDs. We evaluate the model on real-world datasets and validate that our method is efficient in both simulation foggy conditions and real adverse weather conditions. Extensive ablation studies also confirm the practicality of the loss design and data process strategy. Although CMSCNet achieves relatively competitive results, the absolute performance is still limited in some weather conditions such as snow and rainy. We plan to work on performance improvement by integrating some Lidar denoising methods (e.g., defog, derain and desnow), and other modalities such as RGB image, infrared and gated image in the future.

Acknowledgment. This work was supported in part by the National Natural Science Foundation of China under Grant 62306059 and Fundamental Research Funds for the Central Universities under Grant DUT21RC(3)028

References

1. Barnes, D., Gadd, M., Murcutt, P., Newman, P., Posner, I.: The oxford radar robotcar dataset: a radar extension to the oxford robotcar dataset. In: IEEE International Conference on Robotics and Automation (ICRA), pp. 6433–6438 (2020)
2. Bijelic, M., et al.: Seeing through fog without seeing fog: deep multimodal sensor fusion in unseen adverse weather. In: 2020 IEEE/CVF Conference on Computer Vision and Pattern Recognition (CVPR), pp. 11679–11689 (2020)
3. Caesar, H., et al.: nuscenes: a multimodal dataset for autonomous driving. In: Proceedings of the IEEE/CVF Conference on Computer Vision and Pattern Recognition (CVPR), June 2020
4. Chen, X., Ma, H., Wan, J., Li, B., Xia, T.: Multi-view 3d object detection network for autonomous driving. In: Proceedings of the IEEE Conference on Computer Vision and Pattern Recognition (CVPR), July 2017
5. Kim, Y., Choi, J.W., Kum, D.: Grif net: gated region of interest fusion network for robust 3d object detection from radar point cloud and monocular image. In: 2020 IEEE/RSJ International Conference on Intelligent Robots and Systems (IROS), pp. 10857–10864 (2020)
6. Lang, A.H., Vora, S., Caesar, H., Zhou, L., Yang, J., Beijbom, O.: Pointpillars: fast encoders for object detection from point clouds. In: Proceedings of the IEEE/CVF Conference on Computer Vision and Pattern Recognition, pp. 12697–12705 (2019)
7. Li, Y.J., Park, J., O'Toole, M., Kitani, K.: Modality-agnostic learning for radar-lidar fusion in vehicle detection. In: Proceedings of the IEEE/CVF Conference on Computer Vision and Pattern Recognition (CVPR), pp. 918–927, June 2022
8. Major, B., et al.: Vehicle detection with automotive radar using deep learning on range-azimuth-doppler tensors. In: IEEE/CVF International Conference on Computer Vision (ICCV) Workshops, October 2019
9. Nabati, R., Qi, H.: Centerfusion: Center-based radar and camera fusion for 3d object detection. In: Proceedings of the IEEE/CVF Winter Conference on Applications of Computer Vision (WACV), pp. 1527–1536, January 2021
10. Qian, K., Zhu, S., Zhang, X., Li, L.E.: Robust multimodal vehicle detection in foggy weather using complementary lidar and radar signals. In: IEEE/CVF Conference on Computer Vision and Pattern Recognition (CVPR), pp. 444–453 (2021)
11. Ren, S., He, K., Girshick, R., Sun, J.: Faster R-CNN: towards real-time object detection with region proposal networks. In: Advances in Neural Information Processing Systems, vol. 28 (2015)

12. Sheeny, M., De Pellegrin, E., Mukherjee, S., Ahrabian, A., Wang, S., Wallace, A.: Radiate: a radar dataset for automotive perception in bad weather. In: 2021 IEEE International Conference on Robotics and Automation (ICRA), pp. 1–7 (2021)
13. Shi, S., et al.: PV-RCNN++: point-voxel feature set abstraction with local vector representation for 3d object detection. arXiv preprint arXiv:2102.00463 (2021)
14. Shi, S., Wang, X., Li, H.: PointRCNN: 3d object proposal generation and detection from point cloud. In: The IEEE Conference on Computer Vision and Pattern Recognition (CVPR), June 2019
15. Simonyan, K., Zisserman, A.: Very deep convolutional networks for large-scale image recognition. arXiv preprint arXiv:1409.1556 (2014)
16. Wang, Y., Jiang, Z., Gao, X., Hwang, J.N., Xing, G., Liu, H.: RodNet: radar object detection using cross-modal supervision. In: IEEE/CVF Winter Conference on Applications of Computer Vision (WACV), pp. 504–513, January 2021
17. Wu, Y., Kirillov, A., Massa, F., Lo, W.Y., Girshick, R.: Detectron2. https://github.com/facebookresearch/detectron2 (2019)
18. Yang, B., Guo, R., Liang, M., Casas, S., Urtasun, R.: RadarNet: exploiting radar for robust perception of dynamic objects. In: Vedaldi, A., Bischof, H., Brox, T., Frahm, J.-M. (eds.) ECCV 2020. LNCS, vol. 12363, pp. 496–512. Springer, Cham (2020). https://doi.org/10.1007/978-3-030-58523-5_29
19. Yang, B., Luo, W., Urtasun, R.: Pixor: real-time 3d object detection from point clouds. In: Proceedings of the IEEE Conference on Computer Vision and Pattern Recognition (CVPR), June 2018
20. Yang, Z., et al.: Focal and global knowledge distillation for detectors. In: Proceedings of the IEEE/CVF Conference on Computer Vision and Pattern Recognition (CVPR), pp. 4643–4652 (2022)
21. Zhu, X., et al.: Cylindrical and asymmetrical 3d convolution networks for lidar-based perception. IEEE Trans. Pattern Anal. Mach. Intell. **44**(10), 6807–6822 (2021)

Enhancing Active Visual Tracking Under Distractor Environments

Qianying Ouyang[1,2], Chenran Zhao[2,3], Jing Xie[1,2], Zhang Biao[2,3], Tongyue Li[1,2], Yuxi Zheng[1,2], and Dianxi Shi[1,2](✉)

[1] Intelligent Game and Decision Lab(IGDL), Beijing, China
{oyqy,dxshi}@nudt.edu.cn
[2] Tianjin Artificial Intelligence Innovation Center (TAIIC), Tianjin, China
[3] College of Computer, National University of Defense Technology, Changsha, China

Abstract. Active Visual Tracking (AVT) faces significant challenges in distracting environments characterized by occlusions and confusion. Current methodologies address this challenge through the integration of a mixed multi-agent game and Imitation Learning(IL). However, during the IL phase, if the training data of students generated by the teacher lacks diversity, it can lead to a noticeable degradation in the performance of the student visual tracker. Furthermore, existing works neglect visual occlusion issues from distractors beyond the collision distance. To enhance AVT performance, we introduce a novel method. Firstly, to tackle the limited diversity issue, we propose an intrinsic reward mechanism known as Asymmetric Random Network Distillation (AS-RND). This mechanism fosters target exploration, augmenting the variety of states among trackers and distractors, thereby enriching the heterogeneity of the visual tracker's training data. Secondly, to address visual occlusion, we present a distractor-occlusion avoidance reward predicated on the positional distribution of the distractors. Lastly, we integrate a classification score map prediction module to bolster the tracker's discriminative abilities. Experiments show that our approach significantly outperforms previous AVT algorithms in a complex distractor environment.

Keywords: Active Tracking · Reinforcement Learning · Imitation Learning · Dynamic Obstacle Avoidance

1 Introduction

Active Visual Tracking (AVT) controls the tracker's movement to ensure the target is always in the center of the field of view. As an emerging research direction of computer vision, AVT has many application scenarios in autonomous

---

This work was supported by the Science and Technology Innovation 2030 Major Project under Grant No.2020AAA0104802 and the National Natural Science Foundation of China(Grant No.91948303). Supplementary material is checked from URL.

Supplementary Information The online version contains supplementary material available at https://doi.org/10.1007/978-981-99-8435-0_36.

Q. Liu et al. (Eds.): PRCV 2023, LNCS 14427, pp. 452–464, 2024.
https://doi.org/10.1007/978-981-99-8435-0_36

driving, robotics, and other fields. However, executing AVT in dynamic distractor environments presents challenges such as target occlusion, confusion, and unpredictable movements of targets and distractors.

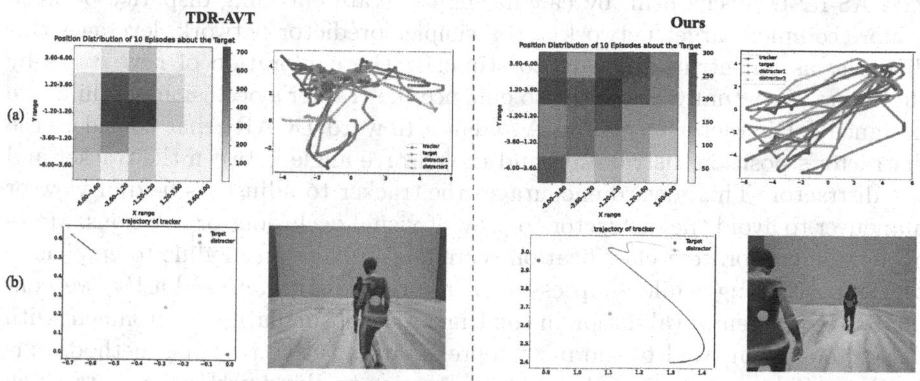

Fig. 1. (a) **Diversity deficit in agent states.** Depiction of target agent positions over ten episodes and agents' trajectories within a single episode. Predominant confinement of the target to specific areas results in other agents experiencing a lack of diverse states. (b) **Visual occlusion from the out-of-range distractor.** Distractor was set between the tracker and stationary target, beyond collision range. We visualized teacher tracker trajectories and student tracker views in this scenario. The results reveal that the teacher tracker of TDR-AVT remains nearly stationary throughout the game, neglecting the handling of occlusions.

To tackle the challenges mentioned above, recent work has made remarkable progress [12,14]. Zhong *et al.* [14](simply as the TDR-AVT) built a multi-agent game wherein the tracker can engage in adversarial reinforcement learning with the distractors and target. More importantly, TDR-AVT introduced cross-modal teacher-student learning from imitation learning [11], which involves training a 2D ground-state tracker as the teacher to guide the learning of the high-dimensional visual tracker. This method dramatically improves the robustness of visual trackers.

However, there are inherent limitations in imitation learning and adversarial learning [11,14]. Firstly, the performance of the visual tracker is directly contingent upon the quality of samples acquired by the teacher network [10]. As shown in Fig. 1 (a), We visualized the position distribution of the target across ten episodes and the trajectory within an episode. The insufficient diversity of target states within the teacher network restricts the variety in adversary tracker [13] and cooperative distractor behavior, leading to homogeneous training data for the student tracker. Additionally, as depicted in Fig. 1 (b), the 2D teacher tracker addresses only distractors within the collision range [14,15], neglecting occlusion caused by distractors beyond this range in 3D scenarios, thereby leaving the visual tracker without a strategy to tackle such issues.

To address these problems, we propose an approach to enhance the robustness of AVT [14]. Firstly, to tackle the issue of training data homogeneity resulting from the lack of diversity in target states, we propose an RND-based [1] intrinsic reward mechanism, the Asymmetric Random Network Distillation (AS-RND). The AS-RND mechanism, by calculating the state encoding disparity between a more complex target network and a simpler predictor network, leverages this disparity as an intrinsic reward to stimulate the exploration of new states by the target perpetually. Secondly, to enhance the tracker's occlusion handling, we designed a Distractor-Occlusion Avoidance Reward(DOAR) that considers the distractor's position distribution and the relative angle between the tracker and the distractor. This reward encourages the tracker to adjust its field of view or maneuver to avoid the distractor to prevent visual occlusions or collisions. Moreover, we incorporate a classification score map prediction module to emphasize target positioning while suppressing non-target disturbances. Lastly, we conducted experimental validation in the UnrealCV [9] simulation environment with distractors. Compared to the most representative active tracking methods currently available, our approach achieved superior tracking performance, tracking time, and success rate. Therefore, our work's contributions can be summarized as follows:

1. We introduced AS-RND, an intrinsic reward method promoting target exploration, leading to diversified training data and improved generalization of trackers.
2. We designed a distractor-occlusion avoidance reward and a classification score map prediction module to reduce occlusions and enhance the state representation of the visual tracker.
3. In normal and unseen environments with distractors, we conducted experimental validation of the efficacy and generalizability of the method.

2 Related Work

2.1 Active Visual Tracking

Active Visual Tracking (AVT) is primarily categorized into two-stage and end-to-end methods. Two-stage methods [4] initially utilize a *passive tracker*, such as DaSiamRPN [16], to detect the target's position, followed by applying a PID controller to manage the tracker's actions. Contrarily, end-to-end methods employ reinforcement learning to map direct visual images to actions directly [6]. To track complex targets more effectively, AD-VAT [13] and E-VAT [2] incorporate adversarial reinforcement learning, conceptualizing the interaction between trackers and targets as an adversarial relationship to enhance tracking performance. In this scenario, the target is trying to learn how to escape the tracker while the tracker is trying to keep up with the target. As for a more complex distracting environment, Unlike Xie *et al.* [12], who emphasized optimizing visual perception, Zhong *et al.* [14] advocate for a greater focus on enhancing tracking strategy. They regard tracker, target, and distractors as a mixed competitive cooperative game and employ imitation learning to accelerate the policy learning of the visual tracker. Despite mitigations in existing methods, improvements remain for handling dynamic target movements, occlusions, and confusion. Therefore, we propose an approach integrating an intrinsic reward for target exploration, a distractor-occlusion reward for avoidance, and a classification map prediction for confusion resistance, aiming to further boost visual tracking performance.

2.2 Curiosity Exploration

Exploration is a crucial reinforcement learning component, encouraging the agent to explore novel states in the environment. ICM [8] introduced the exploration strategy based on prediction error for the first time. Subsequently, RND [1] further evolved this strategy by measuring the novelty of states through the discrepancy between state encoding produced by a prediction network and a fixed target network. In this paper, we apply AS-RND to incentivize target behavior to be more flexible.

3 Method

We propose an enhanced active target tracking method based on cross modal teacher-student learning [14]. The method framework, as shown in Fig. 2, comprises a 2D teacher network trained using reinforcement learning and a student visual tracker. The teacher tracker ($i = 1$) engages in adversarial learning with cooperative targets ($i = 2$) and distractors ($i = 3$) and takes the ground relative positions as input, optimizing the network via A3C [7]. The visual tracker, on the other hand, uses the image frame as an observation input and imitates the actions of the teacher tracker via experience replay.

Based on this framework, we propose three components. (1) AS-RND: To alleviate the issue of a lack of training data diversity for the visual tracker, this mechanism takes the absolute position set of the target as input and uses the

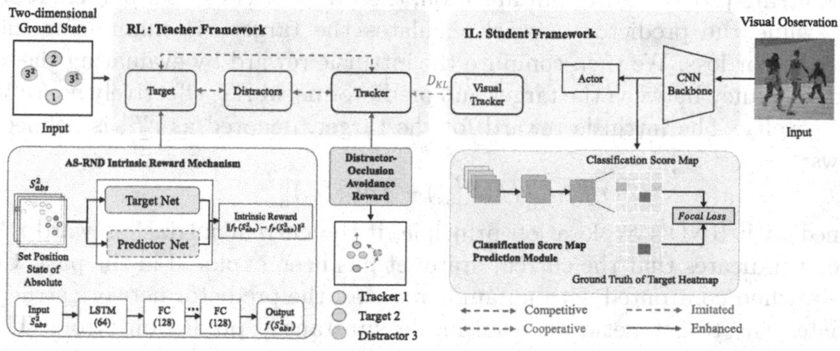

Fig. 2. Overview of the framework.

encoding differences between distinct predictor and target networks as an intrinsic reward to measure state novelty. (2)Distractor-Occlusion Avoidance Reward: Manages occlusions by prompting the tracker's view angle adjustments. (3)Classification Score Map Prediction Module: takes the dimension-reduced feature map as the score map and then uses Focal loss [5] to fit the ground target heatmap, thereby enhancing the discriminative capability of the visual tracker. These three mechanisms, acting in tandem with cross-modal teacher-student learning, collectively enhance the tracker's generalization performance.

3.1 AS-RND Intrinsic Reward Mechanism

We introduced AS-RND, an intrinsic reward mechanism aimed at promoting target exploration. This approach enhances diversity in both target states and behaviors of trackers and distractors, thereby enriching the training data for visual trackers. To begin, we will detail the input for the AS-RND. Unlike the traditional RND [1], AS-RND does not process image inputs but rather the continuous position information of the target. To better measure state novelty, it is recommended to avoid using single-step state algorithm results due to the small differences between continuous states. Thus, we convert the state of the target agent($i = 2$) within n steps into an absolute position state set $S^2_{abs} = \{s^2_{t-n}, s^2_{t-n+1}, \cdots, s^2_t\}$ as input.

Next, we elaborate on the network structure of AS-RND, which encompasses a complex target network f_T featuring three fully connected layers and a simpler predictor network f_P comprising two fully connected layers. The design of networks with different complexity mitigates the premature fitting of the predictor to the target network, preventing late-stage reward decay and thus preserving the exploration of the agent. Furthermore, recognizing the importance of temporal information in contiguous states, we incorporate an LSTM before the fully connected network to process continuous state data. Consequently, the encoder enables target and predictor networks to map S^2_{abs} into respective state encoding: $f_T\left(S^2_{abs}\right)$ and $f_P\left(S^2_{abs}\right)$.

During AS-RND training, we initialize the weights of the target and predictor networks within distinct ranges [3]. This disparity in initial weights ensures a slower fitting process. After initialization, the target network parameters remain fixed, while the predictor network emulates the target via minimizing mean squared error loss. We then compute the intrinsic reward by evaluating the output discrepancy between the target and predictor networks, effectively measuring state novelty. The intrinsic reward for the target, denoted as r^{in}_2, is defined as follows:

$$r^{in}_2 = \left\| f_T(S^2_{abs}) - f_P(S^2_{abs}) \right\|^2, \tag{1}$$

Aligned with RND's exploration principle, if the current intrinsic reward r^{in}_2 is small, it indicates that the current state set has been explored in the past. Conversely, when confronted with unfamiliar states, the predictor network struggles to match the target network, resulting in substantial prediction errors. These errors correspond to high intrinsic rewards, incentivizing the target to explore new states continuously.

3.2 Distractor-Occlusion Avoidance Reward

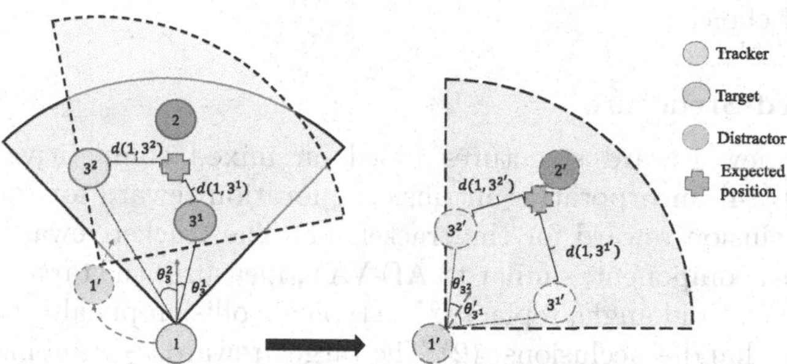

Fig. 3. It depicts a top-down view centered around tracker 1, with green representing the ideal position of the target within the tracker's field of view. The color intensity of the distractor signifies its distance and angle $d(1, 3^j)$ from the ideal target position. Our reward steers tracker 1 to a new tracking perspective at 1' during severe occlusions. (Color figure online)

When distractors appear between the tracker and the target, as shown in Fig. 3, the tracker should actively avoid them to reduce visual obstruction. For simplicity, only the current state distractor position distribution is considered, ignoring future motion impact. In addition, the risk of distractors' emergence obstruction is only considered when they appear within the tracker's field of view $\theta_{vis} = \pm 45°$, and their distance from the tracker is less than that between the tracker and the target. To ensure high-quality tracking while avoiding distractors, we advise the tracker to adjust the relative angle θ_3^j with the distractor primarily, minimizing their presence in the field of view. Complex evasive maneuvers should only be employed when the relative angle is small, indicating severe occlusion. When multiple distractors are present, the tracker should consider their position distribution to prioritize which distractor to avoid first. Inspired by distraction-aware reward, the relative distance and angle $d(1, 3^j)$ [14] between the distractor and expected target position reflects the distractor's impact. The closer the distractor is to the expected position, the more severe the obstruction or confusion. Therefore, combining θ_3^j and $d(1, 3^j)$, the distractor-occlusion reward calculation is:

$$r_1^{avoid} = \sum_{j=1...n} w_3^j * \frac{\left|\theta_3^j - \theta_{vis}\right|}{\theta_{vis}}, \tag{2}$$

w_3^j represents the distractor's impact on the tracker and is followed as:

$$w_3^j = \frac{1}{1 + e^{(-\frac{1}{d(1,3^j)})}} \quad w \sim [0, 1], \tag{3}$$

Utilizing the *sigmoid* function's properties, $d(1, j)$is mapped as an impact factor, effectively reflecting the distractor's influence. In summary, the error term in

r_1^{avoid} of $\frac{|\theta_3^j - \theta_{vis}|}{\theta_{vis}}$ indicates the need for the tracker to change angles to bypass the distractor. w_3^j reflects the distractor's significance, jointly determining the tracker avoid choice.

3.3 Reward Structure

We propose new reward structures based on mixed competitive-cooperative game reward [14], incorporating intrinsic exploration reward for the target and distractor-occlusion reward for the tracker. (1) The tracker reward r_1 tracking quality reward component, similar to AD-VAT, measures the target-to-expected position distance and angle, replacing the typical collision penalty $r_{collison} = -1$ with r_1^{avoid} to handle occlusions. (2) The target reward r_2 combines an escape reward from the zero-sum game with the tracker and an intrinsic reward encouraging new state exploration. (3) The distractor reward r_3^j, inspired by distractor-aware reward, encourages distractors to approach the expected position, distracting the tracker. Thus, the reward function is defined as follows:

$$\begin{aligned} r_1 &= 1 - d(1,2) - r_1^{\text{avoid}}, \\ r_2 &= -r_1 + r_2^{\text{in}} - r_{\text{collision}}, \\ r_3^j &= r_2 - d(1,j) - r_{\text{collision}}. \end{aligned} \tag{4}$$

3.4 Classification Score Map Prediction Module

In the student visual tracker training, it is not only to use the KL loss $\mathcal{L}_{KL}$ [14] to mimic the actions of the teacher tracker but also to adopt measures to reduce visual confusion caused by distractors. Furthermore, we employ the weighted Focal loss $\mathcal{L}_{heat}$ [5] to suppress distractor response points on the classification confidence map, aiming to emphasize the target's center point. Specifically, we use the ground target center $\hat{p} = [x, y]$ and its corresponding low-resolution center $\tilde{p} = [\tilde{p}_x, \tilde{p}_y]$, using the Gaussian kernel to generate the ground heatmap $\hat{P}_{xy} = \exp(-((x - \tilde{p}_x)^2 + (y - \tilde{p}_y)^2)/(2\sigma_p^2))$, where σ is the standard deviation adapted to the target size. We convert the high-dimensional features into a single-channel classification score map P_{xy} using the dimension-reducing network. The classification score map signifies the network's confidence level in determining whether each pixel belongs to the target or other elements (distractors or background). Gradual dimension reduction avoids excessive loss of feature information caused by direct pooling. Finally, we compare the error between the classification score and ground heatmap on a per-pixel basis so that $\mathcal{L}_{heat}$ can be expressed as:

$$\mathcal{L}_{heat} = -\sum_{x,y} \begin{cases} (1 - P_{xy})^{\alpha} \log(P_{xy}), & \text{if } \hat{P}_{xy} = 1 \\ (1 - \hat{P}_{xy})^{\beta} (P_{xy})^{\alpha} \log(1 - P_{xy}), & \text{otherwise} \end{cases}, \tag{5}$$

Finally, the overall loss of the student visual tracker is as follows:

$$\mathcal{L}_{all} = \lambda_1 \mathcal{L}_{KL} + \lambda_2 \mathcal{L}_{heat}, \tag{6}$$

where λ_1,λ_2 are weighing factors.

4 Experiments and Analysis

In this section, we commence by contrasting our approach with the baseline, performing ablation studies on each module, and examining the influence of the intrinsic reward mechanism on the target's exploration and the effectiveness of the distractor-occlusion avoidance reward for the tracker. Environment and evaluation metrics are included in the **Appendix. 2. and Appendix. 3.**

Table 1. Evaluating the active visual trackers on Simple Room with different numbers of scripted distractors

Methods	*Nav−1*			*Nav−2*			*Nav−3*			*Nav−4*		
	AR	EL	SR	AR	EL	SR	AR	EL	SR	AR	EL	SR
DaSiamRPN+PID	58	343	0.45	45	321	0.37	15	302	0.27	3	259	0.23
ATOM+PID	210	423	0.77	193	403	0.63	117	337	0.40	103	318	0.38
DiMP+PID	255	449	0.76	204	399	0.59	113	339	0.38	97	307	0.26
SARL	244	422	0.64	163	353	0.46	112	325	0.46	87	290	0.23
SARL+	226	407	0.60	175	370	0.52	126	323	0.38	22	263	0.15
AD-VAT	221	401	262	171	376	0.50	52	276	0.18	16	223	0.16
AD-VAT+	220	418	0.64	128	328	0.35	96	304	0.34	35	262	0.18
TDR-AVT	357	471	0.88	303	438	0.76	276	427	0.65	250	401	0.54
Ours	**386**	**485**	**0.92**	**351**	**461**	**0.86**	**314**	**443**	**0.78**	**296**	**430**	**0.66**

4.1 Comparison Experiments with Baseline

Basic Environment. In these experiments, we leverage standard AVT metrics including Accumulated Reward (AR), Episode Length (EL), and Success Rate (SR) for evaluation. We compare the tracker's performance with the baseline of two-stage [15,16], and end-to-end [6,13,14] methods, within environments with various quantities of distractors [9]. Among them, TDR-AVT [14], as the state-of-art method at present, is our most crucial comparison method. Despite modifying the reward structure, we maintained the AR calculation from TDR-AVT, ensuring a fair comparison. Table 1 shows our method notably surpassing both two-stage and end-to-end active tracking approaches, underlining the tracker's robustness.

Unseen Environment. We validated the tracker's transferability in unseen environments distinct from the training environment. As shown in Table 2, in the Urban City environment, our tracker performs exceptionally well. Despite the Parking Lot being rife with obstacles and identical-looking distractors, our approach maintains longer tracking times and higher success rates than TDR-AVT. These advantages stem largely from our novel distractor-occlusion reward to reduce the frequency of distractors in the tracker's field of view, and the classification score map, which notably improves the differentiation between distractors, background, and target.

4.2 Ablation Study

We discerned each module's effectiveness in ablation studies within a basic environment. In addition, strategies of the target and two distractors in Meta-2 are trained by the AS-RND method. As shown in Table 3, key findings include: (1) Excluding intrinsic reward for the target diminished the tracker's performance. It highlights the importance of diverse expert data in the imitation learning process. (2) As for the distractor-occlusion avoidance reward(DOAR), when the tracker cannot actively avoid distractors in its visual field and resorts to collision penalties alone for evasion, tracking time is negatively affected because heavy occlusion can easily lead to loss of the target. (3) The classification score map module proved vital in augmenting the tracker's resistance to confusion, especially in high distractor frequency settings such as Meta-2. To illustrate the importance of this module, as shown in Fig. 4, we conducted a visual analysis of activation maps for scenes in the tracker's field of view that simultaneously contain both distractors and targets. Comparatively speaking, our method

Table 2. Result of Tracker's Transferability to Unseen Environments

Methods	Urban City			Parking Lot		
	AR	EL	SR	AR	EL	SR
DaSiamRPN+PID	120	301	0.24	52	195	0.10
ATOM+PID	156	336	0.32	113	286	0.25
DiMP+PID	170	348	0.38	111	271	0.24
SARL	97	254	0.20	79	266	0.15
SARL+	74	221	0.16	53	237	0.12
AD-VAT	32	204	0.06	43	232	0.13
AD-VAT+	89	245	0.11	35	166	0.08
TDR-AVT	227	381	0.51	**186**	331	0.39
Ours	**271**	**420**	**0.66**	183	**343**	**0.42**

Table 3. Ablation Analysis of Visual Policy

Methods	$Nav-4$			$Meta-2$		
	AR	EL	SR	AR	EL	SR
Ours	**296**	**430**	**0.66**	**221**	**403**	**0.60**
w/o AS-RND	262	407	0.56	110	344	0.31
w/o DOAR	270	408	0.59	179	338	0.36
w/o score map	283	418	0.63	180	388	0.57
w/o all	250	401	0.54	86	310	0.24

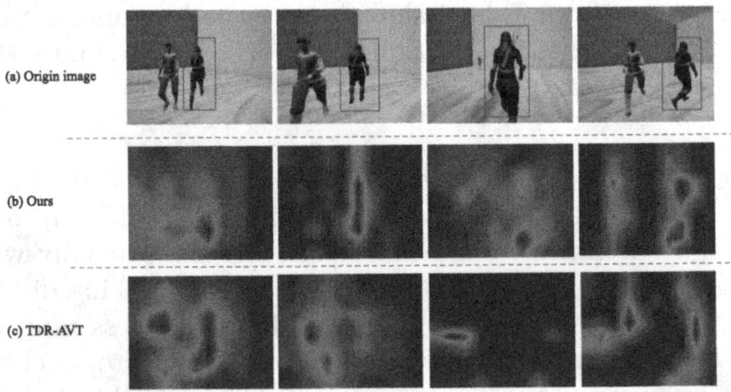

Fig. 4. Activation map visualization

effectively suppresses the feature representation of distractors and enhances the tracker's sensitivity to the target location. (4) When we removed all modules (equivalent to using the baseline [14]), we observed that it struggled to handle stronger targets. This emphasizes that a tracker's performance in adversarial learning often depends on the strength of the target. This indicates that AS-RND not only mitigates the problem of lack of diversity in training data during IL but also guides the tracker to be robust from an adversarial learning perspective.

In fact, experiments reveal that without diverse training data, modules like DOAR Reward and classification score map cannot function properly. This highlights the importance of expert data heterogeneity for IL. If AS-RND serves to increase the diversity of data, then DOAR works to optimize the tracking strategy. The classification score map plays a more vital role when the tracker cannot promptly avoid movable distractors. Together, these experiments form a progressively layered relationship, each addressing a different aspect of the problem.

4.3 Analysis of the Intrinsic Reward Mechanism of AS-RND

Validation of the Target Exploration Ability. Through ablation studies, the target flexibility crucially impacts tracker performance. Thus, we explored this via comparative experiments, using coverage rate (CR) and standard deviation of residence time distribution (RTD) as metrics. The Settings are as follows: (1) Without using the AS-RND mechanism, different action entropy values were set to investigate the impact of action entropy parameters on exploration capability. (2) A comparison was made between SY-RND with the same network structure and AS-RND with different network structures to verify the effectiveness of asymmetric network structures. The results, as shown in Fig. 5, indicate that simply increasing the action entropy is insufficient to enhance target flexibility while using AS-RND demonstrates a higher exploration rate of the targets.

Reward Signal Decay Issue. To address the problem of rapid decay of intrinsic reward signals for 2D targets, we conducted the following comparative experiments: 1) Different fully connected network depths were set to examine the influence of complex(lr=0.005) and simple(lr=0.0001) network structures on the experimental results. 2) Symmetric(lr=0.005) and asymmetric(lr=0.005) network structures were set to verify the necessity of asymmetric networks. 3) Similar and different network parameter initialization sampling distributions [3] were set to evaluate the impact of initialization on decay speed. According to the reward curve in Fig. 6, results showed that complex networks only guarantee higher intrinsic rewards in early training. Asymmetric networks capture state set differences in later stages, encouraging targets to visit less accessed states. Different weight initializations can modulate fitting speed and state differences.

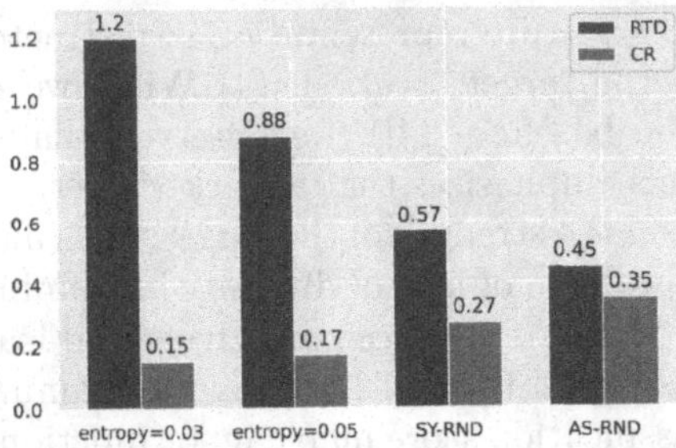

Fig. 5. Metrics over 100 games. RTD represents target residence time deviation (lower is better); CR signifies exploration rate (higher is better).

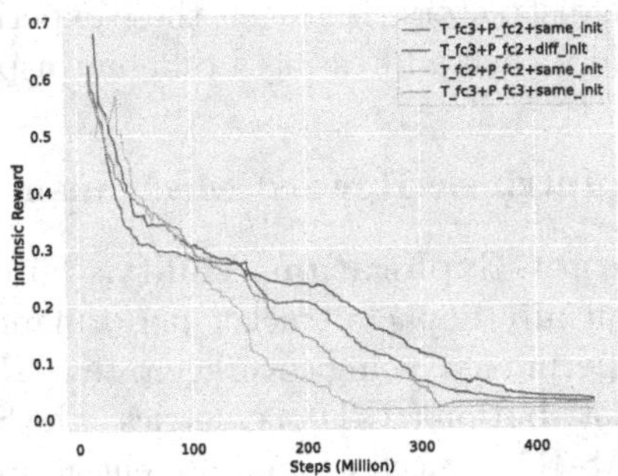

Fig. 6. The training curve for intrinsic reward, denoted as T (target network), P (predictor network), and n (number of fully connected layers

4.4 Analysis Experiment of Distractor-Occlusion Avoidance Reward

To validate the effectiveness of the distractor-occlusion avoidance reward, we conducted the following comparative experiments while training only the tracker: 1) Setting collision penalties with different collision distances, $d_{collide} = 50$ and $d_{collide} = 250$ (expected target distance), to assess the impact of collision distance on avoidance effectiveness and reward. 2) Comparing the use of Distractor-Occlusion Avoidance Reward based solely on Angle (DOAR-A) and considering Position distribution with Angle(DOAR-PA) to examine the influence of position distribution.

From the experiments, as shown in Fig. 7, with only a near-distance collision penalty, distractor frequency remains largely unchanged. Increasing the reward for safe collision distance and using the distractor-occlusion reward both help the tracker avoid the distractors. However, we also observe that although increasing the collision penalty distance results in better avoidance performance, it leads to a more severe decline in the tracker's reward. Trackers utilizing DOAR-A and DOAR-PA achieved similar avoidance results, but DOAR-PA maintained higher rewards compared to DOAR-A. This is due to the influence of the position distribution weight in DOAR-PA, which generally guides the tracker to adjust its orientation on a small scale when stationary to prevent distractors from appearing in the view. Only under severe occlusion, does it receive a larger penalty, promoting evasive actions. However, DOAR-A tends to circumvent distractors for a better view, leading to unnecessary tracking time.

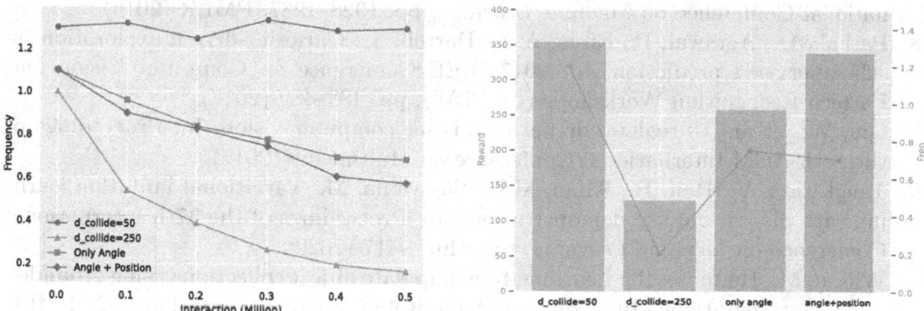

Fig. 7. Left: The frequency curves of distractors during the training. **Right:** The accumulative reward and distractor frequency statistics of the teacher tracker.

5 Conclusion

In this paper, we introduce a method to enhance AVT in distractor environments. This method consists of an intrinsic reward mechanism, AS-RND, to encourage target exploration and mitigate data diversity issues in imitation learning; a distractor-occlusion reward for tackling visual occlusion issues; and a classification map prediction module to enhance the tracker's ability to resist confusion. Experimental results demonstrate the enhanced robustness of our tracker over the baseline in distractor-rich UnrealCV environments. Future work may explore active tracking in more complex scenarios, potentially employing multi-sensor integration or stereo cameras to navigate distractors and obstacles.

References

1. Burda, Y., Edwards, H., Storkey, A.J., Klimov, O.: Exploration by random network distillation. ArXiv abs/1810.12894 (2018)
2. Dionigi, A., Devo, A., Guiducci, L., Costante, G.: E-vat: an asymmetric end-to-end approach to visual active exploration and tracking. IEEE Robot. Autom. Lett. **7**, 4259–4266 (2022)
3. He, K., Zhang, X., Ren, S., Sun, J.: Delving deep into rectifiers: surpassing human-level performance on ImageNet classification. In: Proceedings of the IEEE International Conference on Computer Vision, pp. 1026–1034 (2015)
4. Kristan, M., et al.: A novel performance evaluation methodology for single-target trackers. IEEE Trans. Pattern Anal. Mach. Intell. **38**, 2137–2155 (2015)

5. Lin, T.Y., Goyal, P., Girshick, R.B., He, K., Dollár, P.: Focal loss for dense object detection. In: 2017 IEEE International Conference on Computer Vision (ICCV), pp. 2999–3007 (2017)
6. Luo, W., Sun, P., Zhong, F., Liu, W., Zhang, T., Wang, Y.: End-to-end active object tracking and its real-world deployment via reinforcement learning. IEEE Trans. Pattern Anal. Mach. Intell. **42**(6), 1317–1332 (2019)
7. Mnih, V., et al.: Asynchronous methods for deep reinforcement learning. In: International Conference on Machine Learning, pp. 1928–1937. PMLR (2016)
8. Pathak, D., Agrawal, P., Efros, A.A., Darrell, T.: Curiosity-driven exploration by self-supervised prediction. In: 2017 IEEE Conference on Computer Vision and Pattern Recognition Workshops (CVPRW), pp. 488–489 (2017)
9. Qiu, W., et al.: Unrealcv: virtual worlds for computer vision. In: Proceedings of the 25th ACM International Conference on Multimedia (2017)
10. Tangkaratt, V., Han, B., Khan, M.E., Sugiyama, M.: Variational imitation learning with diverse-quality demonstrations. In: Proceedings of the 37th International Conference on Machine Learning, pp. 9407–9417 (2020)
11. Wilson, M., Hermans, T.: Learning to manipulate object collections using grounded state representations. In: Conference on Robot Learning, pp. 490–502. PMLR (2020)
12. Xi, M., Zhou, Y., Chen, Z., Zhou, W.G., Li, H.: Anti-distractor active object tracking in 3d environments. IEEE Trans. Circ. Syst. Video Technol. **32**, 3697–3707 (2022)
13. Zhong, F., Sun, P., Luo, W., Yan, T., Wang, Y.: Ad-vat: an asymmetric dueling mechanism for learning visual active tracking. In: International Conference on Learning Representations (2019)
14. Zhong, F., Sun, P., Luo, W., Yan, T., Wang, Y.: Towards distraction-robust active visual tracking. In: International Conference on Machine Learning, pp. 12782–12792. PMLR (2021)
15. Zhu, W., Hayashibe, M.: Autonomous navigation system in pedestrian scenarios using a dreamer-based motion planner. IEEE Robot. Autom. Lett. **8**, 3835–3842 (2023)
16. Zhu, Z., Wang, Q., Li, B., Wu, W., Yan, J., Hu, W.: Distractor-aware siamese networks for visual object tracking. In: Proceedings of the European Conference on Computer Vision (ECCV), pp. 101–117 (2018)

Cross-Modal and Cross-Domain Knowledge Transfer for Label-Free 3D Segmentation

Jingyu Zhang[1], Huitong Yang[2], Dai-Jie Wu[2], Jacky Keung[1], Xuesong Li[4], Xinge Zhu[3(✉)], and Yuexin Ma[2(✉)]

[1] City University of Hong Kong, Hong Kong SAR, China
[2] School of Information Science and Technology, ShanghaiTech University, Shanghai, China
mayuexin@shanghaitech.edu.cn
[3] Chinese University of Hong Kong, Hong Kong SAR, China
zhuxinge123@gmail.com
[4] College of Science, Australian National University, Canberra, Australia

Abstract. Current state-of-the-art point cloud-based perception methods usually rely on large-scale labeled data, which requires expensive manual annotations. A natural option is to explore the unsupervised methodology for 3D perception tasks. However, such methods often face substantial performance-drop difficulties. Fortunately, we found that there exist amounts of image-based datasets and an alternative can be proposed, *i.e.*, transferring the knowledge in the 2D images to 3D point clouds. Specifically, we propose a novel approach for the challenging cross-modal and cross-domain adaptation task by fully exploring the relationship between images and point clouds and designing effective feature alignment strategies. Without any 3D labels, our method achieves state-of-the-art performance for 3D point cloud semantic segmentation on SemanticKITTI by using the knowledge of KITTI360 and GTA5, compared to existing unsupervised and weakly-supervised baselines.

Keywords: Point Cloud Semantic Segmentation · Unsupervised Domain Adaptation · Cross-modal Transfer Learning

1 Introduction

Accurate depth information of large-scale scenes captured by LiDAR is the key to 3D perception. Consequently, LiDAR point cloud-based research has gained popularity. However, existing learning-based methods [10,11,37] demand extensive training data and significant labor power. Especially for the 3D segmentation task, people need to assign class labels to massive amounts of points. Such high cost really hinders the progress of related research. Therefore, relieving the annotation burden has become increasingly important for 3D perception.

J. Zhang and H. Yang – Equal contribution.

Q. Liu et al. (Eds.): PRCV 2023, LNCS 14427, pp. 465–477, 2024.
https://doi.org/10.1007/978-981-99-8435-0_37

Recently, plenty of label-efficient methods have been proposed for semantic segmentation. They try to simplify the labeling operation by using scene-level or sub-cloud-level weak labels [24], or utilizing clicks or line segments as object-level or class-level weak labels [19,20]. However, these methods still could not get rid of the dependence on 3D annotations. Actually, annotating images is much easier due to the regular 2D representation. We aim to transfer the knowledge learned from 2D data with its annotations to solve 3D tasks. However, most related knowledge-transfer works [12,23,31] focus on the domain adaption under the same input modality, which is difficult to adapt to our task. Although recent works [27,35] begin to consider distilling 2D priors to 3D space, they use synchronized and calibrated images and point clouds and leverage the physical projection between two data modalities for knowledge transfer, which limits the generalization capability (Fig. 1).

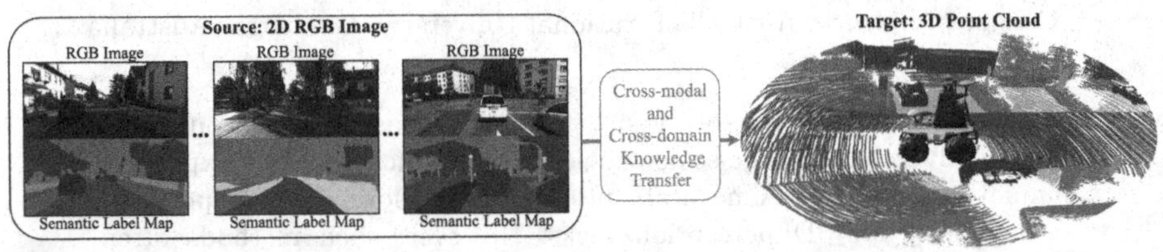

Fig. 1. Our method performs 3D semantic segmentation by transferring knowledge from RGB images with semantic labels from arbitrary traffic datasets.

In this paper, we propose an effective solution for a novel and challenging task, i.e., cross-modal and cross-domain 3D semantic segmentation for LiDAR point cloud based on unpaired 2D RGB images with 2D semantic annotations from other datasets. However, images and point clouds have totally different representations and features, making it not trivial to directly transfer the priors. To deal with the large discrepancy between the two domains, we perform alignment at both the data level and feature level. For data level (i.e., data representation level), we transform LiDAR point clouds to range images by spherical projection to obtain a regular and dense 2D representation as RGB images. For the feature level, considering that the object distributions and relative relationships among objects are similar in traffic scenarios, we extract instance-wise relationships from each modality and align their features from both the global-scene view and local-instance view through GAN. In this way, our method makes 3D point cloud semantic features approach 2D image semantic features and further benefits the 2D-to-3D knowledge transfer. To prove the effectiveness of our method, we conduct extensive experiments on KITTI360-to-SemanticKITTI and GTA5-to-SemanticKITTI. Our method achieves state-of-the-art performance for 3D point cloud semantic segmentation without any 3D labels.

2 Related Work

LiDAR-based Semantic Segmentation: Lidar-based 3D semantic segmentation [5,21] is crucial for autonomous driving. There are several effective representation strategies, such as 2D projection [7,33] and cylindrical voxelization [37]. However, most methods are in the full-supervision manner, which requires accurate manual annotations for every point, leading to high cost and time consumption. Recently, some research has been concentrated on pioneering labeling and training methods associated with weak supervision to mitigate the annotation workload. For instance, LESS [19] developed a heuristic algorithm for point cloud pre-segmentation and proposed a label-efficient annotation policy based on pre-segmentation results. Scribble-annotation [30] and label propagation strategies [28] were also introduced. However, these methods still require expensive human annotation on the point cloud. Several research studies [13,27,35] have explored complementary information to enhance knowledge transfer from images and facilitate 2D label utilization. Our work discards the need for any 3D labels by transferring scene and semantic knowledge from 2D data to 3D tasks, which reduces annotation time and enhances practicality for real-world scenarios.

Unsupervised Domain Adaptation (UDA): UDA techniques usually transfer the knowledge from the source domain with annotations to the unlabeled target domain by aligning target features with source features, which is most related to our task. For many existing UDA methods, the source and target inputs belong to the same modality, such as image-to-image and point cloud-to-point cloud domain adaptation. For image-based UDA, there are several popular methodologies for semantic segmentation, including adversarial feature learning [12,31,34], pseudo-label self-training [9,36] and graph reasoning [17]. For point cloud-based UDA, it is more challenging due to the sparse and unordered point representation. Recent works [2,23] have proposed using geometric transformations, feature space alignment, and consistency regularization to align the source and target point clouds. In addition, xMUDA [13] utilizes synchronized 2D images and 3D point clouds with 3D segmentation labels as the source dataset to assist 3D semantic segmentation. However, xMUDA assumes all modalities to be available during both training and testing, where we relax their assumptions by requiring no 3D labels, no synchronized multi-modal datasets, and during inference, no RGB data. Therefore, existing methods are difficult to extend to solve our cross-domain and cross-modality problem. To the best of our knowledge, we are the first to explore the issue and propose an effective solution.

3 Methodology

3.1 Problem Definition

For the 3D semantic segmentation task, our model associates a class label with every point in a 3D point cloud $X \in R^{N\times4}$, where N denotes the total number of points. Location information in 3D coordinates, and the reflectance intensity, are

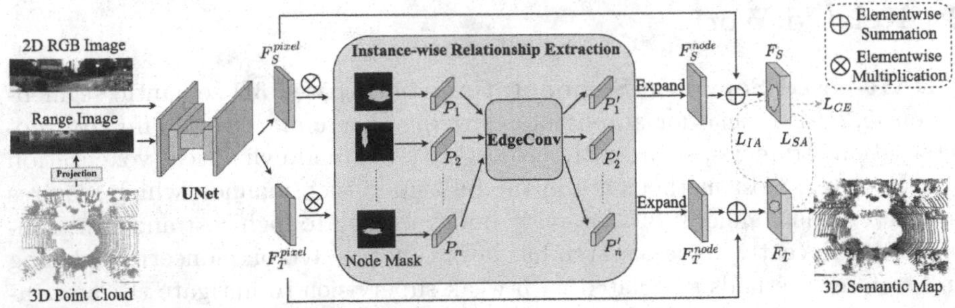

Fig. 2. The pipeline of our method with feature extraction module, instance-wise relationship extraction module, and feature alignment module (L_{IA}, L_{SA}).

provided for each point, denoted by $\{x, y, z, i\}$. Without any 3D annotations, we conduct 3D segmentation by using only 2D semantic information on RGB images from arbitrary datasets in similar scenarios. Unlike existing UDA methods that perform adaptation under the same data modality, our task is more challenging because our source and target data are from different domains and different modalities, and can be named as *Unsupervised Domain-Modality Adaptation (UDMA)*. UDMA transfers knowledge from 2D source domain (images $I_S \in R^{H\times W\times 3}$ with semantic labels $Y_S \in R^{H\times W\times C}$, where H and W are height and width of the image, and C is the number of classes) to 3D target domain (point clouds X_T). For UDMA 3D segmentation, we have $\{I_S, Y_S, X_T\}$ as the input data for training. During inference, only 3D point clouds (projected to range images) are input to the model to produce 3D predictions of the shape R^N (i.e., no RGB data are needed).

3.2 Method Overview

As shown in Fig. 2, our model consists of three important modules: Feature Extraction, Instance-wise Relationship Extraction, and Feature Alignment.

Feature Extraction: This module is used to encode various input data to a certain feature representation. We apply UNet [26] as the feature extraction backbone, where the size of the output feature map retains the same as the input image size, which allows us to obtain pixel-level dense semantic features.

Instance-Wise Relationship Extraction (IRE): This module carries the functionality of learning object-level features that encode local neighborhood information, therefore enhancing object-to-object interaction. The learned features are complementary to the mutually-independent pixel-level features learned by Feature Extraction module. Encoding more spatial information rather than domain- or modality-specific information enhances the efficiency of feature alignment.

Feature Alignment: To solve UDMA tasks (defined in Sect. 3.1), we need to narrow down the 2D-3D domain gap to transfer label information. We achieve this through adversarial learning [8] in both scene level and instance level, which learns to project inputs from different modalities to a common feature space and output domain- and modality-independent features.

3.3 Data Pre-processing

LiDAR Point Cloud Pre-segmentation: Point clouds often contain valuable semantic information that can be used as prior knowledge without the need for

Fig. 3. Visualization of Data-Pre-processing: Top: Point cloud pre-segmentation result. Bottom left: the range image after the spherical projection of the point cloud pre-segmentation. Bottom right: semantic labels in the RGB image. (Color figure online)

any training. We utilize a pre-segmentation method proposed by LESS [19] to group points into components such that each component is of high purity and generally contains only one object. The top image in Fig. 3 displays the pre-segmentation result. The approach involves using the RANSAC [6] for ground points detection. For non-ground points, an adaptive distance threshold is utilized to identify points belonging to the same component. We then utilize component information, such as mean or eigenvalues of point coordinates, to identify component classes. By using this heuristic approach we can successfully identify common classes (e.g., car, building, vegetation) with a high recall rate, and rare classes (e.g., person, bicycle) with a low recall rate. Due to the limited usefulness of low recall rate classes, we finally use three prior categories: car, ground (includes road, sidewalk, and terrain), and wall (building and vegetation) to assist the instance feature alignment in our method. In future work, we will explore the combination of learning-based and heuristic algorithms to improve the quantity and quality of the component class identification. For example, feature matching between unidentified and already identified components.

Range Image Projection: 2D and 3D data have huge a gap in their representations, e.g., structure, dimension, or compactness, which brings difficulties in knowledge transfer. To bridge the gap, we leverage another representation of 3D point cloud, range image [15,21,33]. 2D range images follow similar data representation as RGB (both represented in *height* × *width*). As shown in the bottom two plots in Fig. 3, the road (in gray and olive colors) is always located at the bottom of the picture with cars on top in both range and RGB images. The range image is generated by spherically projecting [33] 3D points onto a 2D plane:

$$\begin{aligned} u &= \frac{1}{2} \cdot (1 - arctan(y \cdot x^{-1})/\pi) \cdot U \\ v &= (1 - (\arcsin(z \cdot r^{-1}) + f_{down}) \cdot f^{-1}) \cdot V, \end{aligned} \tag{1}$$

where $u \in 1, ..., U$ and $v \in 1, ..., V$ indicate the 2D coordinate of the projected 3D point (x, y, z). $r = \sqrt{x^2 + y^2 + z^2}$ is the range of each LiDAR point. We set $U = 2048$ and $V = 64$, respectively, which is ideal for Velodyne HDL-64E laser scan from SemanticKITTI dataset [1]. 2D range images retain the depth information brought by LiDAR and provide rich image-level scene structure information. This helps in scene alignment. During training, we use $[r, r, i]$ for three input channels of range images, where r and i denote the range value defined above and the reflectance intensity, respectively.

3.4 Instance-Wise Relationship Extraction

Based on the consideration that traffic scenarios share similar object location distribution and object-object interaction, we extract relationship features by proposing a novel graph architecture connecting instances as graph nodes. Therefore, it enhances the feature information with object-object interaction details and benefits the following feature alignment. Our Instance-wise Relationship Extraction (IRE) consists of two parts: Instance Node Construction and Instance-level Correlation.

Instance Node Construction: With the motivation of learning object-object relationships in an image, we regard every instance as a node in the input image instead of the whole class as a node. For range images, we use pre-segmentation components as nodes, while for RGB images, we use ground truth to create node masks during training. Initial node descriptors P_i are computed as follows:

$$Agg(F^{pixel} \otimes Mask_i) = P_i, \quad \forall i = 1, ..., n \tag{2}$$

The operation is the same for source and target data. F^{pixel} represents the pixel-level dense feature map extracted by UNet, with the shape $R^{H \times W \times d}$ for source images (F_S^{pixel}) and $R^{V \times U \times d}$ for target images (F_T^{pixel}). Boolean matrices $Mask_i$ denote the binary node mask for the ith component, where the value 1 denotes the pixel belonging to this component. The mask has the shape $H \times W$ in the source domain and $V \times U$ in the target domain. We denote n by the total number of nodes in the current input image. With element-wise matrix multiplication operation $\otimes$, we obtain m_i nonzero pixels features for each node, with shape $R^{m_i \times d}, i = 1, ..., n$. We apply an aggregation function Agg to obtain a compact node descriptor $P_i \in R^d$. Agg is composed of a pooling layer followed by a linear layer. Then, node descriptors are forwarded to EdgeConv to extract instance-level correlation.

Instance-Level Correlation: This submodule aims to densely connect the instance-level graph nodes and enhance recognizable location information for object interaction. After the Instance Node Construction, we embed local structural information into each node feature by learning a dynamic graph $G = (V, E)$ [32] in EdgeConv, where $V = \{1, ..., n\}$ and $E \subseteq V \times V$ are the vertices and edges. We build G by connecting an edge between each node P_i and its k-nearest neighbors (KNN) in R^d.

$$P_i' = \underset{\{j|(i,j)\in E\}}{\square} f_\Theta(P_i, P_j) \tag{3}$$

f_Θ learns edge features between nodes with learnable parameters Θ. All local edge features of P_i are then aggregated through an aggregation operation $\square$ (e.g., max or sum) to produce $P_i' \in R^d$ which is an enhanced descriptor for node i that contains rich local information from neighbors. Instance-level correlation helps the model extract domain- and modality-independent features by focusing more on spatial-distribution features between instances. We then expand the instance features by replicating each P_i' for m_i times to obtain node-level feature map F^{node} that is of the same size as F^{pixel} (note that we omit the subscript here since the operation is the same for source and target data). We concatenate pixel-wise feature F^{pixel} with instance-wise feature F^{node} to form F, which keeps both dense individual information and local-neighborhood information to assist the feature alignment in the next section.

3.5 Feature Alignment

To transfer the knowledge obtained from RGB images to range images, we make the feature of range images approach that of RGB images from two aspects: the global scene-level alignment and local instance-level alignment.

Scene Feature Alignment (SA): Traffic scenarios in range and RGB images follow similar spatial distribution and object-object interaction (e.g., cars and buildings are located above the road). Therefore, we align the global scene features for two modalities. Specifically, we implement the unsupervised alignment through Generative Adversarial Networks (GAN) [8]. Adversarial training of the segmentation network can decrease the feature distribution gap between the source and target domain. Given the ground truth source labels, we train our model with the objective of producing source-like target features. In this way, we can maximize the utilization of low-cost RGB labels by transferring them to the target domain. This can be realized by two sub-process [22]: train the generator (i.e., the segmentation network) with target data to fool the discriminator; train the discriminator with both target and source data to discriminate features from two domains. In Scene Feature Alignment, we use one main discriminator to discriminate global features of the entire scene.

Instance Feature Alignment (IA): Accurate classification for each point is critical for 3D semantic segmentation. We further design fine-grained instance-level feature alignment for two domains to improve the class recognition performance. To perform the instance-level alignment, we build extra category-wise discriminators to align source and target instances features. For target range images, as no labels are available, we use prior knowledge described in Sect. 3.3 to filter prior categories, while for RGB we use ground truth. Then, we compute target features and align them with the corresponding source features of each prior category.

3.6 Loss

Our model is trained with three loss functions, with one supervised Cross-Entropy loss (CE loss) on source domain and two unsupervised feature alignment losses. Formally, we compute the CE loss for each pixel (h, w) in the image and sum them over.

$$L_{CE}(I_S, Y_S) = -\sum_h \sum_w \sum_c Y_S^{(h,w,c)} \log P_S^{(h,w,c)}, \tag{4}$$

where $h \in \{1, ..., H\}$, $w \in \{1, ...W\}$, and class $c \in \{1, ...C\}$. P_S is the prediction probability map of the source image I_S. For Scene Feature Alignment (SA), the adversarial loss is imposed on global target and source features F_T, F_S:

$$L_{SA}(F_S, F_T; G, D) = -\log D(F_T) - \log D(F_S) - \log(1 - D(F_T)). \tag{5}$$

The output of discriminator $D(\cdot)$ is the domain classification, which is 0 for target and 1 for source. In SA, we use one main discriminator to discriminate range and RGB scene global features. For Instance Feature Alignment (IA),

$$L_{IA}(F_S^E, F_T^E; G, D^E) = \sum_{e=1}^{E} -y_T^e \log D^e(F_T^e) - y_S^e \log D^e(F_S^e) - y_T^e \log(1 - D^e(F_T^e)), \tag{6}$$

where F_T^e and F_S^e are the target and source features for each prior category e. $y_S^e = 1$ if category e occurs in the current source image, otherwise $y_S^e = 0$, similarly for y_T^e. Therefore, we only compute losses for the categories that are present in the current image.

We use fine-tuning to further improve the performance. We apply weak label loss [19] (Eq. 7) for the prior categories that contain more than one class and CE loss $L_{CE}(X_T^e, Y_T^e)$ for the prior category that has fine-grained class identification, where $X_T^e \in R^{1 \times n_e \times 3}$ and $Y_T^e \in R^{1 \times n_e \times C}$. In our experiments, the weak label loss is applied to ground and wall, and the CE loss is applied to car,

$$L_{weak} = -\frac{1}{n_e} \sum_{i=1}^{n_e} \log(1 - \sum_{k_{ic}=0} p_{ic}), \tag{7}$$

where n_e indicates the number of pixels in the range image belonging to this prior category. $k_{ic} = 0$ if the class c is not in this prior-category. p_{ic} is the predicted probability of range pixel i for class c. By using weak labels, we only punish negative predictions.

4 Experiments

4.1 Datasets

We evaluate our UDMA framework on two set-ups: *KITTI360* → *SemanticKITTI* and *GTA5* → *SemanticKITTI*. KITTI360 [18] contains more than 60k real-world RGB images with pixel-level semantic annotations of 37 classes. GTA5 [25] is a synthetic dataset generated from modern commercial video games, which consists of around 25k images and pixel-wise annotations of 33 classes. SemanticKITTI [1] provides outdoor 3D point cloud data with 360° horizontal field-of-view captured by 64-beam LiDAR as well as point-wise semantic annotation of 28 classes.

Table 1. The first row presents the results of a modern segmentation backbone [4]. The medium five rows are 2D UDA methods [12,16,29,31,34]. Ours is the result of our trained model with fine-tuning.

(a) KITTI360 → SemanticKITTI.

Method	Road	Sidewalk	Building	Vegetation	Terrain	Car	mIoU
DeepLabV2	17.96	6.79	5.34	33.24	1.07	0.71	10.85
AdvEnt	20.34	10.87	14.48	34.13	0.06	0.21	13.35
CaCo	27.43	8.73	18.05	34.65	6.15	0.43	15.91
CCM	0.01	5.63	**24.15**	20.83	0	7.11	9.62
DACS	19.35	7.44	0	0.13	0	6.64	5.59
DANNet	15.30	0	13.75	31.46	0	0	10.09
Ours	**45.05**	**21.50**	2.18	**45.99**	**4.18**	**28.49**	**24.57**

(b) GTA5 → SemanticKITTI.

Method	Road	Sidewalk	Building	Vegetation	Terrain	Car	mIoU
DeepLabV2	18.87	0	24.19	17.46	0	0.97	10.24
AdvEnt	0.65	0	11.96	12.32	0	1.45	4.40
CaCo	12.06	13.85	17.21	21.69	3.14	2.61	11.76
CCM	0	2.18	20.47	20.14	0.27	0.83	7.32
DACS	23.73	0	17.02	0	0	0	6.79
DANNet	1.67	**18.28**	**28.19**	0	**12.17**	0	10.05
Ours	**37.63**	1.23	27.79	**31.40**	0.03	**30.00**	**21.35**

4.2 Metrics and Implementation Details

To evaluate the semantic segmentation of point clouds, we apply the commonly adopted metric, mean intersection-over-union (mIoU) over C classes of interests:

$$\frac{1}{C}\sum_{c=1}^{C}\frac{TP_c}{TP_c+FP_c+FN_c}, \tag{8}$$

where TP_c, FP_c, and FN_c represent the number of true positive, false positive, and false negative predictions for class c. We report the labeling performance over six classes that frequently occur in driving scenarios on SemanticKITTI validation set.

Implementation Details: We use the original size of KITTI360 and resize GTA5 images to [720,1280] for training. Our segmentation network is trained using SGD [3] optimizer with a learning rate 2.5×10^{-4}. We use Adam [14] optimizer with learning rate 10^{-4} to train the discriminators which are fully connected networks with a hidden layer of size 256.

4.3 Comparison

Comparison with UDA Methods: We evaluate the performance of adapting RGB images to range images under five 2D UDA state-of-the-art frameworks and one modern 2D semantic segmentation backbone (DeepLabV2). As shown in Table 1, our method outperforms all existing methods in terms of mIoU under both set-ups. Compared to AdvEnt, CaCo, and DANNet which perform scene-level alignment through either adversarial or contrastive learning, our method achieves 8.66 $\sim$ 14.48 mIoU performance gain in KITTI360 and 9.59 $\sim$ 16.95 mIoU gain in GTA5 in terms of mIoU. This is attributed to our instance-level alignment and instance-wise relationship extraction modules. As for CCM and DACS, they show inferior results on both source datasets (even compared to other 2D UDA methods). DACS mixes images across domains to reduce the domain gap, which is more compatible when both domains follow the same color system. However, in our case, mixing $[R, G, B]$ pixels with $[r, r, i]$ range images is unreasonable and may distort the feature space. Similarly, the horizontal layout matrix used in CCM may fail in our setting as we have different horizontal field-of-view in range (360°) and RGB images (90°). They focus more on data-level modification. Our method accomplishes both data-level and feature-level alignment. In our case, KITTI360 achieves higher mIoU, this may attribute to the fact that KITTI360 is using real-world images, whereas synthetic data in GTA5 may cause a large domain gap for learning.

Table 2. Comparison with two 3D baselines.

Method	Road	Side-walk	Building	Vegetation	Terrain	Car	mIoU
Cylinder3D	7.63	5.18	**9.30**	19.01	**26.39**	31.81	16.55
LESS	6.63	6.88	9.15	41.15	23.18	**39.42**	21.07
Ours	**45.05**	**21.50**	2.18	**45.99**	4.18	28.49	**24.57**

Table 3. Ablation Study: effectiveness of IRE, SA, IA.

Label	# of WeChat accounts
Identifiable	11 (13.4 %)
Partially anonymous	44 (53.7 %)
Anonymous	23 (28.0 %)
Unclassifiable	4 (4.9 %)

Comparison with 3D Segmentation Methods: We evaluate Cylinder3D [37] and LESS [19] with prior labels obtained from pre-segmentation in Table 2. Cylinder3D is trained using CE loss for car and weak label loss for ground and wall. For LESS, we employ their proposed losses except L_{sparse}. Comparing with them, our method delivers a minimum of 3.50 mIoU increase in performance. 3D baselines may produce suboptimal outcomes as they depend on the number of accurate 3D label annotations, while our approach utilizes rich RGB labels. We note that 3D-based methods provide better results in some classes (e.g., cars), which may be due to that they leverage 3D geometric features and representations more effectively, while our range image projection causes spatial information loss. Future work will explore the possibility of modifying our framework to function directly on the 3D point cloud without the need for projection.

4.4 Ablation Studies

Thorough experiments on KITTI360 → SemanticKITTI are conducted in Table 3 for the effectiveness of network modules and fine-tuning. w/o denotes deleting the corresponding module in our network and * denotes adding fine-tuning. We validate the effectiveness of *IRE*, *SA*, and *IA* based on their 6.95, 3.66, and 2.17 boost in mIoU, respectively (compared to 22.53). Through fine-tuning, we obtain another +2.04 in mIoU.

We also analyze how the number of training data in the source domain affects the result in the target domain in Fig. 4 under the setting KITTI360 → SemanticKITTI.

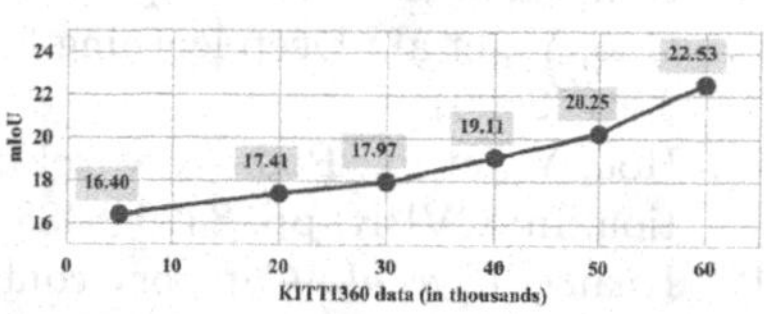

Fig. 4. Analysis for different numbers of training data.

The different number of training data is randomly sampled from KITTI360 with the same training protocol. The mIoU is lifted by 6.13 with $55k$ more source data. More data allow the model to learn more comprehensive scene features capturing object interaction in driving scenarios. This demonstrates the potential efficacy of our proposed method, where further improvement in 3D mIoU is expected to be achieved with more annotated 2D training data.

5 Conclusion

In this work, we propose and address the cross-modal and cross-domain knowledge transfer task for 3D point cloud semantic segmentation from 2D labels to alleviate the burden of 3D annotations. We tackle the challenges via two complementary modules, Instance-wise Relationship Extraction and Feature Alignment. Extensive experiments illustrate the superiority and potential of our method.

Acknowledgments. This work was supported by NSFC (No.62206173), Natural Science Foundation of Shanghai (No.22dz1201900).

References

1. Behley, J., et al.: Semantickitti: a dataset for semantic scene understanding of lidar sequences. In: ICCV, pp. 9297–9307 (2019)
2. Bian, Y., et al.: Unsupervised domain adaptation for point cloud semantic segmentation via graph matching. In: IROS, pp. 9899–9904. IEEE (2022)
3. Bottou, L.: Large-scale machine learning with stochastic gradient descent. In: Lechevallier, Y., Saporta, G. (eds.) Proceedings of COMPSTAT'2010. Physica-Verlag HD, pp. 177–186. Springer, Cham (2010). https://doi.org/10.1007/978-3-7908-2604-3_16

4. Chen, L.C., et al.: DeepLab: semantic image segmentation with deep convolutional nets, Atrous convolution, and fully connected CRFs. TPAMI **40**(4), 834–848 (2017)
5. Cortinhal, T., et al.: SalsaNext: fast, uncertainty-aware semantic segmentation of lidar point clouds for autonomous driving (2020). arXiv:2003.03653
6. Fischler, M.A., et al.: Random sample consensus: a paradigm for model fitting with applications to image analysis and automated cartography. CACM **24**(6), 381–395 (1981)
7. Gerdzhev, M., et al.: Tornado-net: multiview total variation semantic segmentation with diamond inception module. In: ICRA, pp. 9543–9549. IEEE (2021)
8. Goodfellow, I., et al.: Generative adversarial networks. CACM **63**(11), 139–144 (2020)
9. Guo, X., et al.: SimT: handling open-set noise for domain adaptive semantic segmentation. In: CVPR, pp. 7032–7041 (2022)
10. Guo, Y., et al.: Deep learning for 3D point clouds: a survey. TPAMI **43**(12), 4338–4364 (2020)
11. Hou, Y., et al.: Point-to-voxel knowledge distillation for lidar semantic segmentation. In: CVPR, pp. 8479–8488 (2022)
12. Huang, J., et al.: Category contrast for unsupervised domain adaptation in visual tasks. In: CVPR, pp. 1203–1214 (2022)
13. Jaritz, M., et al.: xMUDA: cross-modal unsupervised domain adaptation for 3D semantic segmentation. In: CVPR, pp. 12605–12614 (2020)
14. Kingma, D.P., et al.: Adam: a method for stochastic optimization. arXiv preprint: arXiv:1412.6980 (2014)
15. Langer, F., et al.: Domain transfer for semantic segmentation of LiDAR data using deep neural networks. In: IROS, pp. 8263–8270. IEEE (2020)
16. Li, G., Kang, G., Liu, W., Wei, Y., Yang, Y.: Content-consistent matching for domain adaptive semantic segmentation. In: Vedaldi, A., Bischof, H., Brox, T., Frahm, J.-M. (eds.) ECCV 2020. LNCS, vol. 12359, pp. 440–456. Springer, Cham (2020). https://doi.org/10.1007/978-3-030-58568-6_26
17. Li, W., et al.: SIGMA: semantic-complete graph matching for domain adaptive object detection. In: CVPR, pp. 5291–5300 (2022)
18. Liao, Y., et al.: KITTI-360: a novel dataset and benchmarks for urban scene understanding in 2D and 3D. TPAMI **45**, 3292–3310 (2022)
19. Liu, M., et al.: Less: Label-efficient semantic segmentation for lidar point clouds. In: Avidan, S., Brostow, G., Cisse, M., Farinella, G.M., Hassner, T. (eds.) Computer Vision – ECCV 2022. Lecture Notes in Computer Science, vol. 13699, pp. 70–89. Springer, Cham (2022)
20. Liu, Z., et al.: One thing one click: a self-training approach for weakly supervised 3D semantic segmentation. In: CVPR, pp. 1726–1736 (2021)
21. Milioto, A., et al.: RangeNet++: fast and accurate lidar semantic segmentation. In: IROS, pp. 4213–4220. IEEE (2019)
22. Paul, S., Tsai, Y.-H., Schulter, S., Roy-Chowdhury, A.K., Chandraker, M.: Domain adaptive semantic segmentation using weak labels. In: Vedaldi, A., Bischof, H., Brox, T., Frahm, J.-M. (eds.) ECCV 2020. LNCS, vol. 12354, pp. 571–587. Springer, Cham (2020). https://doi.org/10.1007/978-3-030-58545-7_33
23. Peng, X., et al.: CL3D: unsupervised domain adaptation for cross-LiDAR 3D detection (2022). arXiv:2212.00244
24. Ren, Z., et al.: 3D spatial recognition without spatially labeled 3D. In: CVPR, pp. 13204–13213 (2021)

25. Richter, S.R., Vineet, V., Roth, S., Koltun, V.: Playing for data: ground truth from computer games. In: Leibe, B., Matas, J., Sebe, N., Welling, M. (eds.) ECCV 2016. LNCS, vol. 9906, pp. 102–118. Springer, Cham (2016). https://doi.org/10.1007/978-3-319-46475-6_7
26. Ronneberger, O., Fischer, P., Brox, T.: U-Net: convolutional networks for biomedical image segmentation. In: Navab, N., Hornegger, J., Wells, W.M., Frangi, A.F. (eds.) MICCAI 2015. LNCS, vol. 9351, pp. 234–241. Springer, Cham (2015). https://doi.org/10.1007/978-3-319-24574-4_28
27. Sautier, C., et al.: Image-to-lidar self-supervised distillation for autonomous driving data. In: CVPR. pp. 9891–9901 (2022)
28. Shi, H., et al.: Weakly supervised segmentation on outdoor 4D point clouds with temporal matching and spatial graph propagation. In: CVPR, pp. 11840–11849 (2022)
29. Tranheden, W., et al.: DACS: domain adaptation via cross-domain mixed sampling. In: WACV, pp. 1379–1389 (2021)
30. Unal, O., et al.: Scribble-supervised lidar semantic segmentation. In: CVPR, pp. 2697–2707 (2022)
31. Vu, T.H., et al.: ADVENT: adversarial entropy minimization for domain adaptation in semantic segmentation. In: CVPR, pp. 2517–2526 (2019)
32. Wang, Y., et al.: Dynamic graph CNN for learning on point clouds. TOG **38**(5), 1–12 (2019)
33. Wu, B., et al.: SqueezeSeg: convolutional neural nets with recurrent CRF for real-time road-object segmentation from 3D LiDAR point cloud. In: ICRA, pp. 1887–1893. IEEE (2018)
34. Wu, X., et al.: DanNet: a one-stage domain adaptation network for unsupervised nighttime semantic segmentation. In: CVPR, pp. 15769–15778 (2021)
35. Yan, X., et al.: 2DPASS: 2D priors assisted semantic segmentation on LiDAR point clouds. In: Avidan, S., Brostow, G., Cisse, M., Farinella, G.M., Hassner, T. (eds.) Computer Vision - ECCV 2022. Lecture Notes in Computer Science, vol. 13688, pp. 677–695. Springer, Cham (2022). https://doi.org/10.1007/978-3-031-19815-1_39
36. Zhang, P., et al.: Prototypical pseudo label denoising and target structure learning for domain adaptive semantic segmentation. In: CVPR, pp. 12414–12424 (2021)
37. Zhu, X., et al.: Cylindrical and asymmetrical 3D convolution networks for LiDAR segmentation. In: CVPR, pp. 9939–9948 (2021)

Cross-Task Physical Adversarial Attack Against Lane Detection System Based on LED Illumination Modulation

Junbin Fang[1,2,3], Zewei Yang[4], Siyuan Dai[2,5], You Jiang[1,2,3], Canjian Jiang[1,2,3], Zoe L. Jiang[6,7](✉), Chuanyi Liu[6,7], and Siu-Ming Yiu[8]

[1] Guangdong Provincial Key Laboratory of Optical Fiber Sensing and Communications, Guangzhou 510632, China

[2] Guangdong Provincial Engineering Technology Research Center on Visible Light Communication, and Guangzhou Municipal Key Laboratory of Engineering Technology on Visible Light Communication, Guangzhou 510632, China

[3] Department of Optoelectronic Engineering, Jinan University, Guangzhou 510632, China

[4] International Energy College, Jinan University, Zhuhai 519070, China

[5] Department of Electrical and Computer Engineering, University of Pittsburgh, Pittsburgh, PA 15213, USA

[6] School of Computer Science and Technology, Harbin Institute of Technology(Shenzhen), Shenzhen 518055, China
zoeljiang@hit.edu.cn

[7] Peng Cheng Laboratory, Shenzhen 518000, China

[8] Department of Computer Science, The University of Hong Kong, Hong Kong 999077, China

Abstract. Lane detection is one of the fundamental technologies for autonomous driving, but it faces many security threats from adversarial attacks. Existing adversarial attacks against lane detection often simplify it as a certain type of computer vision task and ignore its cross-task characteristic, resulting in weak transferability, poor stealthiness, and low executability. This paper proposes a cross-task physical adversarial attack scheme based on LED illumination modulation (AdvLIM) after analyzing the decision behavior coexisting in lane detection models. We generate imperceptible brightness flicker through fast intensity modulation of LED illumination and utilize the rolling shutter effect of the CMOS image sensor to inject brightness information perturbations into the captured scene image. According to different modulation parameters, this paper proposes two attack strategies: scatter attack and dense attack. Cross-task attack experiments are conducted on the Tusimple dataset on five lane detection models of three different task types, including segmentation-based models (SCNN, RESA), point detection-based model (LaneATT), and curve-based models (LSTR, BézierLaneNet). The experimental results show that AdvLIM causes a detection accuracy drop of 77.93%, 75.86%, 51.51%, 89.41%, and 73.94% for the above models respectively. Experiments are also conducted in the physical world, and the results indicate that AdvLIM can pose a real and significant security threat to diverse lane detection systems in real-world scenarios.

Q. Liu et al. (Eds.): PRCV 2023, LNCS 14427, pp. 478–491, 2024.
https://doi.org/10.1007/978-981-99-8435-0_38

Keywords: Autonomous driving · lane detection · physical adversarial attack · LED illumination modulation · cross-task

1 Introduction

With the empowerment of deep learning, autonomous driving has achieved rapid development and shown great application prospects. The lane detection system is one of the fundamental technologies in autonomous driving intelligence systems [1], which determines the vehicle's feasible driving area by detecting the lane markings. Lane detection based on deep neural networks (DNNs) has demonstrated superior performance [2], but it also faces many security threats due to the vulnerability of DNNs. As Kurakin et al. [3] revealed, the vulnerability of DNN can be exploited not only for digital adversarial attacks but also for physical adversarial attacks, thus threatening the security of the physical world. Physical adversarial attacks can cause lane detection systems to misidentify the lane markings, leading to serious driving safety threats such as incorrect merge, premature turning, and even entering the opposite lane.

AI vision systems in autonomous driving are often categorized as specific computer vision tasks, e.g., traffic sign recognition and pedestrian detection are considered object detection tasks [4,5]. Adversarial attacks targeting these systems can be easily converted into attacks against their corresponding computer vision tasks, achieving impressive effectiveness. However, lane detection is always modeled as different computer vision tasks, including semantic segmentation, point detection, and curve fitting [6], making it difficult to be unified. Existing adversarial attacks against lane detection ignore its cross-task characteristic, and the designed methods are always crude with poor stealthiness (such as patching or graffitiing the entire road [7,8]), and have weak transferability for diverse lane detection systems. Therefore, their practicality in the physical world is limited, and they cannot help autonomous driving identify potential vulnerabilities.

To address these issues, we analyze the decision behavior coexisting in lane detection models and point out the universal vulnerability in feature extraction. Based on this vulnerability, we propose a physical adversarial attack scheme based on LED illumination modulation (AdvLIM). We generate imperceptible brightness flicker through fast intensity modulation of LED illumination and utilize the rolling shutter effect of the CMOS image sensor to inject brightness information perturbations into the captured scene image, which has the advantages of strong transferability, good stealthiness, and high executability. According to different modulation parameters, we propose scatter attack and dense attack strategies. Cross-task attack experiments are conducted on the Tusimple dataset on five lane detection models of three different task types, including segmentation-based models (SCNN, RESA), point detection-based model (LaneATT), and curve-based models (LSTR, BézierLaneNet). The experimental results show that AdvLIM causes a detection accuracy drop of 77.93%, 75.86%, 51.51%, 89.41%, and 73.94% for the above models respectively. Experiments are also conducted in the physical world, and the results indicate that AdvLIM can pose a real and significant security threat to diverse lane detection systems in real-world scenarios. The main contributions of this paper are as follows:

1. This paper analyzes the decision behavior coexisting in lane detection models, pointing out the universal vulnerability of models in feature extraction.
2. This paper proposes a physical adversarial attack based on LED illumination modulation (AdvLIM), with the advantages of strong transferability, good stealthiness, and high executability. According to different modulation parameters, this paper proposes scatter attack and dense attack strategies.
3. This paper conducts attack experiments on SCNN, RESA, LaneATT, LSTR, and BézierLaneNet, in both the digital and physical world. The experimental results show that AdvLIM has high attack success rates on lane detection models of different task types, posing a real and significant security threat.

2 Related Work

Based on the powerful learning and generalization ability of DNN, researchers have proposed various lane detection models of different task types. Pan et al. [9] proposed SCNN based on the semantic segmentation task, which utilizes slice-by-slice convolution to pass spatial semantic relationships. Zheng et al. [10] introduced a variable stride spatial convolution, which enables the model to capture long-range slice information. Unlike the above-mentioned approach of determining lane area through pixel semantics, Tabelini et al. [11] modeled lane detection as the point detection task, using the angle of the lane and the offset of the sampling points as anchor boxes for regression, greatly improving the detection speed. In addition, Liu et al. [12] achieved lane detection through curve fitting and utilized a transformer to make the model focus more on global context information. Recently, Feng et al. [6] used the Bézier curve to fit the lane and achieved state-of-the-art accuracy on multiple datasets.

Regarding adversarial attacks against lane detection systems, Jing et al. [7] proposed a method to deceive lane detection systems by placing adversarial markings. This method enables the system to detect non-existent lane markings. However, such adversarial markings are unusual in real-world scenarios and can be easily detected by human observers. In addition, Sato et al. [8] proposed Dirty Road Patches. By pasting a large adversarial patch in the driving area, they caused a severe deviation in the detected lane markings. Although the generated dirty road patches resemble common road stains in real-world scenarios, the large adversarial patch is still conspicuous to human observers.

Given that the entrance device of a computer vision system (camera) is essentially an optoelectronic sensor, researchers have proposed some adversarial attacks that utilize illumination devices. Gnanasambandam et al. [13] fooled the image classifier by modulating the structured illumination. Duan et al. [14] successfully attacked the traffic sign recognition system by manipulating the parameters of the laser beam. However, these methods require specific illumination equipment and the perturbation patterns must be computed for different target objects, resulting in low attack executability and high computational cost.

3 Problem Analysis

As previously mentioned, lane detection can be modeled as a variety of computer vision tasks. Based on different modeling tasks, different loss functions have been adopted for the training process of lane detection models, resulting in the acquisition of different input-output mappings. Existing adversarial attacks against lane detection excessively rely on the input-output mappings of models to generate adversarial perturbations, limiting the generalization and transferability of adversarial samples between models of different task types.

To design a cross-task physical adversarial attack against lane detection, we analyze the decision behavior coexisting in models. As shown in Fig. 1, each convolutional layer of the model can be seen as a mask block extracting semantic features from the input sample. Each mask block will filter semantic features of the sample to activate the corresponding feature map while passing it to the next mask block. After the layer-by-layer filtering of mask blocks, the model will obtain the high-level semantic features from the final feature map for decision-making. Moreover, it is worth noting that early mask blocks may focus more on low-level features [15], which is a common characteristic among different models.

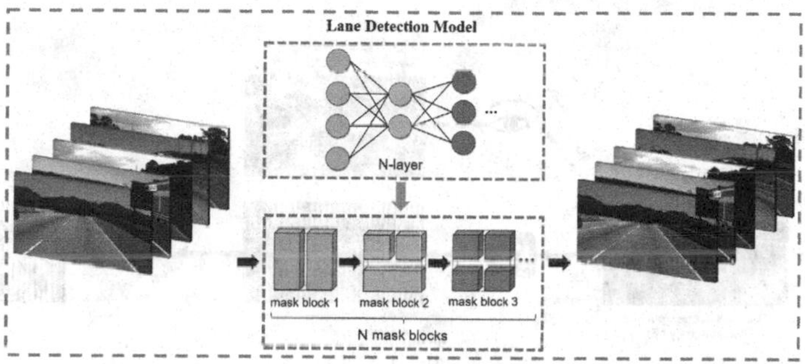

Fig. 1. The decision-making process of the lane detection model

Therefore, by disrupting semantic features in the input sample, thereby interfering with the extraction of features by the mask blocks and affecting the activation of subsequent feature maps, lane detection models of any task type will make erroneous decisions. To achieve this, we propose AdvLIM that can generate adversarial perturbations with both effectiveness and stealthiness based on LED illumination modulation to disrupt semantic features in the sample.

4 The Proposed Scheme

4.1 Threat Model

In our work, we assume that the attacker has zero knowledge of the lane detection system. In this assumption, the attacker has no access to structure or parameters related to the lane detection model but only knows the predicted key points of the lane markings. Furthermore, the attacker cannot manipulate the training process of the model. Instead, the attacker just simply manipulates the physical world setting to deploy an adversarial attack, and the goal of the attacker is to deceive the target model to make deviations in the predicted key points.

4.2 Physical Adversarial Attack Based on LED Illumination Modulation

Based on the vulnerability of lane detection models in feature extraction, we propose a physical adversarial attack based on LED illumination modulation (AdvLIM), as shown in Fig. 2. Firstly, we conduct high-speed on-off keying (OOK) modulation for the LED lamp, causing the environmental brightness to flicker rapidly beyond the perceptual limit of the human eye. Then, based on the rolling shutter effect of the CMOS image sensor, the brightness information perturbations are injected into the scene image during the imaging process, thus resulting in the occurrence of light-dark alternating adversarial stripes and realizing the generation of the adversarial example. The specific steps are as follows:

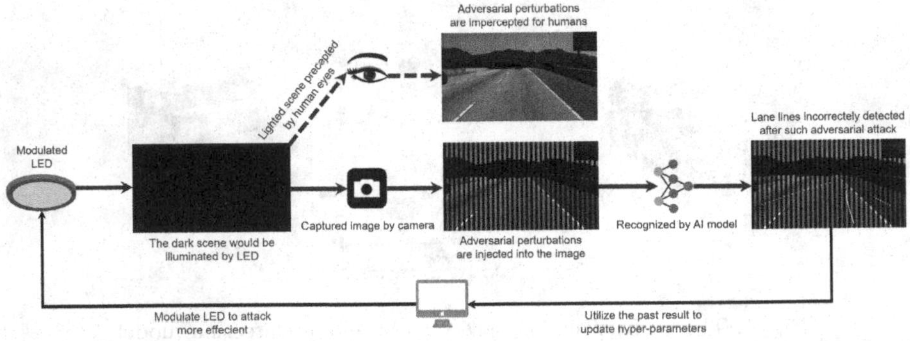

Fig. 2. The framework of the proposed AdvLIM on the lane detection system.

The Modulation of LED Illumination. We modulate the LED illumination at a frequency exceeding 150 Hz. Specifically, we modulate the binary pulse signal to control the on-off switching of the LED lamp, generating the brightness flicker. It is worth noting that the perceptual frequency limit of the human eye is approximately 100 Hz. Therefore, the human eye cannot perceive the brightness flicker at this frequency and can only see continuous and uniform brightness.

The Generation of Adversarial Sample. Based on the rolling shutter effect, the CMOS image sensor exposes the scanned row in sequence with a constant delay, as shown in Fig. 3. Although the human eye cannot perceive the high-frequency brightness flicker, this flicker will be captured by the CMOS image

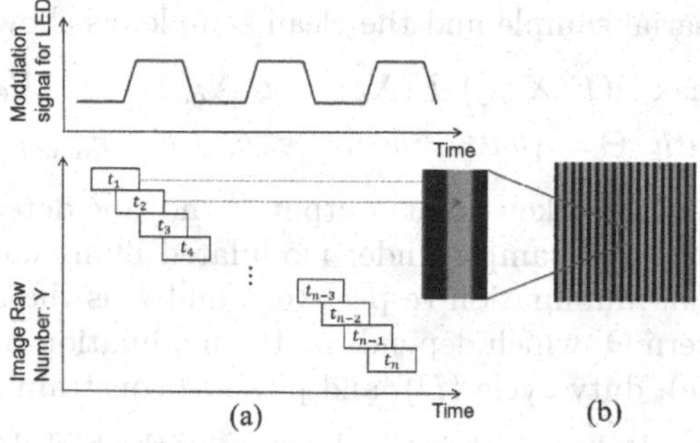

Fig. 3. Based on the rolling shutter effect, (a) and (b) refer to the exposure process under modulated LED illumination and the captured white paper respectively.

sensor with a higher scanning frequency. As shown in Fig. 3(a), at the moment of t_1, the CMOS image sensor exposes the first row of pixels, during which the LED lamp is turned off and dark pixels are stored in the current row of the image. At the moment of t_2, the CMOS image sensor exposes the second row of pixels, during which the LED lamp is turned on and light pixels are stored in the current row. This process continues, and the final captured scene image will carry horizontal light-dark stripes. Similarly, the scene image will carry vertical stripes when the CMOS image sensor images vertically, as shown in Fig. 3(b). Assuming that the pulse repetition period of the control signal for LED is T_p, the periodic width of adversarial stripes in Fig. 3(b) can be expressed as:

$$w_{stripe} = \frac{T_p}{t_d} = \frac{pw}{t_d \cdot D} \tag{1}$$

Where t_d represents the exposure time delay of the CMOS image sensor while pw and D represent the pulse width and duty cycle of the control signal for LED respectively. We can further divide the adversarial stripes within a period into the light and dark stripe, whose widths are expressed by the following formula.

$$w_l = \frac{pw}{t_d}, \quad w_d = \frac{pw \cdot (1 - D)}{t_d \cdot D} \tag{2}$$

From Eqs. (1) and (2), we can see that the pattern of adversarial stripes is related to the modulation parameters. That is, by adjusting the modulation parameters, the adversarial stripes with a specific pattern can be generated targeting the lane detection model. The destruction of semantic features caused by adversarial stripes will accumulate during the forward propagation of each mask block, leading to deviations in the predicted key points of the lane markings.

4.3 Search Space for Adversarial Perturbation Parameters

Given that the attacker can only access the predicted key points of lane markings, we define the goal of AdvLIM as maximizing the matching distance of key points between the adversarial sample and the clean sample, as shown in Eq. (3).

$$\begin{aligned} &\max \ d(F(X_{adv}), F(X)) \quad s.t. \ X_{adv} = X \circ R_{\Theta} \\ &with \ \ \Theta = \{\theta | \theta = [pw, D], \varepsilon_{\min} \leq \theta \leq \varepsilon_{\max}\} \end{aligned} \tag{3}$$

where $F(\cdot)$ is the predicted key points output of the lane detection model, X_{adv} and X are the adversarial sample under modulated illumination and the clean sample under normal illumination respectively, and R is the adversarial stripes with a specific pattern Θ, which depends on the modulation parameters, including pulse width (pw), duty cycle (D), and physical constraints (ε).

Pulse Width (pw). Pulse width is the duration of the high-level voltage within one repetition period. With a fixed duty cycle, a larger pulse width causes a wider periodic width of adversarial stripes, but with fewer periods.

Duty Cycle (D). Duty cycle is the ratio of the duration of the high-level voltage within one repetition period. By adjusting the duty cycle, we can modify the pattern of adversarial stripes, effectively improving the attack success rate.

Physical Constraints (ε). The performance of AdvLIM is subject to certain physical constraints. When modifying the duty cycle in the search space $\left\{D = \frac{1}{i+1} | 1 \leq i \leq 8, step = 1\right\}$, the adversarial stripes are blurry with the pulse widths below 90 μs while they are few in number with the pulse widths exceeding 1100 μs, which may lead to a decline in attack performance. Consequently, we restrict the modulation range of pulse width (pw): $\{pw | 90\,\mu s \leq pw \leq 1100\,\mu s\}$.

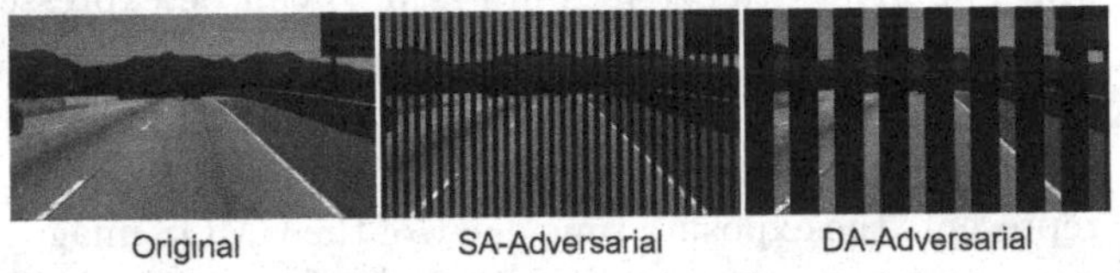

Fig. 4. The lane scene under normal illumination, scatter attack, and dense attack

To enhance the attack efficiency against lane detection systems, we further define the search space of pulse width to guide the deployment of adversarial attacks, proposing two attack strategies: scatter attack and dense attack.

Scatter Attack. Under scatter attack, we modulate the LED illumination to scatter the adversarial stripes extensively across the entire scene, causing significant damage to spatial semantic features. To achieve the aforementioned effect, we modulate the pulse width at a lower level: $\{pw | 90\,\mu s \leq pw \leq 110\,\mu s\}$, generating the adversarial sample shown in Fig. 4(SA-Adversarial).

Dense Attack. Under dense attack, we modulate the LED illumination to cause gradient shaking and occlusion in the local space of the lane scene, densely damaging local semantic features. To achieve the aforementioned effect, we modulate the pulse width at a higher level: $\{pw | 700\,\mu s \leq pw \leq 1100\,\mu s\}$, generating the adversarial sample shown in Fig. 4(DA-Adversarial).

5 Experimental Results and Discussion

5.1 Experimental Setup

Dataset for Digital World Attack. For digital world attack, the widely-used lane detection dataset Tusimple [16] is employed. The dataset primarily comprises images collected from highways with the number of lane markings ranging from two to four and includes 3626 training images and 2782 testing images.

Experimental Environment for Physical World Attack. The experimental scenes for physical world attack are shown in Fig. 5(a). We employ simulated resin roads as the highway and draw white solid or dashed lines above it as the lane markings. Meanwhile, a 17.5 cm diameter 18 W commercial LED lamp is used as the illumination source for the scenes and a high-frequency modulator is integrated into the LED lamp to control the pulse signal, as shown in Fig. 5(b). Furthermore, we use a Huawei P40 smartphone equipped with a CMOS image sensor as the in-vehicle camera to capture the scene images under LED illumination, thus obtaining the test set for the attack experiments.

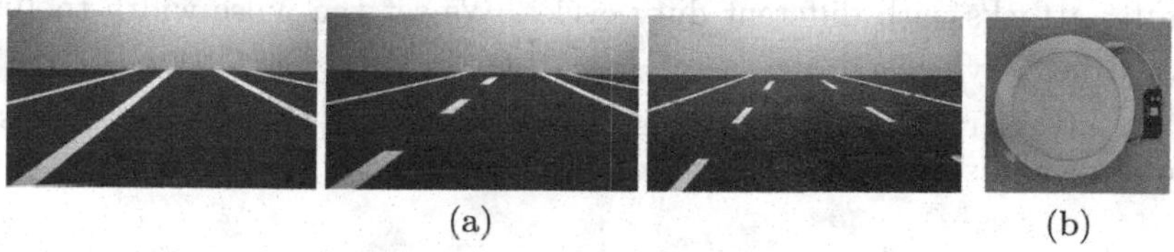

Fig. 5. The experimental environment: (a) refers to three lane scenes in the physical world, and (b) refers to LED lamp and high-speed modulator.

Target Models Under Tests. Five of the most representative lane detection models of three different task types: (1) segmentation-based models: SCNN/RES-A; (2) point detection-based model: LaneATT; (3) curve-based models: LSTR/B-ézierLaneNet are selected as the target models and the detection accuracy is used as the evaluation metric, which is defined as the ratio of the number of correctly predicted key points to the total number of labeled key points. Besides the Tusimple dataset, we collect six images of the three lane scenes under normal illumination to verify the models' detection performance. Table 1 shows the detection accuracy on the Tusimple dataset and the physical sample set of lane scenes. All of our codes are implemented in Pytorch and the testing processes are conducted on an NVIDIA GeForce RTX 3090 platform.

Table 1. The detection accuracy on the Tusimple dataset and the physical sample set

Dataset	Target model				
	SCNN	RESA	LaneATT	LSTR	BézierLaneNet
Tusimple	95.66%	95.76%	94.66%	95.03%	96.29%
Physical sample	92.05%	82.27%	88.64%	81.82%	90.66%

5.2 Experimental Results of Digital World Attack

In the experiments of digital world attack, 2782 testing images of the Tusimple dataset are employed. Firstly, we adjust the modulation parameters of LED illumination under scatter attack and dense attack and utilize a CMOS image sensor to capture the adversarial stripes. Then, by adding light and dark stripes as perturbation information to images, we generate the corresponding adversarial samples with different modulation parameters.

Scatter Attack. To examine the performance of scatter attacks with different pulse widths, we set the duty cycle of the control signal to $1/9$ and vary the pulse width in the search space $\{pw|90\,\mu s \leq pw \leq 110\,\mu s, step = 10\}$ in the first round of scatter attack experiments. In the second round, we focus on the performance of scatter attacks with different duty cycles. We set the pulse width to $100\,\mu s$ and vary the duty cycle in the search space $\left\{D = \frac{1}{i+1} |1 \leq i \leq 8, step = 1\right\}$. The partially adversarial samples generated under scatter attack and the test results of the two rounds of experiments are shown in Fig. 6.

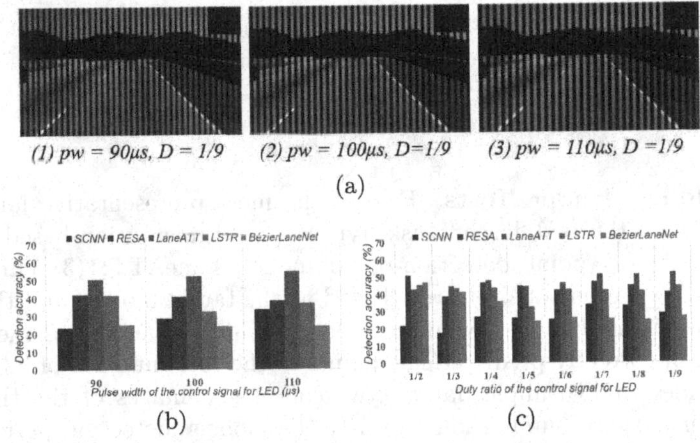

Fig. 6. The experimental results of scatter attacks in the digital world: (a) refers to the partially adversarial samples, while (b) and (c) refer to the detection accuracy under scatter attack with different pulse widths and duty cycles respectively.

As shown in Fig. 6(a), due to the small pulse width, the narrow and numerous adversarial stripes are extensively scattered across the entire scene under scatter attack. From the results in Fig. 6(b), we can observe that scatter attacks with different pulse widths exhibit stable attack performance, achieving a detection accuracy drop of 71.93%, 57.29%, 43.97%, 62.34%, and 72.11% for SCNN, RESA, LaneATT, LSTR, and BézierLaneNet, respectively. Moreover, the detection accuracy of scatter attacks with different duty cycles relatively fluctuates as shown in Fig. 6(c). Notably, the scatter attack with a duty cycle of 1/3 causes a detection accuracy drop of 77.93% and 56.83% for SCNN and RESA, respectively; the scatter attack with a duty cycle of 1/2 causes a drop of 51.51% for LaneATT, while the scatter attack with a duty cycle of 1/5 causes a drop of 62.34% and 71.97% for LSTR and BézierLaneNet, respectively.

Dense Attack. To examine the performance of dense attacks with different pulse widths, we still set the duty cycle of the control signal to 1/9 but vary the pulse width in the search space $\{pw|700\,\mu s \leq pw \leq 1100\,\mu s, step = 200\}$ in the first round of experiments. In the second round, we focus on the performance of dense attacks with different duty cycles. We set the pulse width to 1100 μs and vary the duty cycle in the search space $\left\{D = \frac{1}{i+1}\,|1 \leq i \leq 8, step = 1\right\}$. The partially adversarial samples generated under dense attack and the test results of the two rounds of experiments are shown in Fig. 7.

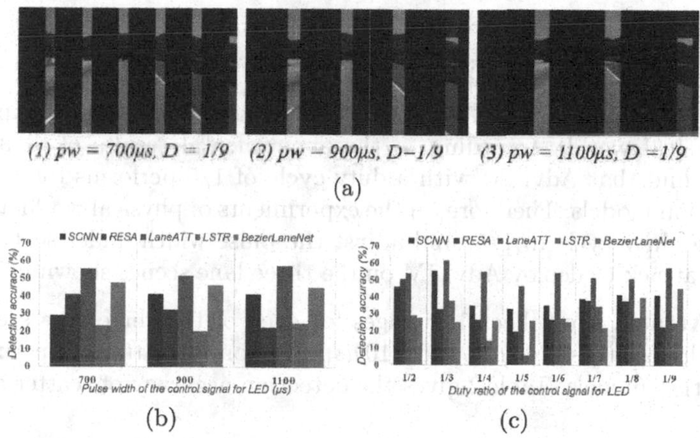

Fig. 7. The experimental results of dense attacks in the digital world: (a) refers to the partially adversarial samples, while (b) and (c) refer to the detection accuracy under dense attack with different pulse widths and duty cycles respectively.

As shown in Fig. 7(a), the width of adversarial stripes is wider than that of scatter attack under dense attack, causing gradient shaking and occlusion in the local space. The results from Fig. 7(b) and Fig. 7(c) indicate that the generated adversarial samples under dense attack can effectively cope with models of different task types. Specifically, for dense attacks with different pulse widths, the dense attack with a pulse width of 700 μs decreases the detection accuracy of SCNN by 66.42%; the dense attack with a pulse width of 900 μs decreases the detection accuracy of LaneATT and LSTR by 43.12% and 74.70%, respectively, while the dense attack with a pulse width of 1100 μs decreases the detection accuracy of RESA and BézierLaneNet by 73.35% and 50.86%, respectively. For dense attacks with different duty cycles, the dense attack with a duty cycle of 1/5 decreases the detection accuracy of SCNN, RESA, LaneATT, and LSTR by 62.72%, 75.86%, 48.14%, and 89.41%, respectively, while the dense attack with a duty cycle of 1/7 decreases the detection accuracy of BézierLaneNet by 73.94%.

From the overall experimental results, we can find that AdvLIM demonstrates significant effectiveness and strong transferability on lane detection models, mainly since the adversarial stripes add a large amount of edge information to the scene, which disrupts the continuity of the original semantic features, thus affecting the early mask block's extraction of low-level features and leading to subsequent mask blocks unable to make accurate decisions. In addition, different attack strategies have distinct impacts on lane detection models. Dense attack exhibits better performance than scatter attack on RESA, LSTR, and BézierLaneNet. However, since SCNN utilizes slice-by-slice convolution to pass spatial semantic relationships between rows and columns, while LaneATT incorporates spatial prior constraints through anchor boxes, the weight of spatial semantic features that these two models focus on may be greater than that of local semantic features, rendering scatter attack more effective than dense attack.

5.3 Experimental Results of Physical World Attack

To verify the executability of AdvLIM in reality, we conduct attack experiments in the physical world. According to the experimental results of digital world attack, we find that AdvLIM with a duty cycle of 1/5 performs the best on all lane detection models. Therefore, in the experiments of physical world attack, we first set the duty cycle to 1/5, and adjust the pulse width under scatter attack and dense attack to deploy AdvLIM on the three lane scenes shown in Fig. 5(a).

Scatter Attack. We collect 10 images for each of the three lane scenes with pulse widths of 90 μs, 100 μs, and 110 μs, and conduct attack experiments on lane detection models. Table 2 gives the detection accuracy of scatter attacks.

Table 2. The experimental results of scatter attacks in the physical world

Pulse width	Target model				
	SCNN	RESA	LaneATT	LSTR	BézierLaneNet
90 μs	0.68%	9.20%	0.00%	8.52%	0.00%
100 μs	0.00%	1.70%	0.00%	15.80%	0.00%
110 μs	0.45%	2.05%	0.00%	10.80%	0.00%

As shown in Table 2, it can be observed that scatter attacks with different pulse widths are effective, significantly decreasing the detection accuracy of lane detection models. The detection accuracy of SCNN, RESA, LaneATT, LSTR, and BézierLaneNet can be decreased from 92.05%, 82.27%, 88.64%, 81.82%, and 90.66% to 0.00%, 1.70%, 0.00%, 8.52%, and 0.00%, respectively.

Dense Attack. We collect 10 images for each of the three lane scenes with pulse widths of 700 μs, 900 μs, and 1100 μs, and conduct attack experiments on lane detection models. Table 3 gives the detection accuracy of dense attacks.

Table 3. The experimental results of dense attacks in the physical world

Pulse width	Target model				
	SCNN	RESA	LaneATT	LSTR	BézierLaneNet
700 μs	12.39%	3.41%	21.48%	6.93%	12.50%
900 μs	13.52%	7.50%	20.34%	11.70%	13.75%
1100 μs	9.20%	6.82%	24.55%	5.57%	14.20%

The results from Table 3 show that dense attacks with different pulse widths greatly decrease the detection accuracy of models. The detection accuracy of SCNN, RESA, LaneATT, LSTR, and BézierLaneNet can be decreased to 9.20%, 3.41%, 20.34%, 5.57%, and 12.50%, respectively, indicating that AdvLIM poses a real and significant threat to lane detection models.

To better illustrate the performance of AdvLIM, we visualize the predicted key points of each model with yellow pixels, as shown in Fig. 8. The visualized results show that AdvLIM has varying degrees of impact on models. In the experiments, segmentation-based models SCNN and RESA detect only a limited number of lane markings. Point detection-based model LaneATT detects lane markings with severe offset, while curve-based models LSTR and BézierLaneNet detect them with different degrees of distortion and deformation, which may cause serious driving safety accidents in real-world scenarios.

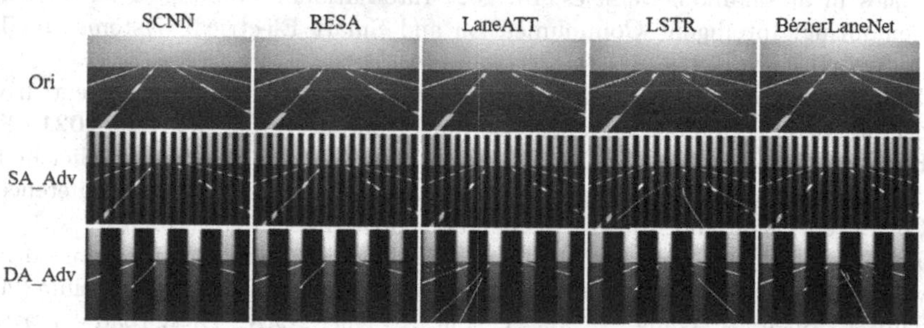

Fig. 8. The visualized test results of some experimental samples

6 Conclusion

In this paper, we analyze the decision behavior coexisting in lane detection models and point out the universal vulnerability of models in feature extraction. Based on this vulnerability, we propose a cross-task physical adversarial attack scheme based on LED illumination modulation (AdvLIM), with the advantages of strong transferability, good stealthiness, and high executability. This scheme does not require any modifications to the detected target or victim model and can generate physical adversarial perturbations that are completely imperceptible to the human eye. We evaluate the practical effectiveness and the performance of AdvLIM in both the digital world and physical world. The experimental results show that the scheme has high attack success rates on lane detection models of different task types, posing a real and significant threat to diverse lane detection systems in real-world scenarios. In future work, we will further develop more effective adversarial attack schemes based on LED illumination modulation, stimulating industry awareness of this security threat and consequentially enhancing the robustness of lane detection technology.

Acknowledgements. This work was partially supported by the National Natural Science Foundation of China (No. 62171202), Shenzhen-Hong Kong-Macao Science and Technology Plan Project (Category C Project: SGDX2021082310353 7030), and Theme-based Research Scheme of RGC, Hong Kong (T35-710/20-R).

References

1. Liang, D., Guo, Y.-C., Zhang, S.-K., Tai-Jiang, M., Huang, X.: Lane detection: a survey with new results. J. Comput. Sci. Technol. **35**, 493–505 (2020)
2. Tang, J., Li, S., Liu, P.: A review of lane detection methods based on deep learning. Pattern Recogn. **111**, 107623 (2021)
3. Kurakin, A., Goodfellow, I.J., Bengio, S.: Adversarial examples in the physical world. In: Artificial Intelligence Safety and Security, pp. 99–112. Chapman and Hall/CRC (2018)
4. Juyal, A., Sharma, S., Matta, P.: Traffic sign detection using deep learning techniques in autonomous vehicles. In: 2021 International Conference on Innovative Computing, Intelligent Communication and Smart Electrical Systems (ICSES), pp. 1–7. IEEE (2021)
5. Tian, D., Han, Y., Wang, B., Guan, T., Wei, W.: A review of intelligent driving pedestrian detection based on deep learning. Comput. Intell. Neurosci. **2021** (2021)
6. Feng, Z., Guo, S., Tan, X., Xu, K., Wang, M., Ma, L.: Rethinking efficient lane detection via curve modeling. In: Proceedings of the IEEE/CVF Conference on Computer Vision and Pattern Recognition, pp. 17062–17070 (2022)
7. Jing, P., et al.: Too good to be safe: tricking lane detection in autonomous driving with crafted perturbations. In: Proceedings of USENIX Security Symposium (2021)
8. Sato, T., Shen, J., Wang, N., Jia, Y., Lin, X., Chen, Q.A.: Dirty road can attack: security of deep learning based automated lane centering under physical-world attack. In: Proceedings of the 30th USENIX Security Symposium (USENIX Security 21) (2021)

9. Pan, X., Shi, J., Luo, P., Wang, X., Tang, X.: Spatial as deep: Spatial CNN for traffic scene understanding. In: Proceedings of the AAAI Conference on Artificial Intelligence, vol. 32 (2018)
10. Zheng, T., et al.: RESA: recurrent feature-shift aggregator for lane detection. In: Proceedings of the AAAI Conference on Artificial Intelligence, vol. 35, pp. 3547–3554 (2021)
11. Tabelini, L., Berriel, R., Paixao, T.M., Badue, C., De Souza, A.F., Oliveira-Santos, T.: Keep your eyes on the lane: Real-time attention-guided lane detection. In: Proceedings of the IEEE/CVF Conference on Computer Vision and Pattern Recognition, pp. 294–302 (2021)
12. Liu, R., Yuan, Z., Liu, T., Xiong, Z.: End-to-end lane shape prediction with transformers. In: Proceedings of the IEEE/CVF Winter Conference on Applications of Computer Vision, pp. 3694–3702 (2021)
13. Gnanasambandam, A., Sherman, A.M., Chan, S.H.: Optical adversarial attack. In: Proceedings of the IEEE/CVF International Conference on Computer Vision, pp. 92–101 (2021)
14. Duan, R., et al.: Adversarial laser beam: Effective physical-world attack to DNNs in a blink. In: Proceedings of the IEEE/CVF Conference on Computer Vision and Pattern Recognition, pp. 16062–16071 (2021)
15. Lee, H., Grosse, R., Ranganath, R., Ng, A.Y.: Convolutional deep belief networks for scalable unsupervised learning of hierarchical representations. In: Proceedings of the 26th Annual International Conference on Machine Learning, pp. 609–616 (2009)
16. Tusimple homepage. https://github.com/TuSimple/tusimple-benchmark. Accessed 17 Oct 2022

RECO: Rotation Equivariant COnvolutional Neural Network for Human Trajectory Forecasting

Jijun Cheng[1], Hao Wang[1], Dongheng Shao[1], Jian Yang[2], Mingsong Chen[1], Xian Wei[1], and Xuan Tang[1](✉)

[1] MoE Engineering Research MOE Center of SW/HW Co-Design Technology and Application, East China Normal University, Shanghai 200062, China
{chengjj,51255902171}@stu.ecnu.edu.cn, shaodongheng@fjirsm.ac.cn, mschen@sei.ecnu.edu.cn, xian.wei@tum.de, xtang@cee.ecnu.edu.cn

[2] School of Geospatial Information, Information Engineering University, Zhengzhou 450052, China
jian.yang@tum.de

Abstract. Pedestrian trajectory prediction is a crucial task for various applications, especially for autonomous vehicles that need to ensure safe navigation. To perform this task, it is essential to understand the dynamics of pedestrian motion and account for the uncertainty and multimodality of human behaviors. However, existing methods often produce inconsistent and unrealistic predictions due to their limited ability to handle different orientations of pedestrians. In this paper, we propose a novel approach that leverages the Euclidean group $C4$ to enhance convolutional neural networks with rotation equivariance. This property enables the networks to learn features that are invariant to rotations, thus maintaining consistent output under various orientations. We present the **R**otation **E**quivariant **CO**nvolutional Neural Network (RECO), a model specifically designed for pedestrian trajectory prediction using rotation equivariant convolutions. We evaluate our model on challenging real-world human trajectory forecasting datasets and show that it achieves competitive performance compared to state-of-the-art methods.

Keywords: Human Trajectory Forecasting · Multimodality · Uncertainty · Rotation Equivariant Convolution

1 Introduction

Historical data often suggest future trends for certain events. Analyzing historical data and predicting future situations is of great practical importance. With the development of deep learning, extracting features from sequence data and using them to predict the future has attracted lots of research efforts in many fields, such as weather prediction [1], traffic forecasting [2], and physical simulation [3]. However, in contrast to inferences with physical laws or specific regulations,

Q. Liu et al. (Eds.): PRCV 2023, LNCS 14427, pp. 492–504, 2024.
https://doi.org/10.1007/978-981-99-8435-0_39

pedestrians are complex agents that engage in various behaviors and actions to achieve their goals [4]. Predicting human behavior, such as motivation, body pose, and speed of motion, still faces significant challenges.

Traditional works use deterministic models to predict pedestrian trajectories [5]. However, in the real world, pedestrian trajectories exhibit distinct multimodal properties, which means that even within the same scene, a pedestrian may follow diverse trajectories [6]. Deterministic models cannot capture the uncertainty of trajectories. This drawback of deterministic models becomes more prominent in long-term trajectory prediction. In recent work, data-driven solutions [7] have provided highly dynamic predictions with impressive results. Nevertheless, these models tend to make inconsistent predictions.

A model that can predict pedestrian trajectories on the road needs to learn and understand the physical and psychological factors that influence pedestrian behaviors and ensure the accuracy of the prediction results. In addition, the model should correctly identify the interaction relationships and states among pedestrians, such as following and avoiding, and then provide equivariant track output for pedestrians in different directions [8]. Data augmentation is useful for rotation equivariance but requires a longer training time. Furthermore, data augmentation requires sampling from an infinite number of possible angles, as rotation is a continuous population. In contrast, using the rotation equivariance method, neural networks can make consistent predictions without increasing parameters and training time (Fig. 1).

Fig. 1. Pedestrian tracks in different directions at the same scene. Models with rotation equivariance should output the same trajectory. The orange line indicates the short-term trajectory and the blue line indicates the long-term trajectory. (Color figure online)

Considering the equivariant nature of pedestrian paths, we propose to encode human interaction uncertainty and predict pedestrian trajectories in a multimodal manner. Concretely, considering that pedestrians in different directions have similar behavioral interaction, inspired by group equivariant neural network [9], we propose to use rotation equivariant for pedestrian trajectory prediction.

We predict more than one trajectory of a pedestrian to describe the uncertainty of the trajectory, which performs better than classical models. Overall, we propose a novel approach for uncertain pedestrian multimodal trajectory prediction using **R**otational **E**quivariant **CO**nvolutional neural network(RECO). We benchmark RECO's performance on the Stanford Drone Dataset [10]. Our method achieves par or better prediction accuracy with better sample efficiency and generalization ability than baseline models. Overall, our contributions are as follows:

- We propose RECO, a multimodality approach to predict short-term and long-term pedestrian trajectory. Our model captures pedestrian similar interaction by rotation equivariant, and uses them to pedestrian trajectory.
- We design RECO considering the invariance of diverse traffic scene especially in different directions of crossroads and construct rotation equivariant neural network, which encodes features in different directions.
- On benchmark Stanford Drone Dataset, RECO achieve competitive accuracy, we show that rotation equivariant can improve the existing pedestrian trajectory methods.

2 Related Work

2.1 Trajectory Analysis and Forecasting

The application of classical dynamic models in vehicle and human trajectory prediction has a long history. For vehicle trajectories, classic models such as the Intelligent Driver model [11] have been used for a long time. Classical prediction methods are mainly designed for primary short-term trajectory prediction. Vehicle trajectory prediction models based on deep learning methods have recently performed significantly. For the ability to learn from time series, Recurrent Neural Networks (RNNs) were studied for vehicle trajectory prediction. For human trajectory forecasting, Social LSTM [7] developed social pooling, a classical framework based on a recurrent neural network that can jointly infer across multiple individuals to predict human trajectories. However, methods based on neural networks are always data-driven, which may generate inconsistent and unrealistic predictions at some times.

2.2 Multimodality Forecast

In the real world, human behavior is inherently multimodal and stochastic. Pedestrians follow different paths in the same scenario. A deterministic model with a single prediction result cannot capture this multimodal property. Previous trajectory prediction methods mainly focused on a single prediction for the future. Recently, generative models have been used to achieve multimodality. Trajectron++ [12] is a graph-structured recurrent model that considers interactions between different agents and forecasts trajectories of agents with dynamics and heterogeneous data. SocialVAE [6] combines a variational autoencoder with a social attention mechanism and uses stochastic recurrent neural networks to encode stochasticity and multimodality of motivation forecast. Another approach of multimodality trajectory forecast is Y-net [13], which proposes to decompose the uncertainty of human motivation into epistemic and aleatoric sources, makes discrete trajectory predictions for an extended period.

2.3 Group Equivariant Deep Learning

Rotation equivariance is a desirable property for neural networks that deal with data that can have different orientations. Without rotation equivariance, data augmentation is widely used to enhance the robustness of neural networks. In contrast, rotation equivariant neural networks can encode rotation symmetry with fewer parameters and achieve excellent results. [14,15] proposed to apply group equivariant to the 3D object detection task. Cohen *et al.* [9] proposed G-CNNs, a seminal work for group equivariant deep learning. G-CNNs use G-convolutions and achieve high-level weight sharing compared to traditional neural networks. For discrete groups such as translations, reflections and rotations, G-CNNs can achieve computational overhead. In subsequent work, equivariant methods have been applied to predict fluid kinetics where Wang *et al.* [16] introduced symmetries into convolutional neural networks to improve accuracy and generalization in fluid flow prediction.

We observe that rotational invariant property is often present in traffic scenarios on roads, especially at intersections. Most previous related works ignore this property. For this reason, we propose RECO for predicting the future trajectories of pedestrians.

3 Approach

Problem Formulation: The formal formulation of the problem of multimodal trajectory prediction is as follows. Let $s_t = (x_t, y_t) \in \mathbb{R}^2$ denote the two-dimensional coordinates of one agent at step t. Given a scene clip $\{I\}$ that contains N agents' observed trajectories $\{X_i\}_{i=1}^{N}$ ($X_i = (s_1, s_2, ..., s_{t_p})^T$ represents i-th agent's observed trajectory in the past $t_p = n_p/FPS$ seconds sampled at the frame rate FPS) during the observation period. Trajectory prediction aims to forecast their possible future coordinates $\{\hat{Y}_i\}_{i=1}^{N}$ ($\hat{Y}_i = (\hat{s}_{t_p+1}, \hat{s}_{t_p+2}, ..., \hat{s}_{t_p+t_f})^T$ denotes one of the prediction) during the corresponding future t_f steps considering their observations and the interactive context.

3.1 Method Overview

Our goal is to predict the future trajectories of pedestrians by incorporating scene information and historical trajectory information of pedestrians. The overall model is shown in Fig. 2. We model the uncertainty of pedestrians' future goals and trajectory by fully considering the rotation and other variable nature of pedestrians on the road, especially at scenes such as intersections.

3.2 Rotation Equivariant Convolutional Network

Inspired by [13], we decompose the human trajectory prediction problem into two components. The first component is goal prediction, there is uncertainty in the target and waypoints of human walking, and we first predict the possible goals and waypoints of pedestrians in the future. Considering that the destinations are likely to have multiple choices, we set a parameter K_g as the number of predicted destinations.The second component is trajectory prediction, which models the uncertainty in the paths that pedestrians take to reach their goals. We assume that the pedestrian's goal is fixed and predict multiple possible trajectories for each goal We set a parameter K_t as the number of predicted trajectory. Thus, we decouple the human trajectory prediction problem into goal prediction and trajectory prediction. We construct the entire trajectory prediction network using rotation equivariant convolution to fully account for the rotation-invariant nature of the traffic scenario. We will describe the trajectory prediction process and each sub-network of the model separately in the following.

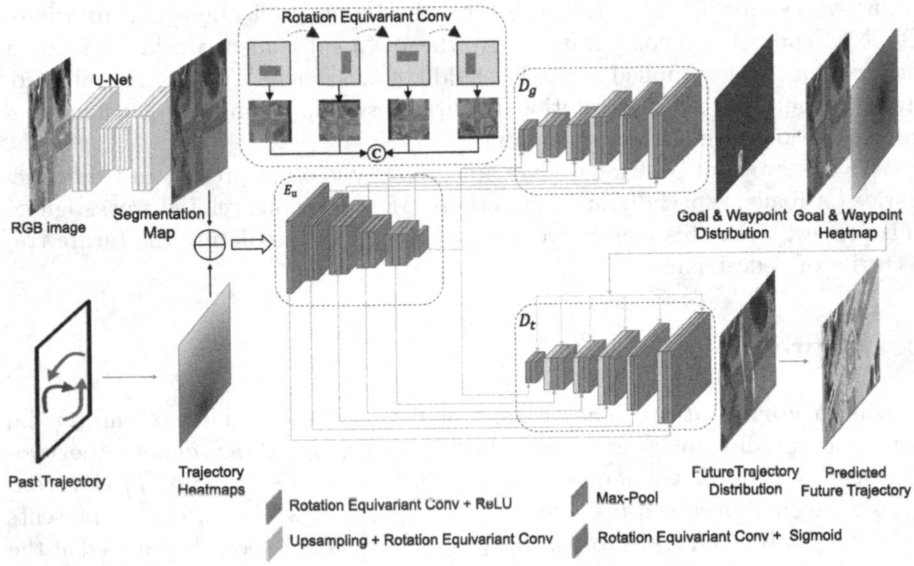

Fig. 2. Model Architecture: The whole pipeline consists of four stages. First, we convert the RGB scene images and observed trajectories into scene segmentation maps and trajectory heat maps, respectively. Second, we concatenate segmentation map and trajectory heatmap and use them as the input of the rotation equivariant network. Third, we apply rotation equivariant convolutions to learn rotation-invariant features from the input. Fourth, we generate a future trajectory heat map from the output of the equivariant network and process it into a future trajectory.

3.2.1 Scene Segmentation and Trajectory Heatmap Generation

In order to make full use of scene information and historical trajectory information for pedestrian future trajectory prediction, we first pre-process the scene image and historical trajectory. Given the RGB image $\mathcal{I}$ of the current scene, we first use U-Net [17] for segmentation to generate a scene segmentation map $\mathcal{S}$ with $\mathcal{I}$, which contains the segmentation results composed of C categories. At the same time, the past trajectory $\{\mathbf{s}_n\}_{n=1}^{n_p}$ of agent p is converted into a tarjectory heatmap $\mathbf{H}$ with the same spatial dimension as scene $\mathcal{I}$. The heat map contains n_p number of channels, and each time step corresponds to a channel respectively, which can be expressed as:

$$\mathbf{H}(n, i, j) = 2 \times \frac{\|(i, j) - \mathbf{s}_n\|}{\max_{(x,y)\in\mathcal{I}} \|(x, y) - \mathbf{s}_n\|} \tag{1}$$

where (i, j) and (x, y) are the pixel space coordinate positions of the heatmap $\mathbf{H}$ and image $\mathcal{I}$, respectively.

The generated heatmap will be stitched with the scene segmentation map in channel dimension to obtain the mixture information $\mathbf{H}_{\mathcal{S}}$, a $H \times W \times (C + n_p)$ tensor, which will be used as input to the subsequent network.

3.2.2 Trajectory Encoder and Decoder

We adopt a U-Net [17] shaped architecture for the encoding and decoding structure of trajectory prediction. To facilitate refinement, we subdivide the main structure into the U-Net shaped encoder E_u, goal and waypoint heatmap decoder D_g and the future trajectory distribution decoder D_t.

The input to E_u is a mixture of scene segmentation and historical trajectory heat map output $\mathbf{H}_{\mathcal{S}}$. E_u consists of Z blocks, each consisting of rotation equivariant convolutional layer, Relu, and a maximum pooling layer with step size 2. After the encoding of Z blocks, the spatial dimension of the input is reduced from $H \times W$ to $H_z \times W_z$ and the depth layer $C + n_p$ is increased to C_z. The final output $\mathbf{H}_z$ of the encoder network and the intermediate features of each block will be used to pass to the two subsequent decoders, respectively.

The decoder D_g will priorly receive the output from the encoder network through two rotation equivariant convolutional layers, and thereafter pass through M blocks in sequence. Each block first uses bilinear upsampling to recover the spatial resolution, then fuses the upsampled features with the intermediate feature representation from the corresponding layer in E_u via rotation equivariant convolutional layer in the middle of the current block. The skip connection between E_u and D_g helps the individual blocks in D_g to take full advantage of the high-resolution features from E_u. For each agent, D_g predicts the non-parametric distribution of the goal $\hat{\mathbb{P}}(\mathbf{s}_{t_p+t_f})$ and N_w waypoints $\hat{\mathbb{P}}(\mathbf{s}_{w_i})$, where $w_i \in \{t_p, t_p + 1, \ldots, t_p + t_f\}\ \forall\ i = 1, \ldots, N^w$. So D_g output $\mathbf{H}_g$ with its shape is $H \times W \times (N_w + 1)$.

D_t has the same network structure as D_g but differs in that D_t receives not only the input from E_u but also the intermediate features from the corresponding layer of D_g at the first layer of each block. For each future time step, D_t predicts an independent probability distribution that eventually produces an output of shape $H \times W \times t_f$ with each channel corresponding to the location distribution at each time step.

It is noteworthy that our encoder and decoder network employs rotation equivariant convolutional layers instead of traditional 2D convolutions, distinguishing it from classical U-net and similar methods. Further details will be provided below.

3.2.3 Rotation Equivariant Convolution

Group Theory: A group $(G, \cdot)$ is a set G together with a binary operation (the group law) $\cdot : G \times G \rightarrow G$ which satisfies the following axioms [18]:

$$\begin{cases} Associativity: \forall a, b, c \in G \quad a \cdot (b \cdot c) = (a \cdot b) \cdot c \\ Identity: \exists e \in G : \forall g \in G \quad g \cdot e = e \cdot g = g \\ Inverse: \forall g \in G \;\; \exists g^{-1} \in G : \quad gg^{-1} = g^{-1}g = e \end{cases} \tag{2}$$

Consider the group G of rotations by multiples of $\frac{\pi}{2}$. This group contains 4 elements and is called the cyclic group of order 4; we usually indicate it with $G = C_4$:

$$C_4 = \{R_0, R_{\frac{\pi}{2}}, R_\pi, R_{3\frac{\pi}{2}}\} = \{R_{r\frac{\pi}{2}} \mid r = 0, 1, 2, 3\} \tag{3}$$

Note that we can identify the elements of C_4 with $\mathbb{Z}/4\mathbb{Z}$, i.e. the integers modulo 4; indeed:

$$R_{r\frac{\pi}{2}} \cdot R_{s\frac{\pi}{2}} = R_{(r+s \mod 4)\frac{\pi}{2}} \tag{4}$$

Due to the road traffic scene, most of the vehicle's rotation is 90°C. Consequently, we decide to choose the rotation operator belongs to the cyclic group $C4$.

Group Equivariant Convolution Layer: Referring to [19], it is helpful to consider the input image of the neural network as a function $f : \mathbb{Z}^2 \rightarrow \mathbb{R}^K$, which maps 2D space to pixel intensities. Group equivariant convolution is a function that generalizes the convolution to groups, and the set $\mathbb{Z}^2$ with translations that usually defines the convolution is a specific type of group. For the group G, the convolution of the filter $\varphi : g \rightarrow \mathbb{R}^K$ with feature map $f : g \rightarrow \mathbb{R}^K$ is defined as the sum of the inner products of all elements in G:

$$(f * \psi)(g) = \sum_k \sum_{h \in G} f_k(h) \psi_k \left(g^{-1}h\right). \tag{5}$$

where, the action of an element g on $h \in G$ is denoted as gh and $g^{-1}h$ is the action of the inverse element of g. For example, if the group is a translation of an element $x \in \mathbb{Z}^2$, then $gx = x + g$ and $g^{-1}x = x - g$ and we will have the standard convolution. Since the output function is a function of G, which indexes not only pixel positions but also rotations, it is possible to preserve that information throughout the network and thus preserve the isotropy for such transformations.

The location of operations within the rotation equivariant convolutional neural network can be seen in the upper middle part of Fig. 2.

3.3 Loss Function

Our method imposes loss constraints on the prediction distributions of the goal, waypoint and trajectory distributions to optimize the network. The predicted probability distributions will be subjected to a binary cross-entropy loss with the corresponding ground truth (Gaussian heatmap $\hat{P}$), respectively.

$$\begin{cases} \mathcal{L}_{\text{goal}} = \text{BCE}(P(\mathbf{s}_{n_p+n_f}), \hat{P}(\mathbf{s}_{n_p+n_f})) \\ \mathcal{L}_{\text{waypoint}} = \sum_{i=1}^{N^w} \text{BCE}(P(\mathbf{s}_{w_i}), \hat{P}(\mathbf{s}_{w_i})) \\ \mathcal{L}_{\text{trajectory}} = \sum_{i=n_p}^{n_p+n_f} \text{BCE}(P(\mathbf{s}_i), \hat{P}(\mathbf{s}_i)) \end{cases} \tag{6}$$

The total loss of the entire network is as follows:

$$\mathcal{L}\& = \mathcal{L}_{\text{goal}} + \mathcal{L}_{\text{waypoint}} + \mathcal{L}_{\text{trajectory.}} \tag{7}$$

4 Experiments

4.1 Dataset and Evaluation Metrics

Stanford Drone Dataset: We evaluate the performance of our proposed model on the widely-used Stanford Drone Dataset [10]. The dataset receives widespread attention from researchers, and higher performance models have been proposed based on it. The dataset consists of over 11,000 unique individuals captured in 20 top-down scenes taken on the campus of Stanford University using a drone. The scenes were captured in a bird's eye view.

Evaluation Metrics: In accordance with established practices, the leave-one-out evaluation methodology is employed for fairness. The observation period for each individual consists of 8 time steps, equivalent to 3.2 s, and the prediction of the trajectory for the next 12 time steps, equal to 4.8 s, is based on the observation period data. Consistent with prior study [20], the performance of various pedestrian trajectory prediction models is evaluated using two error metrics.

1. *Average Displacement Error* (ADE): The average Euclidean distance between the ground-truth trajectory and the predicted one,

$$\text{ADE} = \frac{\sum_{i=1}^{N} \sum_{t=t_o+1}^{t_o+t_p} \left\| Y_i^{(t)} - \hat{Y}_i^{(t)} \right\|_2}{N \times t_p} \tag{8}$$

2. *Final Displacement Error* (FDE): The Euclidean distance between the ground-truth destination and the predicted one,

$$\text{FDE} = \frac{\sum_{i=1}^{N} \left\| Y_i^{(t_o+t_p)} - \hat{Y}_i^{(t_o+t_p)} \right\|_2}{N} \tag{9}$$

4.2 Implementation Details

We train the entire network end to end use Adam optimizer with a learning rate of 1×10^{-4} and batch size of 8. A pre-trained segmentation model is used that is finetuned on the specific dataset.

4.3 Results

As per prior works, in cases where multiple future predictions are available, we report the prediction with the minimum error. Our proposed model, RECO, is compared against a comprehensive list of baselines. The list of baselines includes Social GAN (S-GAN) [21], Conditional Flow VAE (CF-VAE) [20], P2TIRL [22], SimAug [23], and PECNet [24].

Table 1. Short temporal horizon forecasting results on SDD.

	S-GAN	CF-VAE	P2TIRL	SimAug	PECNet	RECO(Ours)
ADE ↓	27.23	12.60	12.58	10.27	9.96	**9.17**
FDE ↓	41.44	22.30	22.07	19.71	15.88	**14.84**

Short Term Forecasting Results: In Table 1, we show the performance of our model for the SDD dataset in short term trajectory prediction. Compared with the prior works which using 20 trajectory samples for evaluation, our model uses $K_t = 1$ and achieved competitive performance. Table 1 shows our proposed model achieving an ADE of 9.17 and FDE of 14.84, which outperforms the previous state-of-the-art performance methods.

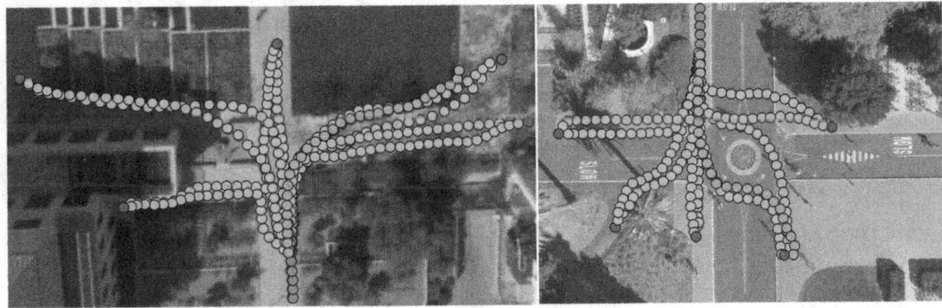

Fig. 3. Multimodal Pedestrian Trajectory Forecasting: We present multimodal trajectory prediction results with uncertainty under various scenarios in the SDD dataset. The example illustrates the prediction of 5 potential goal points for each agent in the future, along with 2 potential trajectories to reach the goals. The blue solid dots represent the observed past trajectories, the green solid dots depict the ground truth trajectories, the red solid dots indicate predicted future goal points, and the orange solid dots illustrate possible predicted trajectories for reaching predicted goal points.

Table 2. Long term trajectory forecasting Results.

	S-GAN	PECNet	R-PECNet	RECO(Ours)		
K_t	1	1	1	1	2	5
ADE ↓	155.32	72.22	261.27	**71.61**	**62.28**	**53.76**
FDE ↓	307.88	118.13	750.42	**91.33**	**91.33**	**91.33**

Long Term Forecasting Results: We propose a long term trajectory forecasting setting with a prediction horizon up to 10 times longer than prior works (up to a minute). Table 2 shows the results of our method compared with other methods for long-term prediction, where K_t denotes the planned number of reachable paths traveling to the target point for the case of determining the target point. In the case that the other methods compared predict only one path, our method achieves multiple path prediction and obtains better performance with lower ADE and FDE. The uncertainty multimodal long term trajectory prediction results under different scenarios are showed in Fig. 3.

Ablation Study: We conducted an ablation study to validate the effectiveness of our proposed rotational isovariant network for predicting road pedestrian trajectories. Specifically, we additionally performed the following experiments: We replaced the rotation equivariant convolutional network in our proposed method with a conventional classical convolutional neural network. We conducted experimental comparisons while ensuring that the other experimental settings were the same. As can be seen in Fig. 4, compared with the conventional 2D convolution, the use of rotation equivariant convolution results in a reduction of 5.85% and 5.83% in ADE and FDE respectively, which shows the effectiveness of our proposed rotational equivariant network.

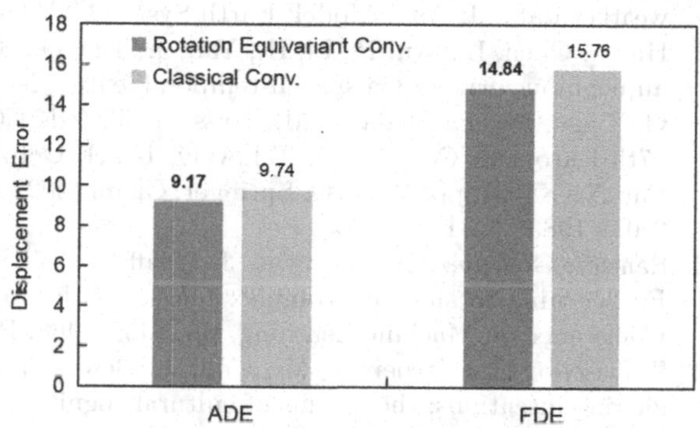

Fig. 4. Benchmarking performance against different convolution.

5 Conclusion

In summary, we propose RECO, a rotation equivariant convolution embedded neural network, which can take advantage of rotation invariant features in traffic scenes to better capture interactions between pedestrians using historical trajectories and predict future trajectories. Considering the uncertainty and multimodality of pedestrian trajectory changes on the roadway, we achieve to predict diverse potential pedestrian trajectories in both the short and long term. Further, we conduct extensive experiments, and the experimental results show that our long-term and short-term predictions are better than other advanced performance methods on the SDD dataset.

Acknowledgements. This work was supported by the Shanghai Munici-pality's "Science and Technology Innovation Action Plan" for the year 2023, Natural Science Fund, General Program (No. 23ZR1419300), Shanghai Trusted Industry Internet Software Collaborative Innovation Center, "Digital Silk Road" Shanghai International Joint Lab of Trustworthy Intelligent Software (No. 22510750100) and the National Natural Science Foundation of China (No. 42130112, No. 41901335).

References

1. Weyn, J.A., Durran, D.R., Caruana, R.: Can machines learn to predict weather? Using deep learning to predict gridded 500-hPa geopotential height from historical weather data. J. Adv. Model. Earth Syst. **11**(8), 2680–2693 (2019)
2. Hu, S., Chen, L., Wu, P., Li, H., Yan, J., Tao, D.: ST-P3: end-to-end vision-based autonomous driving via spatial-temporal feature learning. In: Avidan, S., Brostow, G., Cissé, M., Farinella, G.M., Hassner, T. (eds.) Computer Vision-ECCV 2022: 17th European Conference, Tel Aviv, Israel, October 23–27, 2022, Proceedings, Part XXXVIII, pp. 533–549. Springer, Cham (2022). https://doi.org/10.1007/978-3-031-19839-7_31
3. Sanchez-Gonzalez, A., Godwin, J., Pfaff, T., Ying, R., Leskovec, J., Battaglia, P.: Learning to simulate complex physics with graph networks. In: International Conference on Machine Learning, pp. 8459–8468. PMLR (2020)
4. Tomasello, M., Carpenter, M., Call, J., Behne, T., Moll, H.: Understanding and sharing intentions: the origins of cultural cognition. Behav. Brain Sci. **28**(5), 675–691 (2005)
5. Pellegrini, S., Ess, A., Schindler, K., Van Gool, L.: You'll never walk alone: modeling social behavior for multi-target tracking. In: 2009 IEEE 12th International Conference on Computer Vision, pp. 261–268. IEEE (2009)
6. Xu, P., Hayet, J.B., Karamouzas, I.: SocialVAE: human trajectory prediction using timewise latents. In: Avidan, S., Brostow, G., Cissé, M., Farinella, G.M., Hassner, T. (eds.) Computer Vision-ECCV 2022: 17th European Conference, Tel Aviv, Israel, October 23–27, 2022, Proceedings, Part IV. pp. 511–528. Springer, Cham (2022). https://doi.org/10.1007/978-3-031-19772-7_30
7. Alahi, A., Goel, K., Ramanathan, V., Robicquet, A., Fei-Fei, L., Savarese, S.: Social LSTM: human trajectory prediction in crowded spaces. In: 2016 IEEE Conference on Computer Vision and Pattern Recognition, CVPR 2016, Las Vegas, NV, USA, June 27–30, 2016, pp. 961–971. IEEE Computer Society (2016)
8. Walters, R., Li, J., Yu, R.: Trajectory prediction using equivariant continuous convolution. In: 9th International Conference on Learning Representations, ICLR 2021, Virtual Event, Austria, May 3–7, 2021. OpenReview.net (2021). https://openreview.net/forum?id=J8_GttYLFgr

9. Cohen, T., Welling, M.: Group equivariant convolutional networks. In: International Conference on Machine Learning, pp. 2990–2999. PMLR (2016)
10. Robicquet, A., Sadeghian, A., Alahi, A., Savarese, S.: Learning social etiquette: human trajectory understanding in crowded scenes. In: Leibe, B., Matas, J., Sebe, N., Welling, M. (eds.) ECCV 2016. LNCS, vol. 9912, pp. 549–565. Springer, Cham (2016). https://doi.org/10.1007/978-3-319-46484-8_33
11. Kesting, A., Treiber, M., Helbing, D.: Enhanced intelligent driver model to access the impact of driving strategies on traffic capacity. Philos. Trans. R. Soc. A Math. Phys. Eng. Sci. **368**(1928), 4585–4605 (2010)
12. Salzmann, T., Ivanovic, B., Chakravarty, P., Pavone, M.: Trajectron++: dynamically-feasible trajectory forecasting with heterogeneous data. In: Vedaldi, A., Bischof, H., Brox, T., Frahm, J.-M. (eds.) ECCV 2020. LNCS, vol. 12363, pp. 683–700. Springer, Cham (2020). https://doi.org/10.1007/978-3-030-58523-5_40
13. Mangalam, K., An, Y., Girase, H., Malik, J.: From goals, waypoints & paths to long term human trajectory forecasting. In: 2021 IEEE/CVF International Conference on Computer Vision, ICCV 2021, Montreal, QC, Canada, October 10–17, 2021, pp. 15213–15222. IEEE (2021)
14. Wang, X., Lei, J., Lan, H., Al-Jawari, A., Wei, X.: DuEqNet: dual-equivariance network in outdoor 3D object detection for autonomous driving. In: IEEE International Conference on Robotics and Automation, ICRA 2023, London, UK, May 29 - June 2, 2023, pp. 6951–6957. IEEE (2023)
15. Liu, H., et al.: Group equivariant BEV for 3D object detection. CoRR abs/2304.13390 (2023)
16. Wang, R., Walters, R., Yu, R.: Incorporating symmetry into deep dynamics models for improved generalization. In: 9th International Conference on Learning Representations, ICLR 2021, Virtual Event, Austria, May 3–7, 2021. OpenReview.net (2021). https://openreview.net/forum?id=wta_8Hx2KD
17. Ronneberger, O., Fischer, P., Brox, T.: U-Net: convolutional networks for biomedical image segmentation. In: Navab, N., Hornegger, J., Wells, W.M., Frangi, A.F. (eds.) MICCAI 2015. LNCS, vol. 9351, pp. 234–241. Springer, Cham (2015). https://doi.org/10.1007/978-3-319-24574-4_28
18. Weiler, M., Cesa, G.: General E(2)-equivariant steerable CNNs. In: Wallach, H.M., Larochelle, H., Beygelzimer, A., d'Alché-Buc, F., Fox, E.B., Garnett, R. (eds.) Advances in Neural Information Processing Systems 32: Annual Conference on Neural Information Processing Systems 2019, NeurIPS 2019, December 8–14, 2019, Vancouver, BC, Canada, pp. 14334–14345 (2019)
19. Cohen, T., Welling, M.: Group equivariant convolutional networks. In: Balcan, M., Weinberger, K.Q. (eds.) Proceedings of the 33nd International Conference on Machine Learning, ICML 2016, New York City, NY, USA, June 19–24, 2016. JMLR Workshop and Conference Proceedings, vol. 48, pp. 2990–2999. JMLR.org (2016)
20. Bhattacharyya, A., Hanselmann, M., Fritz, M., Schiele, B., Straehle, C.N.: Conditional flow variational autoencoders for structured sequence prediction. arXiv preprint arXiv:1908.09008 (2019)
21. Gupta, A., Johnson, J., Fei-Fei, L., Savarese, S., Alahi, A.: Social GAN: socially acceptable trajectories with generative adversarial networks. In: Proceedings of the IEEE Conference on Computer Vision and Pattern Recognition, pp. 2255–2264 (2018)
22. Deo, N., Trivedi, M.M.: Trajectory forecasts in unknown environments conditioned on grid-based plans. arXiv preprint arXiv:2001.00735 (2020)

23. Liang, J., Jiang, L., Hauptmann, A.: SimAug: learning robust representations from 3D simulation for pedestrian trajectory prediction in unseen cameras. arXiv preprint arXiv:2004.02022 2 (2020)
24. Mangalam, K., et al.: It is not the journey but the destination: endpoint conditioned trajectory prediction. In: Vedaldi, A., Bischof, H., Brox, T., Frahm, J.-M. (eds.) ECCV 2020. LNCS, vol. 12347, pp. 759–776. Springer, It is not the journey but the destination: endpoint conditioned trajectory prediction (2020). https://doi.org/10.1007/978-3-030-58536-5_45

FGFusion: Fine-Grained Lidar-Camera Fusion for 3D Object Detection

Zixuan Yin[1], Han Sun[1(✉)], Ningzhong Liu[1], Huiyu Zhou[2], and Jiaquan Shen[3]

[1] Nanjing University of Aeronautics and Astronautics, Nanjing, China
sunhan@nuaa.edu.cn

[2] University of Leicester, Leicester, UK

[3] Luoyang Normal University, Luoyang, China

Abstract. Lidars and cameras are critical sensors that provide complementary information for 3D detection in autonomous driving. While most prevalent methods progressively downscale the 3D point clouds and camera images and then fuse the high-level features, the downscaled features inevitably lose low-level detailed information. In this paper, we propose Fine-Grained Lidar-Camera Fusion (FGFusion) that make full use of multi-scale features of image and point cloud and fuse them in a fine-grained way. First, we design a dual pathway hierarchy structure to extract both high-level semantic and low-level detailed features of the image. Second, an auxiliary network is introduced to guide point cloud features to better learn the fine-grained spatial information. Finally, we propose multi-scale fusion (MSF) to fuse the last N feature maps of image and point cloud. Extensive experiments on two popular autonomous driving benchmarks, i.e. KITTI and Waymo, demonstrate the effectiveness of our method.

Keywords: Lidar-Camera Fuison · Fine-grained Fusion · Multi-scale Feature · Attention Pyramid

1 Introduction

3D object detection is a crucial task in autonomous driving [1,8]. In recent years, lidar-only methods have made significant progress in this field. However, relying solely on point cloud data is insufficient because lidar only provides low-resolution shape and depth information. Therefore, researchers hope to leverage multiple modalities of data to improve detection accuracy. Among them, vehicle-mounted cameras can provide high-resolution shape and texture information, which is complementary to lidar. Therefore, the fusion of point cloud data with RGB images has become a research hotspot.

In the early stages of fusion method research, researchers naturally assumed that the performance of fusion methods would be better than that of lidar-only methods, because the essence of fusion methods is to add RGB information as an auxiliary to lidar-only methods. Therefore, the performance of the model should

Q. Liu et al. (Eds.): PRCV 2023, LNCS 14427, pp. 505–517, 2024.
https://doi.org/10.1007/978-981-99-8435-0_40

be at least as good as before, rather than declining [22]. However, this is not always the case.

There are two reasons for the performance decline: 1) a suitable method for aligning the two modal data has not yet been found, 2) the features of the two modalities used in the fusion are too coarse. Regarding the first issue, fusion methods have evolved from the initial post-fusion [3,10] and point-level fusion [22,23] methods to today's more advanced feature fusion [2,12,24] methods. However, the second problem has not yet been solved. Specifically, we know that lidar-only methods are mainly divided into one-stage methods [14,26,31] and two-stage methods [4,18–20]. Usually, the performance of two-stage methods is better than that of one-stage methods because the features extracted by the first stage can be refined in the second stage. However, most current fusion methods focus on how to fuse features more effectively and ignore the process of refining fused features.

To solve the above problems, we utilize fine-grained features to improve the model accuracy and propose an efficient multi-modal fusion strategy called FGFusion. Specifically, since both image and point cloud data inevitably lose detailed features and spatial information during the downscaling process, we design different feature refinement schemes for the two modalities. First, for image data, we exploit a dual-path pyramid structure and designs a top-down feature path and a bottom-up attention path to better fuse high-level and low-level features. For point cloud data, inspired by SASSD [7], we construct an auxiliary network with point-level supervision to guide the intermediate features from different stages of 3D backbone to learn the fine-grained spatial structures of point clouds. In the fusion stage, we select several feature maps of the same number from the feature pyramids of images and point clouds respectively, and fuse them by cross-attention. The fused feature pyramids can then be passed into modern task prediction head architecture [2,28].

In brief, our contributions can be summarized as follows:

- We design different feature refinement schemes for camera image and point cloud data, in order to fuse high-level abstract semantic information and low-level detailed features.
- We design a multi-level fusion strategy for point clouds and images, which fully utilizes the feature pyramids of the two modalities in the fusion stage to improve the model accuracy.
- We verify our method on two mainstream autonomous driving point cloud datasets (KITTI and Waymo), and the experimental results prove the effectiveness of our method.

2 Related Work

2.1 LiDAR-Only 3D Detection

Lidar-only methods are mainly divided into point-based methods and voxel-based methods. Among them, point-based methods such as PointNet [16] and

PointNet++ [17] are the earliest neural networks directly applied to point clouds. They directly process unordered raw point clouds and extract local features through max-pooling. Based on their work, voxel-based and pillar-based methods have been derived. They transform the original point cloud into a Euclidean feature space and then use standard 2D or 3D convolution to calculate the features of the BEV plane. Representative methods include VoxelNet [31], SECOND [25], PointPillars [11], etc.

The development of lidar-only methods later shows two different development trends. Like 2D object detection, they are divided into one-stage and two-stage methods. One-stage methods [14,26,31] directly regress category scores and bounding boxes in one stage, and the network is relatively simple and has fast inference speed. Two-stage methods [4,18–20] usually generate region proposals in the first stage and then refine them in the second stage. The accuracy of two-stage methods is usually higher than that of one-stage methods, because the second stage can capture more detailed and distinctive features, but the cost is a more complex network structure and higher computational cost.

2.2 Fusion-Based 3D Detection

Due to the sparsity of point cloud data and its sole possession of spatial structural information, researchers have proposed to complement point clouds with RGB images. Early methods [3,15] use result-level or proposal-level post-fusion strategies, but the fusion granularity is too coarse, resulting in performance inferior to that of lidar-only methods.

PointPainting [22] is the first to utilize the hard correlation between LiDAR points and image pixels for fusion. It projects the point clouds onto the images through a calibration matrix and enhances each LiDAR point with the semantic segmentation score of the image. PointAugmenting [23] builds on PointPainting and proposes using features extracted from 2D object detection networks instead of semantic segmentation scores to enhance LiDAR points. Feature-level fusion methods points out that the hard association between points and pixels established by the calibration matrix is unreliable. DeepFusion [12] uses cross-attention to fuse point cloud features and image features. TransFusion [2] uses the prediction of point cloud features as a query for image features and then uses a transformer-like architecture to fuse features.

It can be seen that whether using semantic segmentation scores and image features obtained from pre-trained networks, or directly querying and fusing at the feature level, these methods essentially fuse high-level features with the richest semantic information, while ignoring low-level detailed information.

3 FGFusion

3.1 Motivations and Pipeline

The previous fusion methods only exploit high-level features, ignoring the important fact that detailed feature representations are lost in the downsampling process. For example, PointPainting [22] directly makes use of pixel-wise semantic segmentation scores as image features to decorate point cloud data, which only uses the results of last feature map and ignores multi-scale information. PointAugmenting [23] utilizes the last feature map with the richest semantic information to decorate point cloud data, but discards all the others that contain low-level detailed information. DeepFusion [12] is a feature-level fusion method, which improves the accuracy compared to point-level fusion methods such as PointAugmenting, but the essence is the same, as shown in Fig. 1.

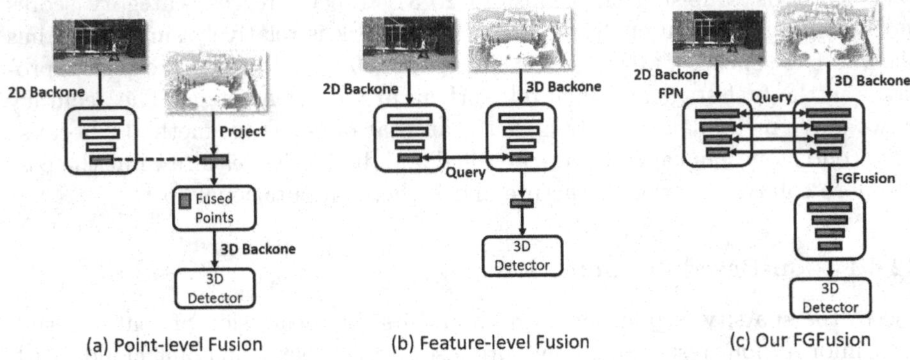

Fig. 1. Most point-level fusion methods [22,23] and feature-level fusion methods [2,12] only use the last layer of image or point cloud features for fusion, while our FGFusion performs fusion at multiple feature scales, fully utilizing low-level detail information to improve model accuracy.

We noticed that in some 2D object detection tasks, such as small object detection and fine-grained image recognition, multi-scale techniques are often used to extract fine-grained features. While in 3D object detection, point cloud data is suitable for capturing the spatial structural features, but it is easy to ignore small targets and fine features due to its sparse characteristic. Therefore, we hope to fuse point cloud and image data in a multi-scale way to make up for the shortcomings of point clouds. To achieve this goal, we fuse the features of point cloud and image at multiple levels instead of only using the last feature map generated by the backbone network. In addition, to extract finer features, we design a dual-path pyramid structure in the image branch and add an auxiliary network to guide convolutional feature perception of object structures in the point cloud branch.

To summarize, our proposed fine-grained fusion pipeline is shown in Fig. 2. For the image branch, we exploit 2D backbone and a dual-path pyramid structure to obtain the attention pyramid. For the point cloud branch, the raw points are fed into the existing 3D backbone to obtain the lidar features, and at the same time, guide the learning of features through an auxiliary network. Finally, we fuse the image and point cloud features at different levels and attach the same designed head to each fused layer of features to obtain the final results.

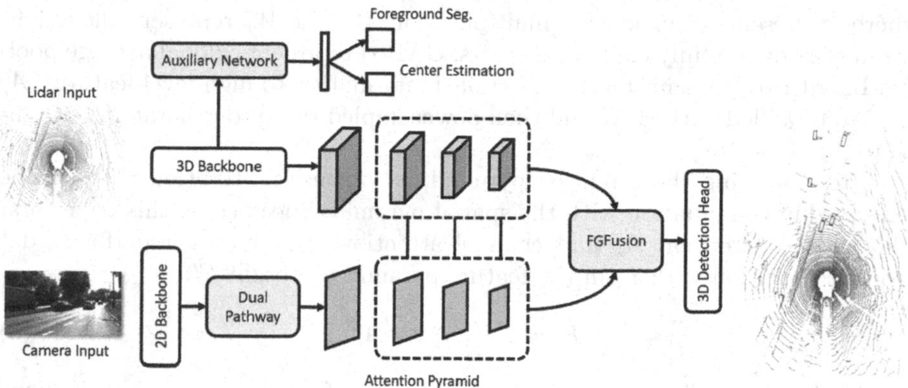

Fig. 2. An overview of FGFusion framework. FGFusion consists of 1) a dual pathway hierarchy structure with a top-down feature pathway and a bottom-up attention pathway, hence learning both high-level semantic and low-level detailed feature representation of the images, 2) an auxiliary network to guide point cloud features to better learn the fine-grained spatial information, and 3) a fusion strategy that can fuse the two modalities in a multi-scale way.

3.2 Camera Stream Architecture

In general, the input image will be processed by a convolutional neural network to obtain a feature representation with high-level semantic information. However, many low-level detailed features will be lost, which is insufficient for robust fusion. In order to retain the fine-grained features, inspired by the FPN network [13], we design a top-down feature path to extract features of different scales.

Let $\{B_1, B_2, ..., B_l\}$ represent the feature maps obtained after the input image passes through the backbone and l represent the number of convolutional blocks. The general method is to directly use the output of the last block B_l for fusion, but we hope to make full use of each B_i. Since it will bring huge cost overheads inevitably if making full use of every blocks of the network, we only select the last N outputs to generate the corresponding feature pyramid. The final feature pyramid obtained can be denoted as $\{F_{l-N+1}, F_{l-N+2}, ..., F_l\}$.

After obtaining the feature pyramid, we design a bottom-up attention path which includes spatial attention and channel attention. Spatial attention is used to locate the identifiable regions of the input image at different scales. It can be represented as:

$$A_i^s = \sigma(K * F_i), \tag{1}$$

where σ is the sigmoid activation function, $*$ represents the deconvolution operation, and K represents the convolution kernel. Channel attention is used to add associations between channels and pass low-level detailed information layer by layer to higher levels:

$$A_i^c = \sigma(W_b \cdot ReLU(W_a \cdot GAP(F_i))), \tag{2}$$

where · represents element-wise multiplication, W_a and W_b represent the weight parameters of two fully connected layers. GAP(·) represents global average pooling. In order to transmit low-level detailed information to high-level features, A_i^c need to be added with A_{i-1}^c and then downsampled twice to generate a bottom-up path.

After obtaining the attention pyramid, a bottom-up attention path can be generated in combination with the spatial pyramid. Specifically, this paper first adds spatial attention A_i^s and channel attention A_i^c, and then performs dot product operation with F_i in the feature pyramid to obtain F_i':

$$F_i' = F_i \cdot (A_i^s + \alpha A_i^c). \tag{3}$$

Finally, $\{F_{l-N+1}', F_{l-N+2}', ..., F_l'\}$ can be obtained for subsequent classification.

3.3 LiDAR Stream Architecture

Our framework can use any network that can convert point clouds into multi-scale feature pyramids as our lidar flow. At the same time, inspired by SASSD [7], we designed an auxiliary network, which contains a point-wise foreground segmentation head and a center estimation head, to guide the backbone CNN to learn the fine-grained structure of point clouds at different stages of intermediate feature learning. It is worth noting that the auxiliary network can be separated after training, so no additional computation is introduced during inference.

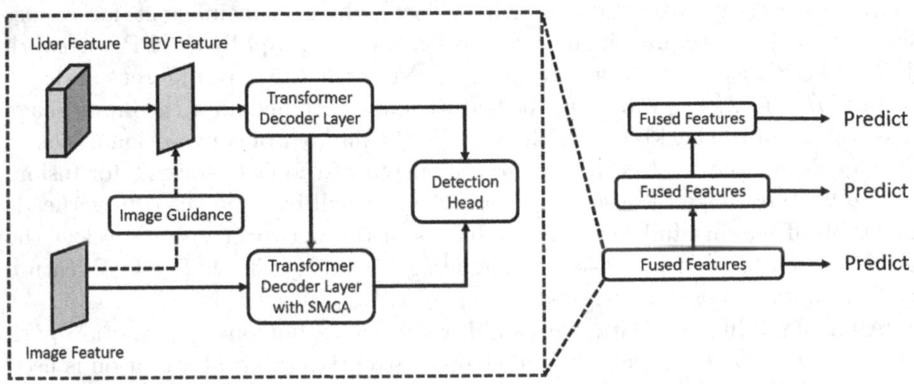

Fig. 3. The multi-scale fusion module first compresses the point cloud features into BEV features, and then uses TransFusion [2] to fuse the last N layers of BEV features and image features separately to obtain the prediction results of each layer. Finally, the post-processing is performed to obtain the final results.

3.4 Multi-scale Fusion Module

Now we have obtained the attention pyramid of the image and the feature pyramid of the point cloud separately. In order to fully fuse the two modalities, we take the last N layers of features of both for fusion, rather than just using the last layer, as shown in Fig. 3. Through the point cloud feature pyramid, we can obtain a multi-scale point cloud BEV feature map $\{F^B_{l-N+1}, F^B_{l-N+2} \cdots, F^B_l\}$. Following TransFusion [2], we use two transformer decoding layers to fuse the two modalities: first decodes object queries into initial bounding box predictions using the LiDAR information, and then performs LiDAR-camera fusion by attentively fusing object queries with useful image features. Finally, each fusion feature can generate corresponding prediction results, and the final prediction is obtained through post-processing.

4 Experiments

We evaluate our proposed FGFuison on two datasets, KITTI [6] and Waymo [21], and conduct sufficient ablation experiments.

4.1 Datasets

The KITTI dataset contains 7481 training samples and 7518 testing samples of autonomous driving scenes. As common practice, we divide the training data into a training set containing 3712 samples and a validation set containing 3769 samples. According to the requirements of the KITTI object detection benchmark, we conduct experiments on three categories of cars, pedestrians, and cyclists and evaluate the results using the average precision (AP) with an IoU threshold of 0.7.

The Waymo Open Dataset contains 798 training sequences, 202 validation sequences and 150 testing sequences. Each sequence has about 200 frames, which contain lidar points, camera images, and labeled 3D bounding boxes. We use official metrics, i.e., Average Precision (AP) and Average Precision weighted by Heading (APH), to evaluate the performance of different models and report the results of LEVEL1 (L1) and LEVEL2 (L2) difficulty levels.

4.2 Implementation Details

For the KITTI dataset, the voxel size is set to (0.05 m, 0.05 m, 0.1 m). Since KITTI only provides annotations for the front camera's field of view, the detection range of the X, Y and Z axes are set to [0, 70.4 m], [−40 m, 40 m], and [−3 m, 1 m], respectively. The image size is set to 448 × 800. For the Waymo dataset, the voxel size is set to (0.1 m, 0.1 m, 0.15 m). The detection range of the X and Y axes is [−75.2 m, 75.2 m], and the detection range of the Z axis is [−2 m, 4 m].

We choose TransFusion-L and the DLA34 of the pre-trained CenterNet as the 3D and 2D backbone networks, respectively. Following TransFusion [2], our training consists of two stages: 1) First we train the 3D backbone with the first decoder layer and FFN for 20 epochs. It only requires point clouds as input, and the last BEV feature map is used to produce initial 3D bounding box predictions. 2) Then we train the LiDAR-camera fusion and image-guided query initialization module for another 6 epochs. In this stage, the last three feature maps of the 3D and 2D backbone are fused separately. The advantage of this two-step training scheme over joint training is that auxiliary networks can be used only in the first stage, as well as data augmentation methods for pure point cloud methods. For post-processing, we use NMS with the threshold of 0.7 for Waymo and 0.55 for KITTI to remove redundant boxes.

4.3 Experimental Results and Analysis

Table 1. Performance comparison on the KITTI *val* set with AP calculated by 40 recall positions.

Method	Modality	mAP	Car			Pedestrian			Cyclist		
			Easy	Mod	Hard	Easy	Mod	Hard	Easy	Mod	Hard
SECOND [25]	L	68.06	88.61	78.62	77.22	56.55	52.98	47.73	80.58	67.15	63.10
PointPillars [11]	L	66.53	86.46	77.28	74.65	57.75	52.29	47.90	80.05	62.68	59.70
PointRCNN [19]	L	70.67	88.72	78.61	77.82	62.72	53.85	50.25	86.84	71.62	65.59
PV-RCNN [18]	L	73.27	92.10	84.36	82.48	64.26	56.67	51.91	88.88	71.95	66.78
Voxel-RCNN [5]	L	–	**92.38**	**85.29**	82.86	–	–	–	–		–
MV3D [3]	L+C	–	71.29	62.68	56.56	–	–	–	–	–	–
AVOD [10]	L+C	–	84.41	74.44	68.65	–	58.80	–	–	49.70	–
F-PointNet [15]	L+C	65.58	83.76	70.92	63.65	70.00	61.32	53.59	77.15	56.49	53.37
3D-CVF [29]	L+C	–	89.67	79.88	78.47	–	–	–	–	–	–
EPNet [9]	L+C	70.97	88.76	78.65	78.32	66.74	59.29	54.82	83.88	65.60	62.70
CAT-Det [30]	L+C	75.42	90.12	81.46	79.15	**74.08**	**66.35**	58.92	87.64	72.82	68.20
FGFusion(Ours)	L+C	**77.05**	**92.38**	84.96	**83.84**	72.63	65.07	**59.21**	**90.33**	**74.19**	**70.84**

KITTI. To prove the effectiveness of our method, we compare the average precision (AP) of FGFusion with some state-of-the-art methods on the KITTI dataset. As shown in Table 1, the mAP of our proposed FGFusion is the highest among all methods. KITTI divides all objects into three difficulty levels: easy, moderate and hard based on the size of the object, occlusion status and truncation level. The higher the difficulty level, the harder it is to detect. Our method leads in different levels of difficulty for multiple categories and has higher accuracy than all other methods in the difficult levels of all three categories, which proves that our method can effectively fuse fine-grained features.

In lidar-only methods, the accuracy of one-stage methods such as SECOND [25] and PointPillars [11] is lower than that of two-stage methods such as PV-RCNN [18]. In the easy and medium difficulty levels of the car category, our FGFusion is competitive with Voxel-RCNN [5], the best-performing method in lidar-only methods, and surpasses 0.98% AP in the difficult level. In fusion methods, early works such as MV3D [3] and AVOD [10] have lower performance than lidar-only methods. However, recently proposed CAT-Det [30] can achieve higher overall accuracy than lidar-only methods in all three categories, and achieve 75.42 in mAP, which is a little lower than that of our method.

Table 2. Performance comparison on the Waymo *val* set for 3D vehicle (IoU = 0.7) and pedestrian (IoU = 0.5) detection.

Method	Modality	Vehicle(AP/APH)		Pedestrian(AP/APH)	
		L1	L2	L1	L2
SECOND [25]	L	72.27/71.69	63.85/63.33	68.70/58.18	60.72/51.31
PointPillars [11]	L	71.60/71.00	63.10/62.50	70.60/56.70	62.90/50.20
PV-RCNN [18]	L	77.51/76.89	68.98/68.41	75.01/65.65	66.04/57.61
CenterPoint [28]	L	–	-/66.20	–	-/62.60
3D-MAN [27]	L	74.50/74.00	67.60/67.10	71.70/67.70	62.60/59.00
PointAugmenting [23]	L+C	67.40/-	62.70/-	75.04/-	70.60/-
DeepFusion [12]	L+C	80.60/80.10	72.90/72.40	**85.80/83.00**	78.70/76.00
FGFusion(Ours)	L+C	**81.92/81.44**	**73.85/73.34**	85.73/82.85	**78.81/76.14**

Waymo. Compared with the KITTI dataset, the Waymo dataset is larger and more diverse in sample diversity, and hence is more challenging. To verify our proposed FGFusion, we also conduct experiments on the Waymo dataset and compare it with some state-of-the-art methods. Table 2 shows that our FGFusion is better than other methods for both car and pedestrian categories in LEVEL2 difficulty, which is the main metric for ranking in the Waymo 3D detection challenge. Compared with the best PV-RCNN [18] in lidar-only methods, FGFusion has improved the APH of vehicle recognition by 4.93% and that of pedestrian recognition by 18.53%, which proves that our fusion method is more advantageous in small object detection.

4.4 Ablation Study

We conduct a series of experiments on Waymo to demonstrate the effectiveness of each component in our proposed FGFusion, including the attention pyramid of the image branch (AP), the auxiliary network of the point cloud branch (AN), and the multi-scale fusion module (MSF).

Effect of Each Component. As shown in Table 3, our FGFusion is 2.92% and 3.2% higher than the baseline in APH for the two categories, vehicles and pedestrians, respectively. Specifically, the multi-scale fusion module brings improvements of 1.74% and 1.93% to the baseline on two categories, which confirms our proposed fine-grained fusion strategy. The attention pyramid or the auxiliary network can further bring improvements of (0.7%, 0.87%) and (0.61%, 0.51%),

Table 3. Effect of each component in FGFusion on Waymo *val* set with APH in L2 difficulty.

MSF	AP	AN	Vehicle	Pedestrian
			70.42	72.94
✓			72.16	74.87
✓	✓		72.86	75.74
✓		✓	72.77	75.38
✓	✓	✓	73.34	76.14

Table 4. Performance comparison on Waymo *val* set with APH in L2 difficulty using different number of features for fusion.

Feature Num	Vehicle	Pedestrian
1	70.42	72.94
2	71.62 (+1.20)	74.05 (+1.11)
3	72.16 (+0.54)	74.81 (+0.76)
4	72.32 (+0.16)	75.01 (+0.20)

respectively. This indicates that the finer the fused features, the higher the model accuracy can achieve, which is consistent with our expectation.

Number of Feature Layers Selected for Fusion. The number of fusion features for point clouds and images is the key hyperparameter of our multi-scale fusion module. In order to determine the optimal value, we conduct experiments on the Waymo dataset without using attention pyramids or auxiliary networks. As shown in Table 4, the more feature layers used, the higher the model accuracy can achieve. This is because high-level features have rich semantic information and low-level features reserve complementary detailed information. The more feature layers used for fusion, the less information lost during downsampling. From the experimental results, it is intuitive that using two or three layers of features for fusion can bring significant improvements to model accuracy. While the number of fusion layers reaches four, the degree of improvement will be greatly reduced. It is worth noting that the more fusion layers used, the more weights the cross-attention model needs to train during fusion. In order to balance between model accuracy and computational cost, we use three layers of features for fusion in our experiments.

5 Conclusion

In this paper, we propose a novel multimodal network FGFusion for 3D object detection in autonomous driving scenarios. We design fine-grained feature extraction networks for both the point cloud branch and the image branch, and fuse features from different levels through a pyramid structure to improve detection accuracy. Extensive experiments are conducted on the KITTI and Waymo datasets, and the experimental results show that our method can achieve better performance than some state-of-the-art methods.

References

1. Arnold, E., Al-Jarrah, O.Y., Dianati, M., Fallah, S., Oxtoby, D., Mouzakitis, A.: A survey on 3D object detection methods for autonomous driving applications. IEEE Trans. Intell. Transp. Syst. **20**(10), 3782–3795 (2019)
2. Bai, X., et al.: Transfusion: robust lidar-camera fusion for 3D object detection with transformers. In: Proceedings of the IEEE/CVF Conference on Computer Vision and Pattern Recognition, pp. 1090–1099 (2022)
3. Chen, X., Ma, H., Wan, J., Li, B., Xia, T.: Multi-view 3D object detection network for autonomous driving. In: Proceedings of the IEEE conference on Computer Vision and Pattern Recognition, pp. 1907–1915 (2017)
4. Chen, Y., Liu, S., Shen, X., Jia, J.: Fast point R-CNN. In: Proceedings of the IEEE/CVF International Conference on Computer Vision, pp. 9775–9784 (2019)
5. Deng, J., Shi, S., Li, P., Zhou, W., Zhang, Y., Li, H.: Voxel R-CNN: towards high performance voxel-based 3D object detection. In: Proceedings of the AAAI Conference on Artificial Intelligence, vol. 35, pp. 1201–1209 (2021)
6. Geiger, A., Lenz, P., Urtasun, R.: Are we ready for autonomous driving? The Kitti vision benchmark suite. In: 2012 IEEE Conference on Computer Vision and Pattern Recognition, pp. 3354–3361. IEEE (2012)
7. He, C., Zeng, H., Huang, J., Hua, X.S., Zhang, L.: Structure aware single-stage 3D object detection from point cloud. In: Proceedings of the IEEE/CVF Conference on Computer Vision and Pattern Recognition, pp. 11873–11882 (2020)
8. Huang, K., Shi, B., Li, X., Li, X., Huang, S., Li, Y.: Multi-modal sensor fusion for auto driving perception: a survey. arXiv preprint arXiv:2202.02703 (2022)
9. Huang, T., Liu, Z., Chen, X., Bai, X.: EPNet: enhancing point features with image semantics for 3D object detection. In: Vedaldi, A., Bischof, H., Brox, T., Frahm, J.-M. (eds.) ECCV 2020. LNCS, vol. 12360, pp. 35–52. Springer, Cham (2020). https://doi.org/10.1007/978-3-030-58555-6_3
10. Ku, J., Mozifian, M., Lee, J., Harakeh, A., Waslander, S.L.: Joint 3D proposal generation and object detection from view aggregation. In: 2018 IEEE/RSJ International Conference on Intelligent Robots and Systems (IROS), pp. 1–8. IEEE (2018)
11. Lang, A.H., Vora, S., Caesar, H., Zhou, L., Yang, J., Beijbom, O.: PointPillars: fast encoders for object detection from point clouds. In: Proceedings of the IEEE/CVF Conference on Computer Vision and Pattern Recognition, pp. 12697–12705 (2019)
12. Li, Y., et al.: DeepFusion: lidar-camera deep fusion for multi-modal 3D object detection. In: Proceedings of the IEEE/CVF Conference on Computer Vision and Pattern Recognition, pp. 17182–17191 (2022)
13. Lin, T.Y., Dollár, P., Girshick, R., He, K., Hariharan, B., Belongie, S.: Feature pyramid networks for object detection. In: Proceedings of the IEEE Conference on Computer Vision and Pattern Recognition, pp. 2117–2125 (2017)
14. Liu, Z., Zhao, X., Huang, T., Hu, R., Zhou, Y., Bai, X.: TANet: robust 3D object detection from point clouds with triple attention. In: Proceedings of the AAAI Conference on Artificial Intelligence, vol. 34, pp. 11677–11684 (2020)

15. Qi, C.R., Liu, W., Wu, C., Su, H., Guibas, L.J.: Frustum pointnets for 3D object detection from RGB-D data. In: Proceedings of the IEEE Conference on Computer Vision and Pattern Recognition, pp. 918–927 (2018)
16. Qi, C.R., Su, H., Mo, K., Guibas, L.J.: PointNet: deep learning on point sets for 3D classification and segmentation. In: Proceedings of the IEEE Conference on Computer Vision and Pattern Recognition, pp. 652–660 (2017)
17. Qi, C.R., Yi, L., Su, H., Guibas, L.J.: PointNet++: deep hierarchical feature learning on point sets in a metric space. In: Advances in Neural Information Processing Systems 30 (2017)
18. Shi, S., et al.: PV-RCNN: Point-voxel feature set abstraction for 3D object detection. In: Proceedings of the IEEE/CVF Conference on Computer Vision and Pattern Recognition, pp. 10529–10538 (2020)
19. Shi, S., Wang, X., Li, H.: PointRCNN: 3D object proposal generation and detection from point cloud. In: Proceedings of the IEEE/CVF Conference on Computer Vision and Pattern Recognition, pp. 770–779 (2019)
20. Shi, S., Wang, Z., Shi, J., Wang, X., Li, H.: From points to parts: 3D object detection from point cloud with part-aware and part-aggregation network. IEEE Trans. Pattern Anal. Mach. Intell. **43**(8), 2647–2664 (2020)
21. Sun, P., et al.: Scalability in perception for autonomous driving: Waymo open dataset. In: Proceedings of the IEEE/CVF Conference on Computer Vision and Pattern Recognition, pp. 2446–2454 (2020)
22. Vora, S., Lang, A.H., Helou, B., Beijbom, O.: PointPainting: sequential fusion for 3D object detection. In: Proceedings of the IEEE/CVF Conference on Computer Vision and Pattern Recognition, pp. 4604–4612 (2020)
23. Wang, C., Ma, C., Zhu, M., Yang, X.: PointAugmenting: cross-modal augmentation for 3D object detection. In: Proceedings of the IEEE/CVF Conference on Computer Vision and Pattern Recognition, pp. 11794–11803 (2021)
24. Wu, X., et al.: Sparse fuse dense: towards high quality 3d detection with depth completion. In: Proceedings of the IEEE/CVF Conference on Computer Vision and Pattern Recognition, pp. 5418–5427 (2022)
25. Yan, Y., Mao, Y., Li, B.: Second: sparsely embedded convolutional detection. Sensors **18**(10), 3337 (2018)
26. Yang, Z., Sun, Y., Liu, S., Jia, J.: 3DSSD: point-based 3D single stage object detector. In: Proceedings of the IEEE/CVF Conference on Computer Vision and Pattern Recognition, pp. 11040–11048 (2020)
27. Yang, Z., Zhou, Y., Chen, Z., Ngiam, J.: 3D-MAN: 3D multi-frame attention network for object detection. In: Proceedings of the IEEE/CVF Conference on Computer Vision and Pattern Recognition, pp. 1863–1872 (2021)
28. Yin, T., Zhou, X., Krahenbuhl, P.: Center-based 3D object detection and tracking. In: Proceedings of the IEEE/CVF Conference on Computer Vision and Pattern Recognition, pp. 11784–11793 (2021)
29. Yoo, J.H., Kim, Y., Kim, J., Choi, J.W.: 3D-CVF: generating joint camera and LiDAR features using cross-view spatial feature fusion for 3D object detection. In: Vedaldi, A., Bischof, H., Brox, T., Frahm, J.-M. (eds.) ECCV 2020. LNCS, vol. 12372, pp. 720–736. Springer, Cham (2020). https://doi.org/10.1007/978-3-030-58583-9_43

30. Zhang, Y., Chen, J., Huang, D.: CAT-Det: contrastively augmented transformer for multi-modal 3D object detection. In: Proceedings of the IEEE/CVF Conference on Computer Vision and Pattern Recognition, pp. 908–917 (2022)
31. Zhou, Y., Tuzel, O.: VoxelNet: end-to-end learning for point cloud based 3D object detection. In: Proceedings of the IEEE Conference on Computer Vision and Pattern Recognition, pp. 4490–4499 (2018)

Author Index

Q. Liu et al. (Eds.): PRCV 2023, LNCS 14427, pp. 519–521, 2024.
https://doi.org/10.1007/978-981-99-8435-0

Printed in the United States
by Baker & Taylor Publisher Services

Printed in the United States
by Baker & Taylor Publisher Services